Advanced New Physical Layer Technologies for Next-Generation Wireless Communications

Advanced New Physical Layer Technologies for Next-Generation Wireless Communications

Guest Editors

Lei Liu

Zhijin Qin

Chongwen Huang

Yuhao Chi

Yang Liu

MDPI

Basel • Beijing • Wuhan • Barcelona • Belgrade • Novi Sad • Cluj • Manchester

Guest Editors

Lei Liu
College of Information Science and Electronic Engineering
Zhejiang University
Hangzhou
China

Zhijin Qin
Department of Electronic Engineering
Tsinghua University
Beijing
China

Chongwen Huang
College of Information Science and Electronic Engineering
Zhejiang University
Hangzhou
China

Yuhao Chi
School of Telecommunications Engineering
Xidian University
Xi'an
China

Yang Liu
College of Communication and Information Engineering
Xi'an University of Science and Technology
Xi'an
China

Editorial Office

MDPI AG
Grosspeteranlage 5
4052 Basel, Switzerland

This is a reprint of the Special Issue, published open access by the journal *Entropy* (ISSN 1099-4300), freely accessible at: https://www.mdpi.com/journal/entropy/special_issues/KY0367C3JX.

For citation purposes, cite each article independently as indicated on the article page online and as indicated below:

Lastname, A.A.; Lastname, B.B. Article Title. *Journal Name* **Year**, *Volume Number*, Page Range.

ISBN 978-3-7258-5117-1 (Hbk)
ISBN 978-3-7258-5118-8 (PDF)
https://doi.org/10.3390/books978-3-7258-5118-8

Contents

About the Editors vii

Preface ix

Lei Liu
Advances in New Physical Layer Technologies for Next-Generation Wireless Communications
Reprinted from: *Entropy* **2025**, *27*, 616, https://doi.org/10.3390/e27060616 1

Michael Peleg and Shlomo Shamai
On the Capacity of the Peak-Limited and Band-Limited Channel
Reprinted from: *Entropy* **2024**, *26*, 1049, https://doi.org/10.3390/e26121049 8

Yuanrui Liu, Xiaoyu Zhao and Wei Chen
User-Perceived Capacity: Theory, Computation, and Achievable Policies
Reprinted from: *Entropy* **2024**, *26*, 914, https://doi.org/10.3390/e26110914 28

Yingge Li and Haihua Hu
Design of Low-Latency Layered Normalized Minimum Sum Low-Density Parity-Check Decoding Based on Entropy Feature for NAND Flash-Memory Channel
Reprinted from: *Entropy* **2024**, *26*, 781, https://doi.org/10.3390/e26090781 51

Li Deng, Zhiping Shi, Yixin Wang, Xiaoxi Yu, Yong Liang Guan and Zhiping Xu
Hybrid Window Decoding for Joint Source Channel Anytime Coding System
Reprinted from: *Entropy* **2024**, *26*, 940, https://doi.org/10.3390/e26110940 67

Xue Li and Qin Huang
Improved PEG-Based Construction of Analog Fountain Codes
Reprinted from: *Entropy* **2024**, *26*, 841, https://doi.org/10.3390/e26100841 81

Zuo Tang, Jing Lei and Ying Huang
EXIT Charts for Low-Density Algebra-Check Codes
Reprinted from: *Entropy* **2024**, *26*, 1118, https://doi.org/10.3390/e26121118 90

Junhu Shao, Zaiping Liu, Yishuo Liu and Tianjiao Xie
Performance Analysis of Troposphere Scattering Communication Channel with Chirp-BOK Modulation
Reprinted from: *Entropy* **2024**, *26*, 1052, https://doi.org/10.3390/e26121052 106

Yu Xin, Jian Hua, Tong Bao, Yaxing Hao, Ziheng Xiao, Xin Nie and Fanggang Wang
Generalized Filter Bank Orthogonal Frequency Division Multiplexing: Low-Complexity Waveform for Ultra-Wide Bandwidth and Flexible Services
Reprinted from: *Entropy* **2024**, *26*, 994, https://doi.org/10.3390/e26110994 120

Jianjian Wu, Chanzi Liu and Qingfeng Zhou
Random Frequency Division Multiplexing
Reprinted from: *Entropy* **2025**, *27*, 9, https://doi.org/10.3390/e27010009 139

Zhuo Han, Wanming Hao, Zhiqing Tang and Shouyi Yang
Optimal Decoding Order and Power Allocation for Sum Throughput Maximization in Downlink NOMA Systems
Reprinted from: *Entropy* **2024**, *26*, 421, https://doi.org/10.3390/e26050421 147

Peizhe Xin, Zihao Fu, Zhiyi Chen, Jing Jiang, Jing Zou, Yu Zhang and Xinyue Hu
Joint User Association, Power Allocation and Beamforming for NOMA-Based Integrated Satellite–Terrestrial Networks
Reprinted from: *Entropy* **2024**, *26*, 1055, https://doi.org/10.3390/e26121055 160

Kangchun Zhao, Yijie Mao and Yuanming Shi
Simultaneously Transmitting and Reflecting Reconfigurable Intelligent Surfaces Empowered Cooperative Rate Splitting with User Relaying †
Reprinted from: *Entropy* **2024**, *26*, 1019, https://doi.org/10.3390/e26121019 . . . . **173**

Yang Liu, Zhao Yang, Bin Wang, Yanhong Xu
Rate Optimization of Intelligent Reflecting Surface-Assisted Coal Mine Wireless Communication Systems
Reprinted from: *Entropy* **2024**, *26*, 880, https://doi.org/10.3390/e26100880 . . . . **200**

Azar Hakimi, Diluka Galappaththige and Chintha Tellambura
A Roadmapfor NF-ISAC in 6G: A Comprehensive Overview and Tutorial
Reprinted from: *Entropy* **2024**, *26*, 773, https://doi.org/10.3390/e26090773 . . . . **214**

Shenghong Lin, Hongyan Wang, Weiyong Li and Jinyuan Wang
Coverage Analysis for High-Speed Railway Communications with Narrow-Strip-Shaped Cells over Suzuki Fading Channels
Reprinted from: *Entropy* **2024**, *26*, 657, https://doi.org/10.3390/e26080657 . . . . **249**

Kai Zhang, Fenglei Zhang, Yongjun Li, Xiang Wang, Zhaohui Yang, Yuanhao Liu, et al.
A Three-Dimensional Time-Varying Channel Model for THz UAV-Based Dual-Mobility Channels
Reprinted from: *Entropy* **2024**, *26*, 924, https://doi.org/10.3390/e26110924 . . . . **268**

Ling-Hui Zhu, Zhen Zhu, Guo-Lin Lv, Chong-Qiang Ye and Xiao-Yu Chen
Robustness of Entanglement for Dicke-W and Greenberger-Horne-Zeilinger Mixed States
Reprinted from: *Entropy* **2024**, *26*, 804, https://doi.org/10.3390/e26090804 . . . . **281**

Bei Liu, Shuting Long and Xin Su
Knowledge-Assisted Actor Critic Proximal Policy Optimization-Based Service Function Chain Reconfiguration Algorithm for 6G IoT Scenario
Reprinted from: *Entropy* **2024**, *26*, 820, https://doi.org/10.3390/e26100820 . . . . **296**

Haonan Zhang, Peng Xu and Bin Dai
Ultra-Reliable and Low-Latency Wireless Hierarchical Federated Learning: Performance Analysis †
Reprinted from: *Entropy* **2024**, *26*, 827, https://doi.org/10.3390/e26100827 . . . . **310**

About the Editors

Lei Liu

Lei Liu is currently a tenure-track Young Professor at Zhejiang University, Hangzhou, China. His research focuses on message-passing algorithms, statistical signal processing, wireless communications, modern channel coding, and information theory. Since 2023, he has been supported by the Excellent Young Scientists Program (Overseas) of the National Natural Science Foundation of China (NSFC). Dr. Liu received his Ph.D. in Communication and Information Systems from Xidian University in 2017, with research visits to Nanyang Technological University (2014-2016). He has held postdoctoral positions at the Singapore University of Technology and Design (2016-2017) and City University of Hong Kong (2017-2019), and served as Assistant Professor at the Japan Advanced Institute of Science and Technology (2019-2023). His notable achievements include receiving the Young Star Award (one of two recipients) and Best Poster Award at the 30th Chinese Institute of Electronics Conference on Information Theory (2023), being named the "6G New Horizon Young Scientist" at Global 6G Conference 2025, wining the Finalist Prize of National Disruptive Technology Innovation Competition 2025, and serving as Publications Co-Chair for the 2021 IEEE Information Theory Workshop. Dr. Liu is a Senior Member of both IEEE and the China Institute of Communications.

Zhijin Qin

Zhijin Qin is currently an Associate Professor at Tsinghua University, where her research focuses on semantic communications and sparse signal processing. Prior to joining Tsinghua, she held research positions at Imperial College London, Lancaster University, and Queen Mary University of London from 2016 to 2022. Dr. Qin has received numerous prestigious awards, including the 2017 IEEE GLOBECOM Best Paper Award, the 2018 IEEE Signal-Processing Society Young-Author Best Paper Award, the 2021 IEEE Communications Society Signal Processing for Communications Committee Early Achievement Award, and the 2022 IEEE Communications Society Fred W. Ellersick Prize. She has made significant editorial contributions as Guest Editor for the *IEEE Journal on Selected Areas in Communications (JSAC)* Special Issue on Semantic Communications and as Area Editor for the JSAC Series. Dr. Qin has also served as Symposium Co-Chair for IEEE GLOBECOM 2020 and 2021, and currently serves as Associate Editor for *IEEE Transactions on Communications*, *IEEE Transactions on Cognitive Networking*, and *IEEE Communications Letters*. A Senior Member of IEEE, Dr. Qin continues to advance research in next-generation communication technologies through both her academic work and professional service.

Chongwen Huang

Chongwen Huang is currently a tenure-track young professor at Zhejiang University, a position he has held since September 2020. His research focuses on Holographic MIMO Surfaces, Reconfigurable Intelligent Surfaces, B5G/6G wireless communications, mmWave/THz communications, and deep learning applications for wireless technologies. Dr. Huang earned his B.Sc. from Nankai University in 2010, followed by an M.Sc. from the University of Electronic Science and Technology of China in 2013, and completed his Ph.D. at the Singapore University of Technology and Design (SUTD) in 2019. He continued his research as a Postdoctoral Fellow at SUTD from October 2019 to September 2020. His outstanding contributions have been recognized through several prestigious awards, including the 2021 IEEE Marconi Prize Paper Award, the 2023 IEEE Fred W. Ellersick Prize Paper Award, and the 2021 IEEE ComSoc Asia–Pacific Outstanding Young Researcher

Award. Since 2021, Dr. Huang has served as an editor for multiple journals, including *IEEE Communications Letters, Elsevier Signal Processing, EURASIP Journal on Wireless Communications and Networking,* and *Physical Communication.* An active IEEE Member, Dr. Huang continues to advance wireless communication technologies through his innovative research and editorial contributions.

Yuhao Chi

Yuhao Chi is currently an Associate Professor at Xidian University, where his research focuses on multiuser coding and detection, message-passing algorithms, MIMO systems, and deep learning applications in communications. He received his B.S. degree in Electronic and Information Engineering from Shaanxi University of Science and Technology in 2012 and earned his Ph.D. in Communication and Information Systems from Xidian University in 2018. During his doctoral studies (2016-2017), Dr. Chi was awarded the China Scholarship Council fellowship to conduct research as a visiting scholar at both Nanyang Technological University and the Singapore University of Technology and Design in Singapore. Before returning to academia, he worked as a Senior Engineer at Huawei Technologies (2018-2021), gaining valuable industry experience in wireless communications. Dr. Chi's research excellence was recognized with the Best Poster Award at the 30th Chinese Institute of Electronics Conference on Information Theory (CIEIT 2023). As an IEEE Senior Member, he actively contributes to advancing communication technologies through his combined expertise in both industrial and academic research.

Yang Liu

Yang Liu is currently an Associate Professor at Xi'an University of Science and Technology. Her research focuses on channel coding and information theory, with applications in deep learning for mine wireless communications and reconfigurable intelligent surface (RIS)-assisted underground communication systems. She has been funded by National Natural Science Foundation of China (NSFC) Program. Dr. Liu received her B.S. in Communication Engineering (2010) and her Ph.D. in Communication and Information Systems (2016) from Xidian University. To further her expertise, she was a Visiting Scholar at Nanyang Technological University (NTU) in Singapore (2021–2022), where she collaborated on next-generation wireless technologies. Her work uniquely bridges theoretical communication principles with real-world industrial challenges, particularly in harsh mine environments.

Preface

The rapid advancement of wireless communication technologies has brought transformative changes to modern networks, driven by demands for higher efficiency, reliability, and adaptability. This Reprint presents significant research contributions on physical-layer innovations, addressing key challenges in next-generation wireless communication systems. The subject of this Reprint encompasses a broad spectrum of physical-layer advancements, including novel modulation schemes, channel coding and decoding techniques, signal-detection algorithms, integrated sensing and communication (ISAC), reconfigurable intelligent surfaces (RISs), and artificial intelligence-driven radio access networks. Each contribution explores theoretical foundations, algorithmic optimizations, and practical implementations to enhance the performance, reliability, and adaptability of wireless systems. This Reprint is motivated by the fundamental need to develop physical-layer technologies, which can support 6G's ambitious performance targets across diverse scenarios. The included research provides both theoretical foundations and practical approaches for overcoming current limitations in high-frequency, high-mobility, and integrated network environments. This Reprint aims to provide researchers, engineers, and industry practitioners with a comprehensive resource that highlights the latest breakthroughs, methodologies, and open challenges in the field. We express our sincere appreciation to the authors whose valuable work forms the foundation of this Reprint, as well as to the reviewers for their constructive feedback that ensured the quality of this collection. We hope this publication will stimulate further innovation and encourage ongoing investigation into the transformative capabilities of physical-layer technologies for future wireless networks.

Lei Liu, Zhijin Qin, Chongwen Huang, Yuhao Chi, and Yang Liu
Guest Editors

entropy

MDPI

Editorial

Advances in New Physical Layer Technologies for Next-Generation Wireless Communications

Lei Liu

College of Information Science and Electronic Engineering, Zhejiang University, Hangzhou 310027, China; lei_liu@zju.edu.cn

1. Introduction

The world of wireless communication is undergoing an exhilarating transformation. As applications advance at an unprecedented pace, communication services have diversified, and deployment environments have grown increasingly complex—ushering in both extraordinary opportunities and formidable challenges for next-generation systems. As the cornerstone of wireless systems, physical layer technologies are designed to combat common impairments in wireless channels—such as fading, delay spread, Doppler shifts, interference, and noise—ensuring reliable and stable data transmission. This, in turn, supports the seamless operation of upper-layer services. To address the escalating demands of next-generation networks, the evolution of physical layer technologies is advancing across several key frontiers, including, but not limited to, the following:

- *Multicarrier Modulation:* Multicarrier modulation is highly effective in enhancing spectral efficiency and mitigating the adverse effects of complex wireless channels. Multicarrier modulation has evolved from traditional orthogonal frequency division multiplexing (OFDM) to more advanced schemes such as orthogonal time frequency space (OTFS) [1], affine frequency division multiplexing (AFDM) [2], and interleaved frequency division multiplexing (IFDM) [3], in order to meet the demands of high-mobility communication. While OFDM offers simplicity and high efficiency in static environments, its performance degrades significantly in time-varying scenarios. OTFS and AFDM achieve diversity gains by separating signals in the delay-Doppler domain but suffer from a lack of efficient detection algorithms. IFDM constructs a dense equivalent channel matrix to ensure statistically stationary transmission of signals. These modulation techniques serve as a cornerstone for enabling high-mobility communication technologies.
- *Channel Estimation and Signal Detection:* Channel estimation and signal detection are essential components of the receiver for ensuring reliable wireless communications, particularly in complex and high-dimensional systems. While classical algorithms like least squares (LS) and minimum mean square error (MMSE) offer baseline performance, they fall short under massive multiple input multiple output (MIMO), low SNR, or non-linear hardware constraints. Recently, approximate message passing (AMP)-based algorithms and their variants—such as orthogonal AMP (OAMP) [4], vector AMP (VAMP) [5], and memory AMP (MAMP) [6]—have emerged as powerful tools for large-scale signal detection. These algorithms exploit the statistical structure of channels and signals to achieve near-optimal performance with low complexity and provable convergence in certain conditions. However, challenges remain in extending AMP to correlated priors, finite-length effects, and ensuring robustness in dynamic or imperfect channel state information (CSI) scenarios. Integrating AMP

with learning-based or model-driven frameworks is a promising direction for next-generation communication systems.

- *Channel Coding and Decoding Design*: With the widespread adoption of the Internet of Things (IoT) and the growing number of wireless device connections, many communication services are placing new demands on low latency, high reliability, and short-packet transmission. These requirements drive the need to revisit channel coding and decoding design under finite-length and multi-user communication constraints. However, unlike long-length codes, finite-length codes still lack a unified design framework, practical tools, and low-complexity, efficient decoding algorithms. Furthermore, most traditional channel coding schemes are tailored for point-to-point channels, making them inadequate for effectively managing interference in multi-user scenarios [7]. Additionally, the development of flexible, variable-rate coding schemes is also essential for the practical deployment of channel coding in real-world systems.
- *Information Theory Evolution:* As practical communication scenarios become increasingly complex, classical information theory—rooted in idealized or simplified assumptions—can no longer accurately characterize the performance limits of emerging systems such as the constrained (e.g., peak-limited and/or band-limited) channel, millimeter-wave, terahertz, ultra-large-scale MIMO, near-field communication, and semantic communication. Consequently, it is essential to develop modern information theory that accounts for practical communication constraints [8]. This modern theory serves as both an evolution and a complement to classical information theory, providing a more accurate theoretical foundation for the design and performance analysis of next-generation communication systems.
- *Integrated Sensing and Communication (ISAC)*: ISAC has emerged as a key enabling technology for next-generation wireless networks, aiming to unify the traditionally separate functions of wireless sensing and data communication [9]. Recent advances in ISAC have focused on joint waveform design, resource allocation, and signal processing algorithms that enable a single hardware and spectral resource to serve both functions efficiently. Cutting-edge research has proposed dual-functional waveforms based on OFDM, OTFS, AFDM, and other modulation schemes, which offer high spectral efficiency and robustness in dynamic environments. ISAC systems have been extended to support large-scale MIMO, near-field communication, and high-mobility scenarios, demonstrating significant improvements in both sensing accuracy and communication reliability. Additionally, theoretical frameworks for analyzing the rate-sensing trade-off and performance bounds are being developed to guide practical system design. Overall, ISAC represents a paradigm shift in wireless system architecture, promising reduced hardware costs, enhanced spectral efficiency, and improved environmental awareness, and is expected to play a central role in future intelligent and ubiquitous wireless infrastructures.
- *Reconfigurable Intelligent Surface (RIS)*: RIS is a key enabler for future 6G networks, capable of reconfiguring the wireless environment to enhance coverage, spectral efficiency, and energy efficiency [10]. By intelligently controlling wave reflection through nearly passive elements, RIS reduces hardware costs and power consumption. Recent progress includes joint beamforming, channel estimation, and integration with MIMO, THz, and ISAC systems. Future research should focus on scalable deployment strategies, real-time control under dynamic environments, and unified theoretical frameworks to guide practical RIS system design.
- *Artificial Intelligence for Radio Access Networks (AI-RAN)*: AI-RAN represents a transformative paradigm that deeply integrates AI technologies [11]—such as machine learning and deep learning—into the design, optimization, and operation of wireless

access networks. By leveraging data-driven models, AI-RAN enables intelligent resource allocation, real-time traffic prediction, adaptive interference management, and proactive fault detection, thereby enhancing the efficiency, flexibility, and reliability of wireless systems. It plays a pivotal role in supporting diverse 5G and future 6G applications, including autonomous driving, industrial IoT, and immersive media. However, AI-RAN still faces several key challenges: ensuring the generalization and robustness of learning models under dynamic and complex wireless environments, achieving low-latency and energy-efficient AI inference at the network edge, and establishing trustworthy and interpretable AI frameworks. Addressing these challenges is critical to fully realizing the potential of AI-RAN in next-generation wireless networks.

- *Communication Technology-Driven Low-Altitude Economy:* Physical layer communication technologies are crucial for the low-altitude economy, enabling reliable and efficient links in environments with high mobility and dynamic channels [12], such as low earth orbit (LEO) satellites and unmanned aerial vehicle (UAV) networks. Key advances include robust modulation and coding schemes (e.g., OTFS, IFDM), spatial techniques like MIMO and IRS, and the use of high-frequency bands for enhanced capacity. Challenges remain in handling rapidly varying channels, interference management, low-power receiver design, and ensuring security under hardware constraints. Overcoming these issues is essential for the continued growth of low-altitude communication systems.

In summary, this Special Issue is organized to provide a focused platform for presenting and discussing the latest advances in physical layer communication technologies. It aims to gather high-quality contributions that address emerging challenges and explore innovative solutions in areas such as channel coding, modulation, channel estimation, detection algorithms, and integrated sensing and communication.

Specifically, in system capacity analysis under practical constraints, a tight lower bound is derived for peak-power-limited (PPL) additive white Gaussian noise (AWGN) channels using a geometry-based approach [13], while a constrained Markov decision process (CMDP)-based strategy improves user-perceived capacity through joint power-rate adaptation [14]. In channel coding and decoding, efforts focus on reducing latency and complexity: entropy-aware layered scheduling enhances low-density parity-check (LDPC) decoding [15], and hybrid window decoding lowers the complexity of joint source–channel anytime codes [16]. Additionally, analog fountain code designs [17] and low-density algebra-check (LDAC) optimization via extrinsic information transfer (EXIT) charts improve coding efficiency in short-block and low-rate scenarios [18]. On modulation waveforms, chirp-BOK modulation is introduced for tropospheric scattering in deep fading environments [19], a generalized filter bank OFDM scheme enhances bandwidth efficiency for 6G THz systems [20], and random frequency division multiplexing (RFDM) integrates compressed sensing and deep neural networks for robust detection in dynamic channels [21]. Emerging technologies are also explored for their potential to jointly improve transmission efficiency. For example, refs. [22,23] propose optimal decoding order and resource allocation strategies for non-orthogonal multiple access (NOMA) and NOMA-based satellite–terrestrial networks. In reconfigurable intelligent surface (RIS)-assisted systems, ref. [24] presents a simultaneously transmitting and RIS (STAR-RIS)-aided cooperative rate splitting (CRS) framework with optimized beamforming and rate allocation, while ref. [25] proposes a three-step intelligent reflecting surface (IRS) deployment strategy tailored for coal mine tunnels. A comprehensive tutorial on near-field integrated sensing and communication (ISAC) [26] highlights the benefits of jointly optimizing sensing and communication using shared RF resources. Furthermore, recent articles tackle advanced challenges including high-speed railway [27] and UAV terahertz channel modeling [28], quantum entanglement robustness [29], deep learning-based service function chain (SFC) reconfiguration for the 6G IoT scenario [30], and secure federated learning under ultra-reliable and low-latency communication (URLLC) [31].

In summary, the aforementioned studies offer innovative models, algorithms, and insights at the physical layer, providing valuable guidance for future research.

2. An Overview of Published Articles

In the study of system capacity analysis, the first article [13] explores peak-power-limited AWGN channels, using a geometry-based approach to provide tighter lower bounds on channel capacity. These bounds improve power efficiency by factors of 3.3 and 8.6, with a 6 dB penalty at high SNRs, offering insights for modulation and coding schemes. The second article [14] studies user-perceived throughput in mobile multimedia services, showing that capacity is limited with only channel state information at the receiver (CSIR), but improves significantly with both CSIR and channel state information at the transmitter through power and rate adaptation. They propose a CMDP-based approach, revealing that the optimal policy follows a threshold rule.

In the study of channel coding and decoding, the third article [15] tackles high decoding latency in LDPC codes, proposing two techniques (serial entropy feature-based layered normalized min-sum and parallel entropy feature-based layered normalized min-sum) to reduce latency by dynamically allocating layered processing units based on bit reliability. Experimental results show significant efficiency improvements. The fourth article [16] reduces decoding complexity in joint source–channel anytime coding (JSCAC) systems using the hybrid window decoding (JHWD) technique, introducing an adaptive decoding structure and an improved density evolution algorithm. The JHWD algorithm is proven to be efficient for low-complexity decoding in JSCAC and similar systems. The fifth article [17] proposes an improved progressive edge-growth construction for analog fountain codes (AFCs), allocating weight coefficients in descending order. Simulation results show it achieves a low block error rate (BLER) in short block lengths. The sixth article [18] explores the use of EXIT charts to optimize LDAC codes and construct a low-rate LDAC code. They also establish an iterative decoding model and demonstrate a low-rate LDAC code with enhanced coding gain through simulations.

In the study of designing modulation waveforms, the seventh article [19] studies chirp-BOK (binary orthogonal keying) modulation in troposphere scattering systems, demonstrating significant BER performance improvements with the optimal configuration. The eighth article [20] introduces the generalized filter bank OFDM waveform for THz communication in 6G networks, improving bandwidth efficiency and reducing complexity compared to cyclic prefix OFDM (CP-OFDM). The ninth article [21] presents an RFDM method for multicarrier modulation in mobile time-varying channels, utilizing compressed sensing (CS) and deep neural networks (DNNs) for signal detection. Simulation results show that the proposed method significantly improves bit error rate (BER) performance, offering a promising approach for enhancing spectrum efficiency in multi-subcarrier, multi-antenna, and multi-user systems.

In the study of technologies for improving transmission efficiency, the tenth article [22] addresses the optimal decoding order and power allocation for sum throughput maximization in downlink NOMA systems over Nakagami-m channels, deriving a closed-form expression for outage probability, analyzing its behavior in high-SNR conditions, and proving the optimal power allocation range for each decoding order. A joint optimization problem for decoding order and power allocation is formulated and solved through an efficient search, validated by Monte Carlo simulations. The eleventh article [23] investigates a NOMA-based integrated satellite–terrestrial network, proposing a two-stage algorithm for resource optimization, with simulation results showing its effectiveness. The twelfth article [24] proposes a STAR-RIS-assisted cooperative rate splitting framework with six transmission modes to maximize the minimum user rate. They introduce a successive convex approximation (SCA)-based alternative optimization (AO) algorithm for joint optimization and a low-complexity algorithm for passive beamforming. Simulation results show significant improvements in user fairness and performance, with the low-complexity method achieving near-AO results at lower computational cost. The thirteenth article [25] investigates a three-step joint rate optimization method for IRS-assisted wireless communication in coal mines. It optimizes the IRS deployment position, transmit power, and phase shifts to maximize achievable rates. To solve the non-convexity of the optimization problem for the IRS deployment position, logarithmic operations, Taylor relaxation, and two auxiliary variables were introduced to transform it into a convex problem. Simulation results show improvements of 12.32% to 54.17% in achievable rates compared to other schemes. The IRS deployment position converges to an optimal location regardless of its initial placement. This paper also examines the impact of tunnel wall roughness, demonstrating superior performance in coal mine scenarios. The fourteenth article [26] provides an overview and tutorial on near-field (NF) ISAC technology, which offers significant advantages over traditional far-field (FF) ISAC systems by enabling simultaneous communication and sensing using shared RF resources. This paper analyzes NF and FF systems, explores their applicability in various scenarios, and discusses different channel models. It also highlights the benefits of ISAC, examines narrow-band and wide-band NF-ISAC systems, and presents case studies and simulations. In addition, it reviews current research, key methodologies, challenges, and future research directions for NF-ISAC.

In the study of complex communication scenarios, the fifteenth article [27] addresses coverage analysis for high-speed railway (HSR) communications, proposing a channel model incorporating path loss, shadowing, and small-scale fading. It derives expressions for edge coverage probability and cell coverage area, validating the results with Monte Carlo simulations. The sixteenth article [28] proposes a 3D time-varying channel model for UAV-based dual-mobility systems in the THz band, considering scattering, reflection, and atmospheric absorption, with simulation results verifying its accuracy. The seventeenth article [29] investigates the entanglement robustness of Dicke–W and GHz mixed states under varying mixing ratios. Simulation results demonstrate that the entanglement robustness increases when the probability ratio of Dicke and W states is greater than or equal to 3/2, while it follows a different trend for ratios below 3/2. The eighteenth article [30] applies deep learning to service function chain reconfiguration to meet the highly dynamic service requirements in the future 6G Internet of Things (IoT) scenario, introducing the knowledge-assisted actor–critic (AC) proximal policy optimization (KA-ACPPO)-based algorithm, which improves flexibility and reduces computing costs while ensuring optimal quality of service (QoS). The nineteenth article [31] discusses physical layer security in wireless hierarchical federated learning (WHFL), proposing a secure finite block-length approach for multi-antenna URLLC-WHFL systems, which provides near-perfect secrecy without compromising learning performance.

3. Conclusions

This Special Issue highlights the pivotal role of physical layer technologies in addressing the evolving demands of modern and future wireless communication systems. Through diverse contributions spanning capacity analysis, coding and decoding, waveform design, and emerging technologies such as NOMA, RIS, ISAC, and THz communications, the featured works collectively advance our understanding of robust, efficient, and adaptive physical layer solutions. By tackling both fundamental challenges and application-driven scenarios, these studies lay a solid foundation for future innovations in wireless communication, offering practical insights and theoretical tools to guide the development of next-generation networks.

Funding: This work was supported in part by the National Natural Science Foundation of China (NSFC) under Grants 62394292, 62301485, and 62201424, in part by the Zhejiang Provincial Natural Science Foundation under Grant LZ25F010002, and in part by the State Key Laboratory of Integrated Services Networks under Grant ISN25-10.

Conflicts of Interest: The author declares no conflicts of interest.

References

1. Hadani, R.; Rakib, S.; Tsatsanis, M.; Monk, A.; Goldsmith, A.J.; Molisch, A.F.; Calderbank, R. Orthogonal Time Frequency Space Modulation. In Proceedings of the 2017 IEEE Wireless Communications and Networking Conference (WCNC), San Francisco, CA, USA, 19–22 March 2017; pp. 1–6.
2. Bemani, A.; Ksairi, N.; Kountouris, M. Affine Frequency Division Multiplexing for Next Generation Wireless Communications. *IEEE Trans. Wirel. Commun.* **2023**, *22*, 8214–8229. [CrossRef]
3. Chi, Y.; Liu, L.; Ge, Y.; Chen, X.; Li, Y.; Zhang, Z. Interleave Frequency Division Multiplexing. *IEEE Wirel. Commun. Lett.* **2024**, *13*, 1963–1967. [CrossRef]
4. Ma, J.; Ping, L. Orthogonal AMP. *IEEE Access* **2017**, *5*, 2020–2033. [CrossRef]
5. Rangan, S.; Schniter, P.; Fletcher, A.K. Vector Approximate Message Passing. *IEEE Trans. Inf. Theory* **2019**, *65*, 6664–6684. [CrossRef]
6. Liu, L.; Huang, S.; Kurkoski, B.M. Memory AMP. *IEEE Trans. Inf. Theory* **2022**, *68*, 8015–8039. [CrossRef]
7. Chi, Y.; Liu, L.; Song, G.; Li, Y.; Guan, Y.L.; Yuen, C. Constrained Capacity Optimal Generalized Multi-User MIMO: A Theoretical and Practical Framework. *IEEE Trans. Commun.* **2022**, *70*, 8086–8104. [CrossRef]
8. Zhu, J.; Wan, Z.; Dai, L.; Cui, T.J. Electromagnetic Information Theory-Based Statistical Channel Model for Improved Channel Estimation. *IEEE Trans. Inf. Theory* **2025**, *71*, 1777–1793. [CrossRef]
9. Mao, W.; Lu, Y.; Chi, C.-Y.; Ai, B.; Zhong, Z.; Ding, Z. Communication-Sensing Region for Cell-Free Massive MIMO ISAC Systems. *IEEE Trans. Wirel. Commun.* **2024**, *23*, 12396–12411. [CrossRef]
10. Huang, C.; Zappone, A.; Alexandropoulos, G.C.; Debbah, M.; Yuen, C. Reconfigurable Intelligent Surfaces for Energy Efficiency in Wireless Communication. *IEEE Trans. Wirel. Commun.* **2019**, *18*, 4157–4170. [CrossRef]
11. Khan, N.A.; Schmid, S. AI-RAN in 6G Networks: State-of-the-Art and Challenges. *IEEE Open J. Commun. Soc.* **2024**, *5*, 294–311. [CrossRef]
12. Huang, H.; Su, J.; Wang, F.-Y. The Potential of Low-Altitude Airspace: The Future of Urban Air Transportation. *IEEE Trans. Intell. Veh.* **2024**, *9*, 5250–5254. [CrossRef]
13. Peleg, M.; Shamai, S. On the Capacity of the Peak-Limited and Band-Limited Channel. *Entropy* **2024**, *26*, 1049. [CrossRef] [PubMed]
14. Liu, Y.; Zhao, X.; Chen, W. User-Perceived Capacity: Theory, Computation, and Achievable Policies. *Entropy* **2024**, *26*, 914. [CrossRef] [PubMed]
15. Li, Y.; Hu, H. Design of Low-Latency Layered Normalized Minimum Sum Low-Density Parity-Check Decoding Based on Entropy Feature for NAND Flash-Memory Channel. *Entropy* **2024**, *26*, 781. [CrossRef]
16. Deng, L.; Shi, Z.; Wang, Y.; Yu, X.; Guan, Y.L.; Xu, Z. Hybrid Window Decoding for Joint Source Channel Anytime Coding System. *Entropy* **2024**, *26*, 940. [CrossRef]
17. Li, X.; Huang, Q. Improved PEG-Based Construction of Analog Fountain Codes. *Entropy* **2024**, *26*, 841. [CrossRef]
18. Tang, Z.; Lei, J.; Huang, Y. EXIT Charts for Low-Density Algebra-Check Codes. *Entropy* **2024**, *26*, 1118. [CrossRef]
19. Shao, J.; Liu, Z.; Liu, Y.; Xie, T. Performance Analysis of Troposphere Scattering Communication Channel with Chirp-BOK Modulation. *Entropy* **2024**, *26*, 1052. [CrossRef]
20. Xin, Y.; Hua, J.; Bao, T.; Hao, Y.; Xiao, Z.; Nie, X.; Wang, F. Generalized Filter Bank Orthogonal Frequency Division Multiplexing: Low-Complexity Waveform for Ultra-Wide Bandwidth and Flexible Services. *Entropy* **2024**, *26*, 994. [CrossRef]
21. Liu, C.; Wu, J.; Zhou, Q. Random Frequency Division Multiplexing. *Entropy* **2025**, *27*, 9. [CrossRef]
22. Han, Z.; Hao, W.; Tang, Z.; Yang, S. Optimal Decoding Order and Power Allocation for Sum Throughput Maximization in Downlink NOMA Systems. *Entropy* **2024**, *26*, 421. [CrossRef] [PubMed]
23. Xin, P.; Fu, Z.; Chen, Z.; Jiang, J.; Zou, J.; Zhang, Y.; Hu, X. Joint User Association, Power Allocation and Beamforming for NOMA-Based Integrated Satellite–Terrestrial Networks. *Entropy* **2024**, *26*, 1055. [CrossRef] [PubMed]

24. Zhao, K.; Mao, Y.; Shi, Y. Simultaneously Transmitting and Reflecting Reconfigurable Intelligent Surfaces Empowered Cooperative Rate Splitting with User Relaying. *Entropy* **2024**, *26*, 1019. [CrossRef]
25. Liu, Y.; Yang, Z.; Wang, B.; Xu, Y. Rate Optimization of Intelligent Reflecting Surface-Assisted Coal Mine Wireless Communication Systems. *Entropy* **2024**, *26*, 880. [CrossRef]
26. Hakimi, A.; Galappaththige, D.; Tellambura, C. A Roadmap for NF-ISAC in 6G: A Comprehensive Overview and Tutorial. *Entropy* **2024**, *26*, 773. [CrossRef] [PubMed]
27. Lin, S.; Wang, H.; Li, W.; Wang, J. Coverage Analysis for High-Speed Railway Communications with Narrow-Strip-Shaped Cells over Suzuki Fading Channels. *Entropy* **2024**, *26*, 657. [CrossRef]
28. Zhang, K.; Zhang, F.; Li, Y.; Wang, X.; Yang, Z.; Liu, Y.; Zhang, C.; Li, X. A Three-Dimensional Time-Varying Channel Model for THz UAV-Based Dual-Mobility Channels. *Entropy* **2024**, *26*, 924. [CrossRef]
29. Zhu, L.H.; Zhu, Z.; Lv, G.L.; Ye, C.Q.; Chen, X.Y. Robustness of Entanglement for Dicke-W and Greenberger-Horne-Zeilinger Mixed States. *Entropy* **2024**, *26*, 804. [CrossRef]
30. Liu, B.; Long, S.; Su, X. Knowledge-Assisted Actor Critic Proximal Policy Optimization-Based Service Function Chain Reconfiguration Algorithm for 6G IoT Scenario. *Entropy* **2024**, *26*, 820. [CrossRef]
31. Zhang, H.; Xu, P.; Dai, B. Ultra-Reliable and Low-Latency Wireless Hierarchical Federated Learning: Performance Analysis. *Entropy* **2024**, *26*, 827. [CrossRef]

entropy

MDPI

Article

On the Capacity of the Peak-Limited and Band-Limited Channel

Michael Peleg * and Shlomo Shamai

Department of Electrical and Computer Engineering, Technion—Israel Institute of Technology, Haifa 3200003, Israel
* Correspondence: peleg.michael@gmail.com

Abstract: We investigate the peak-power limited (PPL) Additive White Gaussian Noise (AWGN) channels in which the signal is band-limited, and its instantaneous power cannot exceed the power P. This model is relevant to many communication systems; however, its capacity is still unknown. We use a new geometry-based approach which evaluates the maximal entropy of the transmitted signal by assessing the volume of the body, in the space of Nyquist-rate samples, comprising all the points the transmitted signal can reach. This leads to lower bounds on capacity which are tight at high Signal to Noise Ratios (SNRs). We find lower bounds on capacity, expressed as power efficiency, that were higher than the known ones by a factor of 3.3 and 8.6 in the low-pass and the band-pass cases, respectively. The gap to the upper bounds is reduced to a power ratio of 1.5. The new bounds are numerically evaluated for FDMA-style signals with limited duration and also are derived in the general case as a conjecture. The penalty in power efficiency due to the peak power constraint is roughly 6 dB at high SNRs. Further research is needed to develop effective modulation and coding for this channel.

Keywords: peak power; capacity; AWGN; band-limited; entropy

1. Introduction

Signals transmitted by wireless communication systems are limited in their power due to the limited capabilities of practical hardware. Two types of power limits were studied already by Shannon [1], namely the average power, which is the long-term average over all the transmitted sequences, and the peak power, which is the maxima of the instantaneous power over the whole transmission. We investigated the peak-power limited (PPL) Additive White Gaussian Noise (AWGN) channels in which the signal is band-limited and its instantaneous power cannot exceed the power P. This model is relevant to many systems in which the peak power is limited by the power amplifier at the transmitter. The model became even more important with the introduction of Digital Pre-Distortion (DPD), e.g., in [2,3], which linearizes the power amplifier up to its maximal transmit power, thus causing it to perform as an ideal soft limiter. Clearly, the capacity limits of this channel are of major practical interest, e.g., the optimization in Section 3.6 of [3], and the interesting peak-power reducing schemes in [4]. Indeed, Shannon already analyzed this channel and presented lower and upper bounds on capacity in [1]. With the exact capacity of the classical Average Power Limited (APL) channel found by Shannon [1] and used widely for tens of years, the PPL channel capacity was studied only sparsely, yielding lower and upper bounds on capacity with a wide gap in between (see [5,6] and Table 1). We think the reason is the difficulty in analyzing this channel as suggested already in [1]. The importance of limiting the peak power is reflected in many works analyzing and reducing the Peak to Average Power Ratio (PAPR), e.g., [7–9]. The impact of the peak power limit has been classical in many communications settings since the start of the wireless communication era, and it is relevant to a variety of practical communications, e.g., modulation and coding over fading channels in [4]. The problem investigated here is related to communication over the Constrained Gaussian Channel (CGC) [10–12], in which a wideband peak-limited signal is fed into a transmit filter in the transmitter. We show below that the capacity of the CGC is an upper bound of the capacity of the PPL channel. The review in [13] presents and categorizes a wide range of modulation schemes with different types of peak limits, including the CGC and the PPL models.

There are two known types of upper bounds on the capacity of the PPL channel. The first one uses the result of [14] on the power spectral density (PSD) of unit processes, which are the inputs to the CGC, to derive the upper bounds of capacity [12] of the CGC which are also valid for the PPL channel. In [15], the approach is specified for the PPL channel, gaining additional insights. The second type of upper bounds releases the constraint on the peak power by applying it only to samples of the signal taken at the Nyquist sampling rate and then computing capacity based on the Nyquist-rate samples being sufficient statistics of the received signals. We denote this approach here as the "sampled discrete analysis". This is introduced in [1] and used in [5] utilizing the capacity of the scalar peak-limited channel derived in [16]. The known lower bounds on capacity are obtained by achievability schemes based on identically and independently distributed (i.i.d.) symbols with optimized pulse shapes (see [1,5,6]).

In this work, we provide numerical evaluation of a lower bound on capacity which is valid for cyclic prefix-assisted Frequency Domain Equalization (CP-FDE) signalling of lengths of up to 100 channel symbols. The CP-FDE signals are not strictly band-limited because they are limited in time; however, they are practically band-limited in the sense of having zero inter-channel interference between users if the rules for cyclic prefixes are adhered to, thus enabling the assignment of adjacent users to channels with no frequency gaps in between. This is applied, for example, in the multiuser uplink of the Long-Term Evolution (LTE) mobile communications system using the Single-Carrier FDMA (SC-FDMA) [17]. Furthermore, we present a lower bound on the capacity of the general PPL channel, which is conjecture due to the two analytical approximations used. We provide lower bounds on capacities improved about 5 dB and more relative to [5,6]. We show that our lower bounds are tight at asymptotically high Signal to Noise Ratios (SNRs), while it is well known [1] that, at very low SNRs, the PPL capacity approaches that of the APL. We investigate first the real-valued channels which model low-pass signals and then the complex-valued channels which model the bandpass signals relevant to radio frequency communications.

The lower bounds in [1,5,6] were obtained using independently and identically distributed (i.i.d.) symbols. Our new approach utilizes dependencies between symbols to increase capacity while meeting the peak power constraint. Modern efforts at PAPR reduction use diverse methods, frequently adapting the transmission per each individual information sequence resembling coding, e.g., [9]. We found that the signals emerging in our new bounds utilize only a small subset of possible symbol sequences selected by the peak power constraint resembling, in a way, coding, in which only a small subset of all possible binary sequences selected by the parity check matrix are valid codewords.

Our new approach is geometry-based; it evaluates the maximal entropy of the transmitted signal by assessing the volume of the body comprising all the points the transmitted signal can reach in the space of Nyquist-rate samples. This is related to the technique introduced in [18] over the CGC. A preprint version of this paper containing additional figures is available in [19].

Notation: Log is the natural logarithm unless stated otherwise. Differential entropy is denoted by h; E denotes the statistical expectation. The N-dimensional vector space of real variables is denoted R^N. The Probability Density Function (PDF) of x is $p_x(x)$ or $p(x)$. Bold italic letters denote vectors. A full list of symbols, variables and abbreviations is provided in Appendix B.

2. System Model

We begin with the real-valued channel. The system is presented in Figure 1. The encoder produces a real-valued low-pass signal $x(t)$ in the frequency band $|f| < B$. The signal is peak-limited, that is, $|x(t)| \leq \sqrt{P}$ for all t. The signal passes an AWGN channel and is decoded. The channel output y is

$$y(t) = x(t) + n(t) \tag{1}$$

where $n(t)$ is a white Gaussian noise with power spectral density N_0 ($0.5N_0$, two-sided) and power $\sigma_n{}^2 = N_0 B$. The Nyquist interval is $T = 0.5/B$. The Signal to Noise ratio is defined as $\rho = \frac{P}{BN_0}$. The Nyquist-rate sample $x(t)$ is denoted by the vector $\boldsymbol{x} = (x_1 \ldots x_n \ldots {}_N)$ of length N. We seek bounds on the capacity C, which is the maximal Mutual Information (MUI) denoted $I(\boldsymbol{x};\boldsymbol{y})$ per Nyquist interval, $C = \frac{1}{N} I(\boldsymbol{x};\boldsymbol{y})$. In the following, we shall analyze the MUI mostly over finite intervals N. As shown in [1], to achieve reliable communication with bitrate approaching the capacity, either coding over many such intervals is required or N has to be taken to its infinite limit.

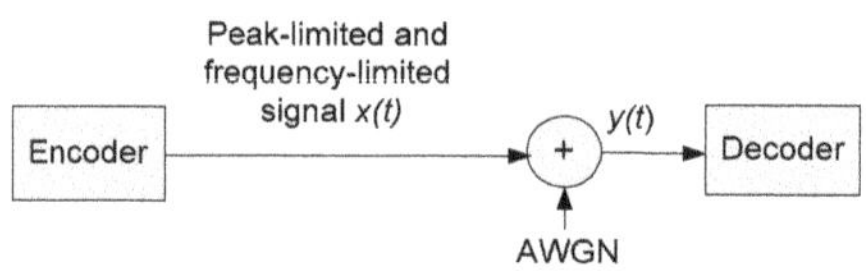

Figure 1. System model of the PPL channel.

The capacity in bits per Nyquist interval of a similar APL system in which the peak power limit P is replaced by the average power limit P is the following famous equation [1]:

$$C_a = 0.5 \log_2\left(\frac{P}{N_0 B} + 1\right) \tag{2}$$

As stated in the Introduction, the capacity of the CGC is an upper bound of the capacity of the PPL channel. The proof is through Lemma 1 of [10], which implies that any signal permitted at the channel input by the PPL model is also valid under the CGC model.

3. Analysis

3.1. General Analysis

The lower bound on capacity can be obtained from the differential entropy $h(\boldsymbol{x})$ of the transmitted signal $\boldsymbol{x}$ via the Entropy Power Inequality (EPI), e.g., [20], as performed, e.g., in [10]. The derivation is presented in Appendix A for completeness. The lower bound γ on the power loss ratio of the PPL channel relative to that of APL channel (2) valid at all SNRs is defined in the sense of (3):

$$C \geq 0.5 \log_2\left(\frac{\gamma \cdot P}{N_0 B} + 1\right) \tag{3}$$

It is shown in Appendix A that the following holds:

$$\gamma = \frac{P^e}{P} \leq 1 \tag{4}$$

where P^e denotes the entropy power of $\boldsymbol{x}$, defined as

$$P^e = \frac{1}{2\pi e} \cdot e^{\frac{2}{N} h(\boldsymbol{x})} \tag{5}$$

The ratio γ is the pre-SNR factor in [6] and it is unity for the APL channel, leading from (3) to (2). The value of γ in (5) is a function of the interval length N. The influence of N is moderate and treated in the sections below.

If the transmitted vector of the N Nyquist-rate samples $\boldsymbol{x}$ is confined to some region of R^N with a volume V_x, then the maximal $h(\boldsymbol{x})$ is the logarithm of the volume V_x and is achieved by the uniform distribution of $\boldsymbol{x}$ over V_x. As shown in Appendix A, under uniform distribution we have

$$P^e = \frac{1}{2\pi e} V_x^{\frac{2}{N}} \tag{6}$$

Combining this with (4) yields

$$\gamma = \frac{V_x^{\frac{2}{N}}}{2\pi e} \tag{7}$$

evaluated for signals with peak power $P = 1$ (see Appendix A).

To provide an upper bound on γ, the peak power limit can be applied on the Nyquist-rate samples only and not on the signal in between, as performed in [5], yielding $\gamma = \frac{2}{\pi e}$; see Appendix A for further explanation.

The lower bounds (3) and (7), which are valid for all SNRs, were shown by [16] to be tight at asymptotically high SNRs for the one-dimensional system analyzed in [16]; they are tight at asymptotically high SNRs in our case too, as shown below in Equation (13).

We follow the method of evaluating the volume V_x presented in [18]. As in [18], all the peak-limited signals form a convex body where convex means that for any pair $\boldsymbol{x}_1, \boldsymbol{x}_2$ in V_x, any linear combination of the two vectors $a\boldsymbol{x}_1 + b\boldsymbol{x}_2$, with a, b being positive and $a_1 + a_2 = 1$, is in V_x. This holds in our case since the absolute value of the linear combination is upper-bounded as

$$|a\boldsymbol{x}_1(t) + b\boldsymbol{x}_2(t)| \leq a|\boldsymbol{x}_1(t)| + b|\boldsymbol{x}_2(t)|;\ a + b = 1,\ a, b > 0.$$

We seek the volume V_x of the N-dimensional convex set $\boldsymbol{x}$, which includes the origin. We denote by r the Distance From the Origin (DFO) to the set surface along the direction of some vector $\boldsymbol{x}$. We denote r as the DFO, while in [21] the term radial function is used. We denote the PDF of r by $p(r)$ for angles θ from the origin to the surface selected randomly and uniformly over the $N-1$-sphere. Then, the volume is

$$V_x = E_\theta\left(V_N^u \cdot r^N\right) \tag{8}$$

where E_θ denotes expectation with respect to θ distributed uniformly over the $N-1$-sphere as defined above and V_N^u denotes the volume of the N-dimensional ball with unit radius. The volume equation is equivalent to proposition 1.13 in [21], where r is denoted radial function. In [21], the expectation is replaced by an integral over all directions. This is equivalent since, in our calculation, the directions are uniformly distributed over the unit $N-1$-sphere. The conversion between (8) here and proposition 1.13 in [21] also utilizes the relation $S_{N-1}^u = N \cdot V_N^u$, where S_{N-1}^u is the area of the unit $N-1$ sphere. The expression in [21] is stated and valid for star-shaped bodies which are a generalization of convex bodies including the origin used in this work.

To estimate the volume, we drew random vectors $\boldsymbol{x}$ with angles θ spread uniformly over the $N-1$ sphere, calculated r for each, substituted them into (8) and computed the average. The random vectors were generated by a method taken from [22], which uses vectors of independent random Gaussian components. The estimation of V_x is illustrated in Figure 2 in two dimensions.

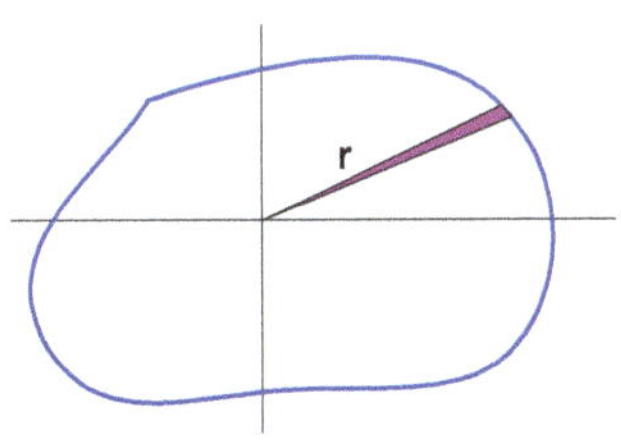

Figure 2. Illustration of the estimation of the volume of a convex set.

This approach is applied to our problem through the following:

1. Generating, numerically or analytically, a signal at a random direction distributed uniformly on the $N-1$ sphere. This is conducted by drawing all the samples independently from a Gaussian distribution.
2. Scaling the generated signal to observe the peak limit of $x(t)$ at any time and sampling it at Nyquist rate to obtain the vector $\boldsymbol{x}$. The resulting length L of $\boldsymbol{x}$ determines the local radius r (DFO) in this direction.
3. Averaging as defined in (8) over many directions yields the volume of the convex body.

The analysis needs to be careful because, in most cases, the peak of the signal before normalization will be very large (maximum of many Gaussian variables) and the volume will be determined mostly by the minority of signals in which the peak is moderate. This minority vanishes with growing N. To address this issue, we shall perform the analysis on blocks of length of NT and then estimate the limit as N approaches infinity.

3.2. Sampled Discrete Analysis

The analysis of maxima of continuous Gaussian processes is difficult and will require approximations [23]. For an initial analysis, we shall peak-limit the signal using only the N Nyquist-rate samples. This is the same as the problem of a sampled system using discrete power-limited symbols treated in [1,5,6] and serves to the evaluate upper bound on γ and to develop our analysis method. The analysis is generalized further below to continuous signals, which are the principal subject of this work. We need the PDF of

$$z = \max_{i \ni \{1 \ldots N\}} |x_i|$$

This is available from the theory of order statistic, e.g., [24,25]. From [24], Equation (2.1.6), we have

$$p_z(z) = N \cdot F_{|x|}(z)^{N-1} \cdot p_{|x|}(z)$$

where p and F denote the PDF and Cumulative Distribution Function (CDF), respectively.

For Gaussian x of unity power P with CDF,

$$F_X(x) = P(x \leq X) = 1 - Q(x)$$

where Q denotes the Q-function. This yields, after few standard steps,

$$p_Z(z) = N[1 - 2Q(z)]^{N-1} \frac{2}{\sqrt{2\pi}} e^{-\frac{z^2}{2}} \tag{9}$$

The distribution *p(z)* is presented for various N values and verified by simulation in Figure 3.

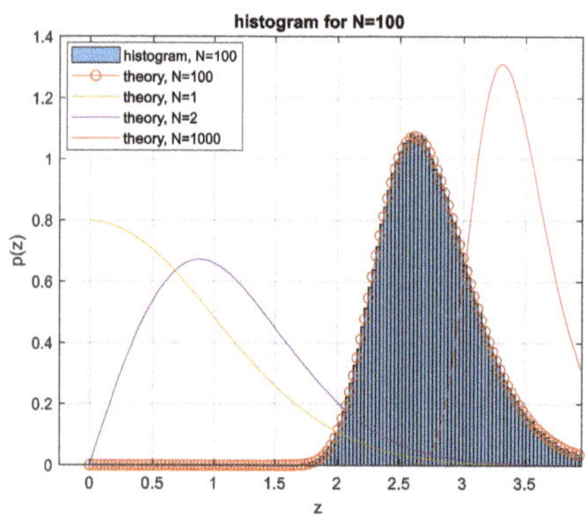

Figure 3. The PDF of the maxima z of absolute values of N Gaussian random variables x_i.

To evaluate the DOF r, we need the PDF of the vector $\boldsymbol{x}$ of the Gaussian variables when the maxima z of the absolute value of its elements is given. This is related to order statistics, e.g., [24,25]; however, such an approach leads to overly complicated analysis. We use the following approximation. The variables x_i are assumed to be i.i.d. Gaussian but limited to the value of the maxima z. This is equivalent to conditioning the Gaussian distribution on $|x| < z$, omitting the index i to simplify presentation in the two following equations.

$$p(x) = \frac{\frac{1}{\sqrt{2\pi}} e^{-\frac{x^2}{2}}}{\int_{-z}^{z} \frac{1}{\sqrt{2\pi}} e^{-\frac{x^2}{2}} dx};\ |x| < z,\ p(x) = 0 \text{ otherwise}$$

So, the variance of x_i is the following function of z:

$$Var(z) = \frac{\int_0^z x^2 e^{-\frac{x^2}{2}} dx}{\int_0^z e^{-\frac{x^2}{2}} dx} \tag{10}$$

We verified (10) by simulation of N-tuples of Gaussian variables for $N = 100$ and found it accurate (see Figure 4).

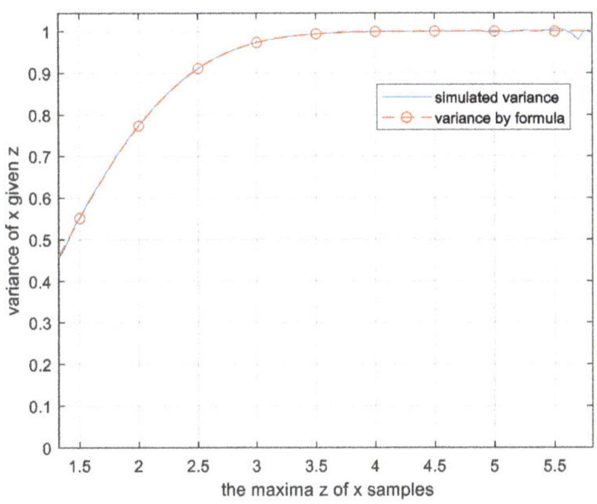

Figure 4. Variance of unordered $N = 100$ Gaussian samples x_i, given their maxima $z = \max(|x_i|)$ with the maximal sample excluded.

The length L of the vector in the analysis below is not a random variable but, rather, the square root of the sum of the variances determined by z. For large N values, this is a good approximation and dropping the randomness forms a lower bound on the entropy, as explained below.

The length L of the normalized vector is then calculated, also accounting for the one sample for which $|x_i| = z$:

$$L(z) = \sqrt{\frac{P \cdot Var(z) \cdot (N-1)}{z^2} + 1}$$

We compute the volume V_x of the peak-limited convex body with a unit power $P = 1$ using (8) and the PDF in (9).

$$V_x = V_N^u \cdot E\left(L^N\right) = V_N^u \cdot \int_0^\infty p_z(z) L(z)^N dz \tag{11}$$

This is a lower bound since $L(z)$ is assumed constant when z is given and only its root mean square is used. The accurate expression would replace in (11) the term $L(z)^N$ with $E[L(z)^N]$.

This would increase V_x by Jensen's inequality since $L(z)$ is positive and raised to a high power in the last equation. Thus, the lower bound evaluated at $P = 1$ is

$$V_x = V_N^u \cdot \int_0^\infty p_z(z)\left(\frac{Var(z)\cdot(N-1)}{z^2}+1\right)^{\frac{N}{2}} dz \quad (12)$$

This is evaluated by numerical integration. The integrand of (12) indicates the range of the maxima z most contributing to the capacity. In Figure 5, we plot the integrand, normalized by its maxima, for each N.

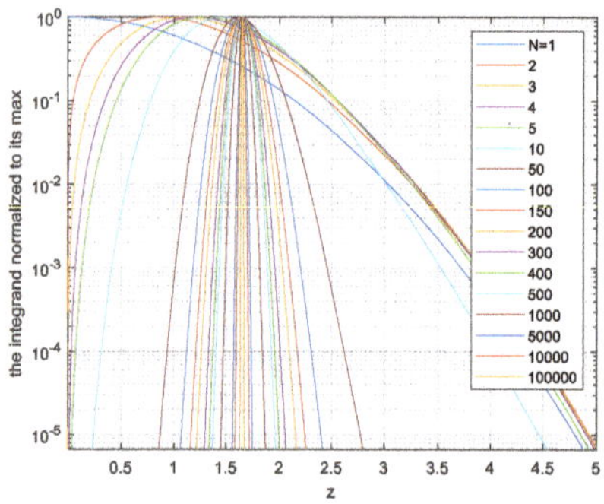

Figure 5. The integrand of Equation (12) for different values of N.

So, for $N > 50$, the system selects signals with relatively low peak values z of about 1.7 and not increasing much with the length N of the signal. This peak occurs, for $N > 100$, with extremely low probability, as seen in Figure 3. This implies that the signals contributing significantly to the integral (12) are rare among signals generated from independent Gaussian-distributed Nyquist-rate samples. The power loss ratio relative to the APL system is obtained by inserting the result of (12) into (7). The power loss is shown in Figure 6.

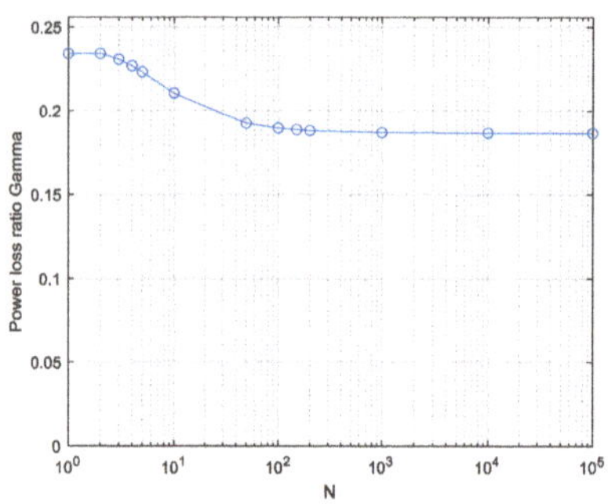

Figure 6. Computed lower bound on the power loss ratio γ; initial analysis limiting the power of Nyquist-rate samples only.

The initial analysis here, which peak-limits only the Nyquist-rate samples, addresses the same discrete symbols problem as [1,5,6], which present a power loss of $\gamma = 0.2342 = 2/\pi e$ (see Appendix A). As explained above, the result in Figure 6 is a lower bound, which explains the gap from our result to the correct $2/\pi e$. Only the values for very low N values approach the exact value. Still, the result is near enough to the exact one to allow qualitative understanding of the numerical evaluation presented further below.

The volume-based evaluation of capacity developed here is tight at asymptotically high SNRs as it is in the one-dimensional case [16] using the following. The capacity is

$$C = \max_{p_x(x)} I(\boldsymbol{x};\boldsymbol{y}) = h(\boldsymbol{y}) - h(\boldsymbol{n}) \quad (13)$$

Our lower bound in (3) and (4) has a higher value than $h(\boldsymbol{x}) - h(\boldsymbol{n})$ (see the derivation in Appendix A); so, to show that it is tight it suffices to show that at high SNRs we have $h(\boldsymbol{x}) \cong h(\boldsymbol{y})$. The volume occupied by all vectors $\boldsymbol{x}$ is (11), which integrates the DFO of $\boldsymbol{x}$, denoted L, raised to the power of N, over all $\boldsymbol{x}$. For any $\boldsymbol{x}$, denoted $\boldsymbol{x}^{i}$, the DFO is $L \geq \sqrt{P}$. Now, $\boldsymbol{y}$ cannot occupy a volume significantly larger than V_x because the added noise increases the DFO in the direction of $\boldsymbol{x}^{i}$ only by a small multiple of σ_n, which is infinitely small relative to $\sqrt{P}$ at asymptotically large SNRs. The volume V_y occupied by all vectors **y** enforces $h(\boldsymbol{y}) \leq \log(V_y)$. Thus, $h(\boldsymbol{x}) \cong h(\boldsymbol{y})$ holds and the lower bound is tight at high SNRs.

3.3. Refinement to Continuous Signals

In this subsection, we replace the peak power limit on the Nyquist-rate samples **x** with the peak power limit on the whole continuous signal $x(t)$. Prasad [26], in his equation (6.4), presents an authoritative and very convenient approximation to the PDF of the PAPR of a complex continuous signal with a duration of N Nyquist rate intervals T. It is derived from (6.3), which is the rigorous PDF of the PAPR of N Nyquist-rate samples, similar to our (9). Prasad carries this out by increasing N by a factor of $\alpha = 2.8$, which is equivalent to replacing the infinity of correlated values in each Nyquist interval by α uncorrelated samples. This yields, in our case (9) of a real-valued signal, the continuous version of CDF and PDF of $z_c = \max_t(|x(t)|)$

$$F_{Z_c}(z_c) = \left[F_{|X|}(z_c)\right]^{\alpha N} \quad (14)$$

And

$$p_{z_c}(z_c) = \alpha N[1 - 2Q(z_c)]^{\alpha N-1}\frac{2}{\sqrt{2\pi}}e^{-\frac{z_c^2}{2}} \quad (15)$$

There are more advanced attempts to approximate $F_{Z_c}(z_c)$, such as [23]; however, those yielded still approximations and very involved expressions. The approximation (15) we use needs verification by simulation. We found the following α to be a good approximation: $\alpha = 2.3$ for N = 101, $\alpha = 2.8$ for N = 1001, $\alpha = 2.9$ for N = 10,001 and N = 100,001. With N = 1001, $\alpha = 2.8$ we obtain Figure 7.

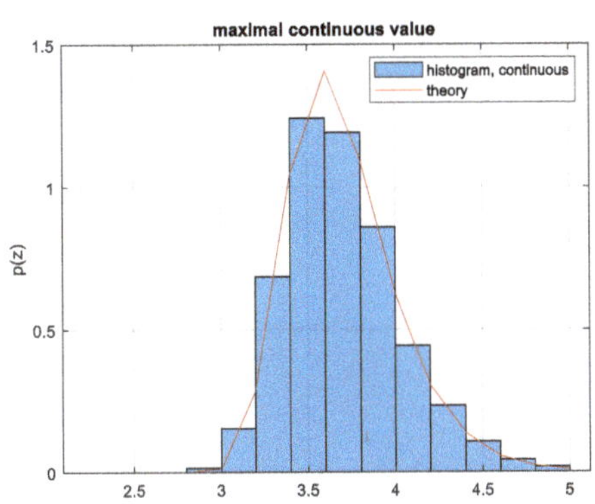

Figure 7. PDF of a maxima of a band-limited Gaussian unit power random process over N = 1001 Nyquist intervals. Histogram by simulation versus (15).

Plugging (15) into (12) and (7) yields the power loss due to the peak limit in Figure 8.

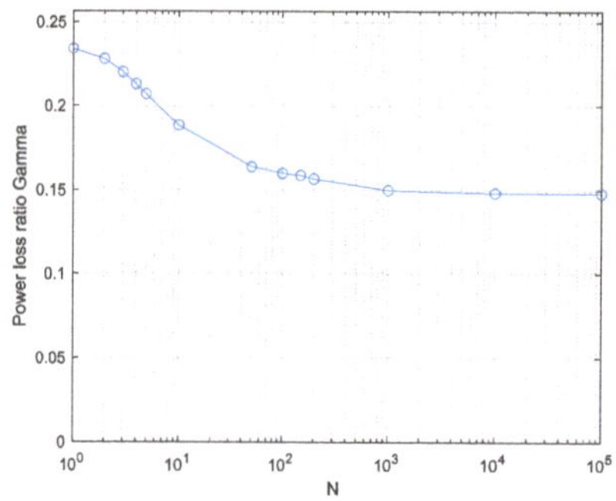

Figure 8. Computed lower bound on the power loss ratio γ, with continuous real-valued PPL signalling.

Conjecture 1. *The power efficiency of 0.15 presented in Figure 8 is a lower bound on γ for continuous low-pass signals.*

Explanation: *The only approximations used were (15), which is similar to [26] and verified numerically, and the approximation (10), verified numerically in Figure 4. Both the approximations seem sound. The result of 0.15 is a lower bound on γ as explained below (11).*

In the next section, to avoid all the approximations used for the analysis above, we shall evaluate the volume V_x, instead of the statistical expectation (11), by generating the vectors x at random and estimating V_x by averaging via (8). We learn from the analysis the following lessons on the number of vectors required in the Monte Carlo evaluation. For $N > 100$, the main contribution to V_x and to capacity is from not overly high values of z which have very low probabilities, as seen by combining Figure 5 with Figure 7. That is, the rare vectors the peak power of which is not too high contribute most to V_x. Generating at random enough vectors, the probability of which is very low, requires a sufficient number of random vectors. As seen in Figures 3 and 7, for $N > 100$, the PDF $p(z)$ of a maxima of Gaussian process decreases sharply with decreasing z. So, it is of interest to evaluate the power ratio γ with the volume in (12) integrated only over $z > z_{min}$ and plotting it as a function of $p\ (z < z_{min})$. Such a plot will enable us to assess the number of simulated sequences required for convergence as follows: select a value of z_{min} that is small enough to enable evaluating γ correctly using Figure 9, in which $p\ (z < z_{min})$ is presented rather than z_{min}, and use number of vectors larger than $1/p\ (z < z_{min})$.

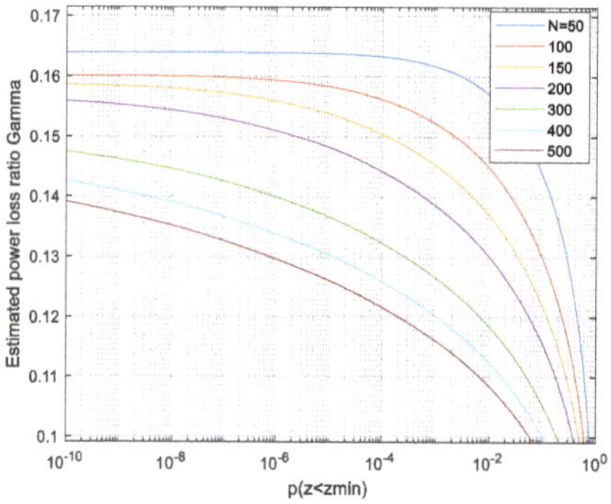

Figure 9. The ratio γ when vectors with a peak smaller then z_{min} are discarded.

Thus, for $N = 50$ about 10^4 simulated sequences are required; for $N = 200$, more than 10^8 should be used. With 10^7 simulated vectors, we can expect reliable results up to an N of about 100. To reach a reliable result, we need to simulate enough vectors to be on the horizontal section of the curve. A useful criterion can be stable results under ten-fold change in p ($z < z_{min}$). A practical method to examine this is discarding the 10 most contributing vectors and permitting only a small change in γ. When using the Monte Carlo evaluation with a too-large N, the estimates will be lower than the true values because the low- probability vectors which contribute most to the capacity will be missed.

4. Monte Carlo Evaluation

The Monte Carlo estimation avoids all the approximations used in the analysis section by generating the vectors $\boldsymbol{x}$ at random and estimating the expectation by averaging as explained above. The evaluation uses importance sampling as presented in Appendix A to accelerate convergence. The signals are evaluated as to be compatible with CP-FDE signalling, that is, generating N Nyquist-rate samples at random, oversampling while keeping the sequence duration at NT, performing Fast Fourier Transform (FFT), brick-wall filtering in the frequency domain, and IFFT. This is the classical transmit side processing of CP-FDE prior to adding the cyclic prefix. Each sequence is scaled as to have $\max(|x(t)|) = 1$; the length r of the scaled and sampled $\boldsymbol{x}$ is calculated and processed by the Monte Carlo equivalent of (8). As a verification, the discrete symbols case, that is, power limiting only the Nyquist-rate samples and not the signal in between, as in [16], evaluates immediately to the correct $\gamma = 2/\pi e$. To evaluate our continuous system, an oversampling ratio of 30 samples per Nyquist interval is used. We begin with a sequence length of N = 101, which is predicted to converge well with the analysis. The results are plotted in Figure 10 as a function of the number N_{sim} of simulated vectors.

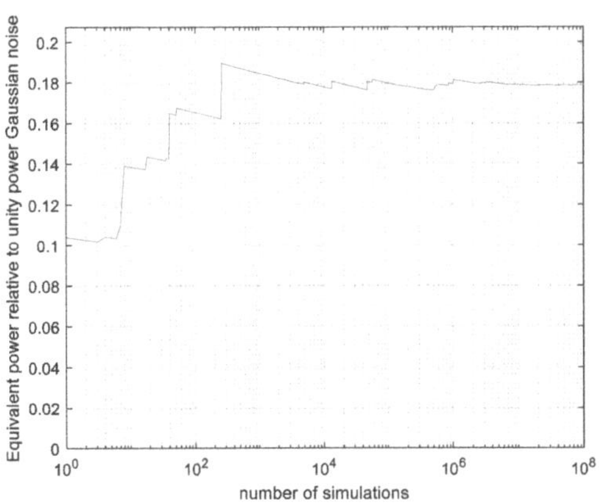

Figure 10. Monte Carlo-evaluated power loss ratio γ; CP-FDE signalling, N = 101, as a function of number of simulated vectors N_{sim}.

The result of $\gamma = 0.18$ is somewhat larger than the lower bound of 0.15 computed by analysis in Figure 8, and there is a convergence after 10^6 simulations. The convergence roughly follows the expectations based on Figure 9. For example, at 10^2 simulations, γ is about 0.02 below its final value as it is approximately for $p\ (z < zmin) = 10^{-2}$ in Figure 9. In Figure 11, we present the results with the same simulation runs if the most contributing runs out of the total 10^8 are discarded.

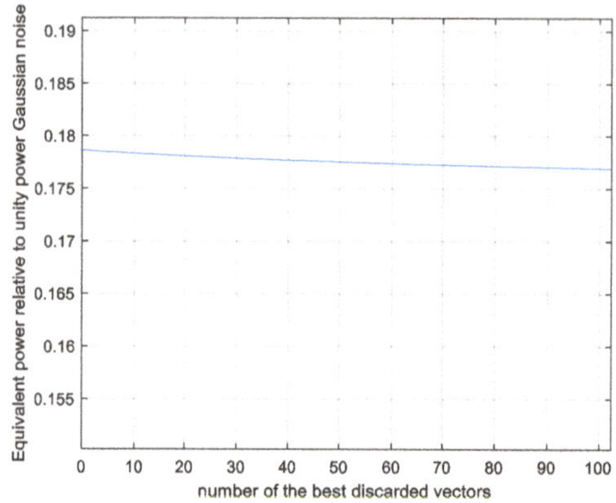

Figure 11. Monte Carlo-evaluated power loss ratio γ; N = 101, as a function of number of the most contributing vectors discarded.

As seen, discarding the ten most contributing runs degrades the performance less than 0.2% and discarding 100 vectors degrades it by 1%, indicating a reliable convergence. This is comparable to the rough expectation based on Figure 9. Next, we estimate γ as a function of the sequence length N using $N_{sim} = 10^8$ simulated vectors per point. The results are plotted in Figure 12.

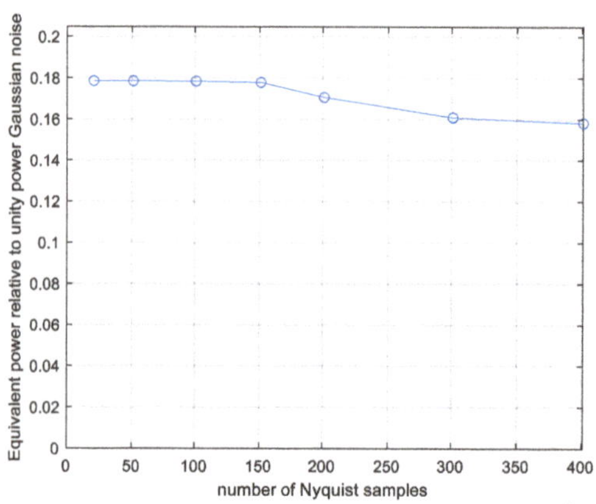

Figure 12. Monte Carlo evaluated power loss ratio γ, CP-FDE signalling, as a function of number of signal duration in Nyquist intervals N.

The γ is above 0.177 for N up to 101 and then starts to decrease. In Figure 13, we show the performance when discarding the p (discard) $\times$ N_{sim} most contributing runs.

The behaviour is qualitatively as predicted in Figure 9, that is, the decrease in the estimated results with rising N is partly due to an insufficient number of simulation runs, with the most contributing vectors becoming rarer as N rises. And, for lengths up to 101 Nyquist intervals, the evaluation is well converged, that is, the results remain the same even if the 10 most contributing runs are discarded as seen in Figure 14.

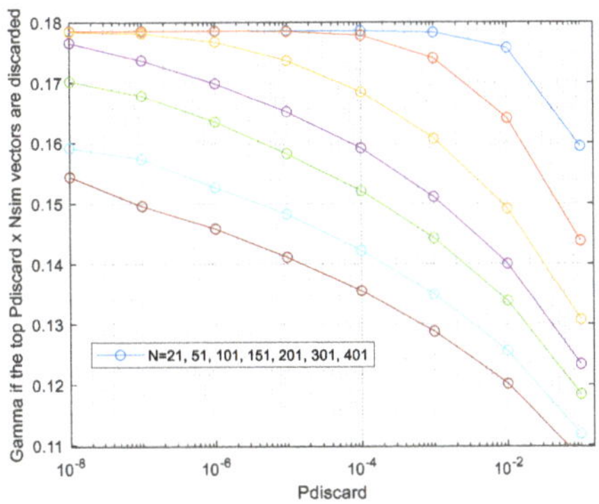

Figure 13. Monte Carlo-evaluated power loss ratio γ; CP-FDE signalling, if the most contributing vectors the probability of which is *Pdiscard* are discarded. The lines correspond to different values of *N*, increasing from top to bottom of the figure.

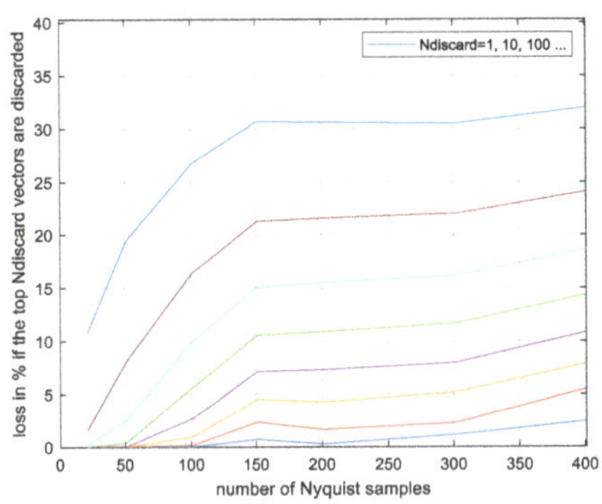

Figure 14. Monte Carlo-evaluated power loss ratio γ; CP-FDE signalling, if the most contributing *Ndiscard* vectors are discarded. The lines correspond to different values of *Ndiscard* increasing fom bottom to top of the figure.

5. Bandpass Signals

We extend the analysis to bandpass signals represented in the complex-valued baseband. We reuse the previous notation with modifications as follows. The encoder produces a complex-valued low-pass signal $x(t)$ in the frequencies $|f| < 0.5B$. The noise is complex-valued with power spectral density N_0 (two-sided). The signal is peak-limited, that is, $|x(t)| \leq \sqrt{P}$ for all t. The Signal to Noise ratio is defined as $\rho = \frac{P}{BN_0}$. The equations are updated as follows (see Appendix A). The classical capacity per Nyquist-rate sample of the APL channel is

$$C_a = \log_2\left(\frac{P}{N_0 B} + 1\right) \tag{16}$$

The lower bound γ on the power loss ratio of the PPL channel relative to the APL channel remains

$$\gamma = \frac{P^e}{P} \leq 1 \tag{17}$$

where P^e denotes the entropy power of $\boldsymbol{x}$ defined now as

$$P^e = \frac{1}{\pi e} \cdot e^{\frac{1}{N} h(x)} \tag{18}$$

It is shown in Appendix A that the following holds:

$$C \geq \log_2\left(\frac{\gamma \cdot P}{N_0 B} + 1\right) \tag{19}$$

In the PPL complex-valued channel, the power ratio γ is shown in Appendix A to be

$$\gamma = \frac{V_x^{\frac{1}{N}}}{\pi e} \tag{20}$$

evaluated with $P = 1$.

To compute the upper bound on capacity by power-limiting only the Nyquist-rate samples and not the signal in between, each complex sample of the entropy-maximizing distribution is uniformly distributed over a disk with a radius of $\sqrt{P}$ yielding $V_x = \left(\pi\sqrt{P}^2\right)^N = (\pi P)^N$, and the ratio γ which is both an upper bound for our continuous signals case and an accurate value for the discrete power-limited problem in [27] is

$$\gamma = \frac{1}{e}, \tag{21}$$

See Appendix A. This value falls correctly between the Smith-based lower and upper bounds in Figure 2 of [27], thus tightening the lower bound of [27] at high SNRs.

Denote the maxima of $|x(t)|$ over N Nyquist intervals as z and $w = z^2$. The work in [26] reminds us that $w_s = |x(t)|^2$ is central chi-square distributed with two degrees of freedom (scaled to a unity mean) with $p_{w_s}(w_s) = e^{-w_s}$ and then provides a good approximation verified by simulations on the CDF of w:

$$F_w(w) = \left(1 - e^{-w}\right)^{\alpha N} \tag{22}$$

with $\alpha = 2.8$. By differentiation,

$$p_w(w) = \alpha N\left(1 - e^{-w}\right)^{\alpha N - 1} e^{-w} \tag{23}$$

Equation (10) is replaced by

$$E(w) = \frac{\int_0^w w_s e^{-w_s} dw_s}{\int_0^w e^{-w_s} dw_s} \tag{24}$$

where E denotes expectation. The last equation was verified numerically by simulation of N-tuples of w_s variables with $N = 100$; see [19] for a plot of the results. The typical length L of the vector normalized to unity $\max(|x(t)|)$ will be, then,

$$L(w) = \sqrt{\frac{E(w)\cdot(N-1)}{w} + 1} \tag{25}$$

The volume V_x is now $V_{2N}^u \cdot E\left(L^{2N}\right)$ because the signals are complex, which doubles the dimension of the convex body. The volume V_x of the peak-limited convex body with a unit power $P = 1$ using the PDF p_w in (23) is

$$V_x = V_{2N}^u \cdot \int_0^\infty p_w(w) L(w)^{2N} dw \tag{26}$$

This is a lower bound as in the real-valued case. It is evaluated by numerical integration and the power loss ratio relative to the APL system is evaluated by (20). In the discrete symbols scenario, that is, power-limiting the Nyquist-rate samples only, evaluated by using $\alpha = 1$ in (23), the exact value is $\frac{1}{e} \cong 0.368$. It is reached only at very low values of N and decreases at larger N values similarly to the real signals case (see the plot in [19]). The lower bound on power efficiency γ with continuous peak power limiting, using $\alpha = 2.8$, is plotted in Figure 15.

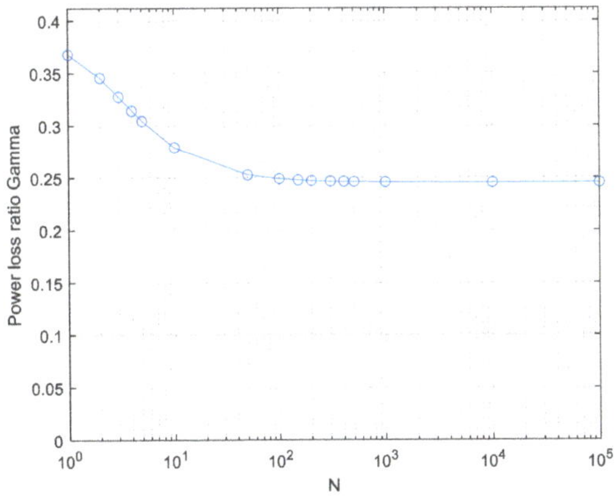

Figure 15. Computed lower bound on power loss ratio γ; continuous PPL signalling. Complex-valued signals.

Conjecture 2. *The power efficiency of 0.245 presented in Figure 15 is a lower bound on γ for continuous bandpass signals.*

Explanation: *The only approximations used were (22), adopted from [26], and the truncation in (24), which was verified numerically. Both the approximations seem sound. The result of 0.245 is a lower bound on γ as explained below (11).*

The Monte Carlo estimation, as in the real-valued case, avoids all the approximations used in the analysis. The signals are evaluated so as to be compatible with CP-FDE signalling. As a verification, the discrete symbols case evaluates immediately to the correct $\gamma = 1/e$. To evaluate our continuous system, an oversampling ratio of 30 samples per Nyquist interval is used. We begin with sequence lengths of $51T$ and $101T$ predicted to converge by the analysis. The results are shown in Figure 16 as a function of the number of simulated vectors N_{sim}.

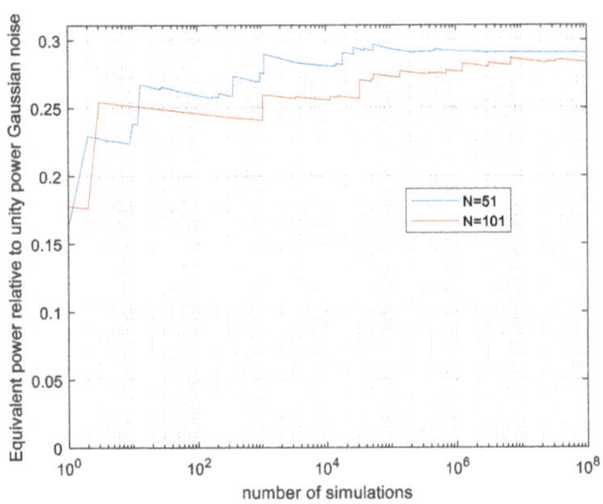

Figure 16. Monte Carlo-evaluated power loss ratio γ; CP-FDE signalling, with $N = 51$ and 101 as a function of number of simulated vectors. Complex-valued signals.

The results N = 51, $\gamma = 0.29$ and N = 101, $\gamma = 0.284$ are slightly larger than the computed lower bound of 0.245 in Figure 15, and there is a convergence after 10^6 and 10^7 simulations for N = 51 and N = 101, respectively. Next, we estimate γ as a function of the sequence length N using $N_{sim} = 10^8$ vectors per point. The results are presented in Figure 17.

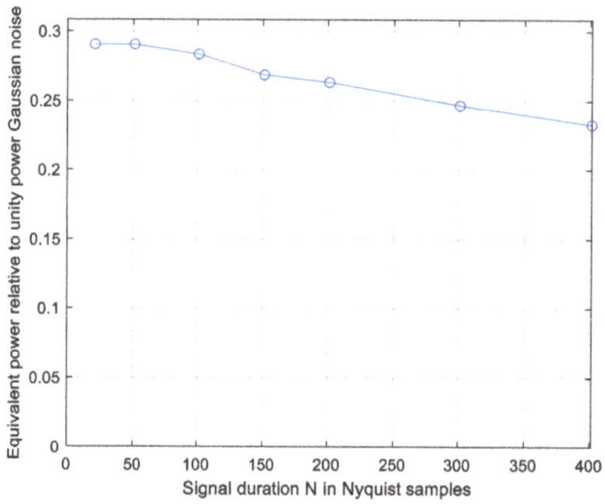

Figure 17. Monte Carlo-evaluated power loss ratio γ; CP-FDE signalling as a function of number of signal durations in Nyquist intervals N. Complex-valued signals.

To assess convergence, figures similar to Figures 9 and 11 and Figures 13 and 14 were constructed for the complex-valued signals; those are available in [19]. The behaviour is similar to the real-valued case. It is evident that more vectors are required in the complex case, e.g., $N = 50$ requires about 10^5 vectors while N = 101 requires about 10^8. The decrease in the estimated results with growing N values is in part due to not enough simulation runs, with the most contributing vectors becoming rarer as N rises. The $N = 51$ result $\gamma = 0.29$ is reliable and is stable even if the 100 most contributing vectors are discarded. The result at N = 101 is probably slightly lower than the true γ; it loses 1.6% if the 10 most contributing vectors are discarded. The result $\gamma = 0.29$ is the γ of the CP-FDE signalling at a signal duration of $51T$ and the values in Figure 17 at $N > 100$ are lower bounds applicable to the CP-FDE signalling.

6. Conclusions

The important problem of the capacity of the PPL channel was investigated. We focussed on the power efficiency γ, which provides a lower bound on capacity, tight at asymptotically high SNRs. The results are summarized in Table 1. We showed that the new lower bounds on γ are about 3.3 and 8.6 times higher than previously known in the low-pass and in the bandpass cases, respectively. The gap to the upper bounds is narrowed to less than 2 dB. The numerical results for γ are valid for practical CP-FDE signalling with limited transmission duration, while the general analytical results are lower bounds on γ as explained below (11) and rely on two approximations, rendering them conjecture. The lower bounds based on γ via (3) are tight at high SNRs and show that the peak power constraint causes, at high SNRs, power loss of about 6 dB in the bandpass case and a little more in the low-pass case.

Future work: The above-stated upper bounds are based on peak-power limiting of only the Nyquist-rate samples which are assumed to be independent. It would be interesting to also account for the correlation which is implied on these samples, as is reflected in [15], to sharpen the upper bounds. This might be addressed via a Multiple Input Multiple Output (MIMO) structure, where both constraints should be addressed. The MIMO setting might also be relevant for a super Nyquist sampling, again accounting for the fact that all samples are peak-power limited; see [15,28,29] for relevant results that might be used.

Table 1. Bounds on power efficiency γ of peak-limited signals.

The Bound	Results of [5]	Results of [6]	This Work, General Result, But Remains Conjecture	This Work	This Work, CP-FDE Signalling, Duration of 101 Nyquist Intervals
Low-pass lower bound	$0.0361 = \pi/32e$	0.04470	$\gamma > 0.15$		$\gamma = 0.18$
Low-pass upper bound	$0.2342 = 2/\pi e$ Also presented in [1].	$0.2342 = 2/\pi e$			
Bandpass lower bound	$\frac{\pi^2}{128e} = 0.0284$		$\gamma > 0.245$		$\gamma = 0.284$
Bandpass upper bound				$\frac{1}{e} \cong 0.368$	

Also, our analysis showed that the capacity achieving signals are a small proportion of Gaussian-bandlimited signals and this proportion vanishes with growing signal duration. The same holds for codewords of error correcting codes; the codewords are a vanishing proportion of all random binary sequences. The peak limit selects the appropriate signals resembling somewhat the effect of the parity check matrix of a binary error correcting code. Thus, further work should seek structures of PPL signals approaching the channel capacity similar to the vast work performed in recent decades on error correcting codes.

Author Contributions: Conceptualization, M.P. and S.S.; Methodology, M.P. and S.S.; Software, M.P.; Formal analysis, M.P. and S.S.; Investigation, M.P. and S.S.; Writing—original draft, M.P. and S.S.; Supervision, S.S. All authors have read and agreed to the published version of the manuscript.

Funding: This work was supported by the German Research Foundation (DFG) via the German-Israeli Project Cooperation (DIP), under Project SH 1937/1-1.

Institutional Review Board Statement: Not applicable since this work did not involve patients, humans, plants or biological samples.

Data Availability Statement: Additional figures are available in [19].

Acknowledgments: We thank the anonymous reviewers for their comments which improved the accuracy and clarity of this paper.

Conflicts of Interest: The authors declare no conflicts of interest.

Appendix A. Bounds

We summarize the bounding technique used, e.g., in [10] for completeness. We shall start with real-valued signals. We denote differential entropy as h. The channel is an AWGN channel with scalar-valued real symbols (1). The noise has a power of $\sigma_n^2 = N_0 B$ and a per-sample entropy of $h(n) = 0.5\log(\sigma_n^2 \cdot 2\pi e)$. The capacity per sample is

$$C = \frac{1}{N}[h(\boldsymbol{y}) - h(\boldsymbol{y}|\boldsymbol{x})] = \frac{1}{N}[h(\boldsymbol{y}) - h(\boldsymbol{n})]$$

where the bold symbols denote vectors and N is the number of Nyquist-rate samples.

Using the EPI,

$$e^{\frac{2}{N}h(\boldsymbol{y})} \geq e^{\frac{2}{N}h(\boldsymbol{x})} + e^{\frac{2}{N}h(\boldsymbol{n})} \tag{A1}$$

$$h(\boldsymbol{y}) \geq \frac{N}{2}\log\left(e^{\frac{2}{N}h(\boldsymbol{x})} + e^{\frac{2}{N}h(\boldsymbol{n})}\right)$$

and

$$e^{\frac{2}{N}h(\boldsymbol{n})} = \sigma_n^2 \cdot 2\pi e$$

So, the per-sample capacity C is

$$C \geq \frac{1}{2}\log\left(e^{\frac{2}{N}h(x)} + \sigma_n^2 \cdot 2\pi e\right) - \frac{1}{2}\log\left(\sigma_n^2 \cdot 2\pi e\right)$$

$$C \geq \frac{1}{2}\log\left(\frac{\frac{1}{2\pi e}\cdot e^{\frac{2}{N}h(x)}}{\sigma_n^2 \cdot} + 1\right) \tag{A2}$$

We use the entropy power of **x** definition $P^e = \frac{1}{2\pi e}\cdot e^{\frac{2}{N}h(x)}$, yielding, in bits per sample,

$$C \geq \frac{1}{2}\log_2\left(\frac{P^e}{\sigma_n^2} + 1\right) \tag{A3}$$

followed by the pre-SNR factor (3), (4). The ratio γ in (4) is the lower bound on the power loss ratio of the PPL channel relative to the APL channel and is valid at all SNRs. It is used in [6] as the pre-SNR factor. Now, $h(x) = \log(V_x)$, where V_x is the volume of the convex body if the distribution of $\boldsymbol{x}$ is the entropy-maximizing uniform one. This yields (6). It follows from the peak power limit P that V_x is proportional to $P^{\frac{N}{2}}$. Then, by (6), P^e is linearly proportional to P and volume-derived γ is invariant with respect to P. Therefore, γ (4) can be evaluated with $P = 1$ with no loss of generality.

To provide an upper bound on γ, the peak power limit can be applied to the Nyquist-rate samples only and not to the signal in between as conducted in [5]. In this case, the signal samples are limited to $-\sqrt{P} < x < \sqrt{P}$, the convex body is an N-cube with volume $(2\sqrt{P})^N$ and (7) yields the power loss γ of $\frac{2}{\pi e}$ presented in [5].

Extension to complex-valued signals:

The symbols are complex. The real and imaginary components of the noise have a power of $\sigma_n^2/2 = N_0B/2$ each and the per-symbol noise entropy is $h(n) = \log(\sigma_n^2 \cdot \pi e)$. The per-symbol capacity is

$$C = \frac{1}{N}[h(\boldsymbol{y}) - h(\boldsymbol{n})]$$

The EPI formula as in (A1) holds for N-dimensional real vectors; for N-dimensional complex vectors, it is

$$e^{\frac{1}{N}h(\boldsymbol{y})} \geq e^{\frac{1}{N}h(\boldsymbol{x})} + e^{\frac{1}{N}h(\boldsymbol{n})}$$

yielding

$$C \geq \log\left(\frac{\frac{1}{\pi e}\cdot e^{\frac{1}{N}h(\boldsymbol{x})}}{\sigma_n^2} + 1\right) \tag{A4}$$

For a Gaussian signal with power P, this turns into the familiar (16). We can denote the entropy power P^e of the complex-valued x as in (18) and define γ by (4) yielding (19). As in the real-valued case, the maximal entropy is $h(\boldsymbol{x}) = \log(V_x)$, where V_x is the volume of the convex body. This yields, for the uniform entropy-maximizing distribution of $\boldsymbol{x}$,

$$P^e = \frac{1}{\pi e}V_x^{\frac{1}{N}} \tag{A5}$$

and

$$C \geq \log\left(\frac{\frac{1}{\pi e}V_x^{\frac{1}{N}}}{\sigma_n^2 \cdot} + 1\right) \tag{A6}$$

and (20) follows.

Appendix B. Importance Sampling in the N-Cube

As explained, generating signals at random on the N-sphere yields Gaussian-distributed samples. This enables analytical results and straightforward simulation; however, there is a problem. As explained above, with no oversampling, the optimal signal is distributed uniformly in the hypercube, the signal volume is concentrated near the N-cube corners and the corner regions are visited by the random signals, which are uniform on the N-sphere, very rarely. Thus, the evaluation by simulation converges very slowly.

We shall accelerate the estimation by importance sampling, generating the random signals uniformly in the hypercube instead on the unit-radius hypersphere. This accelerates the evaluation to some degree even if the signal region in the continuous case is definitely not a hypercube. The principles are as follows:

The signals are generated uniformly distributed in the volume of an N-cube by randomly generating each sample x_i uniformly in the [5] interval. For each signal instance, a probability factor P_c is computed, equal to probability of this particular angle if the signals were generated uniformly over angles, that is, uniformly over the unity radius N-sphere, divided by the probability if the signals were generated uniformly in the N-cube. P_c is the ratio of the volume of the hyperball enclosed in an infinitesimal angle sector around the selected signal to the volume of the hypercube enclosed in the same-width angle sector corrected by the volume ratio of the hypercube 2^N and the hypersphere V_N^u. Thus, denoting by L_c the DFO of the hypercube surface in the chosen direction,

$$P_c = \frac{1}{L_c^N}\frac{2^N}{V_N^u}$$

The volume evaluation (8) is modified according to the importance sampling method to

$$V_Z = \mathrm{E}_\theta\left(V_N^u \cdot r^N \cdot P_c\right) = \mathrm{E}_\theta\left(V_N^u \cdot r^N \cdot \frac{1}{L_c^N}\frac{2^N}{V_N^u}\right) = \mathrm{E}_\theta\left\{\left(\frac{2\cdot r}{L_c}\right)^N\right\} \tag{A7}$$

where the expectation is carried with signals distributed uniformly over the N-cube.

Intuitive check: suppose the signal body is an N-cube with samples $-1 < x_i < 1$. Then, $r = L_c$ is always the case since r is scaled by the peak limiting to touch the hypercube surface and the volume will be 2^N as it should be.

Extension to the complex-valued case:

The signals are now generated uniformly distributed in the volume of a $2N$-dimensional body by randomly generating each of the N complex samples x_i independently and uniformly over an area in the complex plane enclosed by a circle with a unity radius reflecting $|x_i| < 1$ and an area of π.

The last two equations are updated accordingly accounting for the volume of the $2N$-dimensional body of π^N and for the number of dimensions being $2N$ since each x_i is complex. L_c is the DFO of the surface of the $2N$-dimensional body in the chosen direction.

$$P_c = \frac{1}{L_c^{2N}}\frac{\pi^N}{V_{2N}^u}$$

The volume evaluation (8) is modified according to the importance sampling method to

$$V_Z = \mathrm{E}_\theta\left(V_{2N}^u \cdot r^{2N} \cdot P_c\right) = \mathrm{E}_\theta\left\{\pi^N\left(\frac{r}{L_c}\right)^{2N}\right\} \tag{A8}$$

Appendix C. Symbols, Variables and Abbreviations

Symbol or Abbreviation	Meaning
APL	Average Power Limited
AWGN	Additive White Gaussian Noise
B	Bandwidth
$\boldsymbol{b}$, bold italic letters	Vector of Samples of $b(t)$
C	Capacity
C_a	Capacity Over the APL Channel
CDF	Cumulative Distribution Function
CGC	Constrained Gaussian Channel
CP-FDE	Cyclic Prefix-Assisted Frequency Domain Equalization
DFO	Distance From the Origin
DPD	Digital Pre-Distortion
EPI	Entropy Power Inequality
E_θ	The Statistical Expectation with Respect to θ
f	Frequency
F	Cumulative Distribution Function
FDMA	Frequency Domain Multiple Access
FFT	Fast Fourier Transform
h	Differential Entropy
$I(x;y)$	Mutual Information
i.i.d.	Independently and Identically Distributed
IFFT	Inverse Fast Fourier Transform
L	Length of the Signal Vector x
Log	The Natural Logarithm
LTE	Long-Term Evolution
MIMO	Multiple Input Multiple Output
MUI	Mutual Information
N	Duration of Message in Nyquist Intervals
$n(t)$	White Gaussian Noise
N_0	Noise Power Spectral Density
N_{sim}	Number of Simulated Vectors
P	Power and Peak Power
PAPR	Peak to Average Power Ratio
P_c	Probability factor in Importance Sampling
PDF	Probability Density Function
P^e	The Entropy Power
PPL	Peak-Power Limited
PSD	Power Spectral Density
$p_x(x)$ or $p(x)$.	Probability Density Function of x
$Q(x)$	The Q-function
r	Distance From the Origin
R^N	The N-Dimensional Vector Space Of Real Variables
SC-FDMA	Single-Carrier FDMA
SNR	Signal to Noise Ratio
S^u_{N-1}	The Area of the Unit $N-1$ Sphere
t	Continuous Time
T	Nyquist Interval
Var	Variance
V^u_N	The Volume of the N-Dimensional Ball with Unit Radius
V_x	Volume Occupied by all Permissible Vectors $\boldsymbol{x}$
$x = (x_1 \ldots x_n \ldots x_N)$	Vector of Nyquist-Rate Samples of $x(t)$
$x(t)$	Continuous Input Signal
$y(t)$	Continuous Output Signal
z	Maximal Abs. Value of the Nyquist-Rate Samples of x(t)
z_c	Maximal Absolute Value of $x(t)$
α	Correction Factor by Prasad
γ	Lower Bound on the Power Loss Ratio
θ	Angles from the Origin to Points on the Surface of the Convex Body
ρ	Signal to Noise Ratio
${\sigma_n}^2$	Noise power

References

1. Shannon, C.E.; Weaver, W. *The Mathematical Theory of Communication*; University of Illinois Press: Champaign, IL, USA, 1949.
2. Haider, M.F.; You, F.; He, S.; Rahkonen, T.; Aikio, J.P. Predistortion-Based Linearization for 5G and Beyond Millimeter-Wave Transceiver Systems: A Comprehensive Survey. *IEEE Commun. Surv. Tutor.* **2022**, *24*, 2029–2072. [CrossRef]

3. Ugolini, A. Physical and MAC Layer Techniques for Next Generation Satellite Communications. Ph.D. Thesis, Università Degli Studi Di Parma, Parma, Italy, 2016.
4. Miller, S.L.; O'Dea, R.J. Peak power and bandwidth efficient linear modulation. *IEEE Trans. Commun.* **1998**, *46*, 1639–1648. [CrossRef]
5. Shamai, S. On the capacity of a Gaussian channel with peak power and bandlimited input signals. *AEU Arch. Fur Elektron. Und Ubertragungstechnik* **1988**, *42*, 340–346.
6. Wang, Y.; Zhou, J.; Zhang, W. Capacity bounds for bandlimited Gaussian channels with peak-to-average-power-ratio constraint. In Proceedings of the 2018 IEEE Information Theory Workshop, ITW 2018, Guangzhou, China, 25–29 November 2018. [CrossRef]
7. Ochiai, H.; Imai, H. On the distribution of the peak-to-average power ratio in OFDM signals. *IEEE Trans. Commun.* **2001**, *49*, 282–289. [CrossRef]
8. Savaux, V.; Louet, Y. PAPR Analysis as a ratio of two random variables: Application to multicarrier systems with low subcarriers number. *IEEE Trans. Commun.* **2018**, *66*, 5732–5739. [CrossRef]
9. Sayyari, R.; Pourrostam, J.; Ahmadi, H. Efficient PAPR reduction scheme for OFDM-NOMA systems based on DSI & precoding methods. *Phys. Commun.* **2021**, *47*, 101372. [CrossRef]
10. Ozarow, L.H.; Wyner, A.D.; Ziv, J. Achievable rates for a constrained Gaussian channel. *IEEE Trans. Inf. Theory* **1988**, *34*, 365–370. [CrossRef]
11. Neuhaus, P.; Dörpinghaus, M.; Fettweis, G. Zero-crossing modulation for wideband systems employing 1-bit quantization and temporal oversampling: Transceiver design and performance evaluation. *IEEE Open J. Commun. Soc.* **2021**, *2*, 1915–1934. [CrossRef]
12. Shamai, S.; Bar-David, I. Upper bounds on capacity for a constrained Gaussian channel. *IEEE Trans. Inf. Theory* **1989**, *35*, 1079–1084. [CrossRef]
13. Bar-David, I. On the Capacity of Peak Power Constrained Gaussian Channels. In *Performance Limits in Communication Theory and Practice*; NATO ASI Series; Springer: Dordrecht, The Netherlands, 1988. [CrossRef]
14. Shepp, L.A. Covariance of unit processes. In Proceedings of the Working Conference on Stochastic Processes, Santa Barbara, CA, USA; 1967; pp. 205–218.
15. Shamai, S.; Chayat, N. The autocorrelation function of a peak-power-limited process. *Signal Process.* **1991**, *24*, 127–136. [CrossRef]
16. Smith, J.G. The information capacity of amplitude- and variance-constrained scalar gaussian channels. *Inf. Control* **1971**, *18*, 203–219. [CrossRef]
17. Zhang, J.; Yang, L.-L.; Hanzo, L.; Gharavi, H. Advances in Cooperative Single-Carrier FDMA Communications: Beyond LTE-Advanced. *IEEE Commun. Surv. Tutor.* **2015**, *17*, 730–756. [CrossRef]
18. Peleg, M.; Shamai, S. Geometrical bounds on the capacity of the binary bipolar input AWGN channel. In Proceedings of the IEEE COMCAS 2024 Conference, Tel Aviv, Israel, 9–11 July 2024.
19. Peleg, M.; Shamai, S. On the Capacity of the Peak Limited and Band Limited Channel. *arXiv* **2024**, arXiv:2410.04912.
20. Blachman, N. The convolution inequality for entropy powers. *IEEE Trans. Inf. Theory* **1965**, *11*, 267–271. [CrossRef]
21. Klain, D.A. Invariant valuations on star-shaped sets. *Adv. Math.* **1997**, *125*, 95–113. [CrossRef]
22. Marsaglia, G. Choosing a Point from the Surface of a Sphere. *Ann. Math. Stat.* **1972**, *43*, 645–646. [CrossRef]
23. Diebolt, J.; Posse, C. On the density of the maximum of smooth Gaussian processes. *Ann. Probab.* **1996**, *24*, 1104–1129. [CrossRef]
24. David, H.A.; Nagaraja, H.N. *Order Statistics*, 3rd ed.; Wiley: Hoboken, NJ, USA, 2004; ISBN 978-0-471-65401-8.
25. Tong, Y.L. Order Statistics of Normal Variables. In *The Multivariate Normal Distribution*; Springer Series in Statistics; Springer: New York, NY, USA, 1990. [CrossRef]
26. Prasad, R. *OFDM for Wireless Communications Systems*; Chapter 6; Artech House: London, UK, 2004.
27. Shamai, S.; Bar-David, I. The Capacity of Average and Peak-Power-Limited Quadrature Gaussian Channels. *IEEE Trans. Inf. Theory* **1995**, *41*, 1060–1071. [CrossRef]
28. Dytso, A.; Goldenbaum, M.; Poor, H.V.; Shamai, S. Amplitude constrained MIMO channels: Properties of optimal input distributions and bounds on the capacity. *Entropy* **2019**, *21*, 200. [CrossRef] [PubMed]
29. Favano, A.; Ferrari, M.; Magarini, M.; Barletta, L. The Capacity of the Amplitude-Constrained Vector Gaussian Channel. In Proceedings of the 2021 IEEE International Symposium on Information Theory, Melbourne, Victoria, Australia, 12–20 July 2021.

entropy

MDPI

Article

User-Perceived Capacity: Theory, Computation, and Achievable Policies

Yuanrui Liu [1], Xiaoyu Zhao [2,*] and Wei Chen [1]

[1] Department of Electronic Engineering, State Key Laboratory of Space Network and Communications, Beijing National Research Center for Information Science and Technology, Tsinghua University, Beijing 100084, China; liuyr19@mails.tsinghua.edu.cn (Y.L.); wchen@tsinghua.edu.cn (W.C.)

[2] School of Cyber Science and Engineering, Southeast University, Nanjing 211189, China

* Correspondence: xy-zhao@seu.edu.cn

Abstract: User-perceived throughput is a novel performance metric attracting a considerable amount of recent attention because it characterizes the quality of the experience in mobile multimedia services. For instance, it gives a data rate of video streaming with which a user will not experience any lag or outage in watching video clips. However, its performance limit remains open. In this paper, we are interested in the achievable upper bound of user-perceived throughput, also referred to as the user-perceived capacity, and how to achieve it in typical wireless channels. We find that the user-perceived capacity is quite limited or even zero with channel state information at the receiver (CSIR) only. When both CSIR and channel state information at the transmitter (CSIT) are available, the user-perceived throughput can be substantially improved by power or even rate adaptation. A constrained Markov decision process (CMDP)-based approach is conceived to compute the user-perceived capacity with joint power–rate adaptation. It is rigorously shown that the optimal policy obeys a threshold-based rule with time, backlog, and channel gain thresholds. With power adaptation only, the user-perceived capacity is equal to the hard-delay-constrained capacity in our previous work and achieved by joint diversity and channel inversion.

Keywords: user-perceived capacity; ultra-reliable and low-latency communications; real-time wireless communication; hard delay constraint; cross-layer scheduling; finite horizon Markov decision processes

1. Introduction

Future 6G communication aims to provide a revolutionary user experience [1,2], with real-time communication holding high expectations in achieving the goals of future mobile communications. On the other hand, the user-perceived capacity is a novel performance metric that has attracted considerable recent attention because it characterizes the quality of experience in mobile multimedia services [3]. In this paper, our objective is to improve the user-perceived capacity to meet the demands of real-time communication in future wireless communications.

Real-time communication is essential for applications where latency is a critical factor. It ensures that the time taken for data to travel from the transmitter to the receiver is within acceptable limits, enabling seamless and interactive experiences. Study [4] indicated that communication in real-time systems must be predictable, and a scheme was proposed for providing predictable inter-process communication in real-time systems with a point-to-point interconnection network. Study [5] focused on the time sensitive networks, analyzed the key functional parameters affecting the deterministic behavior in real-time communication protocols and, based on the configuration of these parameters, derived the constraints required for computing schedules to ensure the deterministic end-to-end delay of critical communication flows. In [6], the authors concentrated on real-time communication for industrial automation environments, aiming to meet the demands of real-time communication in a scenario where the number of networked devices and computational capabilities are expected to significantly increase. The authors of [7] focused on wireless sensor networks, summarizing and employing various methods based on both hard real-time and soft real-time models. Furthermore, they provided scheduling policies to ensure the requirements of real-time communication. In particular, hard real-time communication implies that missing the deadline can lead to a system-wide failure, necessitating the satisfaction of end-to-end latency constraints within the deadline. On the other hand, soft real-time communication aims to reduce the transmission of packets that exceed the deadline, that is, to minimize the probability of exceeding the deadline. In this paper, we focus on hard latency constraints. This is due to hard latency constraints playing an extremely important role in the ultra-reliable and low-latency communications (URLLC) scenario, ensuring timely and reliable data delivery [8]. Many studies have primarily targeted hard latency constraints, such as [9–17]. In the literature, the authors of [10] considered a communication network under hard real-time constraints, wherein the source node transmits information to the destination node via a directed acyclic graph subject to hard latency constraints. In [9], the authors demonstrated that hard latency constraints can be provided using frequency or spatial diversity and time domain power adaptation, with the condition that diversity gains are at least two. Study [9] also revealed the underlying relationships between the required average power, throughput, and outage probability. The authors of [12] presented a novel power control scheme enhancing the link layer performance in delay-sensitive communications over fading channels. This scheme is particularly adept at managing joint queue length awareness. In [13], the authors introduced an optimal dynamic coding algorithm, along with its low-complexity approximations, designed for deadline-constrained traffic. Additionally, a joint coding and rate control policy proposed by [13] outperformed in time-varying channels, making it particularly beneficial for multimedia streaming and interactive real-time applications. This proposed policy was demonstrated to offer superior performance and stability. However, there is also a considerable amount of research that concentrates on soft real-time communications, specifically addressing the issue of delay violation probabilities over wireless channels [18]. Refs. [19,20] contributed to the field by formulating a CMDP to minimize the queue-length-bound violation probability, utilizing linear programming to obtain the optimal probabilistic policy, balancing violation probability with transmission power. The authors of [21] proposed a framework for cross-layer optimization to ensure the required quality-of-service with finite transmit power, considering the transmission error probability, queueing delay violation probability, and packet dropping probability. Liu et al. proposed a URLLC-centric task offloading and resource allocation framework, incorporating the statistical characteristics' extreme queue length events [22]. In [23], the authors focused on reducing packet delay violation probabilities within periodic traffic models and, by utilizing Markov decision processes and nonlinear Knapsack Problems, presented an asymptotically optimal policy.

In the context of real-time communication, wireless networks are responsible for transmitting substantial amounts of data in a specified time frame, with an increasing emphasis on ensuring high service quality. A significant portion of data has strict latency requirements, which arise from various application scenarios involving real-time communication, such as autonomous driving, health monitoring, and industrial automation. To meet these latency requirements, data transmissions must be completed within a given deadline constraint. Consequently, Zafer and Modiano proposed a BT problem in [24], addressing the challenge of transmitting B units of data within a deadline T. They employed continuous-time stochastic control theory to derive the optimal policy for the BT problem. In addition, Berry et al. provided an overview of research related to packet delay and energy consumption in [25], including the BT problem and delay-constrained issues. For discrete-time point-to-point communication systems, refs. [26,27] investigated the optimal data transmission for BT problems under constraints such as random data arrival and Quality of Service (QoS), without considering time-varying channels. To specifically address the impact of time-varying channel conditions on the BT problem, several studies [11,28,29] extended the BT problem from static channels to fading channels. Furthermore, studies [24,30–32] have considered the BT problem in continuous-time systems, exploring the optimal data transmission policies in the presence of time-varying channels. In addition to single-user scenarios, the academic community has also conducted research on multi-user scenarios. For instance, the authors of [33] investigated the BT problem in a time-shared multi-user system by considering random arrivals and time-varying channels for multiple users. Furthermore, study [34] extended the BT problem to multi-access scenarios and examined power minimization scheduling with deadline constraints in a multi-access channel environment. Moreover, under the framework of the BT problem, studies such as [35–39] further explored deadline-constrained communications in scenarios such as multi-hop networks, computation offloading, lossy networks, and multi-robot task allocation, providing optimal scheduling policies.

Cross-layer design is widely recognized as a crucial methodology for ensuring real-time performance for delay-constrained services. Notably, Collins and Cruz pioneered the integration of network and physical layer designs to optimize low-latency wireless transmissions, resulting in the formulation of a cross-layer scheduling framework [40]. After that, there has been an increasing focus on interactions across different layers [41–43], leading to the proposal of various approaches for cross-layer scheduling in diverse application scenarios. Specifically, Lyapunov optimization and Markov decision processes are commonly used in cross-layer designs to enhance system throughput and reduce latency. Inspired by the existing literature on cross-layer scheduling approaches [40–43], researchers utilized Lyapunov optimization to describe data arrivals at the network layer and channel state information at the physical layer, thereby exploring power-constrained cross-layer scheduling for low-latency wireless communications. In [44], the authors introduced online dynamic control algorithms using Lyapunov optimization theory for multi-hop networks. In [45], Neely provided a systematic summary of Lyapunov optimization theory, focusing primarily on communication and queuing systems. Building upon this, refs. [46–48] demonstrated the throughput optimality of the Lyapunov drift method under the maximum weight rule, the logarithmic rule, and the exponential rule in multi-user scenarios over time-varying wireless channels. Subsequently, the Lyapunov drift method has been widely applied in packet switching [49–51], queuing networks [52], wireless communication [53,54], and satellite communication systems [55]. When the system exhibits Markovian properties, a Markov decision process (MDP) can be employed to achieve a more rigorous analysis and optimization for metrics such as delay and power in wireless communication systems [56,57]. Berry and Gallager conducted an analysis on the average delay and power for a single-user scenario over block-fading channels in discrete-time systems [58]. Building upon this, ref. [59] studied the trade-off among the average power, delay, and the packet drop ratio in a single-user link over fading channels. It provided deterministic policies for unconstrained problems and probabilistic policies for constrained problems using value iteration and linear programming, respectively. Refs. [60,61] further investigated the trade-off between delay and power for random arrivals in AWGN and fading channels based on the constrained MDP (CMDP) framework. Specifically, the authors showed structural properties of the optimal scheduling policy, such as a threshold-based structure and piecewise linearity, and consequently developed an efficient algorithm to search for the optimal delay-power tradeoff. As an extension of [60,61], ref. [62] considered a Markovian arrival process, and showed similar threshold and piecewise linear characteristics for the optimal scheduling policy. Under a CMDP framework with cross-layer designs, refs. [63,63] presented the optimal delay-power trade-off for two-user systems over AWGN and block-fading channels, respectively.

In the aforementioned references, the studies from [40–64] focus on the research of cross-layer design and do not conduct specific studies on real-time communication. Although the references from [4–23] primarily concentrate on summarizing and designing the framework of real-time communication, they do not specifically address the issue of transmitting B units of data within a deadline T. The studies from [24–39], while focusing on the BT problem, do not take into account the user-perceived capacity, that is, they do not consider how to meet the requirement for the continuous reading of the user. Based on this, this paper focuses on real-time communication, utilizing the framework of cross-layer design to improve the user-perceived capacity under the constraint of average power.

This paper primarily investigates single-user communication systems over wireless fading channels. The objective is to maximize the user-perceived capacity under the constraint of average power. Specifically, we consider periodic data arrivals, with B packets of data arriving at the beginning of each period. The user reads the received data at a certain rate at the receiver. To satisfy the user-perceived capacity, we need to transmit B units of data within a deadline T. In particular, the main contributions of the paper are as follows.

- We model the user-perceived capacity-oriented system as a two-dimensional Markov chain with the queue length and time index as the system state. Based on this, we rigorously analyze the power consumption and show conditions for ensuring a target user-perceived capacity.
- Under this Markov chain, we formulate a dual problem of maximizing the user-perceived capacity over fading channels, i.e., minimizing the average power while achieving a constraint on the user-perceived capacity. To address the non-convexity of this dual problem, we convert it into a linear programming problem, equivalently, via a variable combination method.
- Next, we obtain an optimal cross-layer scheduling policy by solving the derived linear programming problem. Subsequently, a binary search method is developed to acquire the maximum user-perceived capacity under an arbitrary average power constraint.
- Furthermore, we represent the single user link as a finite horizon Markov decision process, and demonstrate a series of threshold structures of the optimal policy in terms of queue length, time-slot, and channel state.

The rest of this paper is organized as follows. Section 2 initially provides a detailed definition of user-perceived capacity. Section 3 illustrates the system model. The BT problem with the cross-layer approach is investigated in Section 4. In Section 5, we prove the structure of the optimal policy. Numerical results and concluding remarks are presented in Sections 6 and 7, respectively.

2. User-Perceived Capacity

We define C and $R(t)$ as the user-perceived capacity and transmission rate, respectively. For the time duration $[0, T]$, the user-perceived capacity C satisfies

$$Ct \leq \int_0^t R(\tau)d\tau, \ \ w.p.\,1, \, \forall t \in [0, T]. \tag{1}$$

In discrete-time systems, the integral transforms into the form of a summation $\sum_{t=0}^{T} R(t)$. Specifically, the user-perceived capacity refers to the capacity that enables users to avoid any interruptions during reading. Then, we obtain the following three propositions.

Proposition 1. *In the AWGN channel, the user-perceived capacity C is equal to*

$$C = W \log_2(1+\gamma) \ bit/s, \tag{2}$$

where W and γ represent the bandwidth and the signal-to-noise ratio at the receiver, respectively.

Proposition 2. *In the fading channel with the channel state information at the receiver, each positive constant $c > 0$ satisfies*

$$\Pr\left\{ct > \int_0^t R(t)dt\right\} > 0, \quad \forall t \in [0, T], \tag{3}$$

due to the unbounded power under deep fading. As such, the user-perceived capacity C is equal to zero.

Proposition 3 (Our previous work [9]). *In the fading channel with the channel state information at the transmitter and the receiver, the user-perceived capacity is equal to the hard-delay-constrained capacity with power adaptation only, i.e.,*

$$C = W \log_2(1+\gamma) \text{ bit/s}, \tag{4a}$$

$$P = \mathbb{E}_h\left\{\frac{\gamma\sigma^2}{h^2}\right\}, \tag{4b}$$

where P denotes the expectation of transmission power over the random channel gain h. Meanwhile, σ^2 denotes the noise variance. (The condition for Proposition 3 to hold is that the power P is bounded as stated in Equation (4b). The reader may have observed that Equation (4b) does not hold under Rayleigh fading channels. In response to this case, we have demonstrated in our paper [9] that when the frequency or spatial diversity order exceeds two, Equation (4b) can be satisfied.).

In this paper, leveraging a cross-layer approach, we achieve the user-perceived capacity with the rate adaptation. Our proposed policy yields a user-perceived capacity that is inevitably superior to that presented in Proposition 3.

3. System Model

In this paper, we consider a data transmission from a base station to a user. When a user request is sent to the base station, the base station shall send an amount of data to the user, supporting a target experience rate within a deadline, i.e., meeting the user-perceived capacity. Specifically, a target rate illustrates how fast the user reads the transmitted data at the receiver. Due to the randomness of the wireless channel, a careful channel resource arrangement shall be acquired to avoid any interruption for reading the data.

3.1. Queueing Model

As shown in Figure 1, the time is divided into time-slots, each of which spans τ s and is indexed by $t = 0, 1, 2, \ldots$. We define $a[t]$ as the number of packets that arrive at the base station in the beginning of time-slot t. Meanwhile, we suppose that the data packet arrival follows a periodical process with the arrival period of T time-slots and the number of arriving packets set as B.

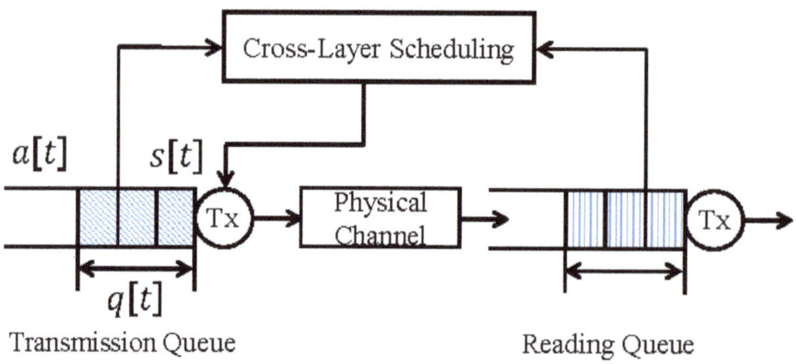

Figure 1. System model

As such, $a[t]$ evolves as

$$a[t] = \begin{cases} B & \text{if } t \bmod T = 0, \\ 0 & \text{if } t \bmod T \neq 0. \end{cases} \tag{5}$$

According to the classic BT problem proposed in [25], these B data packets that arrive periodically must be transmitted within T time-slots.

Let $s[t]$ represent the transmission rate, which is the number of packets transmitted in time-slot t. Owing to the constrained throughput at the base station, we set an upper bound of $s[t]$ by S. This way, the feasible region of $s[t]$ is defined as $\mathcal{S} = \{0,1,2,\ldots,S\}$.

Meanwhile, a buffer of size Q is utilized to store remaining packets at the base station. Let $q[t]$ denote the queue length, i.e., the number of packets backlogged in the buffer, in the t-th slot. As such, we have $q[t] \in \mathcal{Q} = \{0,1,2,\ldots,Q\}$. The queue state $q[t]$ evolves as

$$q[t+1] = \min\{(q[t]-s[t])^+ + a[t+1], Q\}, \tag{6}$$

where $(x)^+ = \max\{x,0\}$.

3.2. Physical Layer Model

In the physical layer, we represent the signal transmitted during the t-th time-slot as $X[t]$, and the corresponding signal received at the receiver as $Y[t]$. The channel gain is denoted as $h[t]$. We express the received signal as follows.

$$Y[t] = h[t]X[t] + Z[t], \tag{7}$$

where $Z[t]$ represents an additive white Gaussian noise process.

Note that a quantized channel state is typically used to determine modulation and coding schemes' real-world systems due to the limited resources for channel estimation. As such, we employ a set of thresholds for the quantization of the channel gain $h[y]$, namely, $h_0^{th} = 0 < h_1^{th} < \cdots < h_L^{th} = +\infty$. Specifically, we set $h[t]$ as $h_l = \frac{1}{2}(h_{l-1}^{th} + h_l^{th})$ for $l = 1,\ldots,L$, if the actual channel gain falls within the interval of $[h_{l-1}^{th}, h_l^{th})$. Consequently, we can model $h[t]$ as an L-state block-fading channel, with $\mathcal{H} = \{h_1, h_2, \ldots, h_L\}$ representing the feasible region of channel states, characterized by $0 < h_1 < h_2 < \cdots < h_L < +\infty$. Furthermore, we define the probability of $h[t] = h_l$ as follows.

$$\eta_l = \int_{h_{l-1}^{th}}^{h_l^{th}} p(x)dx \quad \text{for } l = 1,2,\ldots,L, \tag{8}$$

where $p(x)$ is the probability density function of the channel gain that is conducted from the real-world wireless environment. Additionally, $h[t]$ is independently and identically distributed (*i.i.d.*) across time-slots. We assume that the channel state information $h[t]$ is available at the transceiver and receiver at each timeslot. The assumption of the full knowledge of the channel state is widely adopted in the literature and grounded in the CSI measuring and reporting processes stipulated for 5G NR systems in the 3GPP standard [65]. The effect of channel estimation error is also investigated by numerical simulations.

We denote by $P[t]$ the power consumption in the t-th time-slot. When the transmission rate $s[t] = s$ and the channel state $h[t] = h_l$, the power consumption $P[t] = P(s,l)$. Without the loss of generality, we assume that $P(s,l)$ is convex with respect to s and that it is monotonically increasing. Meanwhile, S is the maximum value of transmission rate $s[t]$.

3.3. Reading Model

The transmitted packets are stored in a read-only buffer at the user side, and subsequently read under a fixed rate. In the context of the classic BT problem, the reading rate of the user is typically set as $\frac{B}{T}$ packets per time-slot. To avoid any interruption to the reading process at the user, we next develop a continuous reading constraint for the transmission of B data packets within an arrival period, i.e., T time-slots. To this end, we first denote by $m[t]$ the index of the current time-slot over its corresponding arrival period, which is represented by

$$m[t] = t \bmod T. \tag{9}$$

Here, we have $m[t] \in \mathcal{M} = \{0, 1, 2, \ldots, T-1\}$. As shown in Figure 2, we can avoid the interruption of the data reading process, as long as the amount of remaining unread data at the user side (red line) is no less than the amount of remaining in-transmitted data at the base station side (blue line). This way, we formulate the continuous reading constraint as

$$\frac{B - q[t]}{m[t]} \geq \frac{B}{T}. \tag{10}$$

To circumvent the influence of the zero point, we introduce

$$q[t]T + m[t]B \leq TB. \tag{11}$$

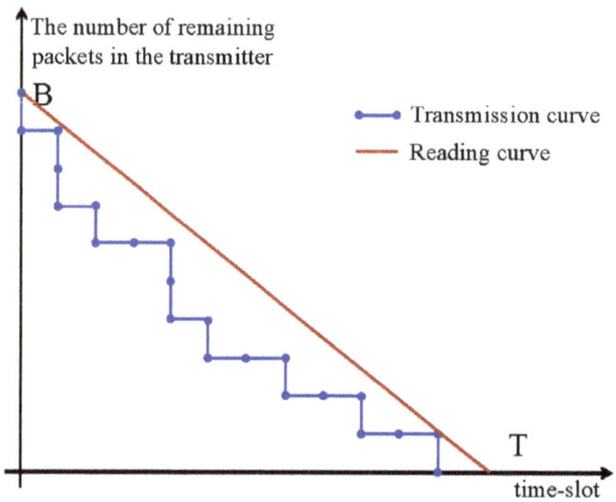

Figure 2. Reading mode. The blue step-change line represents the number of packets remaining as backlog at the base station in each time-slot in a period. The red line represents the number of packets remaining that are unread at the user side.

Subsequently, we find that optimizing the user-perceived capacity directly, under the constraint of average power, is highly challenging. Therefore, we consider optimizing the dual problem of improving the user-perceived capacity, which minimizes the average power while satisfying the user-perceived capacity. Based on this, we formulate an optimization problem to minimize the long-term average transmission power consumption under the continuous reading constraint in Equation (10), i.e.,

$$\min_{\{s[t] \in \mathcal{S}\}} \lim_{N \to +\infty} \frac{1}{N} \sum_{t=1}^{N} P[t] \tag{12a}$$

$$\text{s.t. } q[t]T + m[t]B \leq TB, \tag{12b}$$

$$q[t+1] = \min\{(q[t] - s[t])^{+} + a[t+1], Q\}, \tag{12c}$$

$$m[t] = t \bmod T. \tag{12d}$$

Before leaving this section, we shall emphasize that the system determines the transmission rate $s[t]$ by jointly considering the current values of the queue state $q[t]$, the channel state $h[t]$, and the time index $m[t]$. As such, an energy-efficiency scheduling policy is conducted to achieve the user-perceived capacity.

4. The Cross-Layer Approach

In this section, we prioritize addressing the dual problem of enhancing the user-perceived capacity, that is, minimizing the average power while meeting the user-perceived capacity requirements. Subsequently, we propose a search algorithm capable of determining the achievable user-perceived capacity under power constraints. Firstly, we delve into a detailed exposition of the scheduling policy, aiming to minimize the average power consumption while ensuring a continuous supply of packets for the user. Specifically, we analytically describe the scheduling policy under a probabilistic method, which is comprised of a conditional probability distribution of $s[t]$ under different values of $q[t]$, $h[t]$, and $m[t]$. Based on this, we represent the system as a Markov chain with the system state being the pair of the queue length and time index, i.e., $(q[t], m[t])$. By exploiting the steady-state distribution of the Markov chain and the probabilistic scheduling policy, we re-write problem (12) as a linear programming problem. By solving the derived linear programming problem, we obtain the minimum average power under the continuous reading constraint, as well as the corresponding scheduling policy.

4.1. The Probabilistic Scheduling Policy and Markov Chain

We develop a probabilistic scheduling policy based on the queue state, the time index, and the channel state. To this end, we use $f^s_{q,m,l}$ to characterize the probabilistic scheduling policy, which is given by

$$f^s_{q,m,l} = \Pr\{s[t] = s | q[t] = q, m[t] = m, h[t] = h_l\}. \tag{13}$$

Moreover, we have

$$\sum_{s \in \mathcal{S}} f^s_{q,m,l} = 1, \quad \forall q \in \mathcal{Q}, m \in \mathcal{M}, h_l \in \mathcal{H}. \tag{14}$$

Given a policy $\boldsymbol{F} = \left\{ f^s_{q,m,l} \right\}$, we denote by $\boldsymbol{\pi_F}$ the steady-state distribution of the Markov chain. Then, we have

$$\boldsymbol{\pi_F} = [\pi_{\boldsymbol{F}}(q,m)], \quad q \in \mathcal{Q}, m \in \mathcal{M}, \tag{15}$$

where $\pi_{\boldsymbol{F}}(q,m)$ is the steady-state probability of $q[t] = q$ and $m[t] = m$. Considering the normalization condition, we have

$$\sum_{q \in \mathcal{Q}} \sum_{m \in \mathcal{M}} \pi_{\boldsymbol{F}}(q,m) = 1. \tag{16}$$

For each given scheduling policy $\boldsymbol{F}$, the following lemma presents the transition probability of the Markov chain, defined as

$$\lambda_{(q,m),(\tilde{q},\tilde{m})} = \Pr\{(q[t+1], m[t+1]) = (\tilde{q}, \tilde{m}) | (q[t], m[t]) = (q,m)\}. \tag{17}$$

Lemma 1. *The transition probability is equal to*

$$\lambda_{(q,m),(\tilde{q},\tilde{m})} = \begin{cases} \sum\limits_{h_l \in \mathcal{H}} \sum\limits_{s \in \mathcal{S}} \eta_l f^s_{q,m,l} \mathbb{1}\{\tilde{q} = q - s\} \mathbb{1}\{\tilde{m} = m+1\} & \text{if } m < T-1, \\ \mathbb{1}\{\tilde{q} = B\} \mathbb{1}\{\tilde{m} = 0\} & \text{if } m = T-1, \end{cases} \tag{18}$$

where $\mathbb{1}\{\cdot\}$ is the indicator function.

Proof. When $m < T-1$, the transition to the next state does not involve packet arrival. Thus, the queue state transition depends solely on the transmission rate, leading to a change from q to $q - s$ with probability $f^s_{q,m,l}$. Meanwhile, the time index will inevitably increase by 1. Thus, we have

$$\lambda_{(q,m),(\tilde{q},\tilde{m})} = \sum_{h_l \in \mathcal{H}} \sum_{s \in \mathcal{S}} \eta_l f^s_{q,m,l} \mathbb{1}\{\tilde{q} = q - s\} \mathbb{1}\{\tilde{m} = m+1\}, m < T-1. \tag{19}$$

When $m = T-1$, the queue state will definitely transition to B, and the time index will be reset to zero. □

Based on Lemma 1, the balance equation is expressed as

$$\sum_{q\in\mathcal{Q}}\sum_{m\in\mathcal{M}}\lambda_{(q,m),(\tilde{q},\tilde{m})}\pi_{\mathbf{F}}(q,m)=\pi_{\mathbf{F}}(\tilde{q},\tilde{m}). \tag{20}$$

4.2. The Power Metric and the Continuous Reading Constraint

We next express the long-term average power consumption in terms of the steady-state distribution $\pi_{\mathbf{F}}$ and the scheduling policy $\mathbf{F}$ as follows.

Theorem 1. *The long-term average power consumption* $P_{ave}=\lim\limits_{N\to+\infty}\frac{1}{N}\sum\limits_{t=1}^{N}P[t]$ *is calculated as*

$$P_{ave}=\sum_{q\in\mathcal{Q}}\sum_{m\in\mathcal{M}}\sum_{h_l\in\mathcal{H}}\sum_{s\in\mathcal{S}}P(s,l)\eta_l\pi_{\mathbf{F}}(q,m)f^{s}_{q,m,l}. \tag{21}$$

Proof. We calculate the average power consumption from a probability perspective. The probability of the queue state $q[t]=q$, time index $m[t]=m$, channel state $h[t]=h_l$, and transmission rate $s[t]=s$ is given by $\eta_l\pi_{\mathbf{F}}(q,m)f^{s}_{q,m,l}$, with a power consumption of $P(s,l)$. Subsequently, proving Theorem 1 involves computing the mathematical expectation. □

The continuous reading constraint in Equation (10) can be transformed into

$$\sum_{q\in\mathcal{Q}}\sum_{m\in\mathcal{M}}\pi_{\mathbf{F}}(q,m)\mathbb{1}\{qT+mB>TB\}=0. \tag{22}$$

Therefore, we formulate an optimization problem to minimize the average power while satisfying the continuous reading constraint,

$$\min_{\{\pi_{\mathbf{F}}(q,m),f^{s}_{q,m,l}\}}\sum_{q\in\mathcal{Q}}\sum_{m\in\mathcal{M}}\sum_{h_l\in\mathcal{H}}\sum_{s\in\mathcal{S}}P(s,l)\eta_l\pi_{\mathbf{F}}(q,m)f^{s}_{q,m,l} \tag{23a}$$

$$\text{s.t.}\sum_{q\in\mathcal{Q}}\sum_{m\in\mathcal{M}}\pi_{\mathbf{F}}(q,m)\mathbb{1}\{qT+mB>TB\}=0, \tag{23b}$$

$$\sum_{s\in\mathcal{S}}f^{s}_{q,m,l}=1,\forall q\in\mathcal{Q},m\in\mathcal{M},h_l\in\mathcal{H}, \tag{23c}$$

$$\sum_{q\in\mathcal{Q}}\sum_{m\in\mathcal{M}}\pi_{\mathbf{F}}(q,m)=1, \tag{23d}$$

$$\sum_{q\in\mathcal{Q}}\sum_{m\in\mathcal{M}}\lambda_{(q,m),(\tilde{q},\tilde{m})}\pi_{\mathbf{F}}(q,m)=\pi_{\mathbf{F}}(\tilde{q},\tilde{m}),\forall\tilde{q}\in\mathcal{Q},\tilde{m}\in\mathcal{M}, \tag{23e}$$

$$\pi_{\mathbf{F}}(q,m)\geq 0,f^{s}_{q,m,l}\geq 0. \tag{23f}$$

Note that the derived problem (23) is non-convex, making it challenging to find the optimal solution. We next convert this problem into a linear programming problem by introducing a series of variables, which are defined as

$$y^{s}_{q,m,l}=\pi_{\mathbf{F}}(q,m)\eta_l f^{s}_{q,m,l}. \tag{24}$$

For the new variable, there is an intuitive understanding: the original variable $f^{s}_{q,m,l}$ is the probability of taking action $s[t]=s$ when the queue state $q[t]=q$, time index $m[t]=m$, and channel state $h[t]=h_l$, whereas the new variable $y^{s}_{q,m,l}$ represents the probability of being in the queue state $q[t]=q$, time index $m[t]=m$, channel state $h[t]=h_l$, and taking action $s[t]=s$. Subsequently, we propose Theorem 2.

Theorem 2. *The optimization problem* (23) *is equivalent to the following linear programming problem,*

$$\min_{\{y^{s}_{q,m,l}\}}\sum_{q\in\mathcal{Q}}\sum_{m\in\mathcal{M}}\sum_{h_l\in\mathcal{H}}\sum_{s\in\mathcal{S}}P(s,l)y^{s}_{q,m,l} \tag{25a}$$

$$\text{s.t.} \sum_{q\in\mathcal{Q}}\sum_{m\in\mathcal{M}}\sum_{h_l\in\mathcal{H}}\sum_{s\in\mathcal{S}} y^s_{q,m,l}\mathbb{1}\{qT+mB>TB\}=0, \tag{25b}$$

$$\sum_{q\in\mathcal{Q}}\sum_{m\in\mathcal{M}}\sum_{h_l\in\mathcal{H}}\sum_{s\in\mathcal{S}} y^s_{q,m,l}=1, \tag{25c}$$

$$\sum_{q\in\mathcal{Q}}\sum_{m\in\mathcal{M}}\sum_{h_l\in\mathcal{H}}\sum_{s\in\mathcal{S}} \tilde{\lambda}_{(q,m),(\tilde{q},\tilde{m})} y^s_{q,m,l} = \sum_{h_l\in\mathcal{H}}\sum_{s\in\mathcal{S}} y^s_{\tilde{q},\tilde{m},l}, \forall \tilde{q}\in\mathcal{Q}, \tilde{m}\in\mathcal{M}, h_l\in\mathcal{H} \tag{25d}$$

$$y^s_{q,m,l}\geq 0, \tag{25e}$$

where we define

$$\tilde{\lambda}_{(q,m),(\tilde{q},\tilde{m})} = \begin{cases} \mathbb{1}\{\tilde{q}=q-s\}\mathbb{1}\{\tilde{m}=m+1\} & \text{if } m<T-1, \\ \mathbb{1}\{\tilde{q}=B\}\mathbb{1}\{\tilde{m}=0\} & \text{if } m=T-1. \end{cases} \tag{26}$$

Proof. The proof starts with the observation that the objective functions of optimization problems (23) and (25) can be directly transformed into each other. The constraints (23b)–(23d) and (25b)–(25c) can be mutually converted using Equation (24) with the help of each other. The conversion of the transition equations, however, is more complex. Next, we focus on proving that the transition equations remain linear after the variable combination through Equation (24) and can be mutually converted. Then, when $m<T-1$, we have

$$\lambda_{(q,m),(\tilde{q},\tilde{m})} = \sum_{h_l\in\mathcal{H}} \eta_l f^s_{q,m,l}\mathbb{1}\{\tilde{q}=q-s\}\mathbb{1}\{\tilde{m}=m+1\} \tag{27a}$$

$$= \sum_{h_l\in\mathcal{H}}\sum_{s\in\mathcal{S}} \eta_l f^s_{q,m,l}\mathbb{1}\{\tilde{q}=q-s\}\mathbb{1}\{\tilde{m}=m+1\} \tag{27b}$$

$$= \sum_{h_l\in\mathcal{H}}\sum_{s\in\mathcal{S}} \eta_l f^s_{q,m,l}\tilde{\lambda}_{(q,m),(\tilde{q},\tilde{m})}. \tag{27c}$$

When $m=T-1$, we have

$$\lambda_{(q,m),(\tilde{q},\tilde{m})} = \mathbb{1}\{\tilde{q}=B\}\mathbb{1}\{\tilde{m}=0\} \tag{28a}$$

$$= \sum_{h_l\in\mathcal{H}}\sum_{s\in\mathcal{S}} \eta_l f^s_{q,m,l}\mathbb{1}\{\tilde{q}=B\}\mathbb{1}\{\tilde{m}=0\} \tag{28b}$$

$$= \sum_{h_l\in\mathcal{H}}\sum_{s\in\mathcal{S}} \eta_l f^s_{q,m,l}\tilde{\lambda}_{(q,m),(\tilde{q},\tilde{m})}. \tag{28c}$$

Next, Equations (27) and (28) yield

$$\sum_{q\in\mathcal{Q}}\sum_{m\in\mathcal{M}} \lambda_{(q,m),(\tilde{q},\tilde{m})}\pi_{\mathbf{F}}(q,m) = \sum_{q\in\mathcal{Q}}\sum_{m\in\mathcal{M}}\sum_{h_l\in\mathcal{H}}\sum_{s\in\mathcal{S}} \eta_l f^s_{q,m,l}\tilde{\lambda}_{(q,m),(\tilde{q},\tilde{m})}\pi_{\mathbf{F}}(q,m) \tag{29a}$$

$$= \sum_{q\in\mathcal{Q}}\sum_{m\in\mathcal{M}}\sum_{h_l\in\mathcal{H}}\sum_{s\in\mathcal{S}} \tilde{\lambda}_{(q,m),(\tilde{q},\tilde{m})} y^s_{q,m,l}. \tag{29b}$$

Then,

$$\pi_{\mathbf{F}}(\tilde{q},\tilde{m}) = \pi_{\mathbf{F}}(\tilde{q},\tilde{m})\sum_{h_l\in\mathcal{H}}\sum_{s\in\mathcal{S}} \eta_l f^s_{\tilde{q},\tilde{m},l} \tag{30a}$$

$$= \sum_{h_l\in\mathcal{H}}\sum_{s\in\mathcal{S}} \eta_l f^s_{\tilde{q},\tilde{m},l}\pi_{\mathbf{F}}(\tilde{q},\tilde{m}) \tag{30b}$$

$$= \sum_{h_l\in\mathcal{H}}\sum_{s\in\mathcal{S}} y^s_{\tilde{q},\tilde{m},l}. \tag{30c}$$

Thus, we have proven that the two optimization problems can be mutually converted. □

The solutions $\pi_{\mathbf{F}}(q,m)$ and $f^{s}_{q,m,l}$ of optimization problem (23) can generate the solutions $y^{s}_{q,m,l}$ of optimization problem (25) according to Equation (24). Similarly, $y^{s}_{q,m,l}$ in problem (25) generates $\pi_{\mathbf{F}}(q,m)$ by

$$\pi_{\mathbf{F}}(q,m) = \sum_{h_l \in \mathcal{H}} \sum_{s \in \mathcal{S}} y^{s}_{q,m,l}. \tag{31}$$

Likewise, we have

$$f^{s}_{q,m,l} = \begin{cases} \frac{y^{s}_{q,m,l}}{\pi_{\mathbf{F}}(q,m)} & \text{if } \pi_{\mathbf{F}}(q,m) > 0, \\ \mathbb{1}\{s=0\} & \text{otherwise.} \end{cases} \tag{32}$$

We have demonstrated that if $\left(\pi_{\mathbf{F}}(q,m), f^{s}_{q,m,l}\right)$ constitutes an optimal solution to problem (23), then the corresponding $y^{s}_{q,m,l}$ is an optimal solution to problem (25), and vice versa.

The attentive reader should have noticed that we are optimizing power consumption given B and T. The user-perceived capacity, on the other hand, is the capacity of the wireless communication system under a given power consumption. Therefore, we propose Algorithm 1 to address this issue, that is, to find the user-perceived capacity under a given power consumption. Meanwhile, the complexity of Algorithm 1 is $\mathcal{O}\left(\log\left(\frac{P_{ave}T}{\Delta P_{\min}}\right)\right)$, where $\Delta P_{\min} = \min_{s,h_l} P(s+1,h_l) - P(s,h_l)$.

Algorithm 1 User-Perceived Capacity Search Algorithm

Input: T, P_{ave}.
$B_1 \leftarrow 0$, $B_2 \leftarrow \frac{P_{ave}T}{\Delta P_{\min}}$
$B \leftarrow \left\lfloor \frac{(B_1+B_2)}{2} \right\rfloor$
$P(B) \leftarrow$ the optimal objective function of problem (25) with the frame size as B
$P(B+1) \leftarrow$ the optimal objective function of problem (25) with the frame size as $B+1$
while $P(B) > P_{ave}$ or $P(B+1) < P_{ave}$ **do**
  if $P(B) > P_{ave}$ **then**
    $B_2 \leftarrow B$
  else
    $B_1 \leftarrow B$
  end if
  $B \leftarrow \left\lfloor \frac{(B_1+B_2)}{2} \right\rfloor$
  $P(B) \leftarrow$ the optimal objective function of problem (25) with the frame size as B
  $P(B+1) \leftarrow$ the optimal objective function of problem (25) with the frame size as $B+1$
end while
$\frac{B}{T}$ is the maximum user-perceived capacity.

5. The Structure of the Optimal Policy

In this section, we develop a series of theoretical results for the structural properties of the optimal scheduling policy. Note that B packets arriving at the beginning of each period must be transmitted within one arrival period. Thus, we only need to consider the data transmission for one period, i.e., T time-slots.

Within one period, we first represent the constraint on the user-perceived capacity, i.e., Equation (25b) by introducing the upper and lower bounds of the queue length. In particular, they are expressed as

$$\begin{cases} q[m] \le \left\lfloor B - \frac{Bm}{T} \right\rfloor & \text{upper bound,} \\ q[m] \ge (B - Sm)^{+} & \text{lower bound,} \end{cases} \tag{33}$$

through which we define $q^{\text{lb}}_{m+1} = (B - S(m+1))^+$ and $q^{\text{ub}}_{m+1} = \left\lfloor B - \frac{B(m+1)}{T} \right\rfloor$. Consequently, the feasible region of the transmission rate in time-slot m is given by

$$\mathcal{S}_m(q[m]) = \left\{ s[m] \in \{0,1,\ldots,S\} \mid \max\{q[m] - q^{\text{ub}}_{m+1}, 0\} \le s[m] \le \min\{q[m], S\} \right\}. \quad (34)$$

Furthermore, we represent the data transmission as a finite horizon Markov decision process, which is constructed as follows:

- **State:** The pair of the queue length and channel state $(q[m], h[m])$, where $q^{\text{lb}}_m \le q[m] \le q^{\text{ub}}_m$ and $h[m] \in \{h_1, \ldots, h_L\}$ with $h_1 < h_2 < \cdots < h_L$.
- **Action:** The transmission rate $s[m] \in \mathcal{S}_m(q[m])$.
- **Transition:** The queue length evolves according to Equation (6) and the channel state is *i.i.d.* over time, following the probability distribution $\{\eta_1, \ldots, \eta_L\}$.
- **Value function:**

$$V^*_m(q,l) = \min_{s \in \mathcal{S}_m(q)} P(s,l) + \sum_{l=1}^{L} \eta_l V^*_{m+1}(q-s,l), \quad m = 0,1,\ldots,T-1, \quad (35)$$

  where $V^*_T(q,l) = 0$ for each q and l.
- **Scheduling policy:**

$$s^*_m(q,l) = \arg \min_{s \in \mathcal{S}_m(q)} Q^*_m(q,l,s) = \arg \min_{s \in \mathcal{S}_m(q)} P(s,l) + \sum_{l=1}^{L} \eta_l V^*_{m+1}(q-s,l), \quad (36)$$

  where the item $P(s,l) + \sum_{l=1}^{L} \eta_l V^*_t(q-s,l)$ is referred to as the Q-value function, denoted by $Q^*_m(q,l,s)$. According to theoretical results for finite horizon MDPs [66], considering the scheduling policy in (36) will not lose the optimality for the average minimizing problem (12). Furthermore, the policy in (36) can be re-written in the form of the probabilistic policy $\boldsymbol{F}$ in Section 3.1 by setting $f^s_{q,m,l} = \mathbb{1}\{s = s^*_m(q,l)\}$.

Then, we solve the finite horizon MDP by the dynamic programming in Algorithm 2 with the complexity $\mathcal{O}((B+1)^2 L^2 (S+1) T)$. By analyzing Algorithm 2, we further have the following theorem.

Algorithm 2 Dynamic programming for finite horizon Markov decision processes

$m \leftarrow T-1$
for $q^{\text{lb}}_m \le q[m] \le q^{\text{ub}}_m, h[m] \in \{h_1, \ldots, h_L\}$ **do**
  $V^*_m(q,l) = \min_{s \in \mathcal{S}_m(q)} P(s,l)$
end for
Recursion: For each time step m from $T-2$ to 0, the value function $V^*_m(q,l)$ is updated using the Bellman equation.
for $m \in \{T-2, T-3, \ldots, 0\}$ **do**
  $V^*_m(q,l) = \min_{s \in \mathcal{S}_m(q)} P(s,l) + \sum_{l=1}^{L} \eta_l V^*_{m+1}(q-s,l)$
end for
Policy Extraction: After computing the value functions for all time steps, the optimal policy can be extracted by choosing the action that maximizes the value function at each state and time step.
for $m \in \{T-1, T-2, \ldots, 0\}$ **do**
  for $q^{\text{lb}}_m \le q[m] \le q^{\text{ub}}_m, h[m] \in \{h_1, \ldots, h_L\}$ **do**
    for $s \in \mathcal{S}_m(q)$ **do**
      $Q^*_m(q,l,s) \leftarrow P(s,l) + \sum_{l=1}^{L} \eta_l V^*_{m+1}(q-s,l)$
    end for
    $s^*_m(q,l) \leftarrow \arg\min_{s \in \mathcal{S}_m(q)} Q^*_m(q,l,s)$
  end for
end for

Theorem 3. *The optimal scheduling s_m^*, $m = 0, \ldots, T-1$, can be re-written as a series of thresholds $q_m^{th}(l,s)$, $\forall m, h, s$. That is,*

$$s_m^*(q,l) = s \quad \textit{if}\; q_m^{th}(l,s-1) < q \leq q_m^{th}(l,s), \tag{37}$$

where the thresholds satisfy $q_m^{th}(l,s-1) \leq q_m^{th}(l,s)$ for all $s=0,\ldots,S$. Meanwhile, $q_m^{th}(l,-1)=-1$ and $q_m^{th}(l,S) = B$.

Proof. We prove the threshold-based structure of the optimal policy by induction. Assuming that $V_{m+1}^*(q,l)$ is convex over q, we demonstrate the thresholds of the optimal policy at time-slot m. After that, the threshold-based structure can induce to the convexity of $V_m^*(q,l)$. Specifically, the convexity of $V_{m+1}^*(q,l)$ holds for $m = T-1$ due to $V_T^*(q,l) = 0$. For time-slot m, we first represent the threshold-based structure by the following two assertions:

$$Q_m^*(q,l,s^*) \leq Q_m^*(q,l,s^*-\delta) \Rightarrow Q_m^*(q+1,l,s^*) \leq Q_m^*(q+1,h,s^*-\delta), \tag{38}$$
$$Q_m^*(q,l,s^*) \leq Q_m^*(q,l,s^*+\delta) \Rightarrow Q_m^*(q+1,l,s^*+1) \leq Q_m^*(q+1,h,s^*+1+\delta), \tag{39}$$

where $s^* = s_m^*(q,l) = \arg\min_{s\in\mathcal{S}_m(q)} Q_m^*(q,l,s)$. (For completeness, we set $Q_m^*(q,l,s) = +\infty$ for $s \notin \mathcal{S}_m(q)$ and $V(q,l) = +\infty$ for $q < q_m^{\rm lb}$ or $q > q_m^{\rm ub}$.) Based on the definition of the optimal policy in Equations (36), (38) and (39) implies that the optimal transmission rate $\tilde{s} = \arg\min_{s\in\mathcal{S}_m(q+1)} Q_m^*(q+1,l,s)$ for the state $(q+1,l)$ satisfies $s^* \leq \tilde{s} \leq s^*+1$. That is, the optimal transmission rate is monotonically increasing over q for each given h and m, through which we can derive the threshold-based structure in Theorem 1.

Next, we prove Equations (38) and (39) based on the convexity of $V_{t+1}^*(q,l)$ and $P(s,l)$. For Equation (38), it suffices to show that

$$Q_m^*(q+1,l,s^*) - Q_m^*(q,l,s^*) \leq Q_m^*(q+1,l,s^*-\delta) - Q_m^*(q,l,s^*-\delta) \tag{40}$$

when $Q_m^*(q,l,s^*) \leq Q_m^*(q,l,s^*-\delta)$. Note that $Q^*(q,l,s) = P(s,l) + \sum_{l=1}^{L} \eta_l V_{m+1}^*(q-s,l)$ is given based on the expression of $Q^*(q,l,s)$ in (36). As such, we can re-write (38) as

$$\begin{aligned}&\sum_{l=1}^{L} \eta_l \big[\big(V_{m+1}^*(q-s^*+1,l) - V_{m+1}^*(q-s^*,l)\big)\big]\\ \leq&\sum_{l=1}^{L} \eta_l \big[\big(V_{m+1}^*(q-s^*+\delta+1,l) - V_{m+1}^*(q-s^*+\delta,l)\big)\big].\end{aligned} \tag{41}$$

According to the convexity of $V^*(q,l)$, $V_{m+1}^*(q-s^*+1,l) - V_{m+1}^*(q-s^*,l) \leq V_{m+1}^*(q-s^*+\delta+1,l) - V_{m+1}^*(q-s^*+\delta,l)$ holds for any l. Based on this, Equation (38) has been proven.

Similarly, we demonstrate Equation (39) by considering its sufficient condition:

$$Q_m^*(q+1,l,s^*+1) - Q_m^*(q,l,s^*) \leq Q_m^*(q+1,l,s^*+1+\delta) - Q_m^*(q,l,s^*+\delta). \tag{42}$$

when $Q_m^*(q,l,s^*) \leq Q_m^*(q,l,s^*+\delta)$. Note that $Q^*(q,l,s) = P(s,l) + \sum_{l=1}^{L} \eta_l V_{m+1}^*(q-s,l)$ is given based on the expression of $Q^*(q,l,s)$ in (36). As such, we can re-write (39) as

$$\begin{aligned}&P(s^*+1,l) - P(s^*,l) + \sum_{l=1}^{L} \eta_l \big[\big(V_{m+1}^*(q-s^*,l) - V_{m+1}^*(q-s^*,l)\big)\big]\\ \leq&P(s^*+1+\delta,l) - P(s^*+\delta,l) + \sum_{l=1}^{L} \eta_l \big[\big(V_{m+1}^*(q-s^*-\delta,l) - V_{m+1}^*(q-s^*-\delta,l)\big)\big].\end{aligned} \tag{43}$$

Consequently, it is equivalent to

$$P(s^*+1,l) - P(s^*,l) \leq P(s^*+1+\delta,l) - P(s^*+\delta,l). \tag{44}$$

This assertion is true due to the convexity of $P(s,l)$ over s.

Finally, we prove the convexity of $V_m^*(q,l)$ over q, i.e.,

$$2V_m^*(q,l) \leq V_m^*(q+1,l) + V_m^*(q-1,l). \tag{45}$$

Recalling that $s^* = s_m^*(q,l)$, we have $V_m^*(q,l) = Q(q,l,s^*)$. Likewise, we represent $\tilde{s} = s_m^*(q+1,l)$ and $\breve{s} = s_m^*(q-1,l)$. According to the threshold-based structure, illustrated by Equations (38) and (39), we have

$$s^* \leq \tilde{s} \leq s^* + 1 \quad \text{and} \quad s^* - 1 \leq \breve{s} \leq s^*. \tag{46}$$

This way, four potential cases for the values of s^*, $\tilde{s}$, and $\breve{s}$ are listed as follows: (1) $\tilde{s} = s^*$ and $\breve{s} = s^*$; (2) $\tilde{s} = s^* + 1$ and $\breve{s} = s^*$; (3) $\tilde{s} = s^*$ and $\breve{s} = s^* - 1$; and (4) $\tilde{s} = s^* + 1$ and $\breve{s} = s^* - 1$. For the first three cases, we prove Equation (45) by establishing an upper bound of its left-hand side, i.e.,

$$\begin{aligned} 2V_m^*(q,l) &\leq Q_m^*(q,l,\tilde{s}) + Q_m^*(q,l,\breve{s}) \\ &= P(\tilde{s},l) + P(\breve{s},l) + \sum_{l=1}^{L} \eta_l \left(V_{t+1}^*(q-\tilde{s},l) + V_{m+1}^*(q-\breve{s},l)\right) \end{aligned} \tag{47}$$

Meanwhile, we have

$$V_m^*(q+1,l) + V_m^*(q-1,l) \tag{48}$$

$$= P(\tilde{s},l) + P(\breve{s},l) + \sum_{l=1}^{L} \eta_l \left(V_{m+1}^*(q+1-\tilde{s},l) + V_{m+1}^*(q-1-\breve{s},l)\right). \tag{49}$$

Based on the convexity of $V_{m+1}^*(q,l)$, we have

$$V_{m+1}^*(q-\tilde{s},l) + V_{m+1}^*(q-\breve{s},l) \leq V_{m+1}^*(q+1-\tilde{s},l) + V_{m+1}^*(q-1-\breve{s},l) \tag{50}$$

when $\breve{s} \geq \tilde{s} - 1$. For the last case, i.e., $\tilde{s} = s^* + 1$ and $\breve{s} = s^* - 1$, we prove Equation (45) by

$$\begin{aligned} &2V_m^*(q,l) - V_m^*(q+1,l) - V_m^*(q-1,l) \\ =& 2P(s^*,l) - P(s^*+1,l) - P(s^*-1,l) \\ &+ \sum_{l=1}^{L} \eta_l \left(2V_{m+1}^*(q-s^*,l) - V_{m+1}^*(q+1-(s^*+1),l) - V_{m+1}^*(q-1-(s^*-1),l)\right) \\ =& 2P(s^*,l) - P(s^*+1,l) - P(s^*-1,l) \geq 0 \end{aligned} \tag{51}$$

The equality is given by substituting the definition of $V_m^*(q,l)$ in Equation (45). Meanwhile, the inequality holds due to the convexity of $P(s,l)$ over s.

Overall, we first assumed the convexity of the value function and proved that the optimal policy has a threshold structure. Then, we established the convexity of the value function through mathematical induction. Consequently, we have completed the proof that the optimal policy from the perspective of the queue state possesses a threshold structure. □

Then, we reveal the threshold structure from over the time index $m[t]$ as follows.

Lemma 2. *$V_m^*(q,l)$ monotonically increases over q.*

Proof. We assume $V_{m+1}^*(q+1,l) \geq V_{m+1}^*(q,l)$ inductively. It is true for $m = T-1$ since $V_T^*(q,l) = 0$. Next, we have

$$\begin{aligned} V_m^*(q+1,l) - V_m^*(q,l) &\geq Q_m^*(q+1,l,\tilde{s}) - Q_m^*(q,l,\tilde{s}) \\ &= \sum_{l=1}^{L} \eta_l \left(V_{m+1}^*(q-\tilde{s}+1,l) - V_{m+1}^*(q-\tilde{s},l)\right) \geq 0, \end{aligned} \tag{52}$$

where $\tilde{s} = s_m^*(q+1,l)$. As such, the first equality holds since $V_m^*(q+1,l) = Q_m^*(q+1,l,\tilde{s})$ and $V_m^*(q,l) \leq Q_m^*(q,l,\tilde{s})$. This way, we prove this lemma. □

Theorem 4. *The thresholds of the optimal scheduling policy under different times satisfy*

$$q_{m+1}^{th}(l,s) \leq q_m^{th}(l,s), \quad \forall m = 0,\dots,T-2. \tag{53}$$

Proof. We indicate that the optimal transmission rate increases with time-slot m in this theorem, i.e., $s_m^*(q,l) \leq s_{m+1}^*(q,l)$.

Furthermore, we give an intuitive explanation of how the optimal transmission rate is selected. In particular, we have

$$s_m^*(q,l) = \arg\min_{s\in\mathcal{S}_m(q)} P(s,l) + \bar{V}_{m+1}^*(q-s), \tag{54}$$

where

$$\bar{V}_m^*(q) = \sum_{l=1}^{L} \eta_l V_m^*(q,l). \tag{55}$$

Given the values of q and l, we note that $P(s,l)$ is increasing and convex over s, while $\bar{V}_{m+1}^*(q-s)$ is decreasing and convex over s based on Theorem 1 and Lemma 2. This way, we can represent the optimal transmission rate as

$$s_m^*(q,l) = \max\{s \in \mathcal{S}_m(q) \mid P(s,l) - P(s-1,l) \leq \bar{V}_{m+1}^*(q-s+1) - \bar{V}_{m+1}^*(q-s)\}. \tag{56}$$

As a result, we have $s_{m+1}^*(q,l) \geq s_m^*(q,l)$ when

$$\bar{V}_m^*(q+1) - \bar{V}_m^*(q) \leq \bar{V}_{m+1}^*(q+1) - \bar{V}_{m+1}^*(q), \quad \forall m = 0,\dots,T-2. \tag{57}$$

This statement holds based on two observations: (1) $\mathcal{S}_m(q)$ removes the lower transmission rates when m is increasing, which is due to the monotonicity of q_m^{ub}; (2) for each s satisfying $P(s,l) - P(s-1,l) \leq \bar{V}_{m+1}^*(q-s+1) - \bar{V}_{m+1}^*(q-s)$, we have $P(s,l) - P(s-1,l) \leq \bar{V}_{m+2}^*(q-s+1) - \bar{V}_{m+2}^*(q-s)$ as per (57).

Since $\bar{V}_m^*(q) = \sum_{l=1}^{L} \eta_l V_m^*(q,l)$, we show (57) by the following sufficient condition:

$$\begin{aligned} V_m^*(q+1,l) - V_m^*(q,l) &\leq P(s^*,l) - P(s^*,l) + \sum_{l=1}^{L} \eta_l (V_{m+1}^*(q+1-s^*,l) - V_{m+1}^*(q-s^*,l)) \\ &\leq \sum_{l=1}^{L} \eta_l (V_{m+1}^*(q+1,l) - V_{m+1}^*(q,l)) \\ &= \bar{V}_{m+1}^*(q+1) - \bar{V}_{m+1}^*(q). \end{aligned} \tag{58}$$

where we recall that $s^* = s_m^*(q,l)$. Meanwhile, the last inequality holds due to the convexity of $V_{m+1}^*(q,l)$ over q. □

Then, we reveal the threshold structure from over the channel state $h[t]$ as follows.

Theorem 5. *If the power consumption function $P(s,h)$ satisfies*

$$P(s+1,l^+) - P(s,l^+) \leq P(s+1,l^-) - P(s,l^-), \quad \forall s = 0,\dots,S-1, \tag{59}$$

for the case when $l^+ > l^-$, the order relation of thresholds over different channel states is

$$q_m^{th}(l^+,s) \leq q_m^{th}(l^-,s). \tag{60}$$

Proof. We start the proof by noticing that $q_m^{\text{th}}(l^+,s) \leq q_m^{\text{th}}(l^-,s)$ means that a larger transmission rates is obtained for queue length q when the system suffers a higher channel gain.

That is, $s^*_m(q,l^+) \geq s^*_m(q,l^-)$. To prove this, we represent $\tilde{s} = s^*_m(q,l^+)$ and $\check{s} = s^*_m(q,l^-)$ and give an equivalent expression of $\tilde{s} \geq \check{s}$ as

$$Q^*_m(q,l^-,\check{s}) \leq Q^*_m(q,l^-,\check{s}-\delta) \Rightarrow Q^*_m(q,l^+,\check{s}) \leq Q^*_m(q,l^+,\check{s}-\delta). \tag{61}$$

This implies that the optimal transmission rate for the state (q,l^+) shall be selected from the set $\mathcal{S}_m(q) \cap \{\check{s},\check{s}+1,\ldots,S\}$, i.e., $\tilde{s} \geq \check{s}$. We next show (61) by considering its sufficient condition:

$$\begin{aligned} &Q^*_m(q,l^+,\check{s}) + Q^*_m(q,l^-,\check{s}-\delta) - Q^*_m(q,l^+,\check{s}-\delta) - Q^*_m(q,l^-,\check{s}) \\ =&P(\check{s},l^+) + P(\check{s}-\delta,l^-) - P(\check{s}-\delta,l^+) - P(\check{s},l^-) \leq 0, \end{aligned} \tag{62}$$

where the inequality holds due to (59). □

By exploiting Theorems 3–5, we finally develop an efficient method in Algorithm 3 to search thresholds for the optimal transmission policy. As presented in Algorithm 3, we can reduce the running time to update the optimal transmission rate based on the monotonicity of $s^*_m(q,l)$ over q in line 5. Similarly, given $q^{\text{th}}_m(l,s)$ for specific l and m, we can constrain the transmission rates of states with $\tilde{l} < l$ and $\tilde{m} < m$ in lines 9 and 11. As such, the complexity of Algorithm 3 is $\mathcal{O}(S+1)$ in the best case, while it is equal to $\mathcal{O}((B+1)^2L^2(S+1)T)$ in the worst case, the same as that in Algorithm 2.

Algorithm 3 Algorithm to search thresholds for the optimal transmission policy

```
1: q^th_{T-1}(1,s) ← (B+1)𝟙_{s=S} − 1 for all s
2: for m = T − 1 to 0 do
3:     for l = L to 1 do
4:         for q = q^ub_m to q^lb_m do
5:             s*_m(q,l) ← argmin_{s∈S_m(q)∩{s≤min{š|q≤q^th_m(l,š)}}} Q*_m(q,l,s)
6:             q^th_m(l, s*_m(q,l)) ← q
7:             V*_m(q,l) ← Q*_m(q,l,s*_m(q,l))
8:         end for
9:         q^th_m(l̃,s) ← q^th_m(l,s) for all l̃ < l
10:    end for
11:    q^th_m̃(l,s) ← q^th_m(l,s) for all m̃ < m
12: end for
```

6. Simulation Results

In this section, we will validate the accuracy and effectiveness of our theoretical results through simulation. We assume that the large scale path loss is calculated as $\rho = 28 + 22\log_{10} d + 20\log_{10} f_c$ dB with the carrier frequency f_c as 6 GHz and the distance d as 100 m, grounded in a typical channel model in the 5G NR standard [67]. Next, we calculate the transmission powers over the AWGN channel as $P(0) = 0$ W, $P(1) = 5.19$ mW, and $P(2) = 10.38$ mW by using BPSK and QPSK modulations when transmission rate $s[t] = 1$ and 2, respectively. In particular, the occupied bandwidth is 1 MHz and the one-sided noise power spectral density N_0 is -150 dBm/Hz. Additionally, we model the fading channel with a three-state block-fading distribution, with states $[1,4,9]$ and equal probabilities. The power consumption over fading channels is equal to $P(s,l) = \frac{P(s)}{h_l^2}$.

Figure 3 demonstrates the performance of the policy in the fading channel scenario. Since we adopt discrete time-slots, only the points shown in the figure represent the operating points of our proposed policy. The horizontal axis represents the period size, which is the maximum number of service time-slots for the periodically arriving data packets. The vertical axis represents the average power consumption. We demonstrate the performance of cross-layer scheduling with $B = 6$, 10, and 14, where the frame sizes are quantized by the minimum transmission unit $\delta = 5 \times 10^3$ bits. Meanwhile, we set

$T \in \{7, \ldots, 30\}$, with the timeslot duration as 0.5 ms. We observe that the theoretical results of the proposed policy align well with the simulation results, providing a cross-layer scheduling policy that meets the continuous reading requirement in the fading channel scenario. We also notice that, when the periodically arriving data packets are fixed, the average power consumption decreases as period T increases. There are two reasons for this. First, similar to the AWGN scenario, as the period increases, the average power consumption naturally decreases. Second, a larger period means more time-slots are available for transmission, allowing for the possibility to wait for better channel conditions, thereby reducing power consumption.

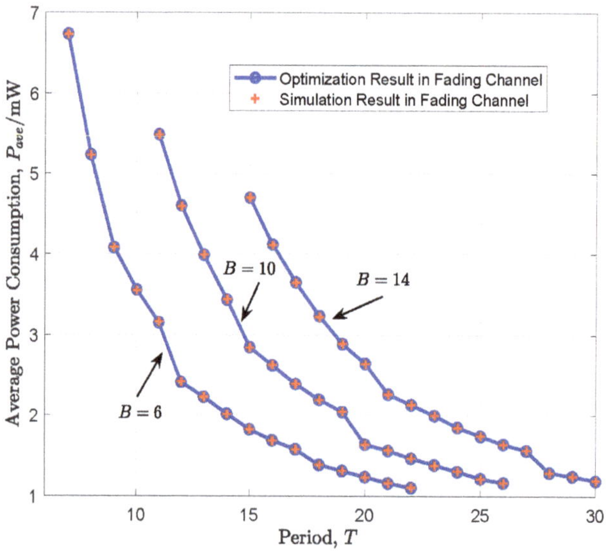

Figure 3. The power–period tradeoff in a fading channel.

Figure 4 illustrates the variation of user-perceived capacity with respect to the average power consumption in a fading channel scenario. The horizontal axis represents the average power consumption, while the vertical axis denotes the user-perceived capacity. We present performance variations for the cases of $T = 15, 20, 30$. It is observed that there is good agreement between theoretical and simulated results. Additionally, it is noted that as the average power increases, the user-perceived capacity continuously increases. However, the rate of increase in user-perceived capacity slows down with the increase in average power. This is consistent with the law of channel capacity, where the marginal effect of the capacity increase diminishes as the average power increases.

We next compare the optimal rate and power adaptation scheme with a heuristic method, referred to as the fixed rate transmission scheme. Specifically, this benchmark serves the user with a fixed rate B/T in each timeslot. In Figure 5a, we set $T = 20$ and evolve the user-perceived capacity under the two schemes. It is evident that the user-perceived capacity achievable by our policy is higher than the fixed rate transmission scheme. Similarly, Figure 5b presents the average power under different values of T with $B = 20$. Since the benchmark dose not consider the time-varying channel state, it suffers from a larger amount of power consumption compared to the optimal scheme.

Figure 6 demonstrates the influence of the imperfect channel state information to the performance of the proposed transmission scheme, where we set $B = 20$ and $T \in \{10, 20\}$. Specifically, ϵ indicates the probability that the actual channel state is not equal to the estimated channel state. In this case, the estimated channel state is randomly selected among all the possible values with equal probabilities. As shown in this figure, the user-perceived capacity will decrease with the increment of ϵ.

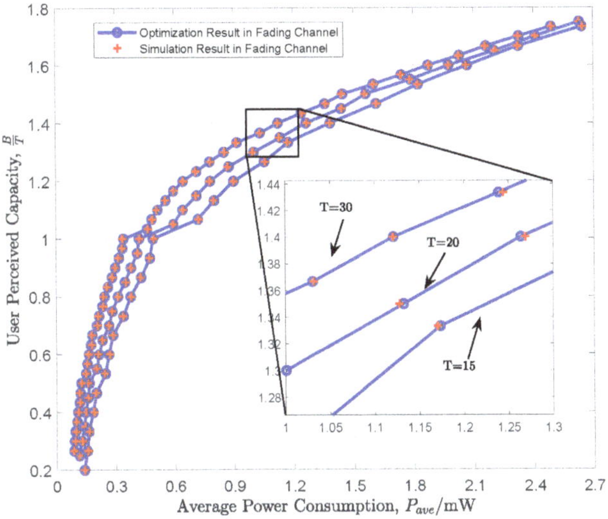

Figure 4. User-perceived capacity in fading channel.

Additionally, Table 1 displays the average running times of Algorithm 2 and the Algorithm 3 under different values of B and T. Notably, Algorithm 3 exhibits a significant reduction in running time compared to the dynamic programming in Algorithm 2.

Table 1. Running times of Algorithms 2 and 3.

B	50				100			
T	25	30	35	40	50	60	70	80
Algorithm 2	0.0106	0.0119	0.0120	0.0119	0.0331	0.0338	0.0462	0.0519
Algorithm 3	0.0033	0.00089	0.00091	0.0011	0.0036	0.0021	0.0030	0.0034

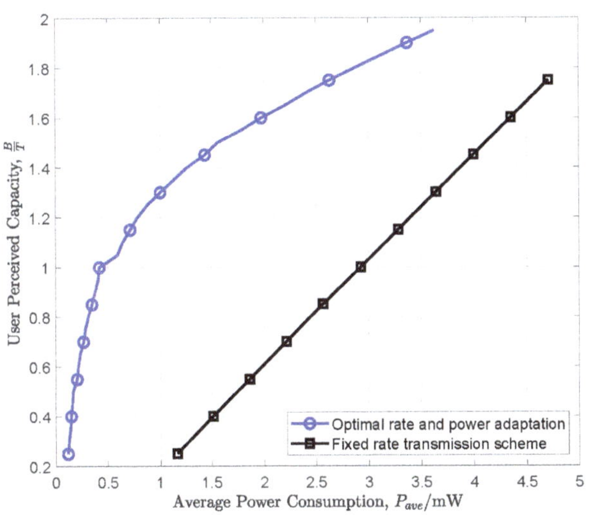

(**a**) User-perceived capacity under different amounts of power with $T = 20$.

Figure 5. *Cont.*

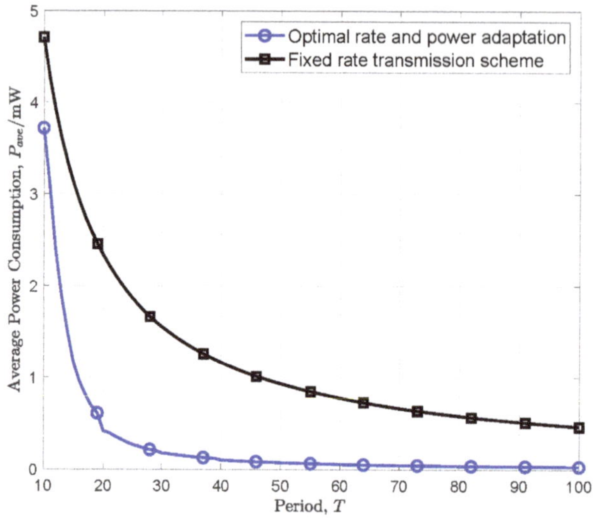

(**b**) Power consumption under different T values with $B = 20$.

Figure 5. Comparisons of the optimal scheme with a fixed rate transmission scheme.

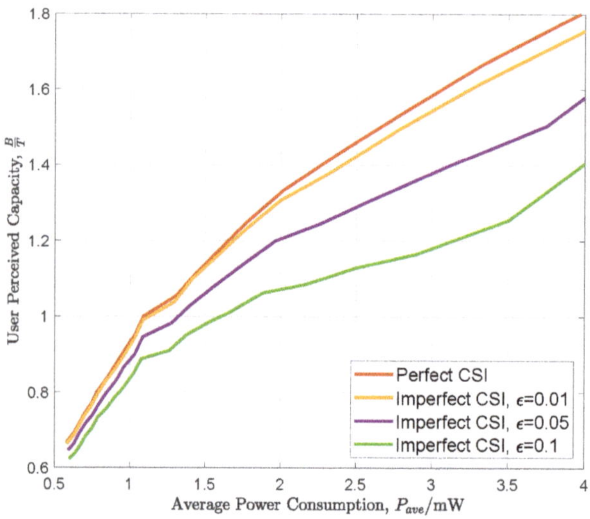

Figure 6. The influence of imperfect CSI for user-perceived capacity.

Figure 7 illustrates the threshold structure of the optimal policy. In this simulation, we set the number of periodically arriving data packets to $B = 10$ and the number of time-slots in a period to $T = 24$. Subplots 1 through 3 represent the optimal scheduling policy under three different channel states, respectively. We observe that within the feasible space $(q, m) \in \{(q, m)|qT + mB \leq TB, B - q \leq mS\}$, the optimal policy exhibits a threshold structure, validating the queue length-based optimal policy structure proposed earlier. Additionally, we note that as the channel condition improves, the transmission rate increases, confirming the channel-based optimal policy structure we proposed.

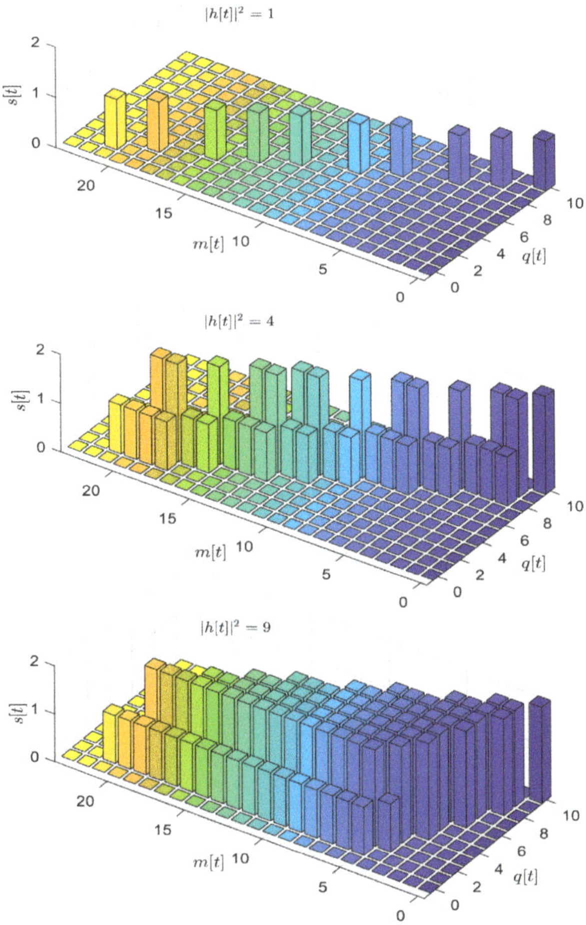

Figure 7. The threshold structure of the optimal policy: the three subfigures present the optimal policy with $|h[t]|^2 = 1, 4$, and 9, successively.

7. Conclusions

This paper has focused on the concept of user-perceived throughput, a crucial performance metric that encapsulates the essence of the user experience in mobile multimedia services. Our study has considered the impact of channel state information, both at the receiver (CSIR) and the transmitter (CSIT), on enhancing the user-perceived capacity. It is observed that the capacity is severely limited when only CSIR is available, potentially reaching a lower bound of zero. In contrast, the integration of CSIT with CSIR facilitates substantial improvements in the user-perceived capacity associated with adaptive power and rate strategies. We have proposed an MDP-based methodology for calculating the user-perceived capacity under a cross-layer design of power and rate adaption. Based on this, we have solved an average power minimization problem with a constraint on the user-perceived capacity by converting it as linear programming. Consequently, the maximum user-perceived capacity under different power consumption amounts has been calculated based on a binary searching method. Furthermore, our findings have indicated that the optimal cross-layer policy adheres to a threshold-based structure dependent on time, backlog, and channel gain. Additionally, when power adaptation is implemented as the sole policy, the user-perceived capacity corresponds to the hard-delay-constrained capacity established in our previous research [9], realized through a combination of diversity and channel inversion techniques. Note that several critical issues, including massive user access, the inter-dependency of data, and secure requirement provisioning, remain to be addressed concerning user-perceived capacity. In future research, we aim to expand this work to achieve immersive, highly dense connectivity, and secure wireless access technologies for emerging human-centric applications in 6G, such as extended reality, digital twins, and remote driving.

Author Contributions: Conceptualization, W.C.; formal analysis, Y.L.; methodology, Y.L. and X.Z.; software, Y.L. and X.Z.; writing—original draft, Y.L. and X.Z.; writing—review and editing, X.Z. and W.C. All authors have read and agreed to the published version of the manuscript.

Funding: This research was funded by the National Natural Science Foundation of China under Grant No. 62401140.

Institutional Review Board Statement: Not applicable.

Informed Consent Statement: Not applicable.

Data Availability Statement: Data are contained within this article.

Conflicts of Interest: The authors declare no conflicts of interest. The funders had no role in the design of the study; in the collection, analyses, or interpretation of data; in the writing of the manuscript; or in the decision to publish the results.

References

1. Saad, W.; Bennis, M.; Chen, M. A Vision of 6G Wireless Systems: Applications, Trends, Technologies, and Open Research Problems. *IEEE Netw.* **2020**, *34*, 134–142. [CrossRef]
2. Lu, Y.; Zheng, X. 6G: A survey on technologies, scenarios, challenges, and the related issues. *J. Ind. Inf. Integr.* **2020**, *19*, 100158. [CrossRef]
3. *3GPP TS 28.554*; 5G End to End Key Performance Indicators (KPI) (Release 18). V18.2.0. 3GPP: Sophia Antipolis, France, 2023.
4. Kandhlur, D.; Shin, K.; Ferrari, D. Real-time communication in multihop networks. *IEEE Trans. Parallel Distrib. Syst.* **1994**, *5*, 1044–1056. [CrossRef]
5. Craciunas, S.S.; Oliver, R.S.; Chmelík, M.; Steiner, W. Scheduling real-time communication in IEEE 802.1 Qbv time sensitive networks. In Proceedings of the 24th International Conference on Real-Time Networks and Systems, Brest, France, 19–21 October 2016; pp. 183–192.
6. Danielis, P.; Skodzik, J.; Altmann, V.; Schweissguth, E.B.; Golatowski, F.; Timmermann, D.; Schacht, J. Survey on real-time communication via ethernet in industrial automation environments. In Proceedings of the 2014 IEEE Emerging Technology and Factory Automation (ETFA), Barcelona, Spain, 16–19 September 2014; pp. 1–8. [CrossRef]
7. Kim, B.S.; Park, H.; Kim, K.H.; Godfrey, D.; Kim, K.I. A survey on real-time communications in wireless sensor networks. *Wirel. Commun. Mob. Comput.* **2017**, *2017*, 1864847. [CrossRef]
8. Popovski, P.; Chiariotti, F.; Huang, K.; Kalør, A.E.; Kountouris, M.; Pappas, N.; Soret, B. A Perspective on Time Toward Wireless 6G. *Proc. IEEE* **2022**, *110*, 1116–1146. [CrossRef]
9. Li, C.; Chen, W.; Poor, H.V. Diversity Enabled Low-Latency Wireless Communications with Hard Delay Constraints. *IEEE J. Sel. Areas Commun.* **2023**, *41*, 2107–2122. [CrossRef]
10. Wang, C.C.; Chen, M. Sending Perishable Information: Coding Improves Delay-Constrained Throughput Even for Single Unicast. *IEEE Trans. Inf. Theory* **2017**, *63*, 252–279. [CrossRef]
11. Lee, J.; Jindal, N. Energy-efficient scheduling of delay constrained traffic over fading channels. *IEEE Trans. Wirel. Commun.* **2009**, *8*, 1866–1875. [CrossRef]
12. Li, X.; Dong, X.; Wu, D. On Optimal Power Control for Delay-Constrained Communication over Fading Channels. *IEEE Trans. Inf. Theory* **2011**, *57*, 3371–3389. [CrossRef]
13. Li, R.; Gangammanavar, H.; Eryilmaz, A. Optimal Dynamic Coding-Window Selection for Serving Deadline-Constrained Traffic Over Time-Varying Channels. *IEEE Trans. Inf. Theory* **2012**, *58*, 6556–6571. [CrossRef]
14. Hou, I.H.; Kumar, P. Queueing systems with hard delay constraints: A framework for real-time communication over unreliable wireless channels. *Queueing Syst.* **2012**, *71*, 151–177. [CrossRef]
15. Singh, R.; Hou, I.H.; Kumar, P.R. Fluctuation analysis of debt based policies for wireless networks with hard delay constraints. In Proceedings of the IEEE INFOCOM 2014—IEEE Conference on Computer Communications, Toronto, ON, Canada, 27 April–2 May 2014; pp. 2400–2408. [CrossRef]

16. Zheng, Q.; Shin, K. On the ability of establishing real-time channels in point-to-point packet-switched networks. *IEEE Trans. Commun.* **1994**, *42*, 1096–1105. [CrossRef]
17. Ewaisha, A.E.; Tepedelenlioğlu, C. Throughput Optimization in Multichannel Cognitive Radios with Hard-Deadline Constraints. *IEEE Trans. Veh. Technol.* **2016**, *65*, 2355–2368. [CrossRef]
18. Bennis, M.; Debbah, M.; Poor, H.V. Ultrareliable and Low-Latency Wireless Communication: Tail, Risk, and Scale. *Proc. IEEE* **2018**, *106*, 1834–1853. [CrossRef]
19. Hu, S.; Chen, W. Minimizing the Queue-Length-Bound Violation Probability for URLLC: A Cross Layer Approach. In Proceedings of the GLOBECOM 2020–2020 IEEE Global Communications Conference, Taipei, Taiwan, 7–11 December 2020; pp. 1–6. [CrossRef]
20. Liu, Y.; Chen, W. Ultra Reliable and Low Latency Non-Orthogonal Multiple Access: A Cross-Layer Approach. In Proceedings of the ICC 2021—IEEE International Conference on Communications, Montreal, QC, Canada, 14–23 June 2021; pp. 1–6. [CrossRef]
21. She, C.; Yang, C.; Quek, T.Q.S. Cross-Layer Optimization for Ultra-Reliable and Low-Latency Radio Access Networks. *IEEE Trans. Wirel. Commun.* **2018**, *17*, 127–141. [CrossRef]
22. Liu, C.F.; Bennis, M.; Debbah, M.; Poor, H.V. Dynamic Task Offloading and Resource Allocation for Ultra-Reliable Low-Latency Edge Computing. *IEEE Trans. Commun.* **2019**, *67*, 4132–4150. [CrossRef]
23. Zhao, X.; Zhang, Y.J.A.; Wang, M.; Chen, X.; Li, Y. Online Multi-User Scheduling for XR Transmissions With Hard-Latency Constraint: Performance Analysis and Practical Design. *IEEE Trans. Commun.* **2024**, *72*, 4055–4071. [CrossRef]
24. Zafer, M.; Modiano, E. Optimal Rate Control for Delay-Constrained Data Transmission over a Wireless Channel. *IEEE Trans. Inf. Theory* **2008**, *54*, 4020–4039. [CrossRef]
25. Berry, R.; Modiano, E.; Zafer, M. *Energy-Efficient Scheduling Under Delay Constraints for Wireless Networks*; Springer Nature: Berlin/Heidelberg, Germany, 2022.
26. Zafer, M.; Modiano, E. A calculus approach to minimum energy transmission policies with quality of service guarantees. In Proceedings of the IEEE 24th Annual Joint Conference of the IEEE Computer and Communications Societies, Miami, FL, USA, 13–17 March 2005; Volume 1, pp. 548–559. [CrossRef]
27. Zafer, M.A.; Modiano, E. A Calculus Approach to Energy-Efficient Data Transmission with Quality-of-Service Constraints. *IEEE/ACM Trans. Netw.* **2009**, *17*, 898–911. [CrossRef]
28. Fu, A.; Modiano, E.; Tsitsiklis, J. Optimal transmission scheduling over a fading channel with energy and deadline constraints. *IEEE Trans. Wirel. Commun.* **2006**, *5*, 630–641. [CrossRef]
29. Lee, J.; Jindal, N. Asymptotically Optimal Policies for Hard-Deadline Scheduling over Fading Channels. *IEEE Trans. Inf. Theory* **2013**, *59*, 2482–2500. [CrossRef]
30. Zafer, M.A. Dynamic Rate-Control and Scheduling Algorithms for Quality-of-Service in Wireless Networks. Ph.D. Thesis, Massachusetts Institute of Technology, Cambridge, MA, USA, 2007.
31. Zafer, M.; Modiano, E. Continuous-time optimal rate control for delay constrained data transmission. In Proceedings of the Allerton Conference, Monticello, IL, USA, 28–30 September 2005.
32. Zafer, M.; Modiano, E. Minimum Energy Transmission Over a Wireless Channel with Deadline and Power Constraints. *IEEE Trans. Autom. Control* **2009**, *54*, 2841–2852. [CrossRef]
33. Tarello, A.; Sun, J.; Zafer, M.; Modiano, E. Minimum energy transmission scheduling subject to deadline constraints. In Proceedings of the Third International Symposium on Modeling and Optimization in Mobile, Ad Hoc, and Wireless Networks (WiOpt'05), Riva del Garda, Italy, 3–7 April 2005; pp. 67–76. [CrossRef]
34. Uysal-Biyikoglu, E.; El Gamal, A. On adaptive transmission for energy efficiency in wireless data networks. *IEEE Trans. Inf. Theory* **2004**, *50*, 3081–3094. [CrossRef]
35. Singh, R.; Kumar, P.R. Throughput Optimal Decentralized Scheduling of Multihop Networks with End-to-End Deadline Constraints: Unreliable Links. *IEEE Trans. Autom. Control* **2019**, *64*, 127–142. [CrossRef]
36. Singh, R.; Kumar, P.R. Adaptive CSMA for Decentralized Scheduling of Multi-Hop Networks with End-to-End Deadline Constraints. *IEEE/ACM Trans. Netw.* **2021**, *29*, 1224–1237. [CrossRef]
37. Sundar, S.; Liang, B. Offloading Dependent Tasks with Communication Delay and Deadline Constraint. In Proceedings of the IEEE INFOCOM 2018—IEEE Conference on Computer Communications, Honolulu, HI, USA, 16–19 April 2018; pp. 37–45. [CrossRef]
38. Zou, Z.; Soldati, P.; Zhang, H.; Johansson, M. Energy-Efficient Deadline-Constrained Maximum Reliability Forwarding in Lossy Networks. *IEEE Trans. Wirel. Commun.* **2012**, *11*, 3474–3483. [CrossRef]
39. Luo, L.; Chakraborty, N.; Sycara, K. Distributed Algorithms for Multirobot Task Assignment with Task Deadline Constraints. *IEEE Trans. Autom. Sci. Eng.* **2015**, *12*, 876–888. [CrossRef]
40. Collins, B.; Cruz, R.L. Transmission policies for time varying channels with average delay constraints. In Proceedings of the Annual Allerton Conference on Communication Control and Computing, Monticello, IL, USA, 22–24 September 1999; pp. 709–717.
41. Shakkottai, S.; Rappaport, T.; Karlsson, P. Cross-layer design for wireless networks. *IEEE Commun. Mag.* **2003**, *41*, 74–80. [CrossRef]
42. Kawadia, V.; Kumar, P. A cautionary perspective on cross-layer design. *IEEE Wirel. Commun.* **2005**, *12*, 3–11. [CrossRef]
43. Srivastava, V.; Motani, M. Cross-layer design: A survey and the road ahead. *IEEE Commun. Mag.* **2005**, *43*, 112–119. [CrossRef]

44. Tassiulas, L.; Ephremides, A. Stability properties of constrained queueing systems and scheduling policies for maximum throughput in multihop radio networks. In Proceedings of the 29th IEEE Conference on Decision and Control, Honolulu, HI, USA, 5–7 December 1990; Volume 4, pp. 2130–2132. [CrossRef]
45. Neely, M. *Stochastic Network Optimization with Application to Communication and Queueing Systems*; Springer Nature: Berlin/Heidelberg, Germany, 2022.
46. Andrews, M.; Kumaran, K.; Ramanan, K.; Stolyar, A.; Vijayakumar, R.; Whiting, P. Scheduling in a queuing system with asynchronously varying service rates. *Probab. Eng. Informational Sci.* **2004**, *18*, 191–217. [CrossRef]
47. Sadiq, B.; Baek, S.J.; de Veciana, G. Delay-Optimal Opportunistic Scheduling and Approximations: The Log Rule. *IEEE/ACM Trans. Netw.* **2011**, *19*, 405–418. [CrossRef]
48. Shakkottai, S.; Stolyar, A.L. Scheduling for multiple flows sharing a time-varying channel: The exponential rule. *Transl. Am. Math. Soc. Ser.* **2002**, *207*, 185–202.
49. Leonardi, E.; Mellia, M.; Neri, F.; Ajmone Marsan, M. Bounds on average delays and queue size averages and variances in input-queued cell-based switches. In Proceedings of the IEEE INFOCOM 2001. Conference on Computer Communications. Twentieth Annual Joint Conference of the IEEE Computer and Communications Society (Cat. No.01CH37213), Anchorage, AK, USA, 22–26 April 2001; Volume 2, pp. 1095–1103. [CrossRef]
50. McKeown, N.; Mekkittikul, A.; Anantharam, V.; Walrand, J. Achieving 100% throughput in an input-queued switch. *IEEE Trans. Commun.* **1999**, *47*, 1260–1267. [CrossRef]
51. Keslassy, I.; McKeown, N. Analysis of scheduling algorithms that provide 100% throughput in input-queued switches. In Proceedings of the Annual Allerton Conference on Communication Control and Computing, Champaign, IL, USA, 23–25 September 1998; Stanford University: Stanford, CA, USA, 2001; Volume 39, pp. 593–602.
52. Kumar, P.; Meyn, S. Stability of queueing networks and scheduling policies. *IEEE Trans. Autom. Control* **1995**, *40*, 251–260. [CrossRef]
53. Tassiulas, L.; Ephremides, A. Dynamic server allocation to parallel queues with randomly varying connectivity. *IEEE Trans. Inf. Theory* **1993**, *39*, 466–478. [CrossRef]
54. Andrews, M.; Kumaran, K.; Ramanan, K.; Stolyar, A.; Whiting, P.; Vijayakumar, R. Providing quality of service over a shared wireless link. *IEEE Commun. Mag.* **2001**, *39*, 150–154. [CrossRef]
55. Neely, M.; Modiano, E.; Rohrs, C. Power allocation and routing in multibeam satellites with time-varying channels. *IEEE/ACM Trans. Netw.* **2003**, *11*, 138–152. [CrossRef]
56. Cui, Y.; Lau, V.K.N.; Wang, R.; Huang, H.; Zhang, S. A Survey on Delay-Aware Resource Control for Wireless Systems—Large Deviation Theory, Stochastic Lyapunov Drift, and Distributed Stochastic Learning. *IEEE Trans. Inf. Theory* **2012**, *58*, 1677–1701. [CrossRef]
57. Han, D.; Chen, W.; Fang, Y. Joint Channel and Queue Aware Scheduling for Latency Sensitive Mobile Edge Computing with Power Constraints. *IEEE Trans. Wirel. Commun.* **2020**, *19*, 3938–3951. [CrossRef]
58. Berry, R.; Gallager, R. Communication over fading channels with delay constraints. *IEEE Trans. Inf. Theory* **2002**, *48*, 1135–1149. [CrossRef]
59. Karmokar, A.; Djonin, D.; Bhargava, V. Optimal and suboptimal packet scheduling over correlated time varying flat fading channels. *IEEE Trans. Wirel. Commun.* **2006**, *5*, 446–456. [CrossRef]
60. Chen, X.; Chen, W.; Lee, J.; Shroff, N.B. Delay-Optimal Buffer-Aware Scheduling with Adaptive Transmission. *IEEE Trans. Commun.* **2017**, *65*, 2917–2930. [CrossRef]
61. Wang, M.; Liu, J.; Chen, W.; Ephremides, A. Joint Queue-Aware and Channel-Aware Delay Optimal Scheduling of Arbitrarily Bursty Traffic over Multi-State Time-Varying Channels. *IEEE Trans. Commun.* **2019**, *67*, 503–517. [CrossRef]
62. Zhao, X.; Chen, W.; Lee, J.; Shroff, N.B. Delay-Optimal and Energy-Efficient Communications with Markovian Arrivals. *IEEE Trans. Commun.* **2020**, *68*, 1508–1523. [CrossRef]
63. Yang, J.; Ulukus, S. Delay-Minimal Transmission for Average Power Constrained Multi-Access Communications. *IEEE Trans. Wirel. Commun.* **2010**, *9*, 2754–2767. [CrossRef]
64. Zhao, X.; Chen, W. Non-Orthogonal Multiple Access for Delay-Sensitive Communications: A Cross-Layer Approach. *IEEE Trans. Commun.* **2019**, *67*, 5053–5068. [CrossRef]
65. *3GPP TS 38.214*; NR; Physical Layer Procedures for Data. V17.4.0. 3GPP: Sophia Antipolis, France, 2023.
66. Puterman, M.L. *Markov Decision Processes: Discrete Stochastic Dynamic Programming*; John Wiley & Sons: Hoboken, NJ, USA, 2014.
67. *3GPP TR 38.824*; Study on Physical Layer Enhancements for NR Ultra-Reliable and Low Latency Case (URLLC). V16.0.0. 3GPP: Sophia Antipolis, France, 2019.

entropy

MDPI

Article

Design of Low-Latency Layered Normalized Minimum Sum Low-Density Parity-Check Decoding Based on Entropy Feature for NAND Flash-Memory Channel

Yingge Li * and Haihua Hu

Information Engineering, Guangdong University of Technology, Guangzhou 510006, China; haihua@mail2.gdut.edu.cn

* Correspondence: 1112003005@mail2.gdut.edu.cn

Abstract: As high-speed big-data communications impose new requirements on storage latency, low-density parity-check (LDPC) codes have become a widely used technology in flash-memory channels. However, the iterative LDPC decoding algorithm faces a high decoding latency problem due to its mechanism based on iterative message transmission. Motivated by the unbalanced bit reliability of codeword, this paper proposes two technologies, i.e., serial entropy feature-based layered normalized min-sum (S-EFB-LNMS) decoding and parallel entropy feature-based layered normalized min-sum (P-EFB-LNMS) decoding. First, we construct an entropy feature vector that reflects the real-time bit reliability of the codeword. Then, the reliability of the output information of the layered processing unit (LPU) is evaluated by analyzing the similarity between the check matrix and the entropy feature vector. Based on this evaluation, we can dynamically allocate and schedule LPUs during the decoding iteration process, thereby optimizing the entire decoding process. Experimental results show that these techniques can significantly reduce decoding latency.

Keywords: entropy feature; LNMS LDPC decoding; cosine similarity; flash-memory channel

1. Introduction

With the rapid development of artificial intelligence, cloud storage, and hyperscale data centers, the speed of data generation and processing continues to accelerate, resulting in a sharp increase in the demand for data storage [1]. In particular, in storage technology, read latency has become a key challenge because it directly affects the response speed and processing efficiency of the system [2,3]. Optimizing the read latency of NAND flash memory channels can not only improve data access speed but also meet the needs of modern applications for high-speed data processing. At the same time, with the increase in storage density, despite the increase in storage capacity, the reliability of NAND flash memory faces challenges [4,5]. In an environment with a high bit-error rate, it becomes particularly important to use multi-threshold soft decision decoding technology, which effectively enhances error correction performance by applying multiple read voltages to achieve fine-grained memory perception, but this may also introduce higher perception latency [6,7]. In addition, the layered normalized minimum sum (LNMS) decoding algorithm has attracted widespread attention due to its high efficiency in the low-density parity-check (LDPC) decoding process. LDPC decoding first calculates the log-likelihood ratio (LLR) of each bit, and then corrects the error bit by iteratively updating the information of the check node (CN) and variable node (VN). This process inherently leads to higher decoding latency [8,9]. In each iteration of LNMS decoding, it is necessary to traverse each row of the check matrix to update the corresponding VN and CN information. The unit that processes each row during one iteration is called a layered processing unit (LPU), which is the most time-consuming step in the decoding process. Therefore, in application environments with strict real-time requirements, it is particularly important to develop an iterative decoding strategy that can dynamically adjust the average number of LPUs executed during decoding.

1.1. Related Work

Significant research has been conducted into the efficient application of LDPC decoding within flash memory systems. A sensing strategy to reduce the raw bit-error rate (RBER) and latency of LDPC is proposed in [10], which comprehensively considers the inter-state, intra-state and cross-layer asymmetric error characteristics of flash memory. The read voltage is optimized to reduce the RBER by utilizing the unbalanced state transition features brought on by changes in the threshold voltage distribution due to retention time and P/E cycles [11]. An optimal read voltage decision-making scheme is proposed in [12] to eliminate the read operations required for read retries, thereby reducing read latency; the scheme uses overlapping error correction codes to determine the optimal best reading voltage. In [13], the nonuniform error characteristics between different states and within the same state are comprehensively quantified, thus the sensing level and the delay can be optimized in a more targeted manner. Given the challenges posed by intra-cell unbalanced bit-error probability and data dependence, in [14], strategies such as interleaving upper and lower page bits have been proposed to enhance min-sum LDPC decoding and reduce decoding latency. A joint decoding strategy is proposed in [15] to achieve latency reduction, which optimizes soft information by exploiting the bit-granularity error rate obtained by combining two bits in a pair of shared pages. Further studies have shown that nonuniform bit-error rates across layers can be mitigated by blending data from error-prone, lower-layer units with more reliable upper-layer units, thereby diluting the concentration of errors and improving decoding speed [16]. Additionally, methods that exploit inter-state error patterns to refine the LNMS LDPC decoding process have been explored, aiming to decrease decoding latency. A bit-error-aware LDPC decoding scheme based on the bit-error characteristics of multi-level cell (MLC) NAND flash pages is proposed in [17] to reduce the number of decoding iterations. A scheme to accelerate decoding by exploiting the error patterns of 3D triple-level cell (TLC) NAND flash pages is proposed in [18], which first decodes two lower RBER pages and stores their corresponding channel LLRs and a posteriori LLRs to optimize the decoding operation of the higher RBER page. A resistance drift-aware LDPC decoding method is proposed in [19], which improves the LDPC decoding performance but also increases the writing time. An efficient LDPC coding scheme has been introduced in [20], which exploits the significant RBER difference between pages to further shorten the decoding iterations for phase-change memory. An improved sliding window decoding algorithm [21] has been proposed based on the joint source-channel coding scheme [22] for dual spatially coupled LDPC codes. Additionally, a joint-grouping shuffling scheduled decoding algorithm [23] has been introduced for the joint source-channel coding scheme in double LDPC code systems. This method considers the overall structure of the Tanner graph for both source and channel coding and applies shuffled decoding. Furthermore, an algorithm has been proposed to jointly optimize the read voltage thresholds [24] across all layers by maximizing mutual information (MMI), thereby improving soft LLR information and reducing the read latency of 3D NAND flash memory. A bilayer LDPC coding scheme [25] has been developed for MLC NAND flash memory, leveraging the inter-page asymmetry of MLC flash memory channels and storing additional parity in the lower pages.

1.2. Contribution

Optimizing access latency by using inter-page or inter-layer error characteristics under flash memory channels to optimize reading reference voltage or LLR information to reduce access latency usually requires sacrificing storage space to obtain additional redundant bits or increasing the latency of read operations, especially in scenarios with high-frequency read and random access, where the potential advantages of such tightly coupled read and decode operations are difficult to fully utilize.

Motivated by this problem, this paper proposes an innovative method based on the bit reliability imbalance in LDPC codewords and develops serial entropy feature-based LNMS (S-EFB-LNMS) and parallel entropy feature-based LNMS (P-EFB-LNMS) LDPC decoding algorithms based on entropy feature matrices. These algorithms aim to reduce access latency, reduce additional space usage, and decouple dense interactions between pages. First, an entropy feature vector is constructed to quantify the reliability of each bit in the codeword. Then, cosine similarity is applied to assess the reliability of the LPU. Finally, an LDPC decoding strategy based on the entropy feature vector is implemented, and its effectiveness is verified through experiments.

The main contributions of this paper are as follows:

1. An LPU reliability assessment method based on the entropy feature vector of codewords is proposed. This method provides a basis for selecting the appropriate LPU for scheduling during the iteration process.
2. Based on the reliability assessment of LPU, S-EFB-LNMS and P-EFB-LNMS LDPC decoding algorithms are proposed for serial and parallel architectures, respectively. These algorithms effectively optimize the transmission of redundant information in the decoding process by adjusting the scheduling strategy of LPU in each iteration, thereby reducing unnecessary calculation and decoding latency.
3. A comprehensive performance evaluation of the proposed algorithm is carried out, which confirms that the algorithm can significantly reduce the average number of LPUs in each iteration and the total number of LPU executed in the decoding process, significantly improving the time efficiency of the decoding process. In addition, through a detailed space overhead analysis, it is proved that the proposed algorithm effectively reduces the additional space occupation. The complexity analysis of the algorithm reveals its linear growth characteristics, indicating that the algorithm shows the advantages of efficiency and practicality when processing large-scale data sets.

The rest of this paper is organized as follows. Section 2 first elaborates on the construction of the entropy feature vector of the codeword and then presents the LPU reliability evaluation method. Section 3 introduces the basic framework of the LDPC LNMS decoding algorithm and then proposes improved S-EFB-LNMS and P-EFB-LNMS LDPC decoding algorithms. Section 4 presents the performance analysis and discussion. Section 5 concludes this paper.

Notations: In this paper, the following parameters are used:

- $H(v)$: Voltage entropy function, where v represents the sensing threshold voltage.
- R_i: The i-th read reference voltage.
- ε_i: The voltage window between R_{i-1} and R_i. Specifically, ε_1 represents a voltage interval of less than R_1 and ε_7 represents a voltage interval greater than R_6.
- LLR_{ch}: LLR obtained from the flash memory channel, where $LLR_{\text{ch},j}$ represents the LLR of the j-th bit in a codeword.
- $\mathbf{f}_{\text{efv}}$: The entropy feature vector of a codeword, where $f_{\text{efv},j}$ represents the entropy feature value of the j-th bit.
- $\mathbf{c}_{\text{cnp}}$: The CN processing unit constraint vector, corresponding to a row of the LDPC matrix $\mathbf{H}$.
- $CS(\mathbf{f}_{\text{efv}}, \mathbf{c}_{\text{cnp}})$: The cosine similarity between $\mathbf{f}_{\text{efv}}$ and $\mathbf{c}_{\text{cnp}}$.
- $CI_{ij}^{(l)}$: Information transmitted from the i-th CN to the j-th VN at the (l)-th decoding iteration, where $l = 1, 2, \ldots, T_{max}$. The initial value $CI_{ij}^{(0)}$ is set to $\mathbf{0}$.
- $VI_{ij}^{(l)}$: Information transmitted from the j-th VN to the i-th CN at the (l)-th decoding iteration.
- $\mathbf{v}^{(l)}$: Posterior information at the l-th decoding iteration, where $v_j^{(l)}$ represents the posterior information of the j-th bit at l-th iteration.
- $c_j^{(l)}$: The j-th bit of the codeword after the l-th decoding iteration.

2. Design of LPU Reliability Assessment Algorithm for Flash-Memory Systems

In this section, we first introduce the construction process of the codeword entropy feature vector in detail, including the relevant theoretical basis and implementation steps. Then, we elaborate on the LPU reliability evaluation method based on the entropy feature vector and explain how this method optimizes the selection and scheduling of LPU by analyzing the reliability information of codewords.

2.1. Design of Entropy Feature Vector for Flash-Memory Channel

For a flash-memory cell, let v represent the sensing threshold voltage; its entropy can be calculated as [26]

$$H(v) = -[\sum_i \frac{P_{s_i}(v)}{\sum_i P_{s_i}(v)} log_2(\frac{P_{s_i}(v)}{\sum_i P_{s_i}(v)})], \quad (1)$$

where $P_{s_i}(v)$ represents the final threshold voltage distribution for cells of state s_i, $i \in 11, 10, 00, 01$. The 1-th bit is denoted as the least significant bit (LSB) and the 2-th bit is denoted as the most significant bit (MSB). An example of voltage entropy function $H(v)$ and voltage distribution is illustrated in Figure 1. It becomes clear that entropy represents the average amount of information before and after passing through the flash-memory channel, or uncertainty. In other words, the greater the entropy, the higher the uncertainty. Then, to represent the uncertainty of the channel information of each bit for the received codeword, the entropy feature vector $\mathbf{f}_{\text{efv}}$ can be written as

$$\mathbf{f}_{\text{efv}} = \left[f_{\text{efv},1}, f_{\text{efv},2}, \cdots, f_{\text{efv},j}, \cdots, f_{\text{efv},n}\right], \quad (2)$$

where n is the length of the received codeword and $f_{\text{efv},j}$ represents the entropy feature value of the j-th bit.

To reduce the overall delay caused by retrying the read operation, this paper sets the maximum number of retries for the read operation to 2 times [27], that is, using 6 soft read reference voltages. By setting the value of $H(v)$ to 0.35 [26], one can get 6 read reference voltages, R_i, where $i = 1, 2, 3, 4, 5, 6$. Then, the voltage window is divided into 7 ranges, ε_1, ε_2, ε_3, ε_4, ε_5, ε_6, ε_7, as shown in Figure 1. According to the quantization range in which each cell falls, the mapping value corresponding to the LLR obtained from the flash memory channel (denoted as LLR_{ch}) for a codeword in the LSB or MSB page is as shown in Table 1. Let $LLR_{\text{ch},j}$ represent the LLR of the j-th bit in a codeword.

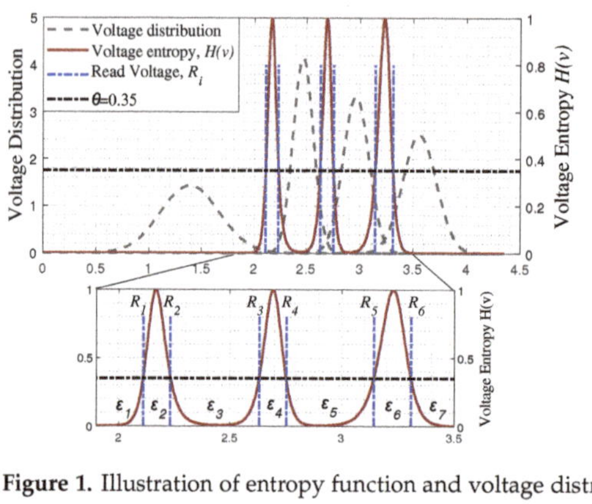

Figure 1. Illustration of entropy function and voltage distribution for 2-bit per cell flash-memory channel where the retention time and number of P/E cycles are set to 5000 and 20,000, respectively.

From Figure 1, one can observe that the entropy of the three regions ε_2, ε_4, and ε_6 is higher. Therefore, the value in the $\mathbf{f}_{\text{efv}}$ of LSB corresponding to the flash-memory cells falling within ε_4 is set to 1, and falling within other ranges is set to 0. The value in the $\mathbf{f}_{\text{efv}}$ of MSB corresponding to the flash-memory cells falling within ε_2 and ε_6 is set to 1, and falling within other ranges is set to 0, as shown in Table 2. It is instructive to note that the entropy value in the ε_1, ε_3, ε_5, and ε_7 ranges is not exactly 0, i.e., $H(v)! = 0$. In other words, the information of a small number of cells falling into these ranges is also unreliable. Note that the mapping we have proposed in Table 2 may not be optimal, and better mappings can be found for further enhancement performance.

Table 1. Mapping between the LLR of LSB and MSB from the flash-memory channel $LLR_{\text{ch},j}$ and the quantization voltage range of the corresponding cell.

	ε_1	ε_2	ε_3	ε_4	ε_5	ε_6	ε_7
$LLR_{\text{ch},j}$ of LSB	−10	−10	−10	0.00001	10	10	10
$LLR_{\text{ch},j}$ of MSB	−10	0.00001	10	10	10	0.00001	−10

Table 2. Mappings between the entropy feature value and the quantization voltage ranges of the corresponding cell.

	ε_1	ε_2	ε_3	ε_4	ε_5	ε_6	ε_7
entropy feature value of LSB	0	0	0	1	0	0	0
entropy feature value of MSB	0	1	0	0	0	1	0

2.2. *Cosine Similarity-Based LPU Reliability Assessment*

Cosine similarity [28] is a measure based on the vector space model that is used to evaluate the similarity between two vectors, which is given by

$$\begin{aligned} CS(\mathbf{x},\mathbf{y}) &= \frac{\mathbf{x}\cdot\mathbf{y}}{\|\mathbf{x}\|\|\mathbf{y}\|} \\ &= \frac{x_1 \times y_1 + x_2 \times y_2 + \cdots + x_n \times y_n}{\sqrt{x_1^2 + x_2^2 + \cdots + x_n^2} \times \sqrt{y_1^2 + y_2^2 + \cdots + y_n^2}}, \end{aligned} \tag{3}$$

where "$\cdot$" represents the dot product of vectors and "$\|\cdot\|$" represents the length of the vector. Under certain conditions, that is, when all components of vectors $\mathbf{x}$ and $\mathbf{y}$ are non-negative, their dot product $\mathbf{x}\cdot\mathbf{y}$ is also non-negative. This is because the dot product is essentially the accumulation of the products of corresponding components, thus ensuring the non-negativity of the result. Compared with Euclidean distance, it shows unique advantages when dealing with sparse vectors, especially when most elements are zero. This feature makes it particularly suitable for sparse data scenarios, such as the processing of LDPC codes, and it can pay more precise attention to the similarity between non-zero element positions. In addition, a significant advantage of cosine similarity compared to Euclidean distance is that its value range is fixed between −1 and 1, independent of the dimension or length of the vector. This means that, no matter how long the vectors are, cosine similarity can provide a standardized similarity assessment, avoiding the changes in the metric due to increasing codeword length.

In this section, we introduce an LPU reliability assessment method based on CS. Each LPU corresponds to a row in the LDPC matrix $\mathbf{H}$, which is denoted as the CN processing unit constraint vector $\mathbf{c}_{\text{cnp}}$. In $\mathbf{c}_{\text{cnp}}$, 1 indicates that the corresponding VN participates in data interaction, while 0 indicates that there is no data interaction. Then, the proposed CS-based detection algorithm aims to effectively detect the reliability of LPU by calculating the cosine similarity between $\mathbf{f}_{\text{efv}}$ (see Section 2.1) and $\mathbf{c}_{\text{cnp}}$, as follows:

$$CS(\mathbf{f}_{\text{efv}},\mathbf{c}_{\text{cnp}}) = \frac{\mathbf{f}_{\text{efv}}\cdot\mathbf{c}_{\text{cnp}}}{\|\mathbf{f}_{\text{efv}}\|\|\mathbf{c}_{\text{cnp}}\|}. \tag{4}$$

The defining characteristic of binary LDPC matrices is that their rows consist solely of 0s and 1s. This not only underscores the sparsity of the matrix but also its distinctive structural properties. Each row in an LDPC matrix can be perceived as a vector composed of non-negative elements. Moreover, given that the entropy feature matrix also incorporates binary elements, it follows that the elements of any two such vectors, $\mathbf{f}_{\text{efv}}$ and $\mathbf{c}_{\text{cnp}}$, are non-negative. Consequently, their dot product, $\mathbf{f}_{\text{efv}} \cdot \mathbf{c}_{\text{cnp}}$, is inherently non-negative, which restrictively bounds the range of their cosine similarity ($CS(\mathbf{f}_{\text{efv}}, \mathbf{c}_{\text{cnp}})$) to [0, 1]. To illustrate this, consider vectors $\mathbf{f}_{\text{efv}} = [0,1,0,0]$ and $\mathbf{c}_{\text{cnp}} = [1,0,1,0]$, where the cosine similarity is computed as follows:

$$\text{CS}(\mathbf{f}_{\text{efv}}, \mathbf{c}_{\text{cnp}}) = 0.$$

For $\mathbf{c}_{\text{cnp}} = [0,1,1,0]$, the cosine similarity is

$$\text{CS}(\mathbf{f}_{\text{efv}}, \mathbf{c}_{\text{cnp}}) = \frac{1}{\sqrt{2}}.$$

Further, considering the definition of $\mathbf{f}_{\text{efv}}$, it is clear that non-zero bits generally indicate unreliability in the prior information of the associated VNs. In cases where $CS(\mathbf{f}_{\text{efv}}, \mathbf{c}_{\text{cnp}})$ is zero, it signifies that no unreliable node information influences the update of the corresponding CN. Conversely, when $CS(\mathbf{f}_{\text{efv}}, \mathbf{c}_{\text{cnp}})$ is non-zero, it implies the involvement of unreliable verification nodes in the update process. Essentially, a higher $CS(\mathbf{f}_{\text{efv}}, \mathbf{c}_{\text{cnp}})$ indicates an increase in participation by unreliable VNs, thus escalating the likelihood of violating verification constraints. Therefore, we can evaluate the reliability of the LPU output information by calculating the $CS(\mathbf{f}_{\text{efv}}, \mathbf{c}_{\text{cnp}})$. The higher the $CS(\mathbf{f}_{\text{efv}}, \mathbf{c}_{\text{cnp}})$, the higher the unreliability of the output information. Subsequently, we can adjust the execution strategy of the LPU during the decoding process by using the $CS(\mathbf{f}_{\text{efv}}, \mathbf{c}_{\text{cnp}})$, thereby reducing the decoding delay. The proposed algorithms can be extended to support multi-level soft quantization, such as 3-level or 4-level quantization voltages. For higher-level quantization, the primary modification lies in the definition of the entropy feature matrix. In the current 2-level case, an entropy value of 1 is assigned to unreliable voltage regions, while in 3-level or 4-level quantization, additional non-zero values are assigned to different unreliable levels. The reliable voltage regions would still be assigned an entropy value of 0. Since the LDPC matrix is binary, the calculation of the cosine similarity between $\mathbf{f}_{\text{efv}}$ and $\mathbf{c}_{\text{cnp}}$ is only affected by the non-zero elements. This extension enables the proposed algorithms to handle increased noise sensitivity while maintaining their effectiveness.

3. Entropy Feature-Based LNMS LDPC Decoding Optimization

In this section, we first introduce the basic framework of the LNMS decoding algorithm for LDPC codes, and then we explain the working principle and key technologies of the algorithm. Then, we propose improved algorithms for serial and parallel architectures and explain in detail how these algorithms adjust their internal operations according to the reliability evaluation results of the LPU to improve decoding efficiency.

3.1. Generalized LNMS Decoding Algorithm

This section provides a detailed description of the iterative LNMS LDPC decoding process utilized in the receiver of the flash storage system. As shown in Figure 2, the system architecture includes several key components: detector, LPU, CN, VN, and decision. The detector is mainly responsible for extracting data from the flash channel output and calculating the LLR information from the channel. This LLR information serves as the key input to the iterative decoding process and provides the necessary signal quality indicators for the subsequent steps.

In iterative decoding, the LPU plays a key role. Each LPU is responsible for processing a single row of the check matrix. For example, the i-th LPU mainly performs the calculation and transmission tasks of information between the i-th CN and the VN connected to it. At the l-th decoding iteration, the operation of the LPU starts by receiving external information from adjacent VNs in combination with the information transmitted to the

VN by the CN in the previous round of iteration ($CI_{ij}^{(l-1)}$), where the initial value of the information transmitted by the CN to the VN ($CI_{ij}^{(0)}$) is set to zero. Then, the LPU integrates the information of all neighboring VNs and calculates and feeds back update information to the VN ($VI_{ij}^{(l)}$), thereby changing the information of the VN. In addition, the LPU is also responsible for calculating the posterior probability information of each bit and performing error detection and correction. The result of each iteration will determine the final decoding output by calculating the posterior information ($\mathbf{v}^{(l)}$) and combining it with the decision logic. This series of iterative operations is crucial to improving the efficiency and accuracy of decoding, which continue until the preset stop condition is met or the maximum number of iterations is reached.

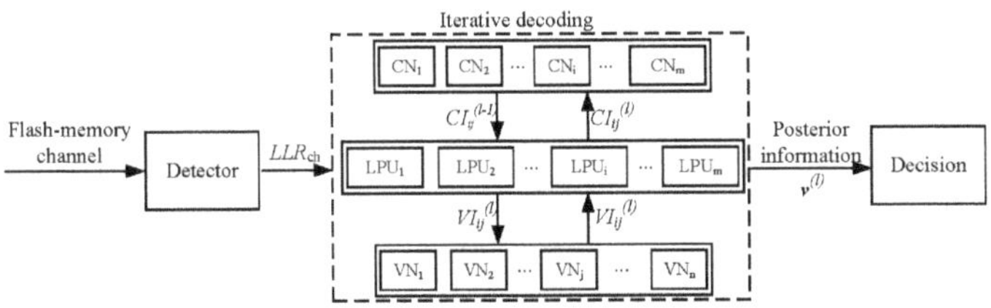

Figure 2. Illustration of the iterative LNMS LDPC decoding process in the receiver of a flash storage system.

One can observe that, based on the principle of iterative decoding, it is obviously critical to reduce the average number of LPUs in each iteration, which directly reduces the latency of the decoding process. In addition, the optimization of resource allocation and scheduling plays a vital role in improving the operating efficiency of flash storage systems. These optimization strategies not only enhance the response speed of decoding but also improve the system's ability to adapt to changing operating conditions, thereby promoting more efficient data processing and storage operations.

Subsequently, we briefly describe the LNMS algorithm presented in [20,29]. Let $\mathbf{c} = \{c_1, c_2, \ldots, c_j, \ldots, c_n\}$ be the received codeword from the flash-memory channel, where c_j represents the j-th bit of the codeword and n is the length of the codeword. Let $M(i)$ represent the set of VNs that are participated in the i-th CN ($i = 1, 2, \ldots, m$). Let $CI_{ij}^{(l)}$ ($i = 1, 2, \ldots, m$, j = 1, 2, ..., n) denote the check-to-variable information at the l-th iteration. Let $VI_{ij}^{(l)}$ ($i = 1, 2, \ldots, m$, j = 1, 2, ..., n) denote the variable-to-check information at l-th iteration. Let $v_j^{(l)}$ ($j = 1, 2, \ldots, n$) denote the posterior information at the l-th iteration. Let $c_j^{(l)}$ denote the j-th bit of the codeword after the l-th decoding iteration.

(1) Initialization:

Initialize the posterior information of the j-th bit to $LLR_{\text{ch},j}$, i.e., $v_j^{(0)} = LLR_{\text{ch},j}$. Clear the check-to-variable information, i.e., $CI_{ij}^{(0)} = 0$.

(2) Iterative decoding:

First, at the l-th iteration, for the i-th LPU, the information of the j-th VN is updated with the incoming message from the i-th CN at the $(l-1)$-th iteration:

$$VI_{ij}^{(l)} = v_j^{(l-1)} - CI_{ij}^{(l-1)}. \quad (5)$$

Subsequently, the information transmitted from the CN is updated with the incoming messages from its neighbor VNs, where α is the normalization factor, and it is set to 0.85 in this work. $M(i) \setminus j$ represents the set of VNs participating in the i-th CN, except for the j-th VN itself.

$$CI_{ij}^{(l)} = \left(\prod_{k \in M(i) \setminus j} sign(VI_{ik}^{(l)}) \right) \times \alpha \times \min_{k \in M(i) \setminus j} |VI_{ik}^{(l)}|. \quad (6)$$

Then, the posterior information of all bits can be calculated through

$$v_j^{(l)} = VI_{ij}^{(l)} + CI_{ij}^{(l)}. \tag{7}$$

(3) Decision:

At the end of each decoding iteration, if $v_j^{(l)} > 0$, $c_j^{(l)} = 1$, otherwise $c_j^{(l)} = 0$. If the code $\mathbf{c}^{(l)}$ is orthogonal to $\mathbf{H}$, i.e., $\mathbf{c}^{(l)} \cdot \mathbf{H}^{\mathrm{T}} = \mathbf{0}$, or l reaches the maximum iteration, the decoding iteration stops, where $[\cdot]^{\mathrm{T}}$ denotes the transposition operation.

3.2. Serial Entropy Feature-Based LNMS (S-EFB-LNMS) LDPC Decoding Optimization Scheme

In this section, we introduce a new S-EFB-LNMS LDPC decoding method based on the similarity between the entropy feature vector and the check matrix (see Section 2.1). This decoding strategy aims to reduce the decoding delay by dynamically adjusting the number of LPUs executed during the iterative decoding process. Specifically, we evaluate the reliability of each LPU by $CS(\mathbf{f}_{\text{efv}}, \mathbf{c}_{\text{cnp}})$. Each LPU corresponds to a row in the check matrix $\mathbf{H}$ and is responsible for processing the information interaction between the CN, represented by the row and the adjacent VNs. According to $\mathbf{f}_{\text{efv}}$, a value of 1 indicates that the LLR of the corresponding bit received from the channel shows high unreliability.

These LLRs are crucial as the initial information of the VNs in the decoding process. If the $CS(\mathbf{f}_{\text{efv}}, \mathbf{c}_{\text{cnp}})$ of an LPU is 0, it means that the LPU does not process LLRs with high unreliability, so the output information of the LPU is considered reliable and classified as a reliable LPU (RLPU). On the contrary, LPUs with non-zero cosine similarity are marked as unreliable (URLPU). It is worth noting that, even within the voltage range defined as the reliable region, there are still voltage intervals with non-zero entropy values, which indicates that the bit LLR reliability in these intervals is low. If only the LPUs marked as unreliable are executed, the information of some unreliable VNs may not be updated, and the errors of the relevant bits cannot be corrected. To solve this problem, the interleaving parameter β is introduced to adjust the execution frequency of reliable and unreliable LPUs. For example, when $\beta = 2$ is set, only RLPUs are executed in one decoding iteration, and only URLPUs are executed in the subsequent decoding iteration. If $\beta = 3$, URLPUs are executed in two consecutive decoding iteration, and then reliable LPUs are executed in the third decoding iteration. This strategy effectively reduces the need to execute all LPUs in each iteration, thereby significantly reducing the decoding latency.

In addition, by monitoring and comparing the bit state changes before and after each iteration, the LPUs that will participate in the next decoding iteration are screened and optimized to improve the decoding performance further. This dynamic adjustment strategy ensures that the entropy feature vector always accurately maps the latest bit reliability data. The detailed implementation of this method has been fully described in Algorithm 1. In Algorithm 1 (line 9), the operator "%" represents the modulus (remainder) operation. It returns the remainder when dividing $(l-1)$ by β, which is commonly used to check divisibility.

Algorithm 1 Decoding Algorithm of S-EFB-LNMS

Input: The LLR of one codeword from the flash-memory channel LLR_{ch}, the entropy feature vector $\mathbf{f}_{\text{efv}}$, the maximum iteration number $T_{\max}$, and the interleaving parameter β.
Output: Decoded bits $\mathbf{c}$.

1: Initialize the posterior information of VN to the LLR from flash-memory channel, i.e., $v_j^{(0)} = LLR_{\text{ch},j}$. Clear the check-to-variable information, i.e., $CI_{ij}^{(0)} = 0$.
2: **if** $LLR_{\text{ch},j} < 0$ **then**
3: $\quad c_j^{(0)} = 1$
4: **else**

Algorithm 1 *Cont.*

$c_j^{(0)} = 0$
end if
Get RLPU and URLPU with $CS(\mathbf{f}_{\text{efv}}, \mathbf{c}_{\text{cnp}})$ calculated by Equation (4).
for l from 1 to $T_{\max}$ **do**
  if $(l-1) \, \% \, \beta == 0$ **then**
    Process the RLPUs.
  else
    Process the URLPUs.
  end if
  Update VN information, CN information, and posterior information calculated by Equations (5)–(7), respectively.
  if $v_j^{(l)} < 0$ **then**
    $c_j^{(l)} = 1$
  else
    $c_j^{(l)} = 0$
  end if
  if $\mathbf{c}^{(l)} \cdot \mathbf{H}^{\mathrm{T}} == 0$ **then**
    break
  else
    Perform an XOR operation on $\mathbf{c}^{(l)}$ and $\mathbf{c}^{(l-1)}$ to find the flipped bit, and subsequently set its corresponding entropy feature value to 0 in the $\mathbf{f}_{\text{efv}}$.
    Refresh RLPUs and URLPUs with $CS(\mathbf{f}_{\text{efv}}, \mathbf{c}_{\text{cnp}})$ calculated by Equation (4).
  end if
end for

3.3. Parallel Entropy Feature-Based LNMS (P-EFB-LNMS) LDPC Decoding Optimization Scheme

In Section 3.2, this paper discusses in detail the use of the entropy feature vector of codewords to optimize the serial decoding framework and reduce the frame decoding latency. Based on this, this section will focus on the development of a parallel decoding framework and its potential advantages, which is named P-EFB-LNMS LDPC decoding.

Parallel processing plays a key role in improving computational efficiency, especially in the LDPC decoding process. Using the entropy feature vector as an efficient analysis tool, this study effectively classifies the LPUs into two groups of high reliability and low reliability, namely, RLPU and URLPU, by calculating $CS(\mathbf{f}_{\text{efv}}, \mathbf{c}_{\text{cnp}})$. This classification strategy ensures that cross-information interactions between VNs are avoided during the decoding process. In the simulations, we design parallel multi-threads, each independently processing LPUs of a different reliability. This parallel strategy not only ensures that both high and low reliability tasks are effectively processed but also optimizes resource allocation by processing similar tasks in a targeted manner. With this parallel design, the system can process multiple LPUs simultaneously, which significantly shortens the decoding latency and improves the overall system performance. In addition, this design also supports flexible adjustment of the workload between threads, so that it can dynamically optimize the execution strategy of threads according to the actual running performance data. The detailed implementation details and performance evaluation of the parallel processing strategy have been fully described in Algorithm 2.

Algorithm 2 Decoding Algorithm of P-EFB-LNMS

Input: Initial channel soft information of each bit of one frame LLR_{ch}, the entropy feature vector $\mathbf{f}_{\text{efv}}$, the cosine similarity matrix $CS(\mathbf{f}_{\text{efv}}, \mathbf{c}_{\text{cnp}})$, and the maximum iteration number $T_{\max}$.
Output: Decoded bits $\mathbf{c}$.
Obtain RLPUs and URLPUs by using the $CS(\mathbf{f}_{\text{efv}}, \mathbf{c}_{\text{cnp}})$ calculated by Equation (4).

Algorithm 2 *Cont.*

Initialize the posterior information of j-th bit to the LLR from flash-memory channel, i.e., $v_j^{(0)} = LLR_{\text{ch},j}$. Clear the check-to-variable information, i.e., $CI_{ij}^{(0)} = 0$.
for l from 1 to $T_{\max}$ **do**
  Process RLPUs and URLPUs in parallel.
  Update VN information, CN information, and posterior information calculated by Equations (5)–(7), respectively.
  if $v_j^{(l)} < 0$ **then**
    $c_j^{(l)} = 1$
  else
    $c_j^{(l)} = 0$
  end if
  if $\mathbf{c}^{(l)} \cdot \mathbf{H}^{\mathrm{T}} == 0$ **then**
    break
  end if
end for

4. Complexity and Performance

This section details the performance evaluation of the proposed S-EFB-LNMS and P-EFB-LNMS LDPC decoders by Monte-Carlo simulations, including decoding performance, decoding complexity, and space overhead, as well as the potential benefits and limitations of these techniques in practical applications.

4.1. Experimental Setup

In this simulations, the standard deviation and write-voltage of the erased state are set to 0.35 and 1.4, respectively, while the standard deviation of the programmed state is 0.05 and the write-voltages are 2.6, 3.2, and 3.93, respectively. For data retention noise, the constant parameters α_i, α_o, A_t, and B_t are set to 0.62, 0.3, 3.5×10^{-5}, and 2.35×10^{-4}, respectively, where the relationship between σ and μ is set to 0.3 [30]. A 2-level soft quantization voltage is used for adjacent states. The LDPC code used is a regular code (4000, 3600) generated by the Progressive Edge Growth algorithm [31,32], with row weight $d_c = 30$ and column weight $d_v = 3$. The maximum number of decoding iterations is set to 15, and the number of error frames collected at each noise point is 30.

4.2. Performance Comparison

This paper presents an in-depth comparative analysis of key performance indicators, including the average number of LPUs processed during decoding (ANLPU), frame error rate (FER), and bit-error rate (BER). The ANLPU is computed by multiplying the average number of iterations during decoding (ANITER) by the average number of LPUs executed per iteration (ANLPUPITER). Specific performance indicators for MSB pages, including ANLPUPITER, ANITER, ANLPU, FER, and BER, measured using LNMS, S-EFB-LNMS, and P-EFB-LNMS LDPC decoding algorithms, are presented in Figures 3–5, respectively.

To verify the decoding latency advantage of the proposed algorithms, we present the results for ANITER and ANLPU. Decoding latency is determined by both the codeword length and the number of decoding iterations. When the number of decoding iterations is similar, the decoding latency is comparable [2,20,27]. For a single decoding iteration, the decoding latency primarily depends on the number of LPUs executed, where each LPU corresponds to one row of the LDPC matrix $\mathbf{H}$ in the LNMS LDPC decoding algorithm. Therefore, reducing the total number of LPUs executed during decoding can directly reduce overall decoding latency.

Figure 3a,b show ANLPUPITER and ANITER, respectively. It can be observed that by adjusting parameter β, the ANLPUPITER of the S-EFB-LNMS LDPC decoding algorithm is significantly reduced compared to the LNMS LDPC decoding algorithm [20]. This adjustment strategy proves to be effective in optimizing the processing frequency of RLPUs and URLPUs. Although the ANITER slightly increases, as shown in Figure 3b, the overall number of LPUs processed is significantly reduced, as shown in Figure 4.

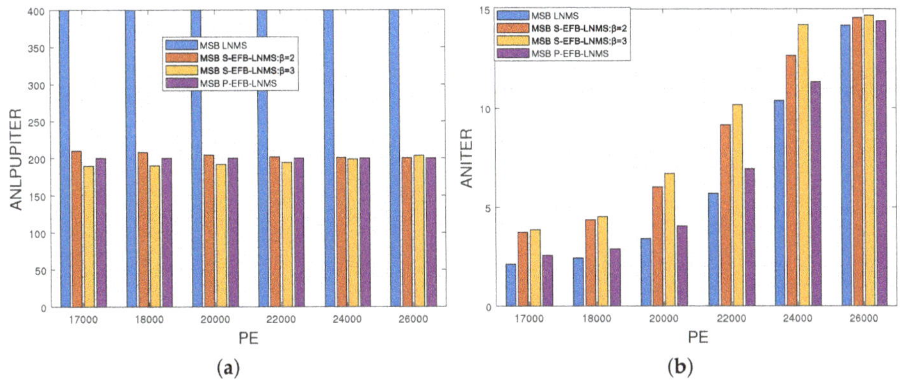

Figure 3. Performance comparison of LNMS, S-EFB-LNMS, and P-EFB-LNMS LDPC decoding algorithms for MSB pages: $T = 5000$, with PE cycles ranging from 17,000 to 26,000. (**a**) Average number of layered processing units executed per Iteration (ANLPUPITER). (**b**) Average number of iterations during decoding (ANITER).

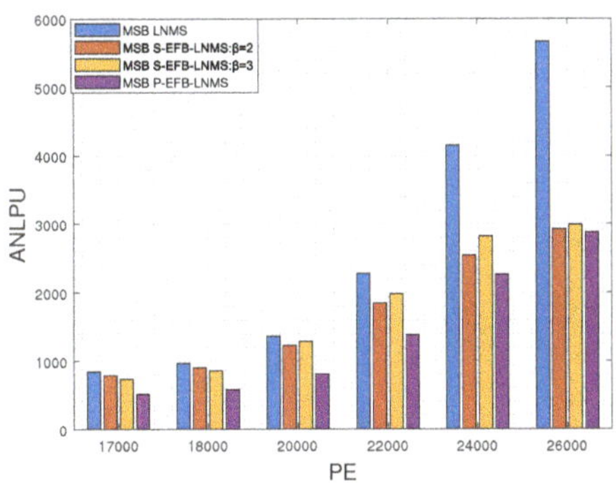

Figure 4. Average number of LPUs processed during decoding (ANLPU) for MSB pages using LNMS, S-EFB-LNMS, and P-EFB-LNMS LDPC decoding algorithms: $T = 5000$, with PE cycles ranging from 17,000 to 26,000. Lower ANLPU values indicate reduced decoding latency.

Figure 4 compares the MSB page's performance of four different algorithms (LNMS, S-EFB-LNMS:$\beta = 2$, S-EFB-LNMS:$\beta = 3$, and P-EFB-LNMS) in terms of ANLPU across varying PE cycles, ranging from 17,000 to 26,000. ANLPU represents the average number of layered processing units required for decoding, with lower ANLPU values indicating higher efficiency in processing overhead. Across all noise levels, LNMS consistently shows the highest ANLPU values, whereas S-EFB-LNMS and P-EFB-LNMS exhibit significantly lower ANLPU values, especially as noise levels increase. Specifically, as the noise level increases, the ANLPU value of LNMS rises from 847.6 to 5670.6, while P-EFB-LNMS only increases from 511.4 to 2881.1. S-EFB-LNMS also shows a substantial reduction in ANLPU under $\beta = 2$ and $\beta = 3$, with the reductions becoming more pronounced at higher noise levels.

These results demonstrate that, compared to the LNMS LDPC decoding algorithm, the S-EFB-LNMS and P-EFB-LNMS LDPC decoding algorithms both significantly improve processing efficiency, particularly under high-noise conditions, by reducing the number of processing nodes. The average reductions in the number of LPUs processed for S-EFB-LNMS with $\beta = 2$, S-EFB-LNMS with $\beta = 3$, and P-EFB-LNMS are 21.63%, 20.47%, and 42.49%, respectively, when compared to the LNMS LDPC decoding algorithm. Consequently, the proposed S-EFB-LNMS and P-EFB-LNMS LDPC decoding algorithms achieve a substantial reduction in decoding latency.

In addition, the P-EFB-LNMS LDPC decoding algorithm enables the simultaneous updating of the posterior probability through its unique internal design optimization by avoiding address conflicts between RLPU and URLPU. This parallel processing strategy not only significantly reduces ANLPU but also ensures high reliability (as shown in Figure 5). These improvements significantly reduce ANLPU, verifying the effectiveness of the proposed optimization strategy in shortening the average time required to decode a frame of data and increasing processing speed. Therefore, even in scenarios where the processor must execute serially, the carefully designed algorithms can effectively reduce decoding latency. In parallel execution scenarios, these algorithms not only accelerate decoding but also maintain high decoding accuracy.

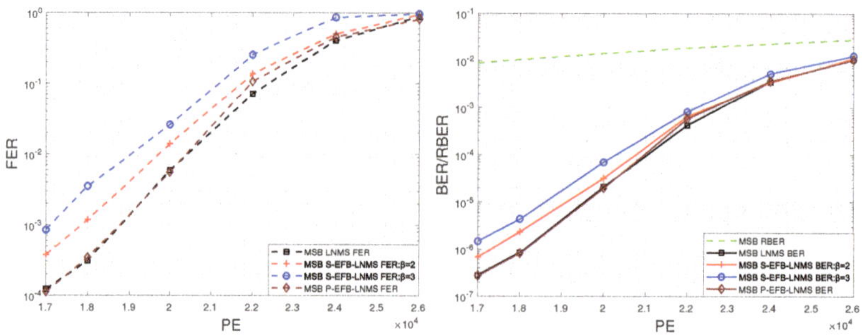

(a) (b)

Figure 5. BER/FER results for MSB pages using LNMS, S-EFB-LNMS, and P-EFB-LNMS LDPC decoding algorithms: $T = 5000$, with PE cycles ranging from 17,000 to 26,000. (**a**) FER. (**b**) BER/RBER.

Moreover, we have evaluated the performance indicators for the LSB page, including ANLPUPITER, ANITER, ANLPU, FER, and BER, as shown in Figure 6a,b; Figure 7; and Figure 8, respectively.

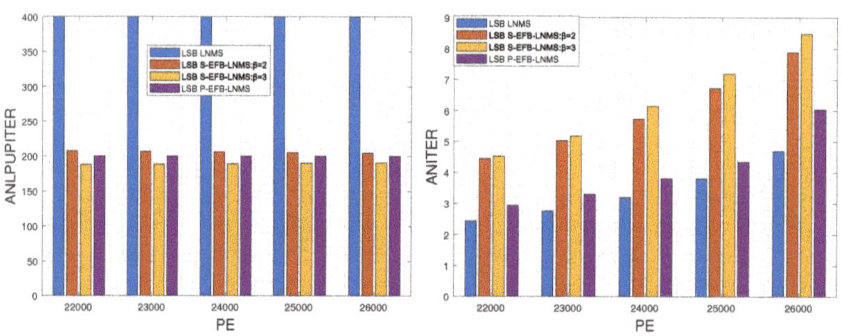

(a) (b)

Figure 6. Performance comparison of LNMS, S-EFB-LNMS, and P-EFB-LNMS LDPC decoding algorithms for LSB pages: $T = 5000$, with PE cycles ranging from 22,000 to 26,000. (**a**) Average number of layered processing units executed per iteration (ANLPUPITER). (**b**) Average number of iterations during decoding (ANITER).

The performance trends for the LSB page are consistent with those observed for the MSB page, with the S-EFB-LNMS and P-EFB-LNMS algorithms demonstrating similar advantages in reducing decoding latency and improving efficiency. Overall, the proposed algorithms maintain their efficiency across both MSB and LSB pages.

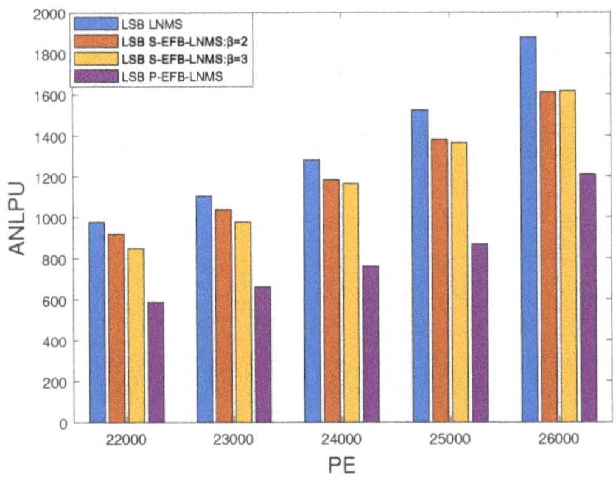

Figure 7. Average number of LPUs processed during decoding (ANLPU) for LSB pages using LNMS, S-EFB-LNMS, and P-EFB-LNMS LDPC decoding algorithms: $T = 5000$, with PE cycles ranging from 22,000 to 26,000. Lower ANLPU values indicate reduced decoding latency.

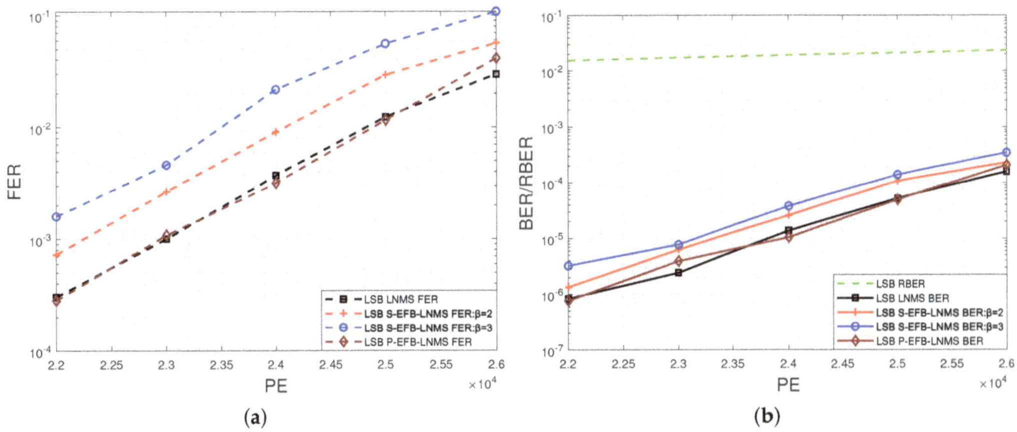

Figure 8. BER/FER results for LSB pages using LNMS, S-EFB-LNMS, and P-EFB-LNMS LDPC decoding algorithms: $T = 5000$, with PE cycles ranging from 22,000 to 26,000. (**a**) FER. (**b**) BER/RBER.

Consequently, the proposed S-EFB-LNMS and P-EFB-LNMS LDPC decoding algorithms significantly reduce the average number of LPUs executed during the decoding process, thereby reducing decoding latency. These algorithms are particularly suitable for NAND flash memory applications that are sensitive to latency.

4.3. Computational Complexity

The computational complexity of the proposed S-EFB-LNMS LDPC decoding algorithm mainly involves the calculation of $\mathbf{f}_{\text{efv}}$ and $CS(\mathbf{f}_{\text{efv}}, \mathbf{c}_{\text{cnp}})$, and the determination of RLPU and URLPU. Assume that the number of rows in the check matrix is m and the number of columns is n. The calculation of $CS(\mathbf{f}_{\text{efv}}, \mathbf{c}_{\text{cnp}})$ involves the dot product of each vector and its length. Calculating the length of a vector and performing dot product operations both have a time complexity of $O(n)$. Therefore, for each row in the check matrix, $CS(\mathbf{f}_{\text{efv}}, \mathbf{c}_{\text{cnp}})$ is calculated, and the total time complexity is $O(mn)$. In addition, the time complexity of updating $\mathbf{f}_{\text{efv}}$ is $O(n)$. Based on the result of $CS(\mathbf{f}_{\text{efv}}, \mathbf{c}_{\text{cnp}})$, the complexity of determining whether the LPU is RLPU or URLPU is $O(m)$. It can be seen that the running time of the algorithm increases linearly with the increase of the number of rows and columns of the check matrix, which shows that the S-EFB-LNMS LDPC decoding algorithm is efficient and practical when processing large-scale data sets. For the proposed P-EFB-LNMS algorithm, although additional bit width is required to calculate $\mathbf{f}_{\mathrm{efv}}$ and $CS(\mathbf{f}_{\mathrm{efv}}, \mathbf{c}_{\mathrm{cnp}})$ during the calculation process, it is notable to point out that the calculation of $\mathbf{f}_{\mathrm{efv}}$ and $CS(\mathbf{f}_{\mathrm{efv}}, \mathbf{c}_{\mathrm{cnp}})$ merely impacts on the computational complexity in the design process but has no impact in the decoding process.

4.4. Space Overhead

The space overhead of this algorithm mainly involves the $\mathbf{f}_{\mathrm{efv}}$ and the reliability determination results of the LPU based on $CS(\mathbf{f}_{\mathrm{efv}}, \mathbf{c}_{\mathrm{cnp}})$. Specifically, the space requirement of $\mathbf{f}_{\mathrm{efv}}$ is directly related to the codeword length, n, and its space complexity is linear. This is because each codeword bit requires 1 bit to store the corresponding entropy feature value. In addition, the classification result of each LPU also only requires 1 bit to store. Compared with the prior LLR and posterior LLR that need to store the entire page in the joint-page decoding optimization methods, the proposed method significantly reduces the storage requirement. This shows that these space overheads are all linear space complexities, indicating that, as the problem size increases, the required storage space grows linearly and controllably. This design ensures the algorithm's high efficiency in storage resource utilization and wide applicability.

5. Conclusions

When dealing with the high-latency problems encountered in high-speed big-data storage systems, the LNMS LDPC decoding algorithm often exhibits insufficient decoding efficiency and response speed due to its inherent iterative processing mechanism. To overcome these limitations, we propose two improved algorithms: the S-EFB-LNMS LDPC decoding algorithm and the P-EFB-LNMS LDPC decoding algorithm. By introducing an $\mathbf{f}_{\mathrm{efv}}$, these algorithms can more accurately locate and handle error patterns in codewords, greatly improving the overall efficiency of the decoding process. In addition, the $\mathbf{f}_{\mathrm{efv}}$ is used in combination with the check matrix to optimize the selection and scheduling of LPUs during the iteration process. This strategy allows for more efficient use of computing resources, significantly reducing the number of required LPUs, thereby speeding up the decoding process and reducing latency. Specifically, this method ensures that each iteration makes a substantial contribution to the decoding result by reducing invalid or redundant iterations, thereby improving the overall decoding efficiency. Experimental results verify the effectiveness of these two algorithms, showing that they can significantly reduce delay in the decoding process.

Author Contributions: Conceptualization, Y.L. and H.H.; methodology, Y.L. and H.H.; software, Y.L.; writing—original draft preparation, Y.L.; writing—review and editing, Y.L. and H.H. All authors have read and agreed to the published version of the manuscript.

Funding: This research received no external funding.

Institutional Review Board Statement: Not applicable.

Informed Consent Statement: Not applicable.

Data Availability Statement: The data used to support the findings of this study are included within the article.

Conflicts of Interest: The authors declare that there are no conflicts of interest.

References

1. Feng, H.; Wei, D.; Gu, S.; Piao, Z.; Wang, Y.; Qiao, L. Random Flip Bit Aware Reading for Improving High-Density 3-D NAND Flash Performance. *IEEE Trans. Circuits Syst. I Reg. Pap.* **2024**, *71*, 2372–2383. [CrossRef]
2. Fang, Y.; Bu, Y.; Chen, P.; Lau, F.C.M.; Otaibi, S.A. Irregular-Mapped Protograph LDPC-Coded Modulation: A Bandwidth-Efficient Solution for 6G-Enabled Mobile Networks. *IEEE Trans. Intell. Trans. Syst.* **2023**, *24*, 2060–2073. [CrossRef]
3. Hu, H.; Han, G.; Wu, W.; Liu, C. Channel Parameter and Read Reference Voltages Estimation in 3D NAND Flash Memory Using Unsupervised Learning Algorithms. *IEEE Trans. Comput.-Aided Design Integr. Circuits Syst.* **2024**, *43*, 305–318. [CrossRef]
4. Hwang, M.; Jee, J.; Kang, J.; Park, H.; Lee, S.; Kim, J. Dynamic Error Recovery Flow Prediction Based on Reusable Machine Learning for Low Latency NAND Flash Memory Under Process Variation. *IEEE Access* **2022**, *10*, 117715–117731. [CrossRef]
5. Mei, Z.; Cai, K.; Shi, L.; Li, J.; Chen, L.; Immink, K.A.S. Deep Transfer Learning-Based Detection for Flash Memory Channels. *IEEE Trans. Commun.* **2024**, *72*, 3425–3438. [CrossRef]
6. Dong, G.; Xie, N.; Zhang, T. On the Use of Soft-Decision Error-Correction Codes in NAND Flash Memory. *IEEE Trans. Circuits Syst. I Reg. Pap.* **2011**, *58*, 429–439. [CrossRef]
7. Li, Q.; Wang, Q.; Xu, Q.; Huo, Z. A Fast Read Retry Method for 3D NAND Flash Memories Using Novel Valley Search Algorithm. *IEICE Commun. Exp.* **2018**, *15*, 20180921. [CrossRef]
8. Yan, Z.; Guan, W.; Liang, L. List-Based Residual Belief-Propagation Decoding of LDPC Codes. *IEEE Commun. Lett.* **2024**, *28*, 984–988. [CrossRef]
9. Li, Y.; Han, G.; Liu, C.; Zhang, M.; Wu, F. Exploiting the Single-Symbol LLR Variation to Accelerate LDPC Decoding for 3-D nand Flash Memory. *IEEE Trans. Comput.-Aided Des. Integr. Circuits Syst.* **2023**, *42*, 5146–5150. [CrossRef]
10. Li, Q.; Shi, L.; Cui, Y.; Xue, C.J. Exploiting Asymmetric Errors for LDPC Decoding Optimization on 3D NAND Flash Memory. *IEEE Trans. Comput.* **2020**, *69*, 475–488. [CrossRef]
11. Zhang, M.; Wu, F.; Yu, Q.; Xie, C. PEAR: Unbalanced Inter-Page Errors Aware Read Scheme for Latency-Efficient 3-D NAND Flash. *IEEE Trans. Device Mater. Reliab.* **2024**, *24*, 49–58. [CrossRef]
12. Li, Q.; Wang, Q.; Yang, L.; Yu, X.; Jiang, Y.; He, J.; Huo, Z. Optimal Read Voltages Decision Scheme Eliminating Read Retry Operations for 3D NAND Flash Memories. *Microelectron. Reliab.* **2022**, *131*, 114509. [CrossRef]
13. Wang, Y.; Wei, D.; Liu, M.; Feng, H.; Qiao, L. EBDN: Entropy-Based Double Nonuniform Sensing Algorithm for LDPC Decoding in TLC NAND Flash Memory. *IEEE Trans. Comput.-Aided Des. Integr. Circuits Syst.* **2024**, *43*, 1914–1918. [CrossRef]
14. Zhao, W.; Sun, H.; Lv, M.; Dong, G.; Zheng, N.; Zhang, T. Improving Min-Sum LDPC Decoding Throughput by Exploiting Intra-Cell Bit Error Characteristic in MLC NAND Flash Memory. In Proceedings of the 2014 30th Symposium on Mass Storage Systems and Technologies (MSST), Santa Clara, CA, USA, 2–6 June 2014; pp. 1–6. [CrossRef]
15. Qiao, L.; Wu, H.; Wei, D.; Wang, S. A Joint Decoding Strategy of Non-Binary LDPC Codes Based on Retention Error Characteristics for MLC NAND Flash Memories. In Proceedings of the 2016 Sixth International Conference on Instrumentation & Measurement, Computer, Communication and Control (IMCCC), Harbin, China, 21–23 July 2016; pp. 183–188. [CrossRef]
16. Yong, K.-K.; Chang, L.-P. Error Diluting: Exploiting 3-D NAND Flash Process Variation for Efficient Read on LDPC-Based SSDs. *IEEE Trans. Comput.-Aided Des. Integr. Circuits Syst.* **2020**, *39*, 3467–3478. [CrossRef]
17. Zhang, M.; Wu, F.; Du, Y.; Liu, W.; Xie, C. Pair-Bit Errors Aware LDPC Decoding in MLC NAND Flash Memory. *IEEE Trans. Comput.-Aided Des. Integr. Circuits Syst.* **2019**, *38*, 2312–2320. [CrossRef]
18. Wu, F.; Zhang, M.; Du, Y.; Liu, W.; Lu, Z.; Wan, J.; Tan, Z.; Xie, C. Using Error Modes Aware LDPC to Improve Decoding Performance of 3-D TLC NAND Flash. *IEEE Trans. Comput.-Aided Des. Integr. Circuits Syst.* **2020**, *39*, 909–921. [CrossRef]
19. Zhang, M.; Wu, F.; Yu, Q.; Liu, W.; Ma, R.; Xie, C. Exploiting Resistance Drift Characteristics to Improve Reliability of LDPC-Assisted Phase-Change Memory. *IEEE Trans. Device Mater. Reliab.* **2021**, *21*, 324–330. [CrossRef]
20. Zhang, M.; Wu, F.; Yu, Q.; Fu, N.; Xie, C. eLDPC: An Efficient LDPC Coding Scheme for Phase-Change Memory. *IEEE Trans. Comput.-Aided Design Integr. Circuits Syst.* **2023**, *42*, 1978–1987. [CrossRef] [CrossRef]
21. Lian, Q.; Chen, Q.; Zhou, L.; He, Y.; Xie, X. Adaptive Decoding Algorithm with Variable Sliding Window for Double SC-LDPC Coding System. *IEEE Commun. Lett.* **2023**, *2*, 404–408. [CrossRef]
22. Xu, Z.; Wang, L.; Hong, S.; Chen, G. Generalized Joint Shuffled Scheduling Decoding Algorithm for the JSCC System Based on Protograph-LDPC Codes. *IEEE Access* **2021**, *9*, 128372–128380. [CrossRef]
23. Chen, Q.; Ren, Y.; Zhou, L.; Chen, C.; Liu, S. Design and Analysis of Joint Group Shuffled Scheduling Decoding Algorithm for Double LDPC Codes System. *Entropy* **2023**, *25*, 357. [CrossRef] [PubMed]
24. Wang, C.; Mei, Z.; Li, J.; Shu, F.; He, X.; Kong, L. Channel Modeling and Quantization Design for 3D NAND Flash Memory. *Entropy* **2023**, *25*, 965. [CrossRef] [PubMed]
25. Kong, L.; Liu, H.; Hou, W.; Meng, C. Bilayer LDPC Codes Combined with Perturbed Decoding for MLC NAND Flash Memory. *Entropy* **2024**, *26*, 54. [CrossRef] [PubMed]
26. Aslam, C.A.; Guan, Y.L.; Cai, K. Read and Write Voltage Signal Optimization for Multi-Level-Cell (MLC) NAND Flash Memory. *IEEE Trans. Commun.* **2016**, *64*, 1613–1623. [CrossRef]
27. Zhang, M.; Wu, F.; Yu, Q.; Liu, W.; Wang, Y.; Xie, C. Exploiting Error Characteristic to Optimize Read Voltage for 3-D NAND Flash Memory. *IEEE Trans. Electr. Devices* **2020**, *67*, 5490–5496. [CrossRef]

28. Zhu, S.; Wu, J.; Xia, G. TOP-K Cosine Similarity Interesting Pairs Search. In Proceedings of the 2010 Seventh International Conference on Fuzzy Systems and Knowledge Discovery (FSKD), Yantai, China, 10–12 August 2010; pp. 1479–1483. [CrossRef]
29. Kim, J.; Cho, J.; Sung, W. A High-Speed Layered Min-Sum LDPC Decoder for Error Correction of NAND Flash Memories. In Proceedings of the 2011 IEEE 54th International Midwest Symposium on Circuits and Systems (MWSCAS), Seoul, Republic of Korea, 7–10 August 2011; pp. 1–4. [CrossRef]
30. Mei, Z.; Cai, K.; He, X. Deep Learning-Aided Dynamic Read Thresholds Design for Multi-Level-Cell Flash Memories. *IEEE Trans. Commun.* **2020**, *68*, 2850–2862. [CrossRef]
31. Nguyen, V. Design of Capacity-Approaching Protograph-Based LDPC Coding Systems. Ph.D. Dissertation, Department of Electrical & Computer Engineering, The University of Texas at Dallas, Dallas, TX, USA, 2012.
32. Lv, L.; Yang, Z.; Fang, Y.; Guizani, M. Adaptive Interleaver and Rate-Compatible PLDPC Code Design for MIMO FSO-RF Systems. *IEEE Trans. Veh. Technol.* **2024**, *in press*. [CrossRef]

entropy

MDPI

Article

Hybrid Window Decoding for Joint Source Channel Anytime Coding System

Li Deng [1,2], Zhiping Shi [1,2,*], Yixin Wang [3], Xiaoxi Yu [4], Yong Liang Guan [4] and Zhiping Xu [5]

[1] National Key Laboratory on Communications, University of Electronic Science and Technology of China, Chengdu 611731, China; dengli@uestc.edu.cn
[2] Yangtze Delta Region Institute (Huzhou), University of Electronic Science and Technology of China, Huzhou 313000, China
[3] The Institute for Infocomm Research (I2R), Agency for Science, Technology and Research (A*STAR), Singapore 138632, Singapore; wang_yixin@i2r.a-star.edu.sg
[4] The School of Electrical and Electronic Engineering, Nanyang Technological University, Singapore 639798, Singapore; xiaoxi001@e.ntu.edu.sg (X.Y.); eylguan@ntu.edu.sg (Y.L.G.)
[5] The School of Ocean Information Engineering, Jimei University, Xiamen 361021, China; zhipingxu@jmu.edu.cn
* Correspondence: szp@uestc.edu.cn

Abstract: Joint source channel anytime coding (JSCAC) is a kind of joint source channel coding (JSCC) systems based on the causal spatially coupled coding and joint expanding window decoding (JEWD) techniques. JSCAC demonstrates greatly improved error correction performance, as well as higher decoding complexity. This work proposes a joint hybrid window decoding (JHWD) algorithm for JSCAC systems, aiming to reduce the decoding complexity while maintaining comparable error correction performance with the state of the art. Unlike the traditional JEWD technique and its variants, the proposed JHWD algorithm utilizes a hybrid window structure. It achieves this by implementing adaptive local expanding window decoding in the sliding window region, guided by syndrome-check-based detection. The hybrid window decoding characteristics of the proposed JHWD algorithm can both effectively reduce the decoding complexity caused by window expanding, and also mitigate the error propagation caused by window sliding. An improved density evolution algorithm is proposed for the asymptotic performance analysis of the proposed JHWD algorithm. Both the analytical and simulation results indicate that the proposed JHWD decoding scheme is a good low-complexity implementation option for JSCAC systems, which is also suitable for other JSCC systems with the spatially coupled coding structure.

Keywords: joint source channel coding (JSCC); anytime coding; adaptive local expanding window decoding

1. Introduction

The separate source-channel coding technique is based on the Shannon separation coding theorem, which assumes that the source entropy is less than the channel capacity under the conditions of unrestricted delay, complexity and code length. That is, the separation coding is asymptotically optimal. Joint source channel coding (JSCC) is a kind of coding technology that designs source/channel encoding and decoding in an interdependent way, so as to achieve the overall optimal communication system [1]. JSCC technology can be used in any system that violates the condition of the separation coding theorem and needs to consider both the source side and the channel side, such as communication systems with limitations of power, bandwidth, complexity or delay, multi-user system with shared channel, and different types of communication environments composed of heterogeneous sources or channels, etc. In recent years, the increasingly mature deep learning technology has brought new design ideas for JSCC technology, and set off a new wave of research [2–4].

The benefits of JSCC are reflected in the following aspects: (a) Improve the transmission efficiency: with comprehensive consideration of the characteristics of the source and channel, JSCC can optimize the coding scheme, reduce the size of the transmitted data, so as to improve the transmission efficiency. This means that more information can be transmitted with the same bandwidth, or less transmission time with the same amount of information. (b) Reduce the error rate: JSCC can adjust the coding mode according to the specific situation of the channel to reduce the bit error rate during transmission. This means that data are less error-prone during transmission, improving the reliability and accuracy of data. (c) Improve the system robustness: JSCC is designed with channel variation and interference in mind, making the system more stable and reliable in the face of different channel conditions. This allows the system to adapt to a wider range of environments, reducing transmission failures due to channel changes.

In terms of JSCC technologies that are not based on deep learning, there are several typical schemes. At the sending side, the channel encoder can provide the channel state information (CSI) to the source encoder for the bit allocation [5,6] and the source code optimization [7]. On the other hand, the source encoder can also provide the source significance information (SSI) for the unequal error protection (UEP) during the channel coding or modulation [8,9]. At the receiving side, the source *a prior* information and *a posterior* information can be used for iterative decoding between the source and channel decoders, which is one of the most popular JSCC solutions and also the focus of this paper. Fresia first proposed a JSCC system that completes the source compression and error correction using two block LDPC codes (D-LDPC), then implements the joint belief propagation (BP) decoding within a global factor graph model framework, including both source and channel coding structures [10]. After that, various codes were introduced into the JSCC system, such as the block double protograph LDPC (DP-LDPC) code [11,12], the block double Polar (D-Polar) code [13], the spatially coupled LDPC (SC-LDPC) code [14–16], and the spatially coupled repeat-accumulate (SC-RA) code [17,18], etc.

Among the above solutions, we focus on the scheme termed SC-RA based joint source channel anytime coding (JSCAC). It is a specific JSCC scheme combined with anytime coding techniques, i.e., causal spatially coupled coding and expanding window decoding [19,20]. The anytime properties of the JSCAC scheme account for both delay and reliability, making it particularly suitable for transmission scenarios demanding both low delay and high reliability. However, the prior-art expanding window decoding strategies have noticeable decoding complexities and memory space requirements [17,18], which affect the practical applications of JSCAC systems.

In order to reduce the decoding complexity while maintaining comparable error correction performance, a hybrid window decoding scheme is proposed. The novelty and contributions of this paper can be summarized as follows.

- A joint hybrid window decoding (JHWD) algorithm is proposed, which adaptively implements the local expanding window decoding in the sliding window region with the constraint of syndrome-check based detection.
- The density evolution algorithm of the JSCAC system in [18] is improved for the proposed JHWD scheme to analyze its asymptotic decoding complexity and anytime properties.
- Some numerical analyses and simulation results are provided to confirm the advantages of the proposed JHWD algorithm over other prior-art decoding schemes for JSCAC systems.

The remainder of this paper is organized as follows. The background of the SC-RA-based JSCAC system is briefly introduced in Section 2. The specific descriptions of the proposed JHWD scheme and improved density evolution algorithm are provided in Section 3. The numerical analyses and simulation results are presented in Section 4, followed by conclusions in Section 5.

2. Preliminaries

In this section, we briefly introduce the coding structure, the prior-art decoding strategies, and the density evolution analysis method of SC-RA-based JSCAC system.

2.1. Source Model

Let $\mathbf{s} = (s_1, \ldots, s_2) \in \{0,1\}^n$ be a binary sequence of length n generated by an independent identically distributed (i.i.d) nonuniform memory-less source with probability distribution $(p, 1-p)$, where p is the probability of the binary bit "1" in a source sequence (named as source probability), and $p = P_r(s_i = 1) \neq 0.5$. The initial log-likelihood-ratio (LLR) of each source bit is denoted by $z^{sc} = \log(\frac{(1-p)}{p})$; and the initial LLRs of the received code word $\mathbf{y}$ after the binary-phase-shift-keying modulation (BPSK) and over the additive white Gaussian noise (AWGN) channel can be represented as $\mathbf{z}^{cc} = \frac{2\mathbf{y}}{\sigma^2}$, where σ^2 is the variance of channel noise.

2.2. Sketch of the SC-RA-Based JSCAC System

Figure 1 shows the encoding and decoding structures of the SC-RA-based JSCAC system [17], which is composed of the upper half part (source code) and the lower half part (channel code). Therein, the white circles represent the parity bit nodes (PNs) with a fixed degree of 2; the black and gray rectangles indicate the source syndrome nodes (SNs) and channel check nodes (CNs) with degrees of A^{sc} and A^{cc}, respectively; while the black and gray circles denote the source and channel information bit nodes (INs) with degrees of Q^{sc} and Q^{cc}, respectively. Each SN or CN is causally and exponentially connected to the former INs with an exponential rate of λ and a coupling length of γ. The exponentially distributed edge connections of source code and channel code are depicted by line thickness in Figure 1, which can also be written as [21]:

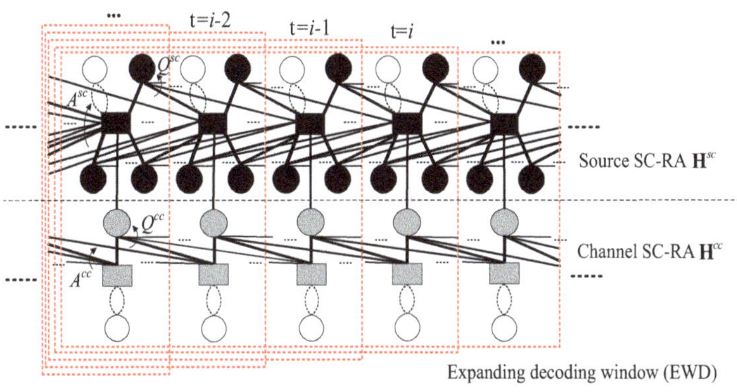

Figure 1. Encoding and decoding structures of the SC-RA-based JSCAC system [17]. Reproduced with permission of [17], *Copyright* ©2020, *IEEE*.

$$P_r(k) = \frac{e^{-k\lambda}(1-e^{-\lambda})}{1-e^{-\gamma\lambda}}, \tag{1}$$

where $k = j - i$ is the distance between the j-th SN (CN) with the connected i-th IN ($j \geq i$). The exponential rate of $\lambda = \lambda^{sc}$ (or λ^{cc}) and the coupling length of $\gamma = \gamma^{sc}$ (or γ^{cc}). The larger the parameter λ, the faster the error probability decays with increasing decoding delay. When $\lambda^{sc} = \lambda^{cc} = 0$, then $P_r(k) = 1$, which is equal to the uniform distribution. The uniform distribution has a slow decline rate over the decoding delay, which is undesired in the anytime transmission. The source code rate and channel code rate can be calculated with the following equation [22]:

$$R = \frac{L}{L + \frac{Q}{A}(L - \gamma + 1 + 2(\gamma - \sum_{i=0}^{\gamma}(\frac{i}{\gamma})^{A}))}, \tag{2}$$

where L is the chain length. For the source code rate R^{sc}, $Q = Q^{sc}$, $A = A^{sc}$, $\gamma = \gamma^{sc}$; and for the channel code rate R^{cc}, $Q = Q^{cc}$, $A = A^{cc}$, $\gamma = \gamma^{cc}$. As $L \to \infty$, the asymptotic code rates of the SC-RA-based JSCAC system are $R^{cc} \to \frac{A^{cc}}{A^{cc}+Q^{cc}}$, $R^{sc} \to \frac{Q^{sc}}{A^{sc}+Q^{sc}}$.

2.3. Encoding Procedure of SC-RA-Based JSCAC System

The source parity-check matrix H^{sc} and the channel parity-check matrix H^{cc} have the same structure as (3)

$$\mathbf{H}_{[0,L-1]} = \begin{bmatrix} \mathbf{H}_{00} & & \\ \vdots & \ddots & \\ \mathbf{H}_{0(\gamma-1)} & \ddots & \\ & \ddots & \mathbf{H}_{(L-1)0} \\ & & \vdots \\ & & \mathbf{H}_{(L-1)(\gamma-1)} \end{bmatrix}, \tag{3}$$

where L is the chain length, the involved sub-matrices for the encoding procedure of the i-th message block are from $\mathbf{H}_{i0}$ to $\mathbf{H}_{(i-\gamma+1)(\gamma-1)}$, $\gamma = \gamma^{sc}$ or (γ^{cc}). The proposed JSCAC follows the concatenated source and channel encoding of block LDPC based JSCC systems.

(1) Syndrome source encoding: Assume the causal source streaming $\mathbf{s}_{[0,L-1]} = \{\mathbf{s}_i | 0 \leq i \leq L-1\}$ produces a N^{sc}-bit message block $\mathbf{s}_i$ at a time instant t and with a chain length of L. The message block $\mathbf{s}_i$ is compressed with the syndrome source coding as $\mathbf{u}_i = \mathbf{s}_i \mathbf{H}_{i0}^{sc^T} + \mathbf{s}_{(i-1)} \mathbf{H}_{(i-1)1}^{sc^T} + \ldots + \mathbf{s}_{(i-\gamma+1)} \mathbf{H}_{(i-\gamma+1)(\gamma-1)}^{sc^T}$ with the block-length of M^{sc}. If $i < \gamma - 1$, then $\mathbf{u}_i = \mathbf{s}_i \mathbf{H}_{i0}^{sc^T} + \mathbf{s}_{(i-1)} \mathbf{H}_{(i-1)1}^{sc^T} + \ldots + \mathbf{s}_0 \mathbf{H}_{0(i-1)}^{sc^T}$.

(2) Channel encoding: The compressed source sequence $\mathbf{u}_{[0,L-1]} = \{\mathbf{u}_i | 0 \leq i \leq L-1\}$ is encoded with the channel encoder to generate the channel codeword $\mathbf{x}_{[0,L-1]} = \{\mathbf{x}_i | 0 \leq i \leq L-1\}$. For each block $\mathbf{x}_i = [\mathbf{u}_i, \mathbf{c}_i]$ with the block-length of N^{cc}, the channel PNs $\mathbf{c}_i$ have a fixed degree of 2, checking both the coupled INs in the early transmitted blocks and local INs, which are denoted as $\mathbf{v}_i = [\mathbf{x}_{i-\gamma+1}, \ldots, \mathbf{x}_{i-1}, \mathbf{u}_i]$ ($\mathbf{v}_i = [\mathbf{x}_0, \ldots, \mathbf{x}_{i-1}, \mathbf{u}_i]$, if $i < \gamma - 1$). The generation of channel PNs $\mathbf{c}_i = \{c_{ij} | 1 \leq j \leq N^{cc} - M^{sc}\}$ is described as

$$c_{ij} = \begin{cases} \mathbf{v}_i \widetilde{\mathbf{H}}^{cc^T}(1,:), & j = 1 \\ \mathbf{v}_i \widetilde{\mathbf{H}}^{cc^T}(j,:) + c_{i(j-1)}, & \text{otherwise} \end{cases}, \tag{4}$$

where $\widetilde{\mathbf{H}}^{cc} = [\mathbf{H}_{(i-\gamma+1)(\gamma-1)}^{cc}, \ldots, \mathbf{H}_{(i-1)1}^{cc}, \mathbf{H}_{i0}^{cc}(:, 1 : M^{sc})]$.

2.4. Decoding Algorithms of SC-RA-Based JSCAC System

The red dashed rectangles in Figure 1 indicate the decoding windows expanding in size as decoding delay increases. The joint BP decoding for D-LDPC based JSCC [10] is implemented in each decoding window, which performs the source BP decoder and channel BP decoder in parallel, exchanging the mutual information through the public edges (red dashed lines in Figure 1). See [10] for more details of joint BP. This classical decoding method for JSCAC systems is the joint expanding window decoding (JEWD) [17], which directly implements the joint BP decoding in the expanding window with high complexities. Authors in [17] proposed a partial joint expanding window decoding (PJEWD) based on the observation of convergence inconsistency between the source and channel decoders, i.e., stop message updating for all channel nodes once the channel decoding convergence is detected. Compared with the JEWD algorithm, the PJEWD algorithm can improve the error performance. However, the reduction ratio of decoding complexity is not significant, as it is affected by the channel conditions.

2.5. Density Evolution (DE) Algorithms for SC-RA-Based JSCAC System

In [18], the authors proposed a density evolution (DE) algorithm for an SC-RA-based JSCAC system with PJEWD scheme. Based on the DE analysis, the anytime properties of the SC-RA-based JSCAC system are provided, i.e., the error probability of any message block can exponentially decay to zero as the decoding delay approaches to infinity. Moreover, the lower bound of the delay exponent α is also provided, which is proportional to the product of λ^{sc} and Q^{sc}. The DE algorithm can also analyze the BP decoding threshold and the decoding complexity of the SC-RA-based JSCAC with PJEWD scheme. See more details of the encoding process, the decoding algorithms, and the DE algorithm of the SC-RA-based JSCAC system in [17,18].

3. Hybrid Window Decoding for JSCAC System

3.1. Hybrid Window Decoding Structure

In this section, we first describe the proposed JHWD algorithm for JSCAC systems in detail, then we provide the improved DE analysis method for the proposed JHWD algorithm.

The proposed hybrid window decoding structure is shown in Figure 2, which is composed of two main parts, i.e., the initial expanding window decoding phase and the sliding window decoding phase. Each part is numerically marked. In order to simplify the description, the source and channel couple lengths are set the same as $\gamma^{sc} = \gamma^{cc} = \gamma$. In the example shown in Figure 2, γ is set as 5. We explain the details of the proposed JHWD scheme as follows.

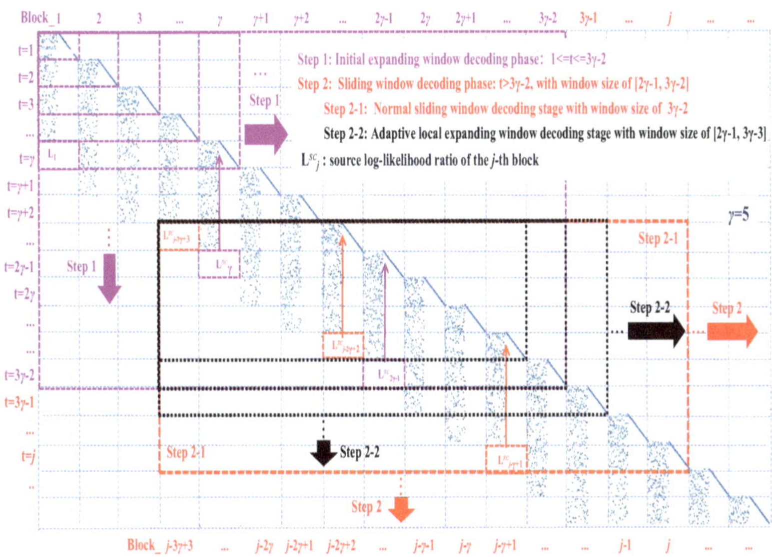

Figure 2. Hybrid window decoding structure of the proposed JHWD algorithm.

Application scenarios: The proposed JHWD scheme is designed for the source streaming with short block length, and with a chain length L far longer than the coupling length γ, i.e., $L \gg \gamma$. Accordingly, as shown in Figure 2, there is not any special decoding process for blocks at the end of the data chain. Moreover, this work focuses on the decoding performances for the source streaming with short block lengths in low-latency and high-reliability communications.

Selection of window size: In the JSCAC system, the coupling length γ is supposed to be finite, and the encoding structure of any block is directly related to its $\gamma - 1$ coupled blocks. However, all blocks in the data chain are causally coupled and connected one by one. Therefore, the decoding performance of any block is still related to the backward-connected blocks beyond the coupling length. However, the mutual influence will gradually weaken as distance increases. In order to trade off the error correction performance and the decoding complexity, we adopt the sliding decoding window size as $3\gamma - 2$ block lengths, covering the coupling range where the object block is located and the two nearest adjacent coupling ranges with the greatest influences. These are called the first, the second and the third coupling interval, respectively.

Emendatory decoding strategies: Combined with the encoding structure of JSCAC system, some emendatory decoding strategies are adopted in the proposed JHWD scheme.

(1) Update the source *a posteriori* log-likelihood ratios (LLRs) at each time slot t after decoding as follows

$$\mathbf{L}_t^{sc} = (\mathbf{L}_{t-1}^{sc} + \mathbf{L}_t^{sc})/2. \tag{5}$$

The reason for updating the source LLRs instead of the channel LLRs is that the source sequence is not actually transmitted through the channel; rather, it is represented by its compressed syndrome. It is found that decoding the object block with the previously well-decoded source information can effectively improve the error correction performance.

(2) In order to alleviate the error propagation caused by sliding window decoding [23], a kind of syndrome-check based detection, shown in (6), is carried out before the local expanding window decoding, where $\hat{\mathbf{S}}_k^{sc}$ and $\hat{\mathbf{U}}_k^{cc}$ are the source and channel decoding results of k-th joint BP, and $\hat{\mathbf{H}}_k^{sc}$ is the involved source parity-check matrix. If the detection result is false, a local expanding window decoding is adaptively carried out within the decoding window until the detection result is true. This syndrome-check-based detection is implemented once, prior to the matrix multiplication in the joint BP decoding stage of the local expanding window decoding. However, it controls the adaptive execution of the local expanding window decoding, and thus can greatly reduce the decoding iterations and message updates.

$$S_{check0} = \begin{cases} 1 & \text{if} \quad \hat{\mathbf{S}}_k^{sc} \cdot \hat{\mathbf{H}}_k^{\mathbf{sc}} = \hat{\mathbf{U}}_k^{cc}, \\ 0 & else. \end{cases} \tag{6}$$

Decoding structure descriptions: The details of the JHWD algorithm are provided in Algorithm 1. The parameters $\mathbf{H}^{sc}$ and $\mathbf{H}^{cc}$ indicate the source and channel parity-check matrix. $I_{\max 1}$ and $I_{\max 2}$ denote the maximum iterations of different decoding stages; and $I_{\max 2}$ is far smaller than $I_{\max 1}$. p is the source probability, σ_{ch} is the variance of channel noise, and T is the maximum observed decoding delay.

There are two main steps in Algorithm 1. First, in the initial expanding window decoding phase of Step 1, the traditional JEWD in [17] is implemented within the time slots range of $t \in [1, 3\gamma - 2]$. Second, when $t > 3\gamma - 2$, the window is slid forward one block and the sliding window decoding is implemented at each time slot in Step 2.

Especially, in the sliding window decoding phase of Step 2, it also includes two sub-steps. (a) Step 2-1: In the normal sliding window decoding stage, the initial source LLR, i.e., $z^{sc} = log\frac{1-p}{p}$, is used for decoding with $I_{\max 1}$, and the decoding range is the whole window length, i.e., $3\gamma - 2$. (b) Step 2-2: In the adaptive local expanding window decoding stage, the source updated *a posteriori* LLRs $\mathbf{L}_t^{sc}$ is used for decoding with $I_{\max 2}$. The decoding range expands with the window size region of $[2\gamma - 1, 3\gamma - 3]$, including the first $2\gamma + k - 2$ blocks $(1 \le k < \gamma)$ of the sliding window. The syndrome-check-based detection is carried out before the local expanding window decoding. As shown in line 16 to line 24 of Algorithm 1, if $S_{check0} = 0$, the local expanding window decoding is carried out within the sliding window region. If $S_{check0} = 1$, the adaptive local expanding window decoding is terminated, and the *a posterior* source LLRs are updated for the involved blocks with (5). After each local expanding window decoding, we update $\mathbf{L}_t^{sc}$ only for blocks in the first coupling interval. The motivation is that in any decoding window range, the LLRs of message blocks in the first coupling interval can converge earlier than the blocks in the other two coupling intervals.

Algorithm 1: Joint hybrid window decoding (JHWD) algorithm

Input: $\mathbf{H}^{sc}, \mathbf{H}^{cc}, I_{\max 1}, I_{\max 2}, T, z^{sc}, \sigma_{ch}$;
Output: the hard decision result $\hat{\mathbf{S}}_{[1,T]}$;

Step 1: Initial expanding window decoding phase
for $t = j, 1 \le j \le 3\gamma - 2$ **do**
  $\hat{\mathbf{H}}_t^{sc} = \mathbf{H}^{sc}[1:j, 1:j]$, $\hat{\mathbf{H}}_t^{cc} = \mathbf{H}^{cc}[1:j, 1:j]$;
  implement joint BP with $\hat{\mathbf{H}}_t^{sc}$, $\hat{\mathbf{H}}_t^{cc}$, $I_{\max 1}$, Z^{sc};
  Update $\mathbf{L}_t^{sc}$ of Block[1:j] with (5);
Step 2: Sliding window decoding phase:
for $t = j, 3\gamma - 2 < j \le T$ **do**
  Step 2-1: Normal sliding window decoding stage
  $\hat{\mathbf{H}}_t^{sc} = \mathbf{H}^{sc}[j - 2\gamma + 2 : j, j - 3\gamma + 3 : j]$,
  $\hat{\mathbf{H}}_t^{cc} = \mathbf{H}^{cc}[j - 2\gamma + 2 : j, j - 3\gamma + 3 : j]$;
  Implement joint BP with $\hat{\mathbf{H}}_t^{sc}$, $\hat{\mathbf{H}}_t^{cc}$, $I_{\max 1}$, Z^{sc};
  Update $\mathbf{L}_t^{sc}$ of Block $[j - 3\gamma + 3 : j]$ with (5);
  Step 2-2: Adaptive local expanding window decoding stage
  Initiate $S_{check0} = 0$;
  for $1 \le k < \gamma$ **do**
    if $S_{check0} == 0$ **then**
      $\hat{\mathbf{H}}_k^{sc} = \mathbf{H}^{sc}[j - 2\gamma + 2 : j - \gamma + k, j - 3\gamma + 3 : j - \gamma + k]$;
      $\hat{\mathbf{H}}_k^{cc} = \mathbf{H}^{cc}[j - 2\gamma + 2 : j - \gamma + k, j - 3\gamma + 3 : j - \gamma + k]$;
      $\mathbf{Z}_k^{sc} = \mathbf{L}_t^{sc}[j - 3\gamma + 3 : j - \gamma + k]$;
      Do joint BP with $\hat{\mathbf{H}}_k^{sc}$, $\hat{\mathbf{H}}_k^{cc}$, $I_{\max 2}$ and $\mathbf{Z}_k^{sc}$;
      Syndrome-check based detection$\rightarrow S_{check0}$;
      if $S_{check0} == 1$ **then**
        Update $\mathbf{L}_t^{sc}$ with (5) for Block $[j - 3\gamma + 3 : j - \gamma + k]$;
        break;
Output: $\hat{\mathbf{S}}_{[1,T]} = -(\mathbf{sign}(\mathbf{L}_T^{sc}) + 1)/2$;
end

3.2. Density Evolution Analysis for JHWD Algorithm

In order to analyze the anytime properties and decoding complexity for the JHWD algorithm, we revised the density evolution algorithm for JSCAC with PJEWD scheme in [18]. In the following density evolution analyses, the specific formulas for message updates are omitted, which are the same as (6) to (15) in [18], and are referred to as "***Message updates***" for simplicity of description.

Step 1: Initialize the parameters of $x_{cv}^0(i_c, t) = x_{cp}^0(j_c, t) = \mu_0 = \frac{2}{\sigma_n^2}, 0 < i_c, j_c \le t$, and $x_{sv}^0(i_s, t) = x_{sp}^0(j_s, t) = \nu_0 = \log\frac{1-p}{p}, 0 < i_s, j_s \le t$, where $x_{cv}^l(i_c, t)$ and $x_{cp}^l(i_c, t)$ are messages exchanged from channel IN_{i_c} and channel PN_{j_c} to connected channel CNs at the l-th iteration and time t; $x_{sv}^l(i_s, t)$ ($x_{sp}^l(j_s, t)$) are messages exchanged from source IN_{i_s} (source PN_{j_s}) to connected source SNs at the l-th iteration and time t. Initialize the number of message updates as $Num_{me} = 0$, and the number of iterations as $l = 0$.

Step 2: Initial expanding window decoding ($1 < t \le 3\gamma - 2$)

(1) Implement "***Message updates***" at the time slot t within the expanding window of $[1:t]$ with $I_{\max 1}$.

(2) Update the following messages as follows:

$$\begin{aligned}
x_{cv}^{t}(i_c,t) &= (x_{cv}^{t-1}(i_c,t) + x_{cv}^{t}(i_c,t))/2, \\
x_{cp}^{t}(j_c,t) &= (x_{cp}^{t-1}(j_c,t) + x_{cp}^{t}(j_c,t))/2, \\
x_{sv}^{t}(i_s,t) &= (x_{sv}^{t-1}(i_s,t) + x_{sv}^{t}(i_s,t))/2, \\
x_{sp}^{t}(j_s,t) &= (x_{sp}^{t-1}(j_s,t) + x_{sp}^{t}(j_s,t))/2.
\end{aligned} \tag{7}$$

(3) Update the number of message updates Num_{me}.

Step 3: Sliding window decoding ($3\gamma - 2 < t \leq T$)

(1) Normal sliding window decoding stage: implement "***Message updates***" at the time slot t within the sliding window of $[t-3\gamma+3:t]$ with $I_{\max 1}$; then update messages with (7), and update Num_{me}.

(2) Adaptive local expanding window decoding stage: The syndrome-check-based detection is adopted by checking $x_{sv}^{t}(i_s,t)$, as below:

$$S_{check1} = \begin{cases} 1 & \text{if} \quad |x_{sv}^{t}(i_s,t) - x_{sv}^{t-1}(i_s,t)| < \epsilon, \\ 0 & else. \end{cases} \tag{8}$$

where ϵ is a threshold with a very small value. If $S_{check1} = 0$, implement "***Message updates***" at the time slot t within the local expanding window of $[t-3\gamma+3:t-\gamma+k]$ ($k=2:\gamma$) with $I_{\max 2}$. If $S_{check1} = 1$, which means that $x_{sv}^{t}(i_s,t)$ could not update anymore compared with $x_{sv}^{t-1}(i_s,t)$, update messages with (7) and Num_{me}.

Step 4: Calculate the error probability with (17) in [18].

4. Simulation and Analysis

In this section, we explore the decoding complexity, anytime properties and error correction performance of the proposed JHWD algorithm with both numerical analyses and simulations. The benchmark algorithms are the joint expanding window decoding (JEWD) algorithm and the partial joint expanding window decoding (PJEWD) algorithm for JSCAC systems in [17,18]. We also compared the proposed algorithm with the traditional joint source channel decoding (JSCD) for block D-LDPC-based JSCC systems [10] and the SC-LDPC-based JSCC scheme with sliding variable-window decoding (SVWD) [16]. The additive white Gaussian noise (AWGN) channel and the binary phase-shift keying (BPSK) modulation are adopted. The simulation parameters are set as follows: $Q^{sc} = 4$, $A^{sc} = 12$, $Q^{cc} = 3$, $A^{cc} = 3$, $R^{sc} = 0.25$, $R^{cc} = 0.5$, $\lambda^{sc} = 0.5$, $\lambda^{cc} = 0.1$, $I_{\max 1} = 30$, $I_{\max 2} = 5$, $p = 0.02$, and $E_b/N_0 = 2$ dB for anytime properties analyses.

We first analyze the asymptotic average message updates of the SC-RA-based JSCAC system as the decoding delay using different decoding schemes. As shown in Figure 3, as the decoding delay increases, the average message updates of all the three decoding schemes show slow and nearly linear growth trends, and their differences become more significant. The right Y-axis of Figure 3 shows the average message updates reduction ratio of PJEWD and the proposed JHWD compared to the traditional JEWD algorithm. As t is greater than 600, the growth rates tend to flatten out. In the decoding delay region of [100, 1000], the message update reduction ratio of PJEWD to JEWD increases from 26% to 48%; while the reduction rate of the proposed JHWD to JEWD increases from 58% to 94%. In other words, the message updates reduction ratio of the proposed JHWD is about two times higher than the PJEWD algorithm.

In Figure 4, the asymptotic anytime properties of SC-RA-based JSCAC with different decoding schemes are analyzed. The anytime properties of a single block in the first sliding window region is observed from $t = 14$. It can be seen that with the same coupling length of $\gamma = 5$, the bit error probability curves of the JEWD and PJEWD algorithm have similar descent rates, while the curve slope of the proposed JHWD algorithm is slightly slower.

Figure 4 also shows the error probabilities of the proposed JHWD with different γ. It can be seen that the smaller the coupling length, the larger the curve slope. However, in the finite length regime (as seen in Figure 5), a smaller coupling length tends to lead a higher error floor. Correspondingly, as shown in Figure 3, the greater the coupling length, the higher the decoding complexity. These results indicate that the anytime property and decoding complexity of the proposed JHWD are related to the coupling length γ, and a proper coupling length should be selected to trade off the error performance and decoding complexity.

In Figure 5, the simulation performances of a single message block is observed as a function of decoding delay within the first sliding window region, from $t = 14$, with a source block length of $N^{sc} = 64$ bits and E_b/N_0 is 2 dB. As shown in Figure 5a, the bit error rate (BER) curve of the proposed JHWD algorithm is located between the JEWD and PJEWD algorithms, which is consistent with the numerical results in Figure 4; and it lastly reaches the error floor at $t = 26$ (also within the sliding window size of $3\gamma - 2$). Moreover, the error floor of the proposed JHWD algorithm is closer to that of the PJEWD algorithm. When the coupling length is larger, i.e., $\gamma = 8$, the slops of the BER curves are slightly smaller than those of $\gamma = 5$ before the time slot $t = 5$, but can achieve lower BER performance. when $t = 21$, the BER of the proposed JHWD algorithm is lower than 10^{-6}. Figure 5b shows the average message updates of a single message block as the decoding delay. The message updates of the JEWD and PJEWD algorithms approximately linearly increase within the observed time slot region. As to the hybrid window decoding of the proposed JHWD algorithm, the adaptive local expanding window decoding strategy with smaller iteration rounds of $I_{\max 2}$ does not result in a significant increase of message updates, which is about 10 times lower than those of the JEWD and PJEWD algorithms at $t = 26$. When the coupling length is $\gamma = 8$, the average message updates of the proposed JHWD algorithm are relatively higher than those of $\gamma = 5$. This experiment indicates that the anytime property of the proposed JHWD algorithm is between the JEWD algorithm and PJEWD algorithm but with lower decoding complexity.

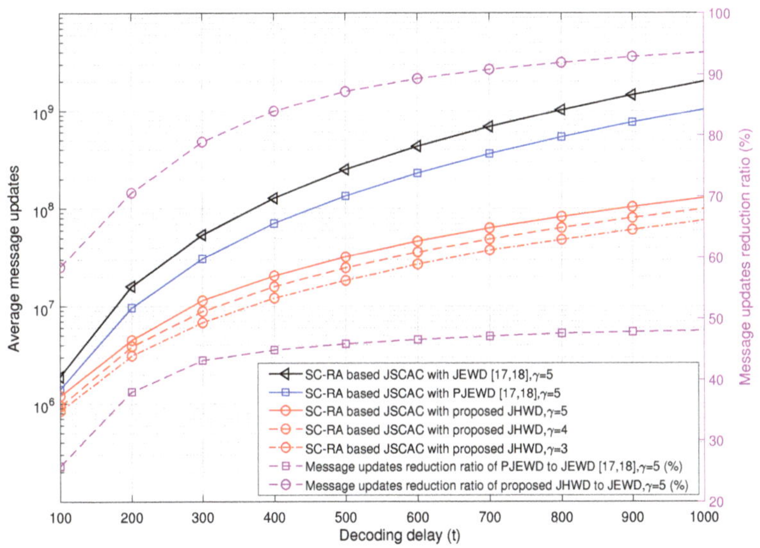

Figure 3. Asymptotic decoding complexity analysis based on the density evolution [17,18].

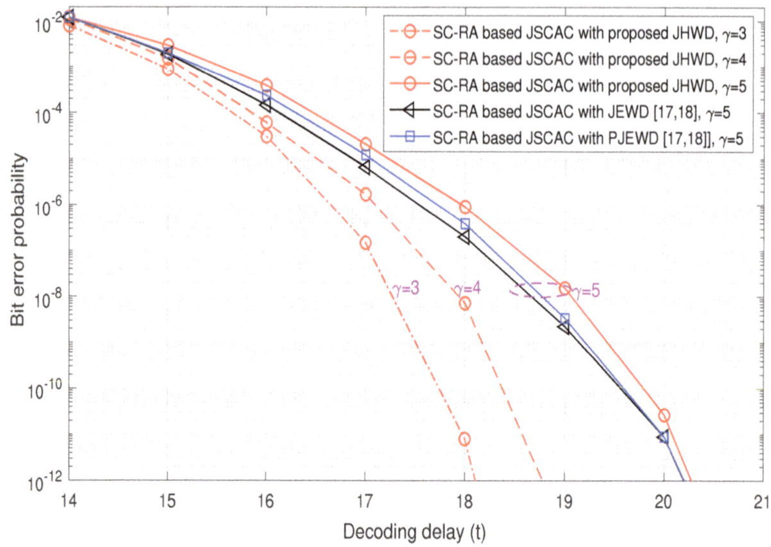

Figure 4. Asymptotic anytime property analysis based on the density evolution [17,18].

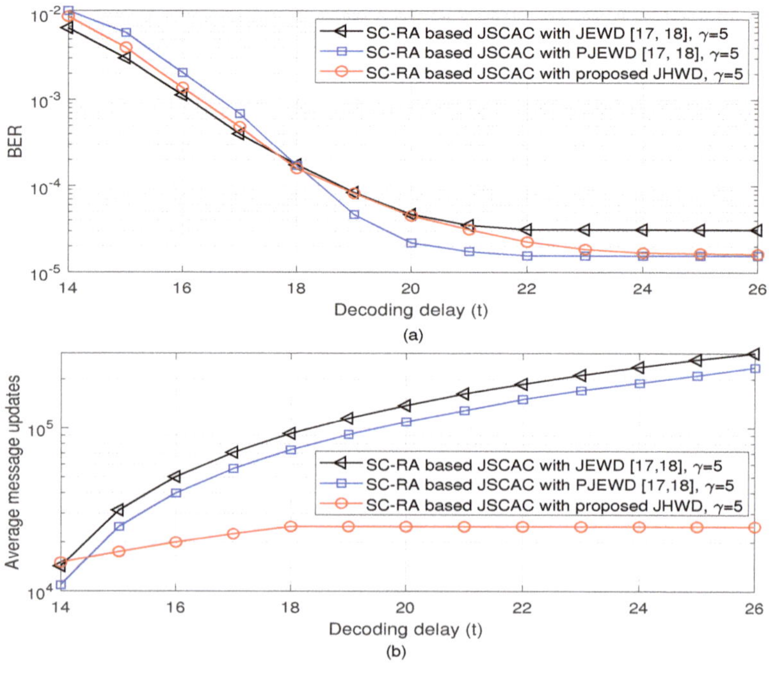

Figure 5. BER performance and decoding complexity of a single message block with $N^{sc} = 64$ bits as the decoding delay within a sliding window region [17,18].

In Figure 6, the simulation performance of the data chain at the time slot $t = 100$ versus E_b/N_0 is observed. In Figure 6a, the proposed JHWD algorithm has the smallest curve slope in the low signal to noise ratio (SNR) region. In the high SNR region, larger than 1.5 dB, the proposed JHWD algorithm can obtain a lower error floor than that of JEWD algorithm and is closer to the BER performance of the PJEWD algorithm. Figure 6b shows the average message updates normalized by that of the JEWD scheme. The decoding complexity of traditional JEWD does not change with SNR. While the decoding complexities of the proposed JHWD and PJEWD scheme are lower than that of the JEWD scheme, especially in the high SNR region. When E_b/N_0 = 3 dB, the average message updates of PJEWD can be decreased about 26.47%, while the proposed JHWD can be decreased about 69.61% compared with the JEWD algorithm. These simulation results show that the proposed JHWD can reduce the decoding complexity to a great extent without sacrificing too much error correction performance.

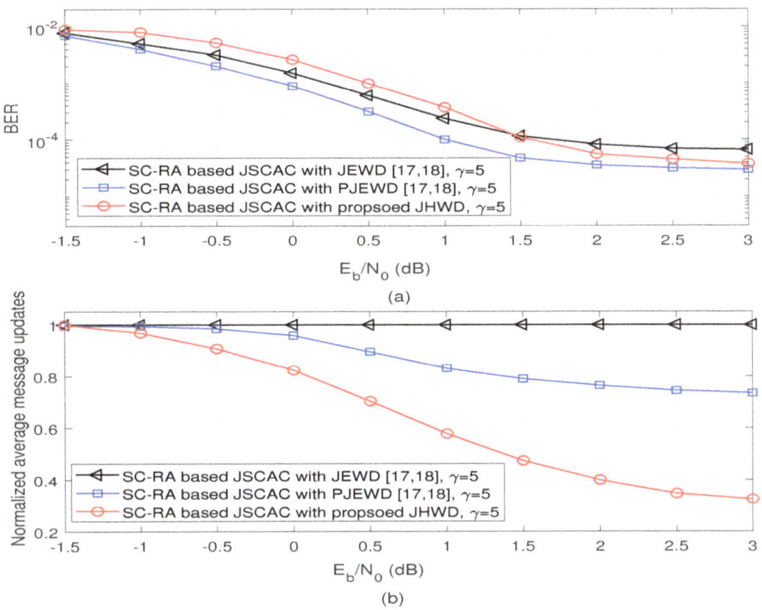

Figure 6. BER performance and decoding complexity of the data chain at the time slot $t = 100$ (L = 100) with $N^{sc} = 64$ bits [17,18].

Figure 7 provides the BER performance comparisons with the state of the art, especially the (3,12)- and (3,6)- SC-LDPC-based JSCC scheme with sliding variable-window decoding (SVWD) of comparable window size [16]. As seen in Figure 7, with the same code rates of R^{sc} = 0.25, R^{cc} = 0.5 and the code length of 6400 bits, the block D-LDPC-based JSCC (LDPC i-8-2-4) [10] has the highest error floor. When the coupling length γ is 3, which is comparable to the syndrome former memory of $m = 2$ in [16] ($\gamma = m + 1$), and the adaptive decoding window size is $2\gamma - 1 = 5$ to $3\gamma - 2 = 7$ (versus the sliding variable-window size of $W_f = 3$ to $W_{max} = 6$ in [16]), the SC-RA-based JSCAC with proposed JHWD has a lower error floor. However, as γ is increased to match the larger window sizes of [16], the preponderance of SC-RA-based JSCAC with the proposed JHWD is gradually reduced. When γ is 5, the JSCAC scheme with the proposed JHWD demonstrates a similar performance to the SC-LDPC JSCC scheme with SVWD of $W_f = 9$. The narrowing gap might be caused by the mismatch between γ and m in [16].

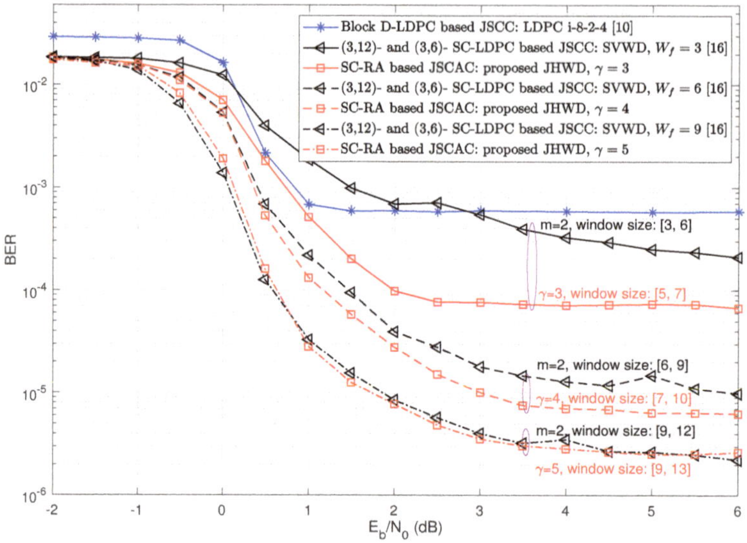

Figure 7. BER performance comparisons with other kinds of JSCC schemes, $R^{sc} = 0.25$, $R^{cc} = 0.5$, $N^{sc} = 640$ bits, $N^{cc} = 320$ bits, and $L = 20$ [10,16].

5. Conclusions

This paper considers a decoding scheme to reduce the decoding complexity without compromising the error correction performance for JSCAC system. We have observed that the improvement of post-decoding BER of the global expanding window decoding in a JSCAC system with finite coupling length gradually decreases with the decoding delay, while the decoding complexity increases approximately linearly. Our key idea, called joint hybrid window decoding (JHWD), is to implement an adaptive local expanding window decoding within the sliding window region. We have also proposed an improved density evolution algorithm to analyze the asymptotic decoding complexity of the JHWD algorithm. The numerical analyses and simulation results have shown that the proposed JHWD algorithm can greatly reduce the decoding complexity while maintaining BER performance similar to that of prior-art decoding schemes, which could be a good alternative for the application of JSCAC systems in low complexity scenarios.

Author Contributions: Conceptualization, L.D. and Y.W.; methodology, L.D. and X.Y.; software, L.D. and Y.W.; validation, Z.S. and X.Y.; formal analysis, Y.L.G.; investigation, Y.W.; resources, Z.X.; data curation, X.Y.; writing—original draft preparation, L.D.; writing—review and editing, Z.S. and Y.L.G.; visualization, Z.X.; supervision, Y.L.G.; project administration, Z.S.; funding acquisition, L.D. and Z.S. All authors have read and agreed to the published version of the manuscript.

Funding: This work was supported in part by the National Natural Science Foundation of China under Grant 62371101 and 62461017; the National Key Laboratory of Wireless Communications Foundation under Grant 2023KP01609; the National Key Research and Development Program under Grant 2022YFC3005702; the ZTE Industry-University-Institute Cooperation Funds under Grant IA20240000001; the Natural Science Foundation of Guangxi under Grant 2022GXNSFBA035508 and 2023GXNSFAA026018.

Institutional Review Board Statement: Not applicable.

Data Availability Statement: The data presented in this study are available on request from the corresponding author

Acknowledgments: The authors thank the anonymous reviewers and the Editor for their helpful comments and review of this article.

Conflicts of Interest: The authors declare no conflict of interest.

Abbreviations

The following abbreviations are used in this manuscript:

AWGN	additive white Gaussian noise
BER	bit error rate
BP	belief propagation
BPSK	binary phase-shift keying
CNs	check nodes
CSI	channel state information
D-LDPC	double low-density parity check
DP-LDPC	double protograph LDPC
JEWD	joint expanding window decoding
JHWD	joint hybrid window decoding
JSCC	joint source-channel coding
JSCAC	joint source-channel anytime coding
LLRs	log-likelihood ratios
PJEWD	partial joint expanding window decoding
PNs	parity bit nodes
INs	information bit nodes
SNR	signal to noise ratio
SNs	syndrome nodes
SC-LDPC	spatially coupled LDPC code
SSI	source significance information
SC-RA	spatially coupled repeat-accumulate code
UEP	unequal error protection

References

1. Massey, J.L. Joint source and channel coding. *Commun. Sys. Rand. Process Theory.* **1978**, *11*, 279–293.
2. Lyu, S.; Dai, J.; Zhang, P. New paradigm of joint source and channel processing: Semantic communications. *ZTE Technol. J.* **2022**, *27*, 2–8. [CrossRef]
3. Xu, J.; Chen, W.; Ai, B. Deep joint source-channel coding based CSI Feedback. *ZTE Technol. J.* **2022**, *27*, 29–33. [CrossRef]
4. Dong, Y.; Dai, J.; Niu, K.; Wang, S.; Yuan, Y. Joint Source-Channel Coding for 6G Communications. *China Commun.* **2022**, *19*, 101–115. [CrossRef]
5. Cheung, G.; Zakhor, A. Bit allocation for joint/channel coding of scalable video. *IEEE Trans. Image Process.* **2000**, *9*, 340–356. [CrossRef] [PubMed]
6. Bystrom, M.; Stockhammer, T. Dependent source and channel rate allocation for video transmission. *IEEE Trans. Commun.* **2004**, *3*, 258–268. [CrossRef]
7. Shirazinia, A.; Chatterjee, S.; Skoglund, M. Joint source-channel vector quantization for compressed sensing. *IEEE Trans. Signal Process.* **2014**, *62*, 3667–3681.
8. Sanchez, V.; Mandal, M. Efficient channel protection for JPEG2000 bitstream. *IEEE Trans. Circuits Syst. Video Technol.* **2004**, *14*, 554–558. [CrossRef]
9. Zhou, C.; Lin, C.; Zhang, X.; Guo, Z. A novel JSCC scheme for UEP-based scalable video transmission over MIMO systems. *IEEE Trans. Circuits Syst. Video Technol.* **2015**, *25*, 1002–1015. [CrossRef]
10. Fresia, M.; Pérez-Cruz, F.; Poor, H.V.; Verdú, S. Joint source and channel coding. *IEEE Signal Process. Mag.* **2010**, *27*, 104–113. [CrossRef]
11. He, J.; Wang, L.; Chen, P. A joint source and channel coding scheme base on simple protograph structured codes. In Proceedings of the IEEE International Symposium on Communications Information Technologies (ISCIT), Gold Coast, Australia, 2–5 October 2012; pp. 65–69.
12. Jiang, X.; Zeng, G.; Chen, W.; Wen, M.; Li, J. Two-Layer LDPC Codes for Low Complexity ML Detection in GSM Systems. *IEEE Wireless Commun. Lett.* **2018**, *7*, 408–411. [CrossRef]
13. Dong, Y.; Niu, K.; Dai, J.; Wang, S.; Yuan, Y. Joint Source and Channel Coding Using Double Polar Codes. *IEEE Commun. Lett.* **2021**, *25*, 2810–2814. [CrossRef]
14. Golmohammadi, A.; Mitchell, D.G.M. Concatenated spatially coupled LDPC codes with sliding window decoding for joint source-channel coding. *IEEE Trans. Commun.* **2022**, *70*, 851–864. [CrossRef]
15. Jiang, X.; Yuang, S.; Huang, P.; Zeng, G. High-Speed Reconciliation for CVQKD Based on Spatially Coupled LDPC Codes. *IEEE Photon. J.* **2018**, *10*, 1–10. [CrossRef]

16. Lian, Q.; Chen, Q.; Zhou, L.; He, Y.; Xie, X. Adaptive decoding algorithm with variable sliding window for double SC-LDPC coding system. *IEEE Commun. Lett.* **2023**, *27*, 404–408. [CrossRef]
17. Deng, L.; Wang, Y.; Yu, X.; Noor-A-Rahim, M.; Guan, Y.; Shi, Z. Joint Source Channel Anytime Coding. In Proceedings of the 2020 IEEE Global Communications Conference (Globecom), Taipei, Taiwan, 7–11 December 2020.
18. Deng, L.; Yu, X.; Wang, Y.; Noor-A-Rahim, M.; Guan, Y.; Shi, Z.; Zhang, Z. Joint Source Channel Anytime Coding based on Spatially Coupled Repeat-Accumulate Codes. *IEEE Trans. Commun.* **2023**, *71*, 6215–6230. [CrossRef]
19. Sahai, A.; Mitter, S. The necessity and sufficiency of anytime capacity for stabilization of a linear system over a noisy communication link-part I: Scalar systems. *IEEE Trans. Inf. Theory* **2006**, *52*, 3369–3395. [CrossRef]
20. Grosjean, L. Practical Anytime Codes. Ph.D. Dissertation, KTH Royal Institute of Technology, Stockholm, Sweden, 2016.
21. Noor-A-Rahim, M.; Khyam, M.; Guan, Y.; Ali, G.; Nguyen, K.; Lechner, G. Delay-Universal Channel Coding with Feedback. *IEEE Access* **2018**, *6*, 37918–37931. [CrossRef]
22. Johnson, S.; Lechner, G. Spatially coupled repeat-accumulate codes. *IEEE Commun. Lett.* **2013**, *17*, 373–376. [CrossRef]
23. Zhu, M.; Mitchell, D.; Lentmaier, M.; Costello, D. Error Propagation Mitigation in Sliding Window Decoding of Spatially Coupled LDPC Codes. *IEEE J. Sel. Areas Inf. Theory* **2023**, *4*, 470–486. [CrossRef]

entropy

MDPI

Article

Improved PEG-Based Construction of Analog Fountain Codes

Xue Li and Qin Huang *

School of Electronic and Information Engineering, Beihang University, Beijing 100191, China; lixue19@buaa.edu.cn

* Correspondence: qhuang.smash@gmail.com; Tel.: +86-010-8231-4663

Abstract: This paper proposes an improved *progressive edge-growth* (PEG) construction of analog fountain codes (AFCs). During edge selection, it simultaneously allocates weight coefficients in descending order. Analysis shows that our proposed construction reduces the probability of large weight coefficients involved in harmful short cycles. Simulation results indicate that it has good block error rate (BLER) in short block length regime.

Keywords: analog fountain codes; progressive edge-growth; weight coefficients

1. Introduction

With the emergence of applications such as intelligent transportation, tele-surgery, and industrial automation, three main service categories are defined in the fifth generation (5G) mobile network. Among these, ultra-reliable low latency communication (URLLC) poses the most significant challenges due to its physical layer design [1]. URLLC requires a short end-to-end latency, along with high reliability evaluated by block error rate (BLER). These strict constraints impose stringent requirements for the design of channel coding in short block length regime.

Recently, analog fountain codes (AFCs) [2–4], a type of rateless codes, have shown their great potential to improve the latency and reliability while keeping bit-level granularity. Several impressive studies [5–12] have explored the coding schemes and constructions of AFCs to make them promising for URLLC. In [5], optimal AFC parameters were analyzed and designed to minimize the bit error rate. In [7], a low-complexity weight-adaptive AFC transmission scheme was proposed based on extrinsic information transfer (EXIT) analysis. In [9], Bose–Chaudhuri–Hocquenghem (BCH) precoders were designed for AFCs to improve performance for short packet communication. In [11], joint design of AFCs and quasi-gray constellation mapping modulation was proposed, approaching the benchmark. In [12], a short partitioned transmission strategy based on AFCs was proposed to further enhance the error rate and latency performance.

AFC generator matrices construction involves two steps: edge selection and weight coefficient allocation. In [2], the edges were selected to maximize the minimum variable node (VN) degree of AFCs. Weight coefficients were allocated randomly. In [6], short AFCs (S-AFCs) reduced error floors by processing these two steps according to a reliability measure. In [8], Online AFC selected edges and allocated weight coefficients according to feedback information, thereby enhancing reliability. Since short cycles deteriorate the performance of the BP decoding, the PEG algorithm [13] was proposed to reduce short cycles of low-density parity-check (LDPC) codes. In [10], an efficient coefficient progressive edge-growth (WC-PEG) algorithm avoided short cycles to obtain good performance. It selected weight coefficients in the same number of times in each column of generator matrix to avoid reducing the rank of generator matrix.

In this paper, we show that large weight coefficients have high mutual information; thus, they contribute most during the BP decoding. Therefore, it is important to avoid them in short cycles, which may severely deteriorate the decoding performance. Note that PEG maximizes the cycle length, and the earlier selected edges are less likely to be involved in short cycles. We propose to add edges and allocate their weight coefficients in descending order simultaneously. To be specific, we select edges with the smallest reliability measure, and allocate larger weight coefficients to earlier selected edges. Our analysis shows that the proposed construction effectively decreases the probability that larger weight coefficients are involved in short cycles. The simulation results show that it has higher reliability over the existing constructions in short block length regime.

This paper is organized as follows. Section 2 presents the necessary background. Section 3 discusses our proposed construction as well as the simulation results. Section 4 concludes the paper.

2. Analog Fountain Codes and the PEG Algorithm

In this section, encoding and decoding processes of AFCs are briefly introduced. Then, PEG applied to AFC construction is reviewed.

2.1. Preliminaries of AFC

Consider an information sequence $\mathbf{b}'$ comprising k information bits $b'_i \in \{0,1\}$, where $i = 0,1,\dots,k-1$. $\mathbf{b}'$ is binary phase shift keying (BPSK) modulated, resulting in $\mathbf{b} = (b_0, b_1, \dots, b_{k-1})$, where $b_i \in \{-1,1\}$. Each AFC codeword $\mathbf{c} = \{c_0, c_1, \dots, c_{n-1}\}$ is then generated as

$$\mathbf{c} = \mathbf{G}\mathbf{b}^T. \tag{1}$$

Here, $\mathbf{G}$ is the generator matrix:

$$\mathbf{G} = \begin{bmatrix} g_{0,0} & g_{0,1} & \cdots & g_{0,k-1} \\ g_{1,0} & g_{1,1} & \cdots & g_{1,k-1} \\ \vdots & \vdots & \ddots & \vdots \\ g_{n-1,0} & g_{n-1,1} & \cdots & g_{n-1,k-1} \end{bmatrix} \tag{2}$$

and $\mathbf{b}^T$ is the transpose of $\mathbf{b}$. The corresponding Tanner graph is shown in Figure 1.

The degree d of AFC is defined as the number of nonzero elements in each row of $\mathbf{G}$. The d nonzero elements correspond to a predefined weight set $\mathrm{W} = \{w_0, w_1, \dots, w_{d-1}\}$, where $w \in \mathrm{W}$ is called the weight coefficient. Without loss of generality, we set $w_0 > w_1 > \cdots > w_{d-1} > 0$.

After being transmitted through an additive white Gaussian noise (AWGN) channel, $\mathbf{r} = \mathbf{c} + \mathbf{n}$ is received, where $\mathbf{n}$ is the noise vector with zero mean and variance σ^2.

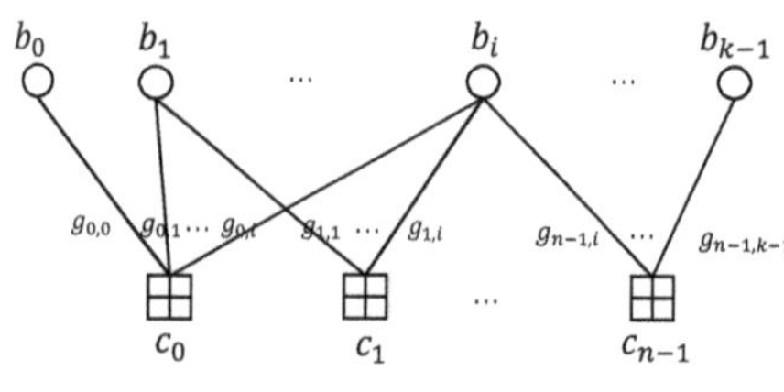

Figure 1. The Tanner graph of an AFC.

At the receiver side, a simplified compressive sensing belief propagation (CS-BP) decoder [14] is employed. The CS-BP decoder is a variant of the BP decoding which can be used in compressive sensing and AFC. The decoding process is briefly reviewed as follows. Let $\mu_{v\to c}(b_i)$ and $\mu_{c\to v}(b_i)$ denote the message transmitted from the VNs b_i to check nodes

(CNs) c_j and vice versa, respectively. After a fixed number of iterations of $\mu_{v\to c(b_i)}$ and $\mu_{c\to v}(b_i)$, the marginal distribution $f(b_i)$ is obtained by

$$f(b_i) = \prod_{j\in\mathcal{J}_i} \mu_{u\to v}(b_i), \tag{3}$$

where $\mathcal{J}_i$ denotes the set of CNs connected to b_i. Then, the decoder outputs the hard decision based on $f(b_i)$. More details of the CS-BP decoder can be reached in [14].

2.2. The PEG Algorithm

Short cycles on Tanner graphs may deteriorate the decoding performance of the BP. Thus, the PEG [13] maximizes the cycle length in the Tanner graph by progressively establishing edges between VNs and CNs in an edge-by-edge manner. VN degree is defined as the number of nonzero elements in each column of G.

Suppose PEG is used to construct an $n \times k$ matrix $\mathbf{G}$ for AFC. For a given c_j, where $0 \le j < n$, we define its neighbor VNs within a depth l as $N_{c_j}^l$, and the remaining VNs are denoted by $\overline{N_{c_j}^l}$. If the subgraph from c_j is fully expanded in depth l, two cases may happen. The cardinality of $N_{c_j}^l$ may stop increasing but $\left|N_{c_j}^l\right| < k$, or $\overline{N_{c_j}^l} \neq \emptyset$ but $\overline{N_{c_j}^{l+1}} = \emptyset$. Let (c_j, b_i) denote the edge connecting c_j and b_i. At last, the newly established edge is (c_j, b_i), where b_i is picked from the set $\overline{N_{c_j}^l}$ with the lowest VN degree. Note that length-r cycle arises if $\overline{N_{c_j}^{l+1}} = \emptyset$, where $r = 2(l+1)$.

When PEG is used to construct generator matrices for AFC, both edge selection and weight coefficients allocation should be considered. In [10], if the edge (c_j, b_i) is selected according to PEG, the weight coefficients are suggested to be allocated according to their appearance in the i-th column of $\mathbf{G}$. Specifically, we calculate the times that each weight coefficient appears among the nonzero elements in the i-th column. The one with the lowest number of times should be allocated in order to balance the weight of columns in $\mathbf{G}$.

3. Improved PEG-Based Construction

In this section, EXIT charts of different weight coefficients at CNs are shown. To avoid large weight coefficients involved in short cycles, we propose to add edges and allocate their weight coefficients simultaneously during edge selection. Analysis shows that the proposed construction efficiently reduces the probability of larger weight coefficients involved in the edges of short cycles.

3.1. EXIT Charts of Different Weight Coefficients

Different weight coefficients contribute differently in the decoding process at CNs. To illustrate that, EXIT charts for various weight coefficients are presented below.

Let $\mathbf{b} = \mathbf{b}_0 \cup \mathbf{b}_1 \cup \cdots \cup \mathbf{b}_{d-1}$ denote the information bits. Without loss of generality, for each CN c_j connected to d VNs $b_{i_0}, b_{i_1}, \cdots, b_{i_{d-1}}$, let $g_{i_s,j} = w_s$, where $s \in \{0, 1, \cdots, d-1\}$. Then $b_{i_s} \in \mathbf{b}_s$. In this way, information bits are divided into d parts, where bits in $\mathbf{b}_s$ are connected to CN through an edge allocated with w_s.

Let $I_{\mathrm{A,C}}$ and $I_{\mathrm{E}_s,\mathrm{C}}$ denote the input mutual information and extrinsic mutual information of the s-th weight coefficient for CN, respectively. For each $\mathbf{b}_s$, the mutual information between $\mathbf{b}_s$ and its corresponding log-likelihood ratio (LLR) value L_s computed at the CNs is represented as

$$I_{\mathrm{E}_s,\mathrm{C}} = \sum_{\mathbf{b}_s\in\{-1,1\}} \int_{-\infty}^{+\infty} p_E(L_s|\mathbf{b}_s)\log_2 \frac{p_E(L_s|\mathbf{b}_s)}{p_E(\mathbf{b}_s)} dL_s. \tag{4}$$

Here, p_E represents the conditional probability distribution function of L_s. As can be seen in Equation (4), each calculation of $I_{\mathrm{E}_s,\mathrm{C}}$ is based on bits $\mathbf{b}_s$ and LLR L_s. It means that after one decoding process at the CNs, CNs output extrinsic mutual information $I_{\mathrm{E}_s,\mathrm{C}}$ to VNs which are connected through weight coefficients w_s. It is difficult to compute p_E directly because of the convolution operation. Thus, the Monte Carlo methods are used to estimate p_E for AFC [7]. Note that $L \sim \mathcal{N}(0, \sigma_L)$, where σ_L can be generated from $I_{A,C}$ as

$$\sigma_L = J^{-1}(\mathrm{I}_{A,C}). \tag{5}$$

The detailed definition of J-function can be reached in [15].

Figure 2 shows the mutual information for four different weight coefficients of AFC, respectively. It can be seen that with the same $I_{A,C}$, the mutual information is larger when weight coefficient is larger. These results indicate that larger weight coefficients contribute higher in the BP decoding.

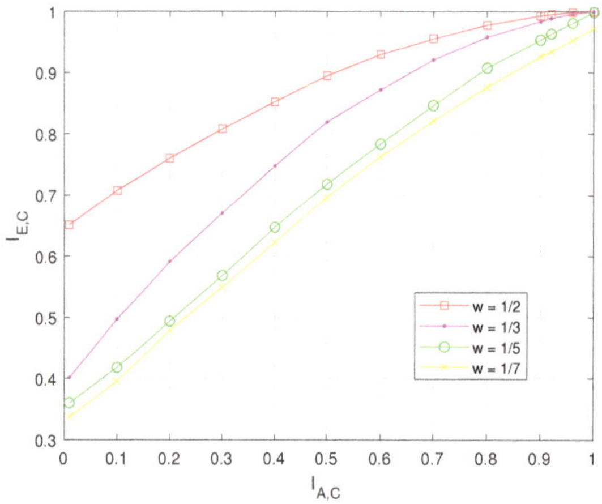

Figure 2. EXIT charts of different weight coefficients, where $d = 4$, $W_s = \{1/2, 1/3, 1/5, 1/7\}$, and signal–noise ratio (SNR) = 18 dB.

3.2. *Improved PEG-Based Construction Method*

It is well known that short cycles may deteriorate the performance of the BP decoding. Large weight coefficients should not be involved in harmful short cycles due to their high contribution in the BP decoding. AFC construction involves two steps: edge selection and weight coefficient allocation. From these two aspects, we propose to construct generator matrix as follows.

During edge selection, we balance the weight of columns in **G** with a reliability measure denoted by $\sum_{p=0}^{i-1} g_{p,q}^2$. The reliability measure represents the sum of squares of weight coefficients for the q-th column of **G** before the edges for c_i are established. Apparently, if a column has a relatively low reliability measure, a corresponding edge should be selected to enlarge its reliability measure. Thus, each time the PEG establishes a candidate set of VN for edge selection, the set is further downsized based on the smallest reliability measure set $Q = \{q | \arg\min_q \sum_{p=0}^{i-1} g_{p,q}^2\}$.

With regard to weight coefficient allocation, recall that the PEG maximizes the cycle length, which means that the edges selected earlier are more likely to be in large cycle. In other words, the edges selected later are more likely to be in the short cycle. Thus, they should not be allocated with larger weight coefficients. It forms the weight coefficient allocation strategy that allocating larger weight coefficients to earlier selected edges.

The proposed construction is presented as follows and the details are summarized in Algorithm 1. Given the parameters k, n, d, and W, an edge-selection procedure is initiated. Let $E_{c_i}^j$ denote the j-th edge selected for c_i, where $0 \leq j < d$. The placement of a new edge $E_{c_i}^0$ on the graph considers the reliability measure of each column. Specifically,

$E_{c_i}^0 \leftarrow \text{edge}(c_i, b_q)$, where q is randomly selected from $Q = \{q | \arg\min_q \sum_{p=0}^{i-1} g_{p,q}^2\}$. Subsequently, the largest weight coefficient w_0 is allocated to $E_{c_i}^0$ as $g_{i,q} = w_0$.

After that, the subgraph from c_i is expanded using the strategy of PEG [13], ending up with the set $\overline{N_{c_i}^l}$. When the expanded stage is finished, $E_{c_i}^j \leftarrow \text{edge}(c_i, b_q)$, where q is randomly selected from a refreshed $Q = \{q | \arg\min_q \sum_{p=0}^{i-1} g_{p,q}^2, b_q \in \overline{N_{c_i}^l}\}$. Then, the next weight coefficient w_j is allocated to $E_{c_i}^j$ as $g_{i,q} = w_j$. This step is repeated until d edges are selected, and then the process moves to the next CN.

Algorithm 1 Improved PEG-based construction algorithm.

Input: k, n, d, W
Output: AFC generator matrix **G**

```
G = zeros(n, k)
for i = 0 to n − 1 do
    for j = 0 to d − 1 do
        if j = 0 then
            Q = {q | argmin_q Σ_{p=0}^{i−1} g²_{p,q}}
            E⁰_{c_i} ← edge(c_i, b_q), where q is randomly selected from Q.
            Update the g_{i,q} = w_0.
        else
            expand a subgraph from c_i up to depth l under the current graph setting such
            that the cardinality N^l_{c_j} stops increasing but |N^l_{c_i}| < k, or N̄^l_{c_i} ≠ ∅ but N̄^{l+1}_{c_i} = ∅.
            Q = {q | argmin_q Σ_{p=0}^{i−1} g²_{p,q}, b_q ∈ N̄^l_{c_i}}
            E^j_{c_i} ← edge(c_i, b_q), where q is randomly selected from Q.
            Update the g_{i,q} = w_j.
        end if
    end for
end for
Return G.
```

3.3. *Analysis of Improved PEG-Based Construction*

Suppose that after applying Algorithm 1, a new initialized CN c_n introduces length-r cycles following the selection of $E_{c_n}^{d-e}$, where e denotes the minimum number of length-r cycles introduced by c_n. During the selection of $E_{c_n}^a$, where $d - e \leq a < d$, a selected edges $E_{c_n}^0, E_{c_n}^1, \cdots, E_{c_n}^{a-1}$ during the selection of c_n, and nd edges established before the initialization of c_n are candidates that can be randomly connected to form a length-r cycle.

Denote the $r-2$ edges of the length-r cycle established before the initialization of c_n as $(c_{j_0}, b_{i_0}), (c_{j_1}, b_{i_1}), \cdots, (c_{j_{r-3}}, b_{i_{r-3}})$, where $0 \leq j < n$ and $0 \leq i < k$. Without loss of generality, denote $E_{c_n}^a \leftarrow (c_n, b_{i_0})$ as the edge closes the cycle, and denote $E_{c_n}^{a'} \leftarrow (c_n, b_{i_1})$ as the edge forming the cycle due to the presence of (c_n, b_{i_0}), where $0 \leq a' < a$.

Moreover, divide W into two subsets according to $E_{c_n}^{d-e}$ as $\text{W} = \text{W}_{\text{large}} \cup \text{W}_{\text{small}}$, where $\text{W}_{\text{large}} = \{w_0, w_1, \cdots, w_{d-e-1}\}$ and $\text{W}_{\text{small}} = \{w_{d-e}, w_{d-e+2}, \cdots, w_{d-1}\}$ represent the sets consisting of relatively large and small weight coefficients, respectively. Based on the aforementioned discussion, the probabilities that a weight coefficient involved in the above edges are analyzed case by case.

Case 1:

$$P(g_{j,i} = w) = 1/d, \; w \in \text{W}, \tag{6}$$

where (c_j, b_i) is one of the $r-2$ edges. The weight coefficient in (c_j, b_i) is allocated before the PEG selects edges for c_n. Thus, it is considered as randomly selected from W.

Case 2:

$$P(g_{n,i_0} = w) = 0, \; w \in \text{W}_{\text{large}}. \tag{7}$$

Considering that length-r cycles arise after selecting $E_{c_n}^{d-e}$, and the weight coefficients are allocated in descending order, the last selected e edges for c_n are allocated with $w \in \mathrm{W}_{\text{small}}$. On the contrary, any $w \in \mathrm{W}_{\text{large}}$ cannot be allocated to g_{n,i_0}.

Case 3:

$$P(g_{n,i_1} = w) = \frac{1}{a} \leq \frac{1}{d-e}, \ w \in \mathrm{W}_{\text{large}}. \tag{8}$$

The corresponding (c_n, b_{i_1}) is randomly connected from the a selected edges. Thus, w can be considered as randomly selected from $\{w_0, w_1, \cdots, w_{a-1}\}$.

Based on the above analysis, $r-2$ edges (c_j, b_i), 1 edge (c_n, b_{i_0}), and 1 edge (c_n, b_{i_1}) have the probabilities in cases 1, 2, and 3, respectively. The probability that w is involved in an edge of the new introduced length-r cycles when $w \in \mathrm{W}_{\text{large}}$ is

$$P(g_{j,i} = w) \leq (\frac{1}{d-e} + \frac{r-2}{d})/r, \ w \in \mathrm{W}_{\text{large}}, \tag{9}$$

where $0 \leq j \leq n$.

Example 1. *Let us consider a* 256×64 *matrix* **G** *with degree* $d = 4$*, which is generated using Algorithm 1. According to the parameter settings and the bound analysis in [10], a few length-4 cycles arise. Thus, we consider* $r = 4$ *and* $e = 1$*.* W *is divided into* $\mathrm{W}_{\text{large}} = \{w_0, w_1, w_2\}$ *and* $\mathrm{W}_{\text{small}} = \{w_3\}$*. Substitute d, e, and r into (9),* $P(g_{j,i} = w) \leq 0.208$*, when* $w \in \mathrm{W}_{\text{large}}$*. Through statistical analysis of a generated* **G***, there are 25 length-4 cycles. The probabilities of weight coefficients involved in the length-4 cycles are shown in Table 1. These results verify our analysis for length-4 cycles.*

Table 1. Probabilities of weight coefficients involved in the length-4 cycles.

Weight Coefficient	Probability
w_0	20%
w_1	23%
w_2	18%
w_3	39%

Example 2. *Let us consider a* 150×128 *matrix* **G** *with* $d = 4$*, which is generated using Algorithm 1. Similarly, we consider* $e = 1$ *and* $r = 6$ *for* **G***.* W *is divided in the same way as in Example 1. Substitute d, e, and r into (9),* $P(g_{j,i} = w) \leq 0.22$*, when* $w \in \mathrm{W}_{\text{large}}$*. Through statistical analysis of a generated* **G***, there are 28 length-6 cycles. The probabilities of weight coefficients involved in the length-6 cycles are shown in Table 2. These results verify our analysis for length-6 cycles.*

Table 2. Probabilities of weight coefficients involved in the length-6 cycles.

Weight Coefficient	Probability
w_0	19%
w_1	21.4%
w_2	21.4%
w_3	38.2%

3.4. Complexity Analysis

The complexity of WC-PEG is $O(nk)$ [13] in the worst case, mainly from expanding the Tanner graph. In addition to expanding the Tanner graph, our proposed algorithm involves calculating the reliability measure. It calculates d weight coefficients introduced by the last CN with complexity $O(nd)$. Then, it sorts the d changed reliability measure with the remainder by binary insertion with complexity $O(nd \log k)$. Thus, its overall complexity is $O(n(k + d \log k))$ in the worst case. All of these algorithms take moderate complexity for constructing generator matrices, which are designed offline.

3.5. Simulation and Discussion

In this subsection, we present the numerical results of the proposed construction and the constructions in the references. Here, we select $d = 8$ and $W = \{1/2, 1/3, 1/5, 1/7, 1/11, 1/13, 1/17, 1/19\}$ for AFC constructions, which are designed in [2]. In addition, $d = 4$ and $W = \{1/2, 1/3, 1/5, 1/7\}$ are also considered. Various construction methods such as random AFC, S-AFC, and WC-PEG are included as comparison. The proposed construction is labeled as IM-PEG. These settings are in line with the previous references. The CS-BP decoder with a maximum iteration number of 15 is applied in the receiver side.

First, we simulate the BLER of different AFC constructions including random AFC, S-AFC, WC-PEG, and the proposed construction. Let $R = k/n$ denote the code rate. Simulations are conducted with fixed rate $R = 2$ and $k = 128$. Various degrees $d = 4$ and $d = 8$ are considered, respectively. The AWGN channel with zero mean and variance δ^2 is employed for the transmission of AFC codewords. Signal–noise ratios (SNRs) range from 16 dB to 24 dB. All of the parameters are set according to [2,6]. Please note that R can be larger than 1, since one AFC symbol involves d bits. It can be seen in Figure 3 that the improved PEG construction has good BLER performance. For example, when $d = 4$, the proposed construction offers about 1 dB gain when BLER is lower than 10^{-4} over WC-PEG and S-AFC.

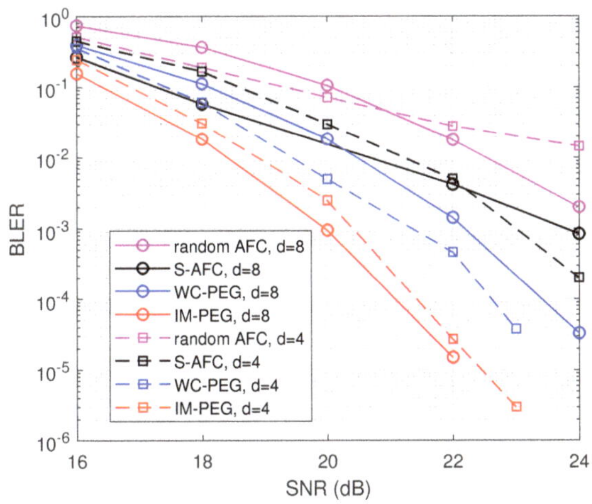

Figure 3. BLER of different AFC constructions with $d = 4$ and $d = 8$.

Then, the achievable rates of different AFC constructions with $k = 128$ are simulated. Similarly, various degrees $d = 4$ and $d = 8$ are considered, respectively. Figure 4 shows that the rates of the proposed construction are higher than others in a large range of SNRs, in both $d = 4$ and $d = 8$.

Moreover, BLER performance with various weight sets is presented in Figure 5. $k = 128, d = 4$, and $R = 1$ are set for the simulation. $W' = \{0.8670, 0.432, 0.2155, 0.1073\}$ [6] is selected for comparison. It can be seen from Figure 5 that the proposed construction provides good error performance with W' as well.

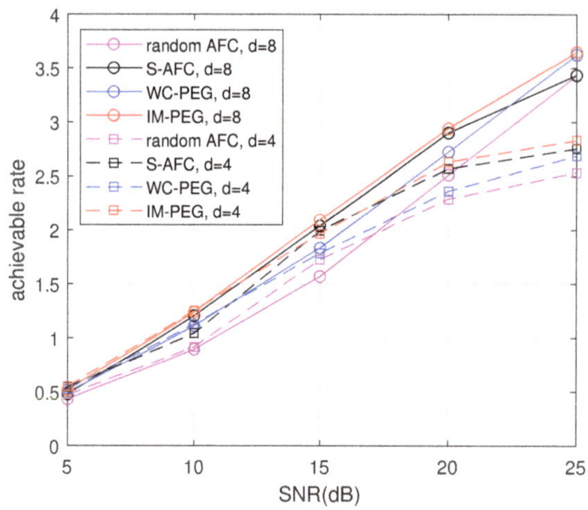

Figure 4. Achievable rates of different AFC constructions with $d = 4$ and $d = 8$.

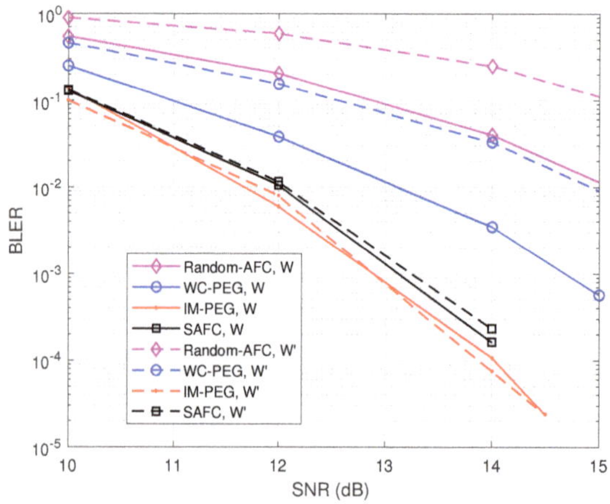

Figure 5. BLER of different weight sets with $d = 4$ and $R = 1$.

4. Conclusions

In this paper, an improved PEG-based construction of AFC is proposed to enhance decoding performance. Specifically, we propose to allocate weight coefficients in descending order simultaneously when PEG selects edges for AFC, avoiding large weight coefficients involved in the short cycles. The analysis demonstrates that the proposed construction efficiently reduces the probability that larger weight coefficients are involved in short cycles. The simulation results show that it has good BLER performance and achievable rates.

Author Contributions: Investigation, X.L. and Q.H.; Writing—original draft, X.L.; Writing—review & editing, Q.H.; Funding acquisition, Q.H. All authors have read and agreed to the published version of the manuscript.

Funding: This research was funded by National Natural Science Foundation of China under Grant 62071026 and 62331002.

Institutional Review Board Statement: Not applicable.

Data Availability Statement: The original contributions presented in the study are included in the article, further inquiries can be directed to the corresponding author.

Conflicts of Interest: The authors declare no conflicts of interest.

References

1. Le, T.K.; Salim, U.; Kaltenberger, F. An Overview of Physical Layer Design for Ultra-Reliable Low-Latency Communications in 3GPP Releases 15, 16, and 17. *IEEE Access* **2021**, *9*, 433–444. [CrossRef]
2. Shirvanimoghaddam, M.; Li, Y.; Vucetic, B. Near-Capacity Adaptive Analog Fountain Codes for Wireless Channels. *IEEE Commun. Lett.* **2013**, *17*, 2241–2244. [CrossRef]
3. Shirvanimoghaddam, M.; Li, Y.; Vucetic, B. Multiple access analog fountain codes. In Proceedings of the 2014 IEEE International Symposium on Information Theory, Honolulu, HI, USA, 29 June–4 July 2014; pp. 2167–2171. [CrossRef]
4. Lim, W.J.; Abbas, R.; Li, Y.; Vucetic, B.; Shirvanimoghaddam, M. Analysis and Design of Short Analog Fountain Codes for the Multiple Access Channel. *IEEE Commun. Lett.* **2022**, *26*, 1454–1458. [CrossRef]
5. Lim, W.J.; Abbas, R.; Li, Y.; Vucetic, B.; Shirvanimoghaddam, M. Analysis and Design of Analog Fountain Codes for Short Packet Communications. *IEEE Trans. Veh. Technol.* **2021**, *70*, 12662–12674. [CrossRef]
6. Abbas, R.; Shirvanimoghaddam, M.; Huang, T.; Li, Y.; Vucetic, B. Novel Design for Short Analog Fountain Codes. *IEEE Commun. Lett.* **2019**, *23*, 1306–1309. [CrossRef]
7. Zhang, K.; Jiao, J.; Huang, Z.; Wu, S.; Zhang, Q. Finite Block-Length Analog Fountain Codes for Ultra-Reliable Low Latency Communications. *IEEE Trans. Commun.* **2020**, *68*, 1391–1404. [CrossRef]
8. Li, X.; Huang, Q. Online Analog Fountain Codes. *IEEE Commun. Lett.* **2023**, *27*, 50–54. [CrossRef]
9. Lim, W.J.; Shirvanimoghaddam, M.; Abbas, R.; Li, Y.; Vucetic, B. On the Design of Analog Fountain Codes for Short Packet Communications in 5G URLLC. In Proceedings of the 2019 IEEE 90th Vehicular Technology Conference (VTC2019-Fall), Honolulu, HI, USA, 22–25 September 2019; pp. 1–5. [CrossRef]
10. Huang, Z.; Jiao, J.; Zhang, K.; Wu, S.; Zhang, Q. A Novel Weight Coefficient PEG Algorithm for Ultra-Reliable Short Length Analog Fountain Codes. *IEEE Wirel. Commun. Lett.* **2019**, *8*, 125–128. [CrossRef]
11. Zhang, K.; Jiao, J.; Wu, S.; Zhang, Q. Short Analog Fountain Code With Quasi-Gray Constellation Mapping Modulation Towards uRLLC. *IEEE Trans. Signal Process.* **2022**, *70*, 4077–4092. [CrossRef]
12. Guo, J.; Zhang, D.; Lee, I.; Li, Y.; Shirvanimoghaddam, M. Partitioned Analog Fountain Codes for Short Packet Communications. *IEEE Commun. Lett.* **2024**, *28*, 1248–1252. [CrossRef]
13. Hu, X.Y.; Eleftheriou, E.; Arnold, D. Regular and irregular progressive edge-growth tanner graphs. *IEEE Trans. Inf. Theory* **2005**, *51*, 386–398. [CrossRef]
14. Baron, D.; Sarvotham, S.; Baraniuk, R.G. Bayesian compressive sensing via belief propagation. *IEEE Trans. Signal Process.* **2009**, *58*, 269–280. [CrossRef]
15. ten Brink, S. Convergence behavior of iteratively decoded parallel concatenated codes. *IEEE Trans. Commun.* **2001**, *49*, 1727–1737. [CrossRef]

entropy

MDPI

Article

EXIT Charts for Low-Density Algebra-Check Codes

Zuo Tang, Jing Lei * and Ying Huang

College of Electronic Science and Technology, National University of Defense Technology, Changsha 410073, China; tz807016280@163.com (Z.T.); inform_huang@sina.com (Y.H.)
* Correspondence: leijing@nudt.edu.cn

Abstract: This paper focuses on the Low-Density Algebra-Check (LDAC) code, a novel low-rate channel code derived from the Low-Density Parity-Check (LDPC) code with expanded algebra-check constraints. A method for optimizing LDAC code design using Extrinsic Information Transfer (EXIT) charts is presented. Firstly, an iterative decoding model for LDAC is established according to its structure, and a method for plotting EXIT curves of the algebra-check node decoder is proposed. Then, the performance of two types of algebra-check nodes under different conditions is analyzed via EXIT curves. Finally, a low-rate LDAC code with enhanced coding gain is constructed, demonstrating the effectiveness of the proposed method.

Keywords: EXIT; LDAC; LDPC; CND; algebraic codes; iterative decoder; channel coding

1. Introduction

In modern communication systems, channel coding techniques play a crucial and fundamental role. Currently, research on channel coding primarily focuses on medium-to-high-rate codes to achieve high-speed and reliable transmission of big data. As society's reliance on information and communication networks grows, the demand for communication in various scenarios rises. Low-rate channel coding technology is of great significance for improving communication capabilities and ensuring Minimum Essential Emergency Communication (MEEC) in extremely adverse channel conditions. For example, in Internet of Things (IoT) applications where sensor nodes transmit small amounts of data with high reliability requirements, or in emergency communication in remote areas with limited bandwidth and harsh environments. Obviously, a low-rate channel coding scheme is an indispensable part of the channel coding system.

However, research on low-rate channel coding techniques remains insufficient. Moreover, the performance of common channel coding schemes with excellent performance in medium-to-high rates, such as turbo code, Low-Density Parity-Check (LDPC) code, and polar code, significantly declines when constructing a low-rate channel code. The LDPC code is widely used in modern communication systems and exhibits excellent performance in medium-to-high-rate codes. However, the construction of low-rate LDPC codes is very difficult. The CCSDS [1] standard uses LDPC codes with a minimum rate of 1/2. The DVB-S2 standard [2,3] improves the coding performance by concatenating LDPC codes with BCH codes and applies LDPC codes with a minimum rate of approximately 1/4. Until now, there has been no uniform standard for channel coding with a rate less than 1/2.

Despite the challenges associated with low-rate LDPC codes, some attempts have been made to develop efficient low-rate channel coding [4–9]. However, these studies failed to break through the LDPC code structure and only achieved rate compatibility. Therefore, the construction of these codes is still under difficult conditions caused by reducing the rate of LDPC codes, often associated with issues such as complex construction and poor flexibility.

Generalized Low-Density Parity Check (GLDPC) codes [10,11] provide another perspective, showing that the check nodes of LDPC codes need not be limited to the Single Parity Check (SPC) and can be extended to other shortcodes with stronger error correction ability. Another code that extends the SPC node is the LDPC-Hadamard code [12], which stems from the parallel iterative decoding structure. Around the same period, the Turbo-Hadamard code [13,14] and the Zigzag-Hadamard code [15], both employing a similar structure, were also proposed, and all exhibit excellent performance at low rates. While the authors only considered approaching the Shannon limit, the length of designed codes was extremely long and difficult to implement in practice. In 2021, a team from Tsinghua University adopted the same structure as LDPC-Hadamard codes to propose LDBCH codes [16]. Subsequently, they further extended it and proposed Generalized Sparse (GS) Codes [17], which further enriched the types of subcodes.

Recently, a novel channel coding scheme named Low-Density Algebra-Check (LDAC) codes has been introduced, which is specifically designed to enhance the coding gain of low-rate codes [18]. In this paper, the Extrinsic Information Transfer (EXIT) chart [19–21] is used to optimize the design of LDAC codes. An iterative decoding model for LDAC codes is established, and the methods for plotting EXIT curves for two types of algebra-check nodes have been proposed. The performance of two types of algebra-check nodes of LDAC codes in different cases is discussed.

2. System Model

2.1. The Structure of LDAC Code

The LDAC code is derived from the LDPC code. To enhance the coding structure, LDAC codes incorporate complex algebraic checks to constrain the check nodes. Algebra-check constraints use algebraic codes as subcodes, which usually have significant algebraic structures, such as BCH codes [22–24], RS codes [25], and Hamming codes [26]. The nodes constrained by algebraic checks are named algebra-check nodes, and the constraint relationship of each algebra-check node can be represented by a set of algebraic equations or matrices. Based on whether all bits of the algebraic code take part in the decoding iteration, the algebra-check nodes are classified into two categories. These two types of algebra-check nodes are different in connection and will show different performance in decoding iteration.

Figure 1 depicts the Tanner graph of LDAC codes. In this figure, a circle represents a variable node, a square represents a check node, and the Tanner graph of two types of algebra-check nodes is presented within the dotted box.

Each check node can be considered a subcode. For instance, an SPC node is related to an SPC code. In both LDAC and LDPC codes, all subcodes are connected in parallel based on the connection shown by the sparse check matrix. LDPC codes consist of SPC codes.

If an LDPC code has M SPC check nodes and its row degree is denoted as $(d_1, d_2, \ldots, d_M)$, then the ith SPC check node corresponds to a $(d_i, d_i - 1)$ SPC subcode.

When the ith row is extended from the SPC check node to the first type of algebra-check node, the subcode related to it has $d_i - 1$ information bits, and its check bits are more than 1. If the subcode length is n_i, this subcode can be represented as an $(n_i, d_i - 1)$ code. The specific structure of the first type of algebra-check node is displayed in algebra-check node 1 of Figure 1. The added check bits of the first type of algebra-check node are corresponded to the variable nodes with $d_v = 1$, as shown in the right of the algebra-check node 1. In order to simplify the encoding and decoding operations of LDAC codes, when constructing the first type of algebra-check nodes, the first d_i variable nodes still satisfy the SPC constraint.

When the ith row is extended from the SPC check node to the second type of algebra-check node, the length of the associated subcode is d_i. If the information bits have a length of k_i, the corresponding subcode can be represented as a (d_i, k_i) code. The specific structure is displayed in algebra-check node 2 of Figure 1.

Compared with LDPC codes, due to the enhanced error correction capabilities of the two types of algebra-check nodes, LDAC codes can construct low-rate to ultra-low-rate codes without introducing short cycles. The utilization of two types of algebra-check nodes provides more flexibility and reliability in LDAC code design, facilitating the construction of ultra-low-rate codes.

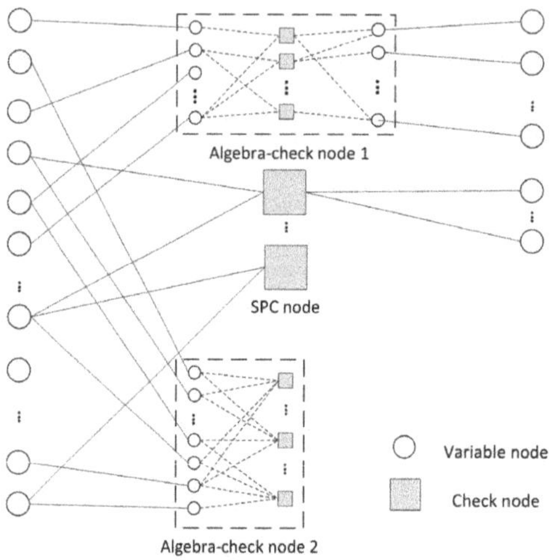

Figure 1. The Tanner graph of LDAC codes.

2.2. The Decoding Algorithm of LDAC Code

The decoding algorithm designed for LDAC codes must be suitable for SPC check nodes and two types of algebra-check nodes. Except for the check nodes, the general structure of the LDAC code closely resembles that of LDPC codes. Consequently, the Log-MAP-SPA decoding algorithm of LDAC codes is a combination of the Maximum a Posteriori Probability (MAP) decoding algorithm and the Log-Sum-Product Algorithm (Log-SPA). The key difference between Log-MAP-SPA and Log-SPA is in the update of check nodes.

Let $\{L_{cv}(i,j)\}_{i\in\{1,\ldots,M\},j\in\{1,\ldots,d_i\}}$ represent the message passing from check nodes to variable nodes, and $\{L_{vc}(i,j)\}_{i\in\{1,\ldots,M\},j\in\{1,\ldots,n_i\}}$ represent the message passing from variable nodes to check nodes. The codeword set of the subcode is denoted by $\mathbf{C}$. At the beginning of the iteration, the message $L_{vc}(i,j)$ is from the extrinsic information received by the channel.

The iterative decoding message update of check nodes is

$$L_{cv}(i,j) = \ln \frac{\sum\limits_{t}^{\phi_{x_k=1}} \left(e^{0.5 \sum\limits_{j=1}^{N} \left(L_{vc}(i,j)\omega\left(c_{t,j}\right) \right)} \right)}{\sum\limits_{t}^{\phi_{x_k=0}} \left(e^{0.5 \sum\limits_{j=1}^{N} \left(L_{vc}(i,j)\omega\left(c_{t,j}\right) \right)} \right)} - L_{vc}(i,j) \tag{1}$$

where $\phi_{x_k} = 1/0$ denote the subsets of $\mathbf{C}$ with the jth is 1 or 0.

The iterative decoding message update of variable nodes can be written as

$$L_{vc}(i,j) = \sum_{i' \in VN(j)\setminus\{i\}} L_{cv}(i',j) \tag{2}$$

where $VN(j)$ represents the set of the check nodes positions connected to the jth variable node.

The range of j in L_{cv} and L_{vc} is different. The index j of message $L_{cv}(i,j)$ ranges from 1 to d_i, while the index j of message $L_{vc}(i,j)$ ranges from 1 to n_i. For the SPC check node, or the second type of algebra-check node, since $n_i = d_i$, all the variable nodes can be updated according to Formula (2). However, when it comes to the first type of algebra-check node, since $n_i > d_i$, it implies that the $n_i - d_i$ variable nodes cannot be updated. These $n_i - d_i$

variable nodes have $d_v = 1$. The message $L_{vc}(i,j)$ with $j \in \{d_i + 1, \ldots, n_i\}$ (when $n_i > d_i$) is the extrinsic information from the channel and remains unchanged during the iterative decoding process.

2.3. Iterative Decoder

EXIT charts are utilized to analyze the input–output mutual information interactions between the Check Nodes Decoder (CND) and the Variable Nodes Decoder (VND), which is an effective method to predict the convergence of the iterative decoding process. Specifically, the CND updates the decoding message in accordance with Formula (1), while the VND updates the decoding message based on Formula (2).

The LDAC codes employ an iterative decoding method, which is similar to LDPC codes. Therefore, the model of the LDAC iterative decoder can be derived from the model of the LDPC iterative decoder. The update message between the CND and VND is quantified by mutual information. As shown in Figure 2, the iterative decoding of LDPC codes is conceptualized as the reciprocal transmission of mutual information between the CND and the VND. Of which, I_{ch} denotes the mutual information originating from the channel. The subscripts A/E represent input/output, while C/V represent check nodes/variable nodes, respectively. The relationships $I_{EV} = I_{AC}$ and $I_{EC} = I_{AV}$ are maintained throughout this process.

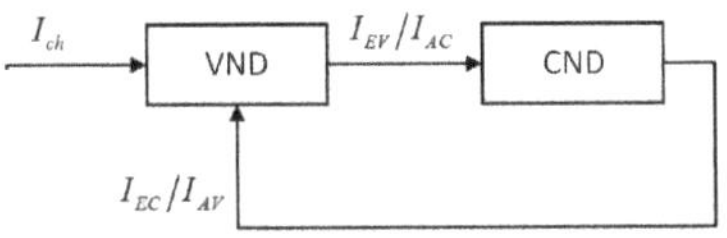

Figure 2. Block diagram of the iterative decoder for LDPC codes.

For the VND of LDPC codes, I_{EV} is a function of degree d_v, input mutual information I_{AV}, and channel mutual information I_{ch}.

$$I_{EV} = f(I_{AV}, I_{ch}, d_v) \tag{3}$$

For the CND of LDPC codes, I_{EC} is a function of degree d_c and input mutual information I_{EV}.

$$I_{EC} = g(I_{AC}, d_c) \tag{4}$$

However, a CND with an algebra-check extension cannot be described by Function (4). The output I_{EC} is a function of input mutual information I_{EV}, channel mutual information I_{ch}, and degree d_c.

$$I_{EC} = g(I_{AC}, I_{ch}, d_c) \tag{5}$$

Depending on the selection of the check nodes, the input channel mutual information I_{ch} may be zero. The block diagram of the iterative decoder for LDAC codes is presented in Figure 3, and the meaning of symbols in the figure refers to Figure 2.

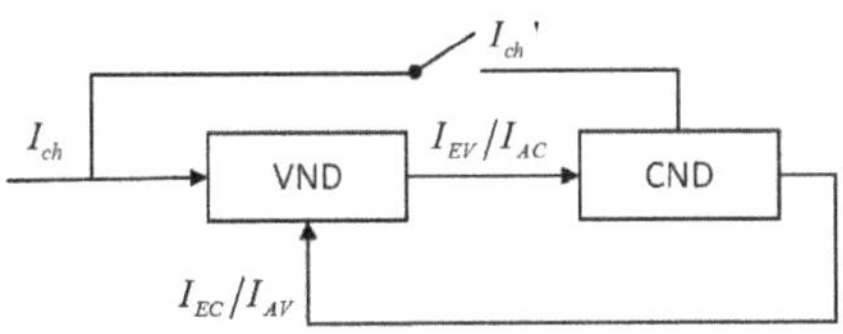

Figure 3. Block diagram of the iterative decoder for LDAC codes.

2.4. Calculation of Mutual Information

The mutual information can be calculated according to the log-likelihood ratio (LLR) of the transmitted bits. Consider the AWGN channel with BPSK modulation. Assume that the transmitted bit is x, and the information received from the channel is y. The calculation of LLR is presented as follows:

$$l = \log \frac{p(x=1|y)}{p(x=0|y)} = \frac{2}{\sigma_n^2} y \tag{6}$$

σ_n^2 is noise variance. Then, the conditional probability density function of the LLR is

$$f_L(l|X=x) \tag{7}$$

Define mutual information I_L:

$$I_L = I(X;L) \tag{8}$$

Through derivation, we can obtain

$$I_L = \frac{1}{2} \sum_X \int_{-\infty}^{+\infty} f_L(l|X=x) \cdot \log_2 \frac{2 f_L(l|X=x)}{f_L(l|X=0) + f_L(l|X=1)} dl \tag{9}$$

Among which, X is the transmitted bit, and L is the LLR of extrinsic information.

If L satisfies symmetry and consistency:

$$f_L(l|X=1) = f_L(-l|X=0) \tag{10}$$

$$f_L(-l|X=x) = e^{-l} f_L(l|X=x) \tag{11}$$

and the conditional probability density function of L follows a Gaussian distribution. The corresponding distribution function is

$$f_L(l|X=1) = \frac{1}{\sqrt{2\pi}\sigma} e^{-\frac{\left(l-\sigma^2/2\right)^2}{2\sigma^2}} \tag{12}$$

$$f_L(l|X=0) = \frac{1}{\sqrt{2\pi}\sigma} e^{-\frac{\left(l+\sigma^2/2\right)^2}{2\sigma^2}} \tag{13}$$

Substitute (12) and (13) into (9) to obtain

$$I_L = 1 - \frac{1}{\sqrt{2\pi}\sigma} \int_{-\infty}^{\infty} e^{-\frac{\left(l-\sigma^2/2\right)^2}{2\sigma^2}} \log_2\left(1 + e^{-l}\right) dl \tag{14}$$

Definition function:

$$J(\sigma) = 1 - \frac{1}{\sqrt{2\pi}\sigma} \int_{-\infty}^{\infty} e^{-\frac{\left(y-\sigma^2/2\right)^2}{2\sigma^2}} \log_2\left(1 + e^{-y}\right) dy \tag{15}$$

Obviously, $\sigma = J^{-1}(I)$. The functions J and J^1 can be segmented for several approximate and simplified calculations as cited in [20].

3. EXIT Curves of LDAC Codes

In LDAC codes, the EXIT curves of VND are identical to those of the LDPC code. However, the CND of LDAC codes contains three possible cases. In this section, the EXIT curves for both the VND and the three different types of CND are introduced, respectively.

3.1. VND

Similar to LDPC codes, the VND of LDAC codes can be regarded as a repetition (REP) code decoder. We define the normalized signal-to-noise ratio (SNR) as $E_b/N_0 = 1/(2R\sigma_n^2)$, where R is the rate of the LDAC code. The mutual information between the input and output of the VND at the same degree d_v can be calculated by Formula (9).

$$I_{EV} = J\left(\sqrt{(d_v - 1)[J^{-1}(I_{AV})]^2 + [J^{-1}(I_{ch})]^2}\right) \tag{16}$$

$$\sigma_{ch}^2 = \left[J^{-1}(I_{ch})\right]^2 = 8R \cdot E_b/N_0 \tag{17}$$

R is the rate of the LDAC code. The EXIT curves of VND with different degrees d_v are shown in Figure 4.

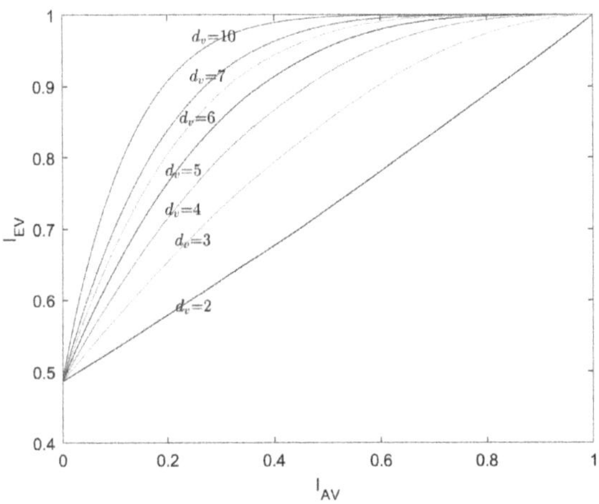

Figure 4. The EXIT curves of VND with $E_b/N_0 = 2$ dB, $R = 1/2$.

3.2. CND

The check nodes of LDAC codes include the SPC node, the first type of algebra-check node, and the second type of algebra-check node.

The SPC can be considered as the dual of repetitive codes, and the EXIT curve of the SPC node is only related to the degree d_c of the check node. The relationship between the input and output mutual information of the VND with only one degree d_c is as follows:

$$I_{EC}^{SPC} = 1 - I^{REP} = 1 - J\left(\sqrt{(d_c - 1)[J^{-1}(1 - I_{AV})]^2}\right) \tag{18}$$

The EXIT curves of the SPC nodes decoder with different degrees of d_c are shown in Figure 5.

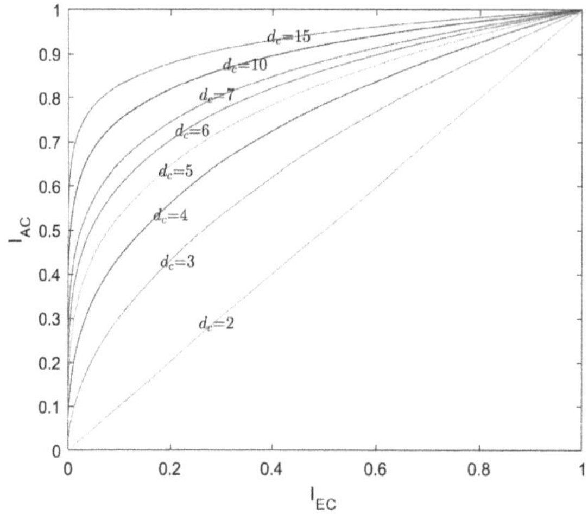

Figure 5. The EXIT curves of the SPC nodes decoder.

In LDAC codes, the two types of algebra-check nodes are distinct from the SPC nodes. The decoding output of an algebra-check node decoder is related to the specific algebra-check constraint. The calculation is more complex than the SPC decoder, and it is difficult to obtain the specific expression of each constraint. Therefore, for the algebra-check nodes decoder, the Monte Carlo method is employed to plot the EXIT curves.

For a given input mutual information value I_{AC}, according to the experimental results in Reference [20], when the number of experiments is large enough, the conditional probability functions $P(L_A|X=0)$ and $P(L_A|X=1)$ can be considered to follow a Gaussian distribution. Therefore, under the circumstances of fulfilling the symmetry and consistency conditions, when the variance is σ_A^2, the mean can be calculated as $\sigma_A^2/2$. The expression of the conditional probability functions is as follows:

$$P(L_A = l|X = x) = \frac{1}{\sqrt{2\pi}\sigma_A} e^{-\frac{\left(l-\left(\sigma_A{}^2/2\right)\omega(x)\right)^2}{2\sigma_A{}^2}} \tag{19}$$

Substitute Formula (19) into the mutual information calculation Formula (9) to obtain

$$I_{AC} = J(\sigma_A) \tag{20}$$

When the input mutual information value I_{AC} is known, obtain

$$\sigma_A = J^{-1}(I_{AC}) \tag{21}$$

When the output LLR sequence L_E is known and the sequence size is sufficiently large, the empirical distribution of the output can be obtained via the Monte Carlo method, which is regarded as an approximation of the probability distribution. As a result, the conditional probability density function $P(L_E = l|X = x)$ can be obtained, and the output mutual information I_{EC} can be calculated according to Formula (9). In addition, according to Formula (17), the variance of channel information $\sigma_{ch}{}^2$ can be obtained. Two types of algebra-check nodes of LDAC codes are analyzed below.

According to the decoding algorithm of LDAC codes [18], if the ith check node is the first type of algebra-check node, only the first d_i variable nodes connected to it participate in the decoding iteration. The latter $n_i - d_i$ variable nodes only provide the extrinsic information from the channel but remain unchanged during the iterative decoding process. Therefore, for the first type of algebra-check node, the output mutual information

I_{EC} of the CND is a function of the input mutual information I_{AC} and the extrinsic mutual information from the channel I_{ch}. And the vector input of the CND is the vector $(L_{A,1}, L_{A,2}, \ldots, L_{A,d_c}, L_{ch,1}, L_{ch,2}, \ldots, L_{ch,n-d_c})$. If the ith check node is the second type of algebra-check node, the output mutual information I_{EC} of the CND is only related to the input mutual information I_{AC}, and every variable node connected to the check node updates during the iterative decoding. Therefore, for the second type of algebra-check node, the vector input of the CND is the vector $(L_{A,1}, L_{A,2}, \ldots, L_{A,d_c})$.

For the first type of algebra-check node, the process for plotting the EXIT curves of the CND by the Monte Carlo method is as follows:

1. Set the value of input mutual information I_{AC} ranging from 0 to 1, and set the target rate R and E_b/N_0;
2. Calculate σ_{ch} according to R and E_b/N_0, and randomly generate the channel information L_{ch};
3. Calculate σ_A according to I_{AC} and randomly generate the input L_A of the CND.
4. Input L_{ch} and L_A into the CND and calculate the output message L_{EC} according to Formula (1);
5. Accumulate sufficient output messages L_E for empirical distribution. Consider it as a probability distribution to obtain the conditional probability density function of L_E and calculate the output mutual information I_{EC};
6. Plot the EXIT curves of the first type of algebra-check node decoder, according to the input mutual information I_{AC} and output mutual information I_{EC}.

For the second type of algebra-check node, the process for plotting the EXIT curves of the CND by the Monte Carlo method is as follows:

1. Set the value of input mutual information I_{AC} ranging from 0 to 1;
2. Calculate σ_A according to I_{AC} and randomly generate the input L_A of the CND;
3. Input L_A into the CND and calculate the output message L_{EC} according to Formula (1);
4. Accumulate sufficient output messages L_E for empirical distribution. Consider it as a probability distribution to obtain the conditional probability density function of L_E and calculate the output mutual information I_{EC};
5. Plot the EXIT curves of the second type of algebra-check node decoder, according to the input mutual information I_{AC} and output mutual information I_{EC}.

According to the above steps, for the two types of algebra-check nodes, the EXIT curves of the CND are simulated respectively, as shown in Figures 6 and 7. The curves indicated by $d_c = 3$, $d_c = 4$, and $d_c = 5$ are the EXIT curves of the SPC node decoder with the corresponding degree. The curves labeled by $sub_1H(31,6)$, $sub_1H(15,5)$, and $sub_1H(7,4)$ are the EXIT curves of the first type of algebra-check node with the modified BCH codes. Specifically, the code lengths are 31, 15, and 7, respectively. Associated with the information bit length are 6, 5, and 4, respectively. Corresponding to the row degree, $d_c = 7$, $d_c = 6$, and $d_c = 5$, respectively. The curves labeled by $sub_2H(7,4)$, $sub_2H(6,4)$, and $sub_2H(5,3)$ are the EXIT curves of the first type of algebra-check node with the Hamming codes. Specifically, the code lengths are 7, 6, and 5, respectively. Associated with the information bit lengths are 4, 4, and 3, respectively. Corresponding to the row degree, $d_c = 7$, $d_c = 6$, and $d_c = 5$, respectively.

In Figures 6 and 7, the shape of the EXIT curves of the algebra-check nodes significantly differs from that of the SPC nodes. Especially for the first type of algebra-check nodes, the EXIT curves of the CND do not pass through the origin but intersect with the X-axis. This is because there is mutual information from the channel inputted into the CND.

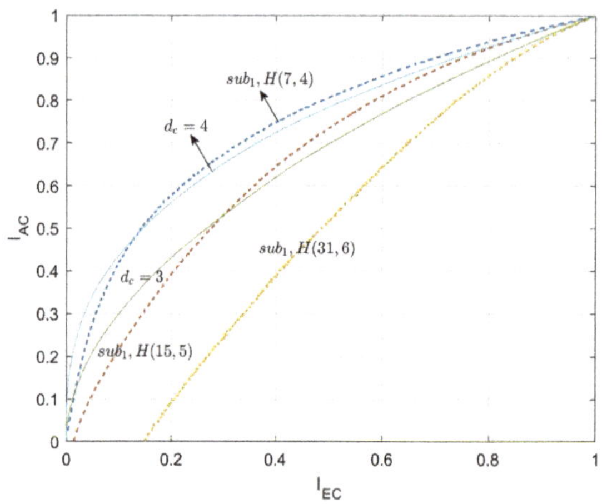

Figure 6. The EXIT curves of the first type of algebra-check node decoder.

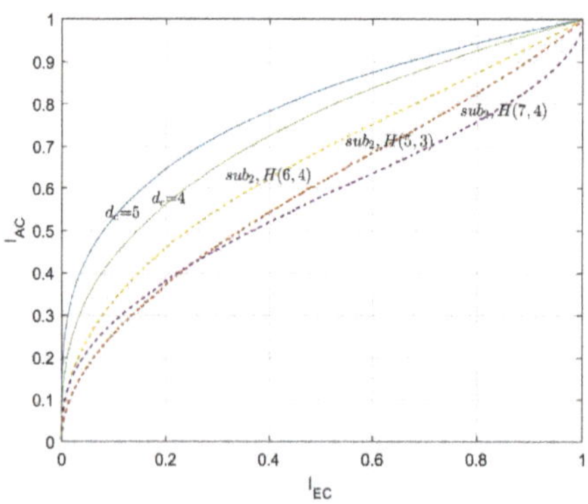

Figure 7. The EXIT curves of the second type of algebra-check node decoder.

4. Simulation Results and Analysis

4.1. The Analysis of the VNDs and SPC Node

Figure 8 shows the VND EXIT curves with $d_v = 2$ and $d_v = 5$ given different code rates R and E_b/N_0. Comparing the EXIT curves about the rate of 1/2 and 1/12, it is observed that the intersection of the EXIT curves with the Y-axis approaches the origin as the code rate decreases. Moreover, when the bit rate is 1/12, increasing the E_b/N_0 from 0.001 dB to 2 dB, the EXIT curves are significantly smaller than the EXIT curves with the rate of 1/2. It is observed that the impact of varying E_b/N_0 on the VND EXIT curves diminishes as the code rate decreases. This indicates that reducing the code rate makes it increasingly challenging to enhance the output performance of the variable node decoder by merely increasing E_b/N_0.

To achieve the iterative decoding, the VND curve in the EXIT chart must surpass the CND curve, ensuring that the two curves remain non-intersecting and establishing a decoding channel. Figure 9 depicts the VND EXIT curves for $d_v = 2$ at different bit rates. Additionally, the figure presents CND EXIT curves with different degrees d_c. It can be directly observed that when $R = 1/2$ and $E_b/N_0 = 2$dB, the CND EXIT curves with $d_c \leq 7$ are below the VND EXIT curve with $d_v = 2$. When $R = 1/2$ and $E_b/N_0 = 0.001$ dB, the CND EXIT curves with $d_c \leq 5$ are below the VND EXIT curve with $d_v = 2$. When $R = 1/12$ and $E_b/N_0 = 2$ dB, only the CND EXIT curves with $d_c = 3$ and $d_c = 2$ are below the VND

EXIT curve with $d_v = 2$. When $R = 1/12$ and $E_b/N_0 = 0.001$ dB, only the CND EXIT curve with $d_c = 2$ is below the VND EXIT curve with $d_v = 2$.

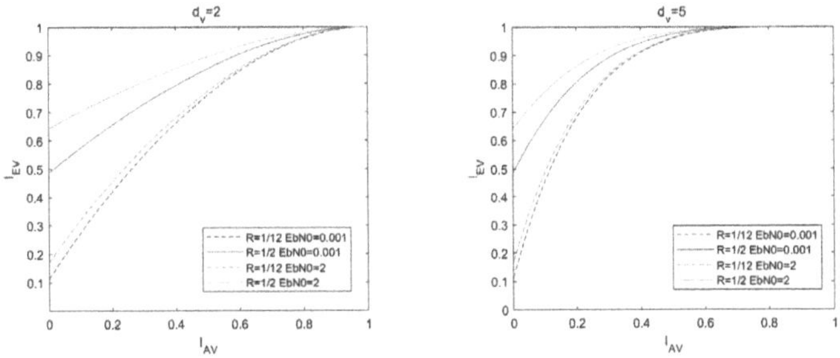

Figure 8. EXIT curves of VND under different conditions.

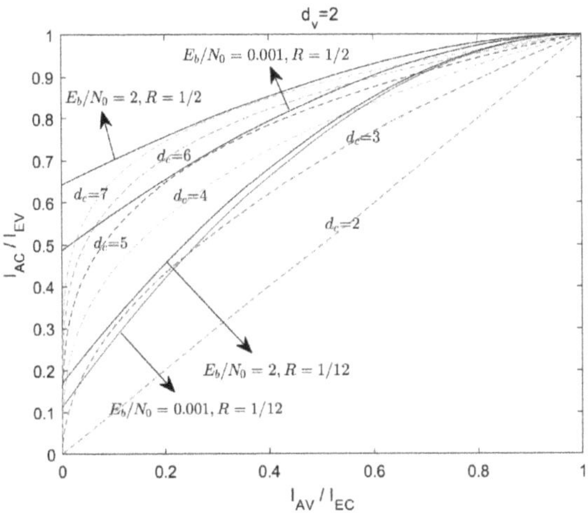

Figure 9. EXIT curves of VND and SPC node decoder at different rates.

The analysis reveals that when the code rate reduces, the variable node degree d_v necessary for achieving iterative decoding increases, while the check node degree d_c decreases. This implies that the matrix of the low-rate codes must be structured with the higher column degrees and the lower row degrees. However, it is impossible to decrease the row degree d_c and increase the column degree d_v at the same time, because the sum of row degrees is always equal to the sum of column degrees. If the row degree d_c decreases, the CNDs will become insufficiently connected, leading to poor performance during decoding. Conversely, if the column degree d_v decreases, the number of short cycles tends to increase, which also adversely affects the decoding performance.

The construction of low-rate LDPC codes presents considerable challenges, and the introduction of algebra-check nodes provides a new ideal for constructing low-rate codes. By modifying the structure of check nodes, its decoding capabilities can be enhanced. The change in the shape of CND EXIT curves can optimize iterative decoding performance without modifying d_v and d_c. This offers another option for constructing a low-rate code.

4.2. The Comparison of Algebra-Check Nodes

Initially, to evaluate and contrast the performance of the two types of algebra-check nodes, the impact on the rate of LDAC codes following the substitution of these check nodes is analyzed. The detailed derivation process is as follows.

Suppose the increase of the code length due to the first type of algebra-check nodes is b. The reduction of the information bits due to the second type of algebra-check nodes is a.

The LDAC code's basis matrix is H_0, with a code length of N_{H_0} and a rate of R_0. The final rate of the LDAC code is R.

$$R = \frac{R_0 N_{H_0} - a}{N_{H_0} + b} \tag{22}$$

If $a = Rb$, then

$$R = \frac{R_0 N_{H_0} - Rb}{N_{H_0} + b} \tag{23}$$

When $R > a/b$, there is

$$\frac{R_0 N_{H_0} - a}{N_{H_0} + b} > \frac{R_0 N_{H_0} - Rb}{N_{H_0} + b} \tag{24}$$

When $R < a/b$, there is

$$\frac{R_0 N_{H_0} - a}{N_{H_0} + b} < \frac{R_0 N_{H_0} - Rb}{N_{H_0} + b} \tag{25}$$

Thus, the following relationships can be obtained:

- If $R > a/b$, substituting the first type of algebra-check nodes leads to a more significant rate loss compared to replacing the second type.
- When $R < a/b$, replacing the first type of algebra-check nodes incurs a relatively smaller rate loss than substituting the second type.
- In the case of $R = a/b$, the influence of both types of algebra-check nodes on the rate of the constructed LDAC code is comparable.

For a specific analysis, the comparison of algebra-check nodes requires the combination with parameters R, E_b/N_0, and the ratio a/b. Here, the cases with code rates $R = 1/12$ and $R = 1/4$ are discussed as examples.

(1) The constructed LDAC code with the target rate $R = 1/12$.

Figure 10 shows the EXIT curves of different algebra-check nodes with the target rate $R = 1/12$ and $E_b/N_0 = 0.001$ dB. The symbols sub_1 and sub_2 denote the first and the second type of algebra-check nodes, respectively. The EXIT curves of the first type of algebra-check nodes are all restricted by the modified BCH codes, while the EXIT curves of the first type of algebra-check nodes are constrained by the modified BCH codes and Hamming codes.

When the rate of the basis matrix is $R_{H_0} = 1/2$, the degrees of the variable nodes are mainly 3 and 4. Thus, the EXIT curve of variable node decoders with $d_v = 3$ and $d_v = 4$ is presented in the figure as a reference.

For the second type of algebra-check nodes, the selected algebra-subcodes are the (7,4) Hamming code and the (7,4) modified BCH system code. The EXIT curve of the (7,4) Hamming code is always below the VND EXIT curve of the $d_v = 4$, yet intersects with the VND EXIT curve of $d_v = 3$. While the EXIT curve of the (7,4) modified BCH code below both the VND EXIT curves of $d_v = 3$ and $d_v = 4$. The simulation results indicate that the second type of algebra-check node exhibits superior iterative convergence performance compared to the SPC node, and the (7,4) modified BCH code outperforms the Hamming code.

For the first type of algebra-check node, When the subcode of the information bit $K^{\text{sub1}} = 6$ and the length of code $N^{\text{sub1}} < 30$, the first type of algebra-check node has a rate advantage compared to the second type of algebra-check node with a (7,4) code. The dotted lines in Figure 10 display the EXIT curves for the first type of algebra-check node with $K^{\text{sub1}} = 6$. The subcode lengths N^{sub1} range from 12 to 31, increasing from left to right. The curves closest to $d_v = 3$ and $d_v = 4$ are the curves of the (15,6) code and the (19,6) code, with code rates of 0.4 and 0.316, corresponding to $a/b = 1/4$ and $a/b = 1/6$, respectively.

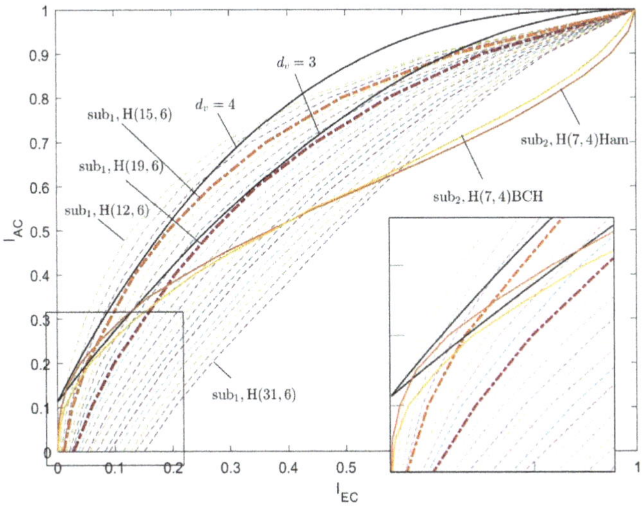

Figure 10. The EXIT curves of two types of algebra-check nodes with the target rate $R = 1/12$.

In summary, if the rate of LDAC code is 1/12, for a check node with row degree $d_c = 7$, the first type of algebra-check nodes with a code length greater than 15 and the second type of algebra-check nodes with the (7,4) code can be considered as candidates.

(2) The constructed LDAC code with the target rate $R = 1/4$.

The second type of algebra-check nodes that participate in the comparison is the (6,4) code and (6,3) code, corresponding to $a = 1$ and $a = 2$, respectively. The two EXIT curves in Figure 11 are located below the EXIT curve of $d_v = 3$ and are significantly lower than the VND EXIT curve of $d_v = 4$.

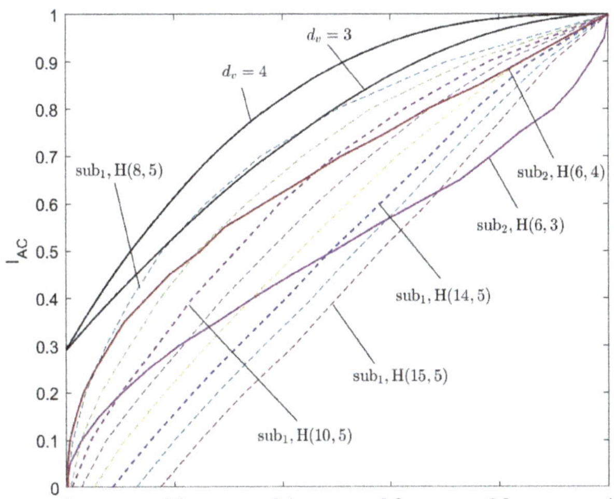

Figure 11. The EXIT curves of two types of algebra-check nodes with the target rate $R = 1/4$.

The first type of algebra-check node with $K^{\text{sub1}} = 5$ and $N^{\text{sub1}} < 14$ has an advantage over the second type of algebra-check node with the (6,3) code. And the first type of algebra-check node with $K^{\text{sub1}} = 5$ and $N^{\text{sub1}} < 10$ has an advantage over the second type of algebra-check node with the (6,4) code.

The dotted lines in Figure 11 display the EXIT curves for the first type of algebra-check node with $K^{\text{sub1}} = 5$. The subcode lengths N^{sub1} increase from left to right, starting at 8 and ending at 15. Except the curves of (8,5) code intersect with the VND EXIT curve of $d_v = 3$, the other curves do not intersect with the VND EXIT curves of $d_v > 2$.

In summary, for the LDAC code construction of the target rate $R = 1/4$, both types of algebra-check nodes show great performance.

4.3. Examples of Optimized Constructions of LDAC Codes

In response to the severe channel conditions with extremely low SNR, we have constructed LDAC codes with a code rate of $R = 1/12$. The selection of check nodes is detailed in Table 1.

Table 1. Selection of check nodes.

Degree d_c	The Type of Check Node	Subcode Length n_i	Information Bit Length k_i
5	sub1	7	4
6	sub1	15	5
7	sub1	31	6
7	sub2	7	4

The EXIT chart for the LDAC code at $R = 1/12$ and $E_b/N_0 = 0.001$ dB is presented in Figure 12.

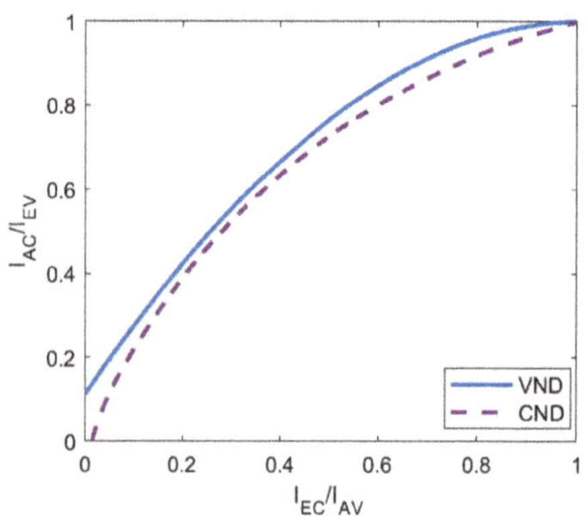

Figure 12. EXIT chart of LDAC code with rate $R = 1/12$.

The LDAC decoding matrix **H** at $R = 1/12$ is shown in Figure 13.

Figure 14 shows the block error rate (BLER) of the LDAC code constructed based on the above parameters in the AWGN channel, comparing it with the 1/2-LDPC repetition code, 1/3-Turbo repetition code, and the 5G LDPC code [27]. The simulation results show that under QPSK modulation, the BLER of the LDAC code with $R = 0.0834$ is significantly lower than the 1/2-LDPC repetition code and outperforms the 5G LDPC code and 1/3-Turbo repetition code, demonstrating superior performance.

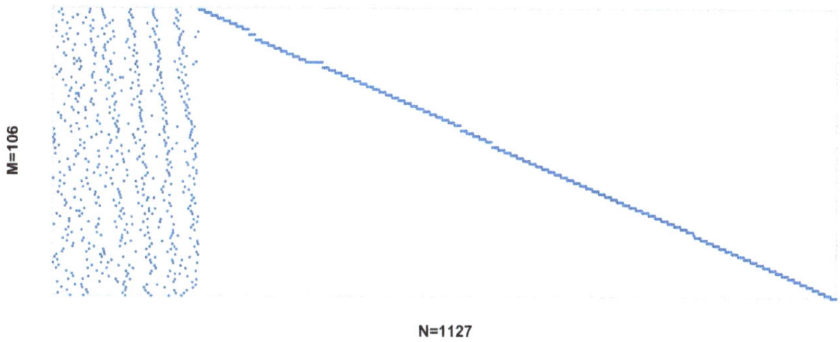

Figure 13. Decoding matrix of LDAC code with code rate $R = 0.0834$.

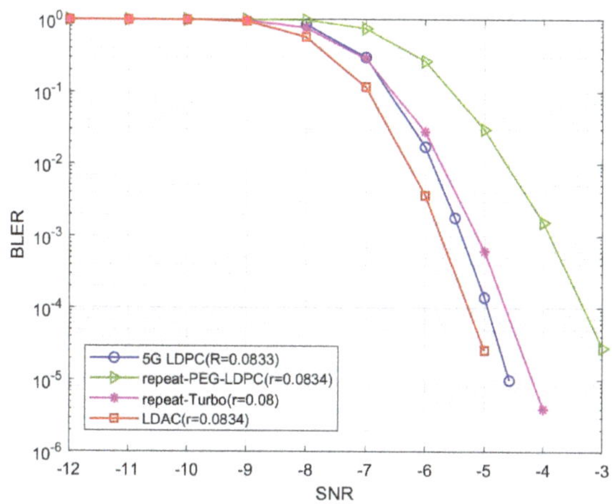

Figure 14. The BLER of LDAC code with $R = 0.0834$.

5. Conclusions

In this paper, the optimization design of LDAC codes based on the EXIT chart is studied. An iterative decoder model for LDAC codes is proposed, along with calculation methods for the input–output mutual information curves of two types of algebra-check node decoders. The paper simulates various EXIT curves of VNDs and CNDs with different conditions, investigates the limitations of low-rate LDPC codes, and analyzes the performance of the two algebra-check nodes. Finally, based on the EXIT chart, the optimitive constructed LDAC code based on the EXIT chart. The simulation results show that the optimized LDAC code greatly reduces the BLER compared with the 1/2 LDPC repetition code and has enhanced coding gain compared with 5G LDPC code and 1/3-Turbo repetition code.

Author Contributions: Investigation, Z.T.; methodology, Z.T.; project administration, J.L., Y.H. and Z.T.; software, Z.T. All authors have read and agreed to the published version of the manuscript.

Funding: This research was supported in part by the Hunan Province Natural Science Foundation under Grant 2021JJ30777.

Institutional Review Board Statement: Not applicable.

Data Availability Statement: The data presented in this study are available on request from the corresponding author.

Conflicts of Interest: The authors declare no conflicts of interest.

Abbreviations

The following abbreviations are used in this manuscript:

EXIT	Extrinsic Information Transfer
LDPC	Low-Density Parity-Check
LDAC	Low-Density Algebra-Check
GLDPC	Generalized Low-Density Parity Check
SPC	Single Parity Check
CND	Check Nodes Decoder
VND	Variable Nodes Decoder
MEEC	Minimum Essential Emergency Communication
GS	Generalized Sparse
SNR	signal-to-noise ratio
BLER	block error rate

References

1. CCSDS. *TM Synchronization and Channel Coding*; Consultative Committee for Space Data Systems: Beltsville, MD, USA, 2011.
2. Digital Video Broadcasting (DVB). *Second Generation Framing Structure, Channel Coding and Modulation Systems for Broadcasting, Interactive Services, News Gathering and Other Broadband Satellite Applications*; Part 2: DVB-S2 Extensions (DVB-S2X); DVB: Geneva, Switzerland, 2015.
3. Digital Video Broadcasting (DVB). *Second Generation Framing Structure, Channel Coding and Modulation Systems for Broadcasting, Interactive Services, News Gathering and Other Broadband Satellite Applications*; Part 1: DVB-S2; DVB: Geneva, Switzerland, 2014.
4. Chu, T.; Jiang, X.-Q.; Hou, J.; Wang, H.-M.; Kong, L. Construction of multiple-rate LDPC codes using modified PEG. In Proceedings of the 2017 9th International Conference on Wireless Communications and Signal Processing (WCSP), Nanjing, China, 11–13 October 2017; pp. 1–5.
5. Park, S.I.; Wu, Y.; Kim, H.M.; Hur, N.; Kim, J. Raptor-Like Rate Compatible LDPC Codes and Their Puncturing Performance for the Cloud Transmission System. *IEEE Trans. Broadcast.* **2014**, *60*, 239–245. [CrossRef]
6. Wu, X.; Jiang, M.; Zhao, C.; Ma, L.; Wei, Y. Low-Rate PBRL-LDPC Codes for URLLC in 5G. *IEEE Wirel. Commun. Lett.* **2018**, *7*, 800–803. [CrossRef]
7. He, Z.; Peng, K.; Zhang, C.; Song, J. Low-Rate LDPC Code Design for DTMB-A. *IEEE Trans. Broadcast.* **2024**, *70*, 739–746. [CrossRef]
8. Park, H.; Kwak, H.-Y.; Hong, S.; No, J.S.; Shin, D.J. Variable-Weight Block Dual-Diagonal Structure for Low-Rate QC LDPC Codes With Low Error Floors. *IEEE Trans. Commun.* **2020**, *68*, 1344–1357. [CrossRef]
9. Gonsioroski, L.H.; dos Santos, A.B.; Fernandes, N.C.; Castellanos, P.V.G.; de Matos, L.J.; da Silva Mello, L.A. Evaluation of the Field-Test Results of Advanced ISDB-T System with Low-Rate LDPC Codes. *IEEE Trans. Broadcast.* **2023**, *69*, 675–690. [CrossRef]
10. Tanner, R. A recursive approach to low complexity codes. *IEEE Trans. Inf. Theory* **1981**, *27*, 533–547. [CrossRef]
11. Lentmaler, M.; Zigangirov, K. On generalized low-density parity-check codes based on Hamming component codes. *IEEE Commun. Lett.* **1999**, *3*, 248–250. [CrossRef]
12. Yue, G.; Ping, L.; Wang, X. Generalized Low-Density Parity-Check Codes Based on Hadamard Constraints. *IEEE Trans. Inf. Theory* **2007**, *53*, 1058–1079. [CrossRef]
13. Ping, L.; Leung, W.K.; Wu, K.Y. Low-rate turbo-Hadamard codes. *IEEE Trans. Inf. Theory* **2003**, *49*, 3213–3224. [CrossRef]
14. Wu, Y.-J.; Ping, L. On the limiting performance of turbo-Hadamard codes. *IEEE Commun. Lett.* **2004**, *8*, 449–451. [CrossRef]
15. Leung, W.; Yue, G.; Ping, L.; Wang, X. Concatenated zigzag hadamard codes. *IEEE Trans. Inf. Theory* **2006**, *52*, 1711–1723. [CrossRef]
16. Sun, Z.; Chen, Z.; Yin, L.; Lu, J. Design of LDBCH Codes for Ultra Reliable Low Latency Communications. *IEEE Commun. Lett.* **2021**, *25*, 2800–2804. [CrossRef]
17. Chen, Z.; Sun, Z.; Pei, Y.; Yin, L. Generalized sparse codes for non-Gaussian channels: Code design, algorithms, and applications. *Fundam. Res.* **2022**, *2*, 284–295. [CrossRef] [PubMed]
18. Tang, Z.; Lei, J.; Lai, K.; Huang, Y. Low-density algebra-check codes for ultra reliable low latency communications. *Electron. Lett.* **2024**, *60*, e13179. [CrossRef]
19. Ten Brink, S. Convergence Behavior of Iteratively Decoded Parallel Concatenated Codes. *IEEE Trans. Commun.* **2001**, *49*, 1727. [CrossRef]
20. Ten Brink, S.; Kramer, G.; Ashikhmin, A. Design of low-density parity-check codes for modulation and detection. *IEEE Trans. Commun.* **2004**, *52*, 670–678. [CrossRef]
21. MacKay D. Good error-correcting codes based on very sparse matrices. *IEEE Trans. Inf. Theory* **1999**, *45*, 399–431. [CrossRef]
22. Hocquenghem, A. Codes correcteurs D'erreurs. *Chiffres* **1959**, *2*, 147–156.
23. Bose, R.; Ray-Chaudhuri, D. On a class of error correcting binary group codes. *Inf. Control* **1960**, *3*, 68–79. [CrossRef]
24. Bose, R.; Ray-Chaudhuri, D. Further results on error correcting binary group codes. *Inf. Control* **1960**, *3*, 279–290. [CrossRef]
25. Reed, I.S.; Solomon, G. Polynomial Codes Over Certain Finite Fields. *J. Soc. Ind. Appl. Math.* **1960**, *8*, 300–304. [CrossRef]

26. Hamming, R.W. Error Detecting and Error Correcting Codes. *Bell Syst. Tech. J.* **1950**, *29*, 147–160. [CrossRef]
27. Channel Coding Techniques for URLLC. TSG RAN WG1 93 Meeting, 3rd Generation Partnership Project (3GPP), 2018: Document R1–1806006. Available online: https://www.3gpp.org/dynareport?code=TDocExMtg--R1-93--18784.htm (accessed on 29 October 2024).

entropy

Article

Performance Analysis of Troposphere Scattering Communication Channel with Chirp-BOK Modulation

Junhu Shao [1], Zaiping Liu [1], Yishuo Liu [1] and Tianjiao Xie [2,*]

[1] Xi'an Key Laboratory of Wireless Optical Communication and Network Research, School of Automation and Information Engineering, Xi'an University of Technology, Xi'an 710048, China; jhshao@xaut.edu.cn (J.S.); 2220321179@stu.xaut.edu.cn (Z.L.); 2220320071@stu.xaut.edu.cn (Y.L.)
[2] China Academy of Space Technology (Xi'an), Xi'an 710100, China
* Correspondence: xiexietianjiao@163.com

Abstract: By utilizing chirp-BOK (binary orthogonal keying) modulation into a troposphere scattering communication system, a lower demodulation threshold can be achieved with excellent linear frequency modulation properties in a strong noise and weak signal environment. Firstly, the bit error rate (BER) formula of chirp-BOK modulation over a Rayleigh fading channel was derived theoretically. Then, the BER performance with different chirp-BOK parameters were numerically calculated. In order to investigate the performance of chirp-BOK over deeping fading troposphere scattering link, a seven-path equal-delay Rayleigh fading model was employed. Finally, the system BER performance was simulated under different tap delay and time–bandwidth product parameters. The results demonstrate that when BER reaches 10^{-4}, the optimal configuration of the system achieves a gains approximately from 1.7 dB to 10 dB compared to non-optimized configuration under different Path-Gain-Vector with varying tap delays.

Keywords: troposcatter; chirp signal; multipath channel; bit error rate

1. Introduction

Troposphere scattering communication is a form of wireless link that utilizes atmospheric inhomogeneity to scatter or reflect ultra-short and short waves, which makes over-the-horizon (OTH) communication possible [1]. This method has become an essential component of the military communication technique due to its characters, such as long transmission range, low susceptibility to channel medium damage and good anti-jamming capabilities. However, a major drawback is the large signal attenuation and deep fading during the troposcatter channel,resulting in adverse multipath interference on the system performance.

Chirp spread spectrum (CSS) technology is an optional scheme in the IEEE802.15.4a physical layer standard [2]. It is a non-single-frequency wave modulation technique that uses chirp signals with different frequencies to carry information. These signals have good cross-correlation characteristics and are less sensitive to environmental noise. Additionally, chirp signals can have an effect of energy compression after matched filtering, which leads to a good performance against multipath fading and Doppler shift even at very low transmit power levels.

Winklery firstly applied CSS technology to the field of wireless communications by using chirp up-sweep and down-sweep signals to transmit binary data [3], i.e., binary orthogonal keying (BOK) modulation. Lee et al. applied the chirp-BOK scheme to transmit information in the ultrasonic-frequency band and achieved reliable long-distance transmission in severely frequency-selective fading channels [4]. Currently, CSS communication has been widely studied within two categories: modulating baseband information into the parameters of the chirp signal itself [3,5–7], or directly utilizing the chirp signal as a carrier for digital modulation [8,9]. Refs. [10,11] introduced the modulation and demodulation process of chirp signal, but they lacked the description of rigorous mathematical signal processing. Ref. [7] presented the mathematical formulas for modulation and demodulation process of frequency shift chirp modulation (FSCM), providing a theoretical derivation of optimal receiver with low complexity based on FFT. Ref. [12] analyzed the technical characteristics and performance of FSCM demodulation under the residual frequency offset. Ref. [13] proposed a fractional chirp rate grounded on CSS (FCR-CSS) as a physical layer-inspired approach to enhance the capacity of CSS modulation based on IoT networks. Ref. [14] developed a novel OFDM-chirp modulation scheme with a low peak-to-average power ratio (PAPR). An analysis method was introduced to evaluate the corresponding symbol error rate (SER) under carrier frequency offset (CFO) in a Rayleigh fading channel. Ref. [15] designed a non-coherent distributed transmission (NCDT) CSS system for low-power communication. Ref. [16] proposed a two-dimensional modulation method chirp-MFCSK for the LoRa system. Ref. [17] raised phase shift keying based on a CSS (PSK-CSS) multiple-access system with a single chirp rate. On the ground of the analysis results under multiple access interference, the PSK-CSS system can support synchronous and asynchronous transmissions simultaneously by multiplying different random phases to the data symbols of different users.

Owing to the special demands and characteristics of the tropospheric scattering communication system, the channel modeling and the adaptive reliable transmission technique have always been a research hotshot. Ref. [18] provide a reference for the theoretical characteristics of communication channels. Ref. [19] presented the troposcatter propagation principles along with several mathematical models, demonstrating integrated theoretical analysis with experimental measurements. Ref. [20] tested the fading characteristics of scattered signals and established a signal model with radar parameters, finally proposing a suitable beamforming and detection algorithm. In ref. [21], equalization techniques were incorporated into a scattering communication system to mitigate inter code interference and achieved high-speed transmission. Ref. [22] investigated the impact of atmospheric conditions on signal transmission and provided insights into uncertainty estimation, serving as a reference for an alternative statistical channel model for bit error rate (BER) analysis. Ref. [23] proposed an innovative anti-fading method for tropospheric scattering communication based on frequency hopping, which also explored synchronization technologies to improve the system performance.

In this paper, by utilizing a chirp modulation scheme into a troposphere scattering communication system, the mathematic formulation of a chirp-BOK signal was introduced to describe the characteristics of the signal passing through troposcatter channels. Then, the averaged BER formula was theoretically deduced over a Rayleigh flat fading channel and frequency-selective fading channel, respectively. In order to investigate the suitable chirp-BOK parameters to deal with different fading parameters, a seven-path fading channel model was established with tap delays and multipath gains. Finally, the BER performance of the chirp-BOK system was simulated and analyzed, with the results showing that an adaptive optimal chirp configuration could offer some significant performance gains.

The main contributions of this work are twofold. Firstly, the theoretical averaged BER formula of chirp-BOK modulation over a troposcatter fading channel was deduced and shown, and it was consistent with the corresponding simulation results over the established seven-path fading model. Secondly, for a troposcatter communication environment with a Doppler shift, multipath gain vectors and delays, optimal parameter configurations for chirp-BOK modulation were achieved to obtain some significant performance gains.

2. System Model with Chirp-BOK Modulation

The schematic diagram of tropospheric scatter propagation is shown in Figure 1, with the transmitter and receiver far apart. In Figure 1, the "scatterers" in the troposphere can form a scattering effect on electromagnetic waves in the microwave band, thereby realizing OTH communication. The transmitter sends the signal toward the troposphere without being blocked by terrestrial obstacles. Although most of the transmitted signal passes out of the troposphere, the remainder scatters at the troposphere, which we call "common volume". In order to set up a tropospheric scatter communication link, an antenna facing a horizontal line should be placed with an elevation angle as low as possible to minimize the path length. As shown in Figure 1, a common volume is formed at the intersection of the antenna beam widths, which can scatter the radio waves to the receiver.

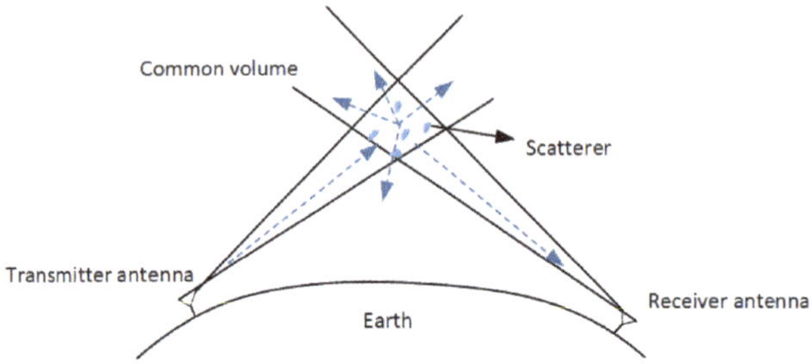

Figure 1. Troposcatter propagation path.

2.1. Chirp Signal Model

Within a symbolic period, a chirp signal can be expressed as

$$s(t) = a(t)e^{j\left(2\pi f_0 t + \pi\mu t^2\right)}, 0 \leq t \leq T_b, \tag{1}$$

where $a(t)$ represents the chirp signal envelope, which is usually a rectangular pulse; f_0 is the initial frequency; μ is the chirp rate, which denotes the instantaneous frequency change rate of the chirp signal; T_b denotes the duration of the chirp signal; and $B = |\mu|T_b$ is used to represent the chirp signal bandwidth. When $\mu > 0$ or $\mu < 0$, the frequency of the chirp signal increases or decreases linearly during the sweep time, which is called the up-chirp signal or down-chirp signal.

A matched filter is a linear filter in which the ratio of signal power to noise power at the output is maximized, and receivers use matched filters to detect received signals. The optimal matched filter for an up-chirp signal is a down-chirp filter, and vice versa. Taking the up-chirp signal as an example, the output waveform of the signal after the matched filter is

$$y_1(t) = \sqrt{BT_b} \cdot \frac{\sin\left[\pi BT_b\left(1 - \frac{|t|}{T_b}\right)\right]}{\pi BT_b} \cdot \cos(2\pi f_0 t), \tag{2}$$

and the output waveform of the signal after the unmatched filter is

$$y_2(t) = \frac{1}{\sqrt{BT_b}} \exp\left(\frac{j\pi\mu t^2}{2}\right)[C(X) + jS(X)], \tag{3}$$

where $X = \frac{\pi}{2}\left(\sqrt{BT_b} - \sqrt{\mu}|t|^2\right)$, $C(\cdot)$, and $S(\cdot)$ are the standard Fresnel cosine and sine integrals, respectively, and are given by

$$C(x) = \int_0^x cos(\frac{\pi}{2}t^2)dt, \; S(x) = \int_0^x sin(\frac{\pi}{2}t^2)dt. \tag{4}$$

As shown in Equation (2), the output of the matched filter has similar characteristics to the sinc function, which can convert the input chirp signal with low peak power into an output signal with high peak power and concentrated energy in a very short period of time. The feature is beneficial for detecting chirp signals. Finally, the information bit can be restored by sampling and judging the output of the matched filter.

2.2. Troposcatter Channel Model

The main characteristics of tropospheric scattering propagation encompass transmission loss and fading. Transmission loss is influenced by various factors, such as the distance between the transmitter and receiver, the geographical location, and the scattering properties of scatterers. There are some classical methods for estimating scattering transmission losses, mainly the NBS-101 model and ITU-R model.

NBS-101 is a widely used calculation method to estimate transmission losses in tropospheric scatter communication links. The NBS-101 prediction method considers the impact of nonlinear variations in the atmospheric refraction index when utilizing the equivalent earth radius. It appropriately adjusts this factor during the calculation process to align with actual link conditions, resulting in a prediction of scattering transmission loss that is more consistent with measured values. However, this method involves complex calculations and requires consultation of numerous charts, making its engineering application relatively challenging.

The ITU-R P.617 model proposes the data and propagation prediction techniques required for the design of a radio relay system beyond the horizon, as well as a transmission loss prediction model for troposcatter signals. Many changes have been made since the P.617 recommendation was first proposed in 1992. From P.617-1 in 1992 to P.617-3 in 2013, this model was mainly used to predict the performance of a radio relay system from 50% to 99.999% of the time percentage. As it was updated to P.617-4 in 2017 and P.617-5 in 2019, the model was improved to predict performance for a time percentage of 0.001% to 99.999%. This pattern can be expressed as follows [24]:

$$L = F + 22\log f + 35\log\theta + 17\log d + L_c - Y_p, \tag{5}$$

In Equation (5), L denotes the median of transmission loss (dB), F is weather factor (dB), f stands for frequency (MHz), θ is the scatter angle (mrad), d is the circuitry length (km), L_c is the surface medium coupling loss (dB), and Y_p is the conversion factor that does not exceed the estimated p time percentage.

The calculation process of the ITU-R617 prediction method is relatively simple, and most of the unknowns can be obtained through the corresponding calculation formula, accurately calculating the transmission loss of tropospheric scattering communication, which is more in line with the practical applications in China.

Using the tranmission loss, as computed in Equation (5), the receiver signal-to-noise ratio (SNR) γ can be expressed as [25]

$$\gamma = 10\log_{10}P_t - L - L_F - NF - 10\log_{10}(KT_1B), \tag{6}$$

where P_t is the transmitter power, L_F is the transmitter and receiver feeder losses, NF is the receiver noise figure, K is Boltzman's constant, T_1 is the noise temperature in Kelvin, and B is the transmission bandwidth.

Fading is a typical characteristic of tropospheric scattering signals, which can be classified into fast fading and slow fading according to the duration of the fading period [26]. Slow fading reflects the attenuation caused by changes in the average value of a received level over a medium range of hundreds of wavelengths, primarily associated with climate conditions. Statistical findings indicate that the slow fading signal power obeys a log-normal distribution. Fast fading comprises frequency-selective fading and time-selective fading, which reflects the fast amplitude changes over a short period of time. The statistical experimental results show that fast fading always follows the Rayleigh distribution [27]. The received SNR γ will be used to analyze and simulate modulation performance over a troposcatter fading channel.

The troposcatter channel is commonly referred to as Wide-Sense Stationary Uncorrelated Scattering (WSSUS), which follows Rayleigh distributions [28]. It is a linear time-varying (LTV) system that can be characterized as a continuous time–impulse response $\tilde{c}(\tau, t)$ [29]. Assuming that the bandwidth of the low-pass input of the channel is limited to the order of $r/2$, and under the band-pass condition, it is equal to $B = r$, where r represents the symbol rate. Therefore, the low-pass input signal is sampled at a minimum sampling rate of r points per second, with a sampling interval represented by $T = 1/r$, i.e., tap delay.

With transmit signal $\tilde{x}(t)$ input, the output $\tilde{y}(t)$ of the troposcatter multipath channel with the tap delay line model (TDL) can be described as

$$\tilde{y}(t) = T\sum_{n=0}^{N-1} \tilde{x}(t - nT)\tilde{c}(nT, t) = \sum_{n=0}^{N-1} \tilde{x}(t - nT)\tilde{g}_n(t), \tag{7}$$

where $\tilde{g}_n(t) = T\tilde{c}(nT, t)$, N denotes the number of taps in the TDL model, i.e., the number of multipaths in the scattering channel.

For the Rayleigh fading channel, $\tilde{g}_n(t)$ is the tap gain, which is a complex Gaussian process with a zero mean [30]. Under the WSSUS assumption, the gain of each tap is uncorrelated, their power spectral density is determined by the Doppler spectrum, and the magnitude of the Doppler shift reflects the rate at which the tap coefficient changes.

In consideration of the aforementioned characteristics of the scattering channel, and its multipath structure that is always consistent with the Rayleigh distribution, an equal-delay multipath Rayleigh fading model is employed for the simulation of non-anomalous propagation within the scattering channel. In this paper, the seven-path Rayleigh fading channel model with equal delay is adopted for the purposes of simulation analysis.

2.3. Chirp-BOK System Model

Chirp-BOK uses a quasi-orthogonal chirp signal with up- and down-frequency to carry different information bits. The up-chirp signal represents the bit "1", and the down-chirp signal represents the bit "0". After passing through the communication channel, the receiver uses a down-chirp and up-chirp matched filter to detect and judge for demodulation. The schematic diagram of the chirp-BOK modulation system is shown in Figure 2 and Algorithms 1 and 2.

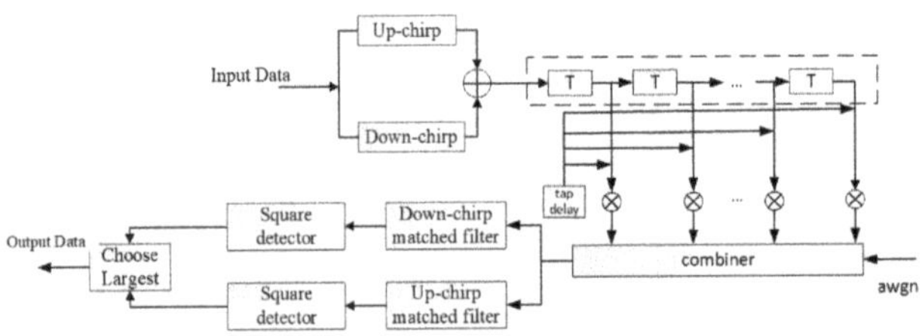

Figure 2. Scattering communication system based on chirp-BOK modulation.

Algorithm 1: Chirp-BOK modulation algorithm.

Input: $d = [d_1, d_2, \ldots, d_n]$ (data series), μ (frequency modulation slope)
Result: $s(t)$
Data: initial phase ϕ_0

for $i = 1$ **to** n **do**
    $s_i(t) \leftarrow A \cdot \cos(2\pi f_0 t + \phi(d_i))$
    if $d_i = 1$ **then**
        $\phi(d_i) \leftarrow \phi_0 + \pi\mu t^2$ /* up-chirp signals */
    else
        $\phi(d_i) \leftarrow \phi_0 - \pi\mu t^2$ /* down-chirp signals */
$s(t) \leftarrow \sum_{i=1}^{n} s_i(t)$
Output: Final modulated signal $s(t)$

Algorithm 2: Chirp-BOK demodulation algorithm.

Input: Received signal(the signal transmitted through the channel after $s(t)$)
Result: Decoded binary data
/* Match the filtering and get the *up_chirp_filtering* result */
up_chirp_filtering = MatchedFilter(*received_signal*, *up_chirp*)
/* Match the filtering and get the *up_chirp_filtering* result */
down_chirp_filtering = MatchedFilter(*received_signal*, *down_chirp*)
/* Compare the *up_chirp_filtering* and *down_chirp_filtering* results to determine the binary value of the output */
if *up_chirp_filtering* > *down_chirp_filtering* **then**
  Output "1"
else
  Output "0"

3. Theoretical Analysis for Bit Error Rate

3.1. Rayleigh Flat Fading

The average bit error rate (BER) ($\overline{p}_e$) of chirp-BOK modulation over a flat fading channel can be calculated by the instantaneous BER integrated over an AWGN channel $p_e(\gamma)$ with a signal-to-noise ratio (SNR) γ as follows:

$$\overline{p}_e = \int_0^{\infty} p_e(\gamma) f_\gamma(\gamma) d\gamma, \tag{8}$$

where $p_e(\gamma)$ is given by [31]

$$P_e(\gamma) = Q\left(\sqrt{\frac{E_b}{N_0}(1-\rho_{ij})}\right) = \frac{1}{2} erfc\left(\sqrt{\frac{1}{2}\gamma}\right). \tag{9}$$

where E_b is symbol energy and N_0 is noise power spectral density. The Gaussian Q-function is defined by $Q(x) = \frac{1}{\sqrt{2\pi}} \int_x^{\infty} e^{-t^2/2} dt$, and ρ_{ij} is the normalized cross-correlation coefficient between the up/down-chirp signals, as follows:

$$\rho_{ij} = \frac{1}{\sqrt{BT_b}} \sqrt{(C(\frac{\pi}{2}BT_b))^2 + (S(\frac{\pi}{2}BT_b))^2}. \tag{10}$$

For Rayleigh fading, the probability density function (PDF) of the received SNR γ and the cumulative distribution function (CDF) of the received SNR γ can be described, respectively, as

$$f_\gamma(\gamma) = \frac{1}{\overline{\gamma}} exp(\frac{-\gamma}{\overline{\gamma}}), \tag{11}$$

$$F_\gamma(\gamma) = 1 - exp(\frac{-\gamma}{\overline{\gamma}}), \tag{12}$$

where $\gamma = \frac{|h|^2 E_b (1-\rho_{ij})}{N_0}$ is the instantaneous SNR per bit, h is the fading coefficient with the Rayleigh distribution, and $\overline{\gamma} = \frac{|\overline{h}|^2 E_b (1-\rho_{ij})}{N_0}$ is the average SNR per bit. Additionally, since Equation (9) is a complementary error function, its original function cannot be easily expressed by a simple elementary function, so the integral by parts method is used to calculate Equation (8). Thus, Equation (8) can be deduced by

$$\overline{p}_e = -\int_0^{\infty} p_e'(\gamma) F_\gamma d\gamma, \tag{13}$$

where

$$p_e'(\gamma) = -\sqrt{\frac{1}{8\pi}}(\gamma)^{-\frac{1}{2}}exp(-\frac{1}{2}\gamma). \tag{14}$$

By substituting Equations (12) and (14) into Equation (13), we can get

$$\overline{p}_e = \frac{1}{2}\left(1 - \sqrt{\frac{\bar{\gamma}}{2+\bar{\gamma}}}\right) \tag{15}$$

Hence, the average BER of the chirp-BOK modulation over the Rayleigh flat fading channel can be written in a closed form as

$$\overline{p}_e = \frac{1}{2}\left(1 - \sqrt{\frac{\bar{h}^2 E_b(1-\rho_{ij})/N_0}{2+\bar{h}^2 E_b(1-\rho_{ij})/N_0}}\right). \tag{16}$$

3.2. Frequency-Selective Fading

The impulse response of baseband signals on frequency-selective channels can be expressed as

$$h(t) = \sum_{l=1}^{L} \alpha_l e^{-j\theta_l}\delta(t-\tau_l), \tag{17}$$

where l is the channel index, L is the number of resolvable paths, $\delta(\cdot)$ is the delta function, and $\{\alpha_l\}_{l=1}^{L}$, $\{\theta_l\}_{l=1}^{L}$, $\{\tau_l\}_{l=1}^{L}$ are the random channel amplitudes, phases and delays, respectively. The received signal can be expressed as

$$\begin{aligned} y(t) &= h(t) * s(t) + n(t) + i_{ISI}(t) \\ &= \sum_{l=1}^{L} \alpha_l e^{-j\theta_l} x(t-\tau_l) + n(t) + i_{ISI}(t), \end{aligned} \tag{18}$$

where $n(t)$ denotes Additive White Gaussian Noise (AWGN) and $i_{ISI}(t)$ is the inter symbol interference (ISI) component. Assuming that the fading amplitude is a random variable with a mean-square value $\Omega = \overline{\alpha^2}$. The instantaneous SNR per bit of the Lth channel is given as $\gamma_l = \Omega_l E_b(1-\rho_{ij})/N_0$ and the average SNR per bit of the Lth channel can be expressed as $\bar{\gamma}_l = \bar{\Omega}_l E_b(1-\rho_{ij})/N_0$.

The mean-square value Ω_l is related to power delay profile of the channel, which is considered as an exponentially decaying power delay profile (PDP).

$$\Omega_l = \Omega_1 e^{-(l-1)\delta}, \delta \geq 0, l = 1, 2, \ldots, L, \tag{19}$$

where Ω_1 is the average fading power of the first path, and the parameter δ is the power decay factor. The variance of the ISI component is [32] $\sigma_{ISI}^2 = (\Omega_T - 1)\Omega_1 E_b/2P_g$, where $\Omega_T = \sum_{l=1}^{L} e^{-(l-1)\delta}$ is the normalized total average fading power, and p_g is the processing gain. Therefore, the equivalent interference plus noise power spectral density can be written as

$$\frac{N_e}{2} = \frac{\Omega_T - 1}{2P_g}\Omega_1 E_b + \frac{N_0}{2} \tag{20}$$

The average signal-to-interference plus noise ratio is

$$\bar{\gamma}_e = \left(\frac{2}{\bar{\gamma}_1} + \frac{\Omega_T - 1}{2P_g}\right)^{-1}. \tag{21}$$

Following the same procedure as in Section 3, the average BER over frequency-selective Rayleigh fading in a closed form is obtained as

$$\bar{p}_e = \frac{1}{2}\left[1 - \sqrt{\frac{\bar{\gamma}_e}{1+\bar{\gamma}_e}}\right]. \tag{22}$$

4. Simulation and Discussion

In the subsequent simulation, the setting parameters of the multipath Rayleigh fading channel are shown in Table 1.

Table 1. Parameters for simulation.

Parameter	Value
Maximum Doppler Shift	$f_d \in [1\text{ Hz}, 100\text{ Hz}]$
Delay-Vector	$[0\ 1\ 2\ 3\ 4\ 5\ 6] \times T(\text{ns})$
Path-Gain-Vector1	$[-10\ -3\ 0\ -2\ -5\ -8\ -10]$ (dB)
Path-Gain-Vector2	$[10\ -3\ 0\ -2\ -5\ -8\ -10]$ (dB)
Time–Bandwidth Product	$D = BT_b$

Equation (14) states the average BER output from the troposcatter channel with chirp-BOK modulation in the absence of ISI (i.e., Rayleigh flat fading), which can be expressed as $P_e = \frac{1}{2}\left(1 - \sqrt{\frac{\bar{\gamma}}{2+\bar{\gamma}}}\right)$, where $\bar{\gamma}$ denotes the average SNR. A comparison between the simulation result and theoretical result is shown in Figure 3; the simulation results were verified through numerous simulations, revealing that when the parameter D is set to 6, there is a high degree of consistency with the theoretical results.

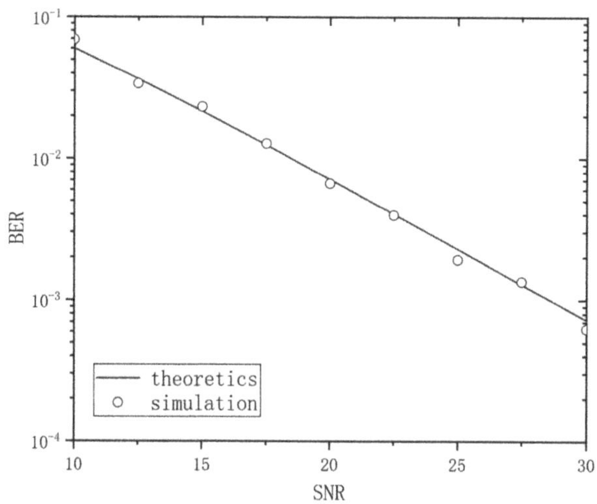

Figure 3. Comparison of theoretical and simulated values.

When tap delay T is equal to 10 ns and 50 ns, respectively, with Path-Gain-Vector1, the influence of different Doppler shifts f_d on the system BER performance is shown in Figure 4. As shown in Figure 4, the BER curves are basically consistent in the range of $f_d \in [1\text{ Hz}, 100\text{ Hz}]$, and the influence of f_d on the BER is relatively small. The subsequent simulation and analysis are based on the assumption that $f_d = 10$ Hz.

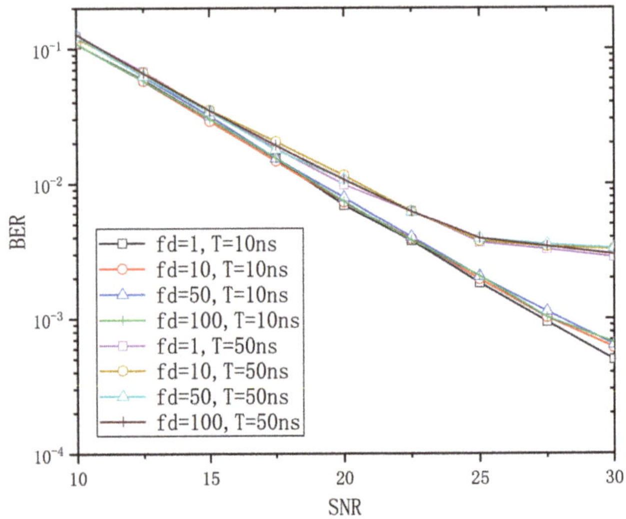

Figure 4. The BER curve of different f_d in $T = 10$ ns and $T = 50$ ns.

The effect of varying tap delay vector T on BER performance of the system with D equal to 6 and 60 is illustrated in Figure 5a and Figure 5b, respectively.

As illustrated in Figure 5a, under the premise of a small D, the BER curve exhibits a flattening trend as the value of T increases. When $T = 1$ ns and $T =$ ns, the two BER curves are essentially identical, indicating that the ISI in the system is small in the presence of a relatively short time interval. This can be regarded as frequency flat fading. As the value of T increases to 10 ns, 30 ns, and 50 ns, the BER performance of the system declines continuously. This is due to the fact that as T increases, the ISI phenomenon in the channel becomes more pronounced, resulting in a transition from a frequency flat fading channel to a frequency-selective fading channel.

Figure 5b shows that the BER performance improves with an increase in T. This is due to the fact that a larger time–bandwidth product results in a more rectangular spectral characteristic of the chirp signal, rendering it more resistant to frequency-selective fading. Consequently, the system is more effectively capable of suppressing frequency-selective fading caused by different paths when T continuously increases, thus enhancing the reception quality and BER performance.

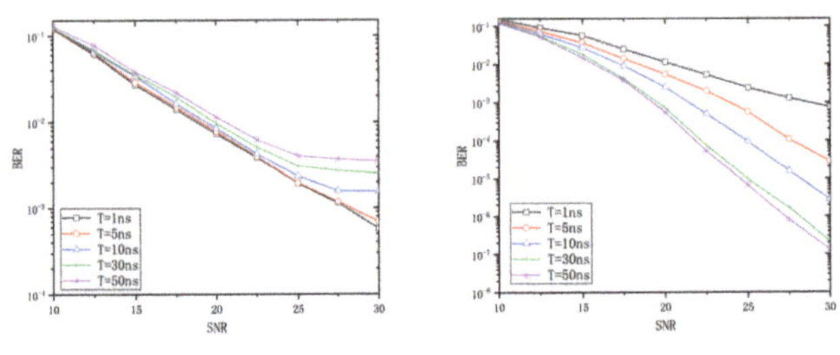

(**a**) D = 6 (**b**) D = 60

Figure 5. The BER curve of different T.

The system BER performance for tap delays T of 5 ns, 10 ns, 30 ns, and 50 ns is analyzed in Figures 6–9 when D is 8, 20, 40, 60, and 120, respectively. From these figures, it can be observed that the system performance attains its optimal level when D is 20, 40, 60, and 120, respectively, with Path-Gain-Vector1. Similarly, the system performance is optimal when D is 20 and 40 with Path-Gain-Vector2.

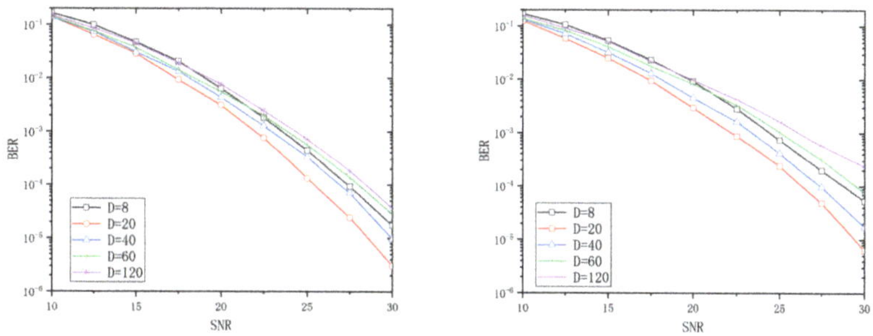

(**a**) Path-Gain-Vector1 (**b**) Path-Gain-Vector2

Figure 6. The BER curve of different D in $T = 5$ ns.

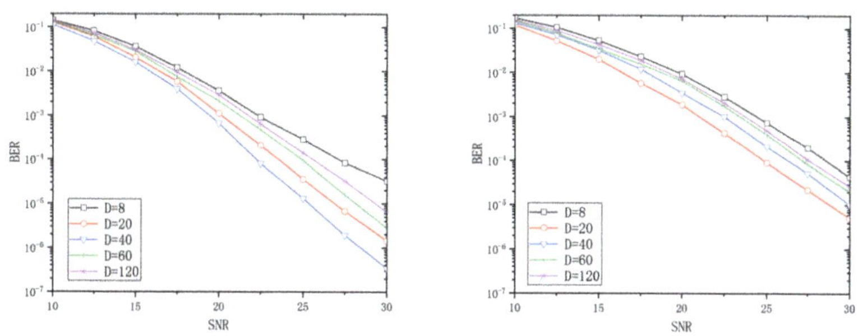

(**a**) Path-Gain-Vector1 (**b**) Path-Gain-Vector2

Figure 7. The BER curve of different D in $T = 10$ ns.

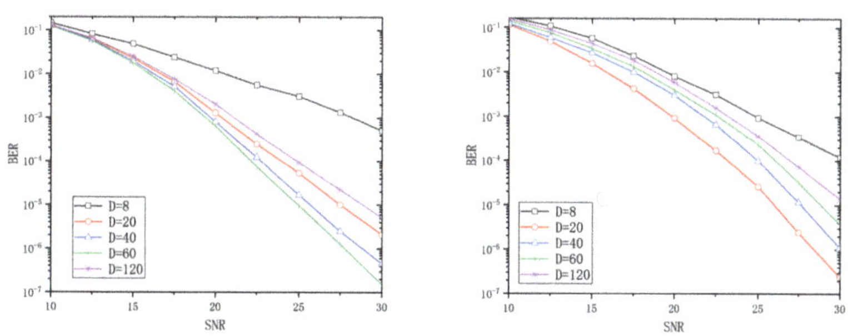

(**a**) Path-Gain-Vector1 (**b**) Path-Gain-Vector2

Figure 8. The BER curve of different D in $T = 30$ ns.

Tables 2 and 3 show the threshold values of the SNR for different channel delay vectors and different D when the BER is 10^{-4} with Path-Gain-Vector1 and Path-Gain-Vector2, respectively.

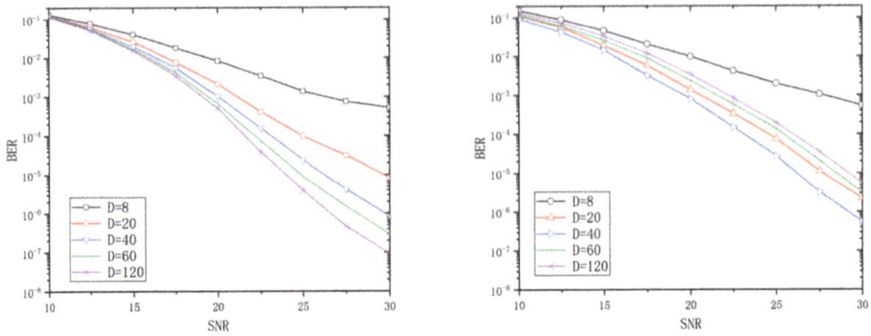

(**a**) Path-Gain-Vector1 (**b**) Path-Gain-Vector2

Figure 9. The BER curve of different D in $T = 50$ ns.

Table 2. SNR (dB) for Path-Gain-Vector1 when the BER is 10^{-4}.

Path-Gain-Vector1	$T = 5$ ns	$T = 10$ ns	$T = 30$ ns	$T = 50$ ns
$D = 8$	27.2	27.5	31.9	32.5
$D = 20$	25.5	24.0	23.5	25.0
$D = 40$	26.8	22.5	22.8	22.7
$D = 60$	28.2	25.0	22.4	22.5
$D = 120$	28.8	25.8	25.0	21.4

Table 3. SNR (dB) for Path-Gain-Vector2 when the BER is 10^{-4}.

Path-Gain-Vector2	$T = 5$ ns	$T = 10$ ns	$T = 30$ ns	$T = 50$ ns
$D = 8$	28.7	28.7	30.4	32.7
$D = 20$	26.0	24.8	22.7	23.4
$D = 40$	27.5	26.9	25.1	22.9
$D = 60$	29.1	27.4	26.3	25.4
$D = 120$	32.6	27.7	27.4	25.6

Assume that target bit error rate is 10^{-4} and the reference $D = 8$ is used as a basis for the gain calculation. In Path-Gain-Vector1, the optimal selection to achieve the target BER is 1.7 dB less than the SNR required for $D = 8$ when the tap delay T is 5 ns. Similarly, the optimal selections to achieve the target BER are 5 dB, 9.5 dB and 10.0 dB less than $D = 8$ when the tap delay T is 10 ns, 30 ns and 50 ns, respectively. In Path-Gain-Vector2, the optimal configuration to reach the target BER is 1.7 dB less than the SNR required for $D = 8$ when the tap delay T is 5 ns. Similarly, the optimal configurations to reach the target BER are 3.9 dB, 7.7 dB, and 9.8 dB less than $D = 8$ when the tap delay T is 10 ns, 30 ns and 50 ns, respectively.

Figure 10 is a BER performance comparison between chirp-BOK, BPSK, QPSK and 16-QAM modulation schemes. During the chirp-BOK process of simulation, the BER trends were found to be consistent under four different channel conditions. Therefore, only two cases are listed here, $T = 30$ ns and various path gains, with results showing that chirp-BOK outperforms BPSK, 16-QAM and QPSK in troposcatter deep fading channel conditions. Although chirp-BOK modulation can have higher sensitivity, the spectral efficiency is relatively low. To settle this problem, the chirp signal spread method can be used with a typical digital modulation scheme and Orthogonal Frequency Division Multiplexing (OFDM) technique.

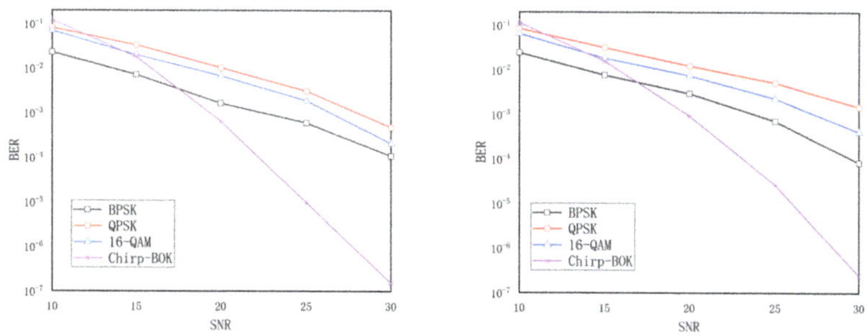

(**a**) Path-Gain-Vector1 (**b**) Path-Gain-Vector2

Figure 10. The BER curve of different modulation schemes in $T = 30$ ns.

5. Conclusions

In this paper, we presented a BER performance analysis and parameter optimization of chirp-BOK modulation over a troposcatter channel, theoretically and numerically. Firstly, the averaged BER performance of chirp-BOK modulation over a Rayleigh flat fading channel was derived theoretically. Then, a seven-path equal-delay Rayleigh fading channel model was established to simulate the chirp-BOK modulation BER performance under various conditions. Furthermore, the impact of the time–bandwidth product and channel tap delay was optimized and analyzed with different chirp-BOK parameters. These results demonstrate that when the path gain remains constant and the tap delay increases, the value of D needs to be increased in order to achieve optimal performance. It is worth noting that an increase in the primary path gain will not necessarily result in an improvement in BER performance, when the same time–bandwidth product and tap delay conditions are considered. As observed in Part 4, when optimized with appropriate parameters, chirp-BOK modulation can exhibit a significantly lower BER compared to other modulation techniques under deep fading channel conditions. On the one hand, in a future study, an actual chirp signal experiment should be tested to characterize chirp-BOK modulation parameters. On the other hand, further enhancement may focus on the combination of chirp modulation with some advanced error correction codes to achieve higher data transmission reliability.

Author Contributions: Conceptualization, J.S. and Z.L.; software, J.S.; validation, Z.L., Y.L. and T.X.; writing—original draft preparation, J.S., Z.L. and T.X.; writing—review and editing, J.S., Z.L. and Y.L.; visualization, J.S.; supervision, J.S. and T.X. All authors have read and agreed to the published version of the manuscript.

Funding: This research was supported by the National Natural Science Foundation of China (62371390).

Institutional Review Board Statement: Not applicable.

Data Availability Statement: Data are contained within this article.

Conflicts of Interest: Author Tianjiao Xie is employed by the company China Academy of Space Technology. The remaining authors declare that the research was conducted in the absence of any commercial or financial relationships that could be construed as a potential conflict of interest.

Abbreviations

The following abbreviations are used in this manuscript:

CSS	Chirp spread spectrum
BOK	Binary orthogonal keying
ISI	Inter-symbol interference
BER	Bit error rate
SNR	Signal-to-noise power ratio
PDF	Probability density function
CDF	Cumulative distribution function

References

1. Li, C.; Chen, X.; Liu, X. Cognitive Tropospheric Scatter Communication. *IEEE. Trans. Vehic. Technol.* **2017**, *67*, 1482–1491. [CrossRef]
2. *IEEE. Std. 802.15.4a-2007;* IEEE Standard for Information Technology—Local and Metropolitan Area Networks—Specific Requirements—Part 15.4: Wireless Medium Access Control (MAC) and Physical Layer (PHY) Specifications for Low-Rate Wireless Personal Area Networks (WPANs): Amendment 1: Add Alternate PHYs. IEEE: Piscataway, NJ, USA, 2007; pp. 1–210. [CrossRef]
3. Winkler, M. Chirp signals for communications. In Proceedings of the 1962 IEEE Western Electronic Show and Convention Conference (WESCON), Los Angeles, CA, USA, 21–24 August 1962; pp. 14–17.
4. Lee, H.; Kim, T.H.; Choi, J.W. Chirp signal-based aerial acoustic communication for smart devices. In Proceedings of the 2015 IEEE Conference on Computer Communications(INFOCOM), Hong Kong, China, 26 April–1 May 2015; pp. 2407–2415. [CrossRef]
5. Zhao, Q.M.; Zhang, Q.Y.; Zhang, N.T. Multiple chirp-rate modulation based on fractional fourier transform. In Proceedings of the 2010 First International Conference on Pervasive Computing, Signal Processing and Applications (PCSPA), Harbin, China, 17–19 September 2010; pp. 688–691. [CrossRef]
6. Kaihan, Z.; Keyu, C.; Fei, Y.; Jinwang, Y. Chirp FSK based on FRFT for underwater acoustic communication. In Proceedings of the 2017 7th IEEE International Conference on Electronics Information and Emergency Communication (ICEIEC), Macau, China, 21–23 July 2017; pp. 49–52. [CrossRef]
7. Vangelista, L. Frequency shift chirp modulation: The LoRa modulation. *IEEE. SPL* **2017**, *24*, 1818–1821. [CrossRef]
8. Yuyang, J. Performance Study of LFM-Based Shared Signal for Radar and Communications. Master's Thesis, Huazhong University of Science and Technology, Wuhan, China, 2019. (In Chinese) [CrossRef]
9. Jianxun, H. Research on Radar Communication Signal Integration and Beamforming. Master's Thesis, Harbin Engineering University, Harbin, China, 2020. (In Chinese) [CrossRef]
10. Reynders, B.; Meert, W.; Pollin, S. Range and coexistence analysis of long range unlicensed communication. In Proceedings of the International Conference on Telecommunications(ICT), Thessaloniki, Greece, 16–18 May 2016; pp. 1–6. [CrossRef]
11. Reynders, B.; Pollin, S. Chirp spread spectrum as a modulation technique for long range communication. In Proceedings of the 2016 Symposium on Communications and Vehicular Technologies (SCVT), Mons, Belgium, 22–22 November 2016; pp. 1–5. [CrossRef]
12. Tang, X.; Li, H.; Zhang, Y.; Zhao, X. Performance Analysis of LoRa Modulation with Residual Frequency Offset. In Proceedings of the 2018 IEEE 4th International Conference on Computer and Communications (ICCC), Chengdu, China, 7–10 December 2018; pp. 835–839. [CrossRef]
13. Jia, W.; Wang, W.Q.; Cheng, J.; Hau, Y.; Kang, Z. Low PAPR OFDM-Chirp Modulation Signaling Scheme. In Proceedings of the 2021 IEEE International Geoscience and Remote Sensing Symposium IGARSS, Brussels, Belgium, 11–16 July 2021; pp. 7963–7966. [CrossRef]
14. Jia, Z.; Zheng,W.; Yuan, F. A Two-Dimensional Chirp-MFCSK Modulation Method for Underwater LoRa System. *IEEE. Intern. Thin. J.* **2022**, *9*, 24388–24397. [CrossRef]
15. Zhao, Z. The Multiple Access PSK-Chirp System and the Demodulation Based on FrFT. In Proceedings of the 2022 IEEE 22nd International Conference on Communication Technology (ICCT), Nanjing, China, 11–14 November 2022; pp. 77–83. [CrossRef]
16. Zhang, R.; Zhang, S.; Wang, X.; Zhu, Y. Fractional Chirp Rate Based Non-Orthogonal Transmission for Chirp Spread Spectrum Modulation. In Proceedings of the 2023 International Conference on Intelligent Communication and Networking (ICN), Changzhou, China, 10–12 November 2023; pp. 111–115. [CrossRef]
17. Gupta, A.; Agarwal, S.; Chakravarty, S. Non-Coherent Distributed Transmission with Chirp Spread Spectrum Modulation. In Proceedings of the 2024 IEEE International Conference on Communications Workshops (ICC Workshops), Denver, CO, USA, 9–13 June 2024; pp. 51–56. [CrossRef]
18. Bacioccola, A.; Cicconetti, C.; Eklund, C.; Lenzini, L.; Li, Z.; Mingozzi, E. IEEE 802.16: History, status and future trends. *Comput. Commun.*, **2010**, *33*, 113–123. [CrossRef]
19. Xuemei, S.; Xudong, Z. Research on tropospheric scattering for radar signal beyond line-of-sight transmission. *Modern Radar* **2011**, *33*, 9–12. [CrossRef]

20. Mengnan, W. Research on Beyond-Line-of-Sight Wideband Array Signal Beamforming Technology. Master's Thesis, Changsha, China, 2013. (In Chinese) [CrossRef]
21. Garg, A.; Das, S.S. Design of troposcatter broadband link based on SCFDE. In Proceedings of the 2017 IEEE International Conference on Advanced Networks and Telecommunications System(ANTS), Bhubaneswar, India, 17–20 December 2017. [CrossRef]
22. Biscarini, M.; De Sanctis, K.; Di Fabio, S.; Montopoli, M.; Milani, L.; Marzano, F.S. Assessment and Uncertainty Estimation of Weather-Forecast Driven Data Transfer for Space Exploration at Ka- and X-Band. *IEEE. Trans. Antenn. Propag.* **2019**, *67*, 3308–3322. [CrossRef]
23. Binsong, S.; Jiancun, Q.; Wencheng, R. A new scattering communication method based on frequency hopping and simulation research. *J. Jilin Univ.* **2020**, *50*, 1870–1875. [CrossRef]
24. ITU-R Recommendation P.617-5, Propagation Prediction Techniques and Data Required for the Design of Trans-Horizon Radio-Relay Systems. 2019 Available online: https://1f8a81b9b0707b63-19211.webchannel-proxy.scarabresearch.com/rec/R-REC-P.617/en (accessed on 4 September 2024).
25. Annavajjala, R.; Zagami, J. Design and Performance Evaluation of High-Capacity Mobile Troposcatter Links Under Mobility, Frequency Selectivity, and Antenna Pointing Errors. In Proceedings of the 2018 IEEE Military Communications Conference (MILCOM), Los Angeles, CA, USA, 29–31 October 2018. [CrossRef]
26. Liu, Y.; Wang, Z.; Dong, H.; Liu, H. Anti-disturbance filter design for a class of stochastic systems with fading channels. *Sci. China Inf. Sci.* **2020**, *63*, 327–329. [CrossRef]
27. Minggao, Z. *Troposphere Scatter Propagation*; Publishing House of Electronics Industry: Beijing, China, 2004; ISBN 9787121003820.
28. Tranter, W.; Shanmugan, K.; Rappaport, T.; Kosbar, K. *Principles of Communication Systems Simulation with Wireless Applications*; Prentice Hall: London, UK, 2003; ISBN 0134947908.
29. Hashemi, H. The indoor radio propagation channel. *Proc. IEEE* **1993**, *81*, 943–968. [CrossRef]
30. Kailath, T. Channel characterization: Time-variant dispersive channels. *Lect. Commun. Syst. Theor.* **1961**, 95–123.
31. Alsharef, M.; Rao, R.K. M-ary chirp modulation for coherent and non-coherent data transmission. In Proceedings of the 2015 IEEE 28th Canadian Conference on Electrical and Computer Engineering (CCECE), Halifax, NS, Canada, 3–6 May 2015; pp. 213–219. [CrossRef]
32. Eng, T.; Milstein, L.B. Coherent DS-CDMA perfomance in Nakagami multipath fading. *IEEE Trans. Commun.* **1995**, *43*, 1134–1143. [CrossRef]

entropy

MDPI

Article

Generalized Filter Bank Orthogonal Frequency Division Multiplexing: Low-Complexity Waveform for Ultra-Wide Bandwidth and Flexible Services

Yu Xin [1,2], Jian Hua [1,2,*], Tong Bao [1,2], Yaxing Hao [3], Ziheng Xiao [3], Xin Nie [3] and Fanggang Wang [3,*]

1 State Key Laboratory of Mobile Network and Mobile Multimedia Technology, Shenzhen 518055, China; xin.yu@zte.com.cn (Y.X.); bao.tong@zte.com.cn (T.B.)
2 ZTE Corporation, Nanshan District, Shenzhen 518055, China
3 State Key Laboratory of Advanced Rail Autonomous Operation, School of Electronic and Information Engineering, Beijing Jiaotong University, Beijing 100044, China; yaxinghao@bjtu.edu.cn (Y.H.); 23120134@bjtu.edu.cn (Z.X.); 23125045@bjtu.edu.cn (X.N.)
* Correspondence: hua.jian2@zte.com.cn (J.H.); wangfg@bjtu.edu.cn (F.W.)

Abstract: Terahertz (THz) communication is a crucial technique in sixth generation (6G) mobile networks, which allow for ultra-wide bandwidths to enable ultra-high data rate wireless communication. However, the current subcarrier spacing and the size of fast Fourier transform (FFT) of the orthogonal frequency division multiplexing (OFDM) in 5G NR are insufficient regarding the bandwidth requirements of terahertz scenarios. In this paper, a novel waveform is proposed to address the ultra-wideband issue, namely the generalized filter bank orthogonal frequency division multiplexing (GFB-OFDM) waveform. The main advantages are summarized as follows: (1) The K-point IFFT is implemented by two levels of IFFTs in smaller sizes, i.e, performing M-point IFFT in N times and performing N-point IFFT in M times, where $K = N \times M$. (2) The proposed waveform can accommodate flexible subcarrier spacings and different numbers of the subbands to provide various services in a single GFB-OFDM symbol. (3) Different bandwidths can be supported using a fixed filter since the filtering is performed on each subband. In contrast, the cyclic prefix orthogonal frequency division multiplexing (CP-OFDM) in 4G/5G requires various filters. (4) The existing detection for CP-OFDM can be directly employed as the detector of the proposed waveform. Lastly, the comprehensive simulation results demonstrate that GFB-OFDM outperforms CP-OFDM in terms of the out-of-band leakage, complexity and error performance.

Keywords: 6G; terahertz; waveform; ultra-wide bandwidth

1. Introduction

The development of the sixth generation (6G) wireless communication system has been accompanied by the introduction of new technologies, among which terahertz (THz) technology, also referred to as one of the ten technologies that will innovate the future world, is a breakthrough spectrum technology [1,2]. It offers a wide transmission bandwidth to cater to the increasing demand for higher transmission rates, and thus it has emerged as a crucial component in 6G communication.

Terahertz communication typically involves the utilization of the electromagnetic waves within a frequency range spanning from 0.1 THz to 10 THz and a wavelength range from 3 mm to 30 μm. Its wavelength lies between the microwave and the far-infrared regions, providing an abundance of spectrum resources that remain unexplored [3]. One of the characteristics of the THz communication is ultra-wide bandwidth. Due to the abundant spectrum of resources brought by the very high frequency, the expected bandwidth of THz communication ranges from several GHz to tens of GHz, surpassing that of the current 5G bandwidth. Therefore, it can achieve an ultra-high rate of up to 1 Tbps, which is

10 to 100 times of the current 5G peak rate [4]. The other characteristics encompass remarkable penetrability [5], the capacity to traverse non-metallic materials without inflicting any damage, and the low energy levels that minimize harm to living organisms, thereby facilitating non-destructive testing. Furthermore, it provides finer time resolution for capturing subtle momentary forms and ensures utmost confidentiality.

In THz communication, achieving an ultra-wide bandwidth involves the following challenges. Hardware cost and implementation complexity set limitations for the maximum subcarrier spacing and the size of the fast Fourier transform (FFT) of orthogonal frequency division multiplexing (OFDM), which may hinder the utilization of the ultra-wide bandwidth in THz scenarios. Currently, the maximum subcarrier spacing is limited to 960 kHz and the maximum FFT size is restricted to 4096 points, resulting in a bandwidth limitation of around 4 GHz when considering oversampling and guard intervals. To support a larger bandwidth in OFDM baseband processing, there are typically two solutions, i.e., increasing the subcarrier spacing or the FFT size. However, increasing the subcarrier spacing leads to a shorter symbol duration, while a sufficiently long cyclic prefix is still required to combat multipath delay. This solution results in a higher proportion of cyclic prefix overhead and reduces the spectral efficiency. Typically, the complexity of N-point IFFT is $N\log N$ [6]. The current hardware faces challenges in regard to accommodating the large size of the FFT. Therefore, it prompts us to propose a new waveform that enables the realization of ultra-wideband THz communication.

The waveforms predominantly employed in the current fifth generation (5G) standards encompass cyclic prefix-orthogonal frequency division multiplexing (CP-OFDM) and discrete Fourier transform spread orthogonal frequency division multiplexing (DFT-s-OFDM). In recent years, a plethora of investigations have been conducted on waveforms [7], including some new waveforms based on OFDM: (1) Filter-bank multi-carrier (FBMC) [8] is a form of real number modulation that separates the real and imaginary components of the modulated symbols, utilizing filters to achieve real orthogonality. (2) Generalized frequency division multiplexing (GFDM) [9,10] adopts a cyclic filter for each OFDM subcarrier and introduces block modulation technology; however, there is interference between subcarriers. (3) Filter bank orthogonal frequency division multiplexing (FB-OFDM) [11–13] utilizes polyphase filters for each subcarrier, with the subcarriers being approximately orthogonal. (4) Universal filtered multi-carrier (UFMC) [14] performs filtering on a set of consecutive subcarriers. Due to the long time-domain filter function, there is inter-symbol interference in the time domain. (5) Filter orthogonal frequency division multiplexing (F-OFDM) [15] filters the entire subband. (6) Orthogonal time frequency space (OTFS) [16–19] is a waveform based on a specific two-dimensional Fourier transform, i.e., a symplectic Fourier transform, which loads the modulated symbols into the delay-Doppler domain. These waveforms encounter various problems when supporting the communication in large bandwidths of the THz communication band.

Recently, we proposed a new waveform called generalized filter bank orthogonal frequency division multiplexing (GFB-OFDM) in [20–22] for THz communication. The waveform utilizes the two-level IFFTs and the polyphase filters to jointly process of the multiple subbands. In [21], it shows that the out-of-band leakage of GFB-OFDM is lower than that of CP-OFDM without error performance loss. In [22], more comparisons were provided in different subcarrier spacings. However, there is still a lack of solid proof regarding an equivalence between the GFB-OFDM transmitter and the CP-OFDM transmitter. Moreover, the complexity analysis and a more detailed evaluation are required to validate its advantages.

In this paper, we present the complexity analysis, solid proof of the equivalence between the two transmitters, and a comprehensive performance evaluation of the coded GFB-OFDM. By decomposing the large size of an IFFT into two levels of IFFTs in smaller sizes, the GFB-OFDM facilitates efficient handling of the large IFFT size and the subband filtering, which reduces the computational complexity and enables the ultra-wideband communication. The GFB-OFDM allows for a flexible configuration of the system parameters, including the number of subbands, the number of subcarriers in each subband, the subcarrier spacing, and the filter parameters, etc. This capability can improve the transmission stability in high-mobility scenarios by suppressing the effects of Doppler spread and satisfy various application requirements in 6G communication, especially in terms of the high bandwidth and different services. Furthermore, the GFB-OFDM can be also configured to reduce power consumption regarding the energy efficiency of 6G networks. Overall, the GFB-OFDM provides a versatile and efficient solution for the complex demands of 6G communication, particularly in supporting ultra-wideband THz communication, improving spectral efficiency, and offering adaptable configurations for diverse applications and services.

2. GFB-OFDM Transceiver

We first introduce the GFB-OFDM transmitter, and then the polyphase filter and GFB-OFDM receiver are illustrated, respectively. Finally, we provide the theoretical derivation.

2.1. Transmitter

The GFB-OFDM transmitter is briefly exhibited in Figure 1. The modulated symbols are divided into N groups, and the symbols in each group are loaded onto its corresponding subband. Then, the first-level (or subcarrier-level) IFFT is performed over each individual subband, and the size of the IFFT is flexible, which aligns with the number of the loaded symbols in the specific subband. The output of each IFFT process is stacked, and the second-level (or subband-level) IFFT is performed over all the subband outputs; thus, the size of the second-level IFFT is the sum of the sizes of all subband IFFT. Then, the polyphase filter is adopted to window the baseband signal. Finally, the digital-to-analog converter (DAC) and the radio frequency (RF) is employed to transmit it out. In summary, the process of dividing modulated symbols into subbands and performing the first-level IFFT is referred to as the subcarrier-level processing module. The process of the second-level IFFT and the polyphase filter are referred to as the subband-level processing module.

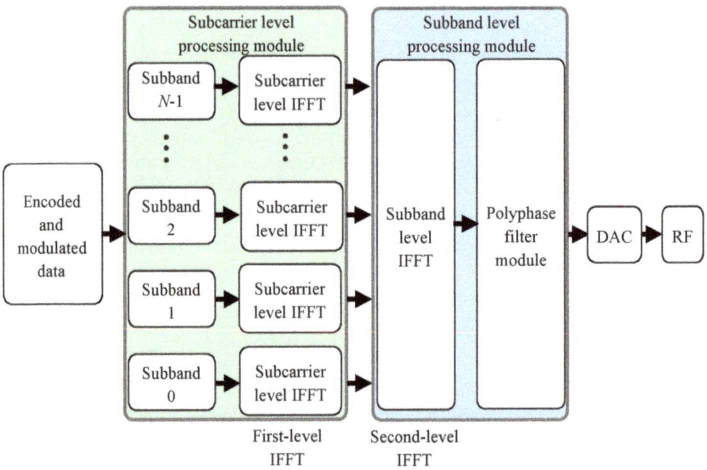

Figure 1. GFB-OFDM transmitter diagram. The encoded and modulated data are first divided into N subbands, with each subband undergoing a subcarrier-level IFFT. The resulting data is then processed through a subband-level IFFT along the subband dimension, followed by a polyphase filter. The output is transmitted after passing through the digital-to-analog converter and radio frequency module.

By using the two levels of IFFT and the polyphase filters, the large size of the IFFT can be decomposed into two levels of IFFT of smaller sizes. This approach effectively addresses the challenges in the issues of the computational complexity and the hardware implementation.

In the following, we formulate the two-level IFFT and the polyphase filtering in the GFB-OFDM transmitter. There are K modulated symbols to be transmitted over the K subcarriers. They are divided into N subbands, and each subband has an identical bandwidth (each subband is able to alter its subcarrier spacing according to different services. For a brief illustration, we assume that each subband has identical subcarrier spacing.). In each subband, M symbols are mapped to the subcarrier, i.e., $K = MN$, which can be represented as $\boldsymbol{X}[l,k], l = 0,1,2,\ldots,M-1, k = 0,1,2,\ldots,N-1$. Then, the output of the first-level IFFT $\boldsymbol{x}[m,k]$ of N subbands is calculated as

$$\boldsymbol{x}[m,k] = \frac{1}{M}\sum_{l=0}^{M-1} \boldsymbol{X}[l,k]e^{j2\pi\frac{ml}{M}}, \quad m = 0,1,2,\ldots,M-1 \tag{1}$$

where $j = \sqrt{-1}$. Let $\boldsymbol{A}$ be the inverse Fourier transform matrix; then, Equation (1) can be rewritten in the matrix form as

$$\boldsymbol{x}_{M\times N} = \frac{1}{M}\boldsymbol{A}_{M\times M}\boldsymbol{X}_{M\times N}. \tag{2}$$

Then, the cyclic prefix is added for each subband as

$$\boldsymbol{x}_{M\times N} \xrightarrow{\text{Add CP}} \boldsymbol{x}_{(N_{\text{cp}}+M)\times N} \tag{3}$$

where N_{cp} is the length of the cyclic prefix. Then, the second-level IFFT is performed row-wise over the matrix $\boldsymbol{x}$. For each row, N samples are used as the input of the N_2 points double bandwidth oversampled IFFT, i.e., $N_2 = 2N$, and the input is given as

$$\boldsymbol{Y}_{(N_{\text{cp}}+M)\times N_2} = \left[0_{(N_{\text{cp}}+M)\times N_s}, \boldsymbol{x}_{(N_{\text{cp}}+M)\times N}, 0_{(N_{\text{cp}}+M)\times N_s}\right] \tag{4}$$

where $N_s = \frac{N_2-N}{2}$ is the number of zero padding on one side. The operation is equivalent to oversampling in the time domain. After the $N_{\text{cp}}+M$ second-level IFFT, the $N_{\text{cp}}+M$ rows are obtained, and each row contains N_2 samples. Each row of a sample is a subsymbol data sequence, and the time-domain length T of a subsymbol data sequence is the reciprocal of the interval between the adjacent subband in the frequency domains, which is also the reciprocal of the width of the subband in the frequency domain. The second-level IFFT process can be written in matrix form as

$$\boldsymbol{y}_{(N_{\text{cp}}+M)\times N_2} = \frac{1}{N_2}\boldsymbol{Y}_{(N_{\text{cp}}+M)\times N_2}\boldsymbol{B}_{N_2\times N_2} \tag{5}$$

where $\boldsymbol{B}$ is the inverse Fourier transform matrix. The output of the second-level IFFT $\boldsymbol{y}$ is repeated L times column-wise, which is given as

$$\boldsymbol{y}_{(N_{\text{cp}}+M)\times N_2} \xrightarrow{\text{Repeat } L \text{ times}} \boldsymbol{y}_{(N_{\text{cp}}+M)\times LN_2} \tag{6}$$

Then, the output is filtered by the polyphase filter $\mathbf{g}$, which is written as

$$\boldsymbol{z}_{(N_{\text{cp}}+M)\times LN_2} = \boldsymbol{y}_{(N_{\text{cp}}+M)\times LN_2} \odot [\mathbf{g}, \mathbf{g}, \ldots, \mathbf{g}]_{(N_{\text{cp}}+M)\times LN_2} \tag{7}$$

where "$\odot$" is the Hadamard product operation. The column vector $\mathbf{g}$ is the discrete form of the polyphase filtering function with the length of LN_2. Since each column of $\boldsymbol{y}$ is filtered by the same $\mathbf{g}$, the filtering matrix is obtained by repeating $\mathbf{g}$ in $N_{\text{cp}}+M$ times row-wise.

Then, the $N_{\text{cp}}+M$ subsymbol is summed over and the mth subsymbol is placed with the lag of the $N(m-1)$ samples. The sum can be elaborated by Figure 2. Finally, the sum of the time-domain samples can be transmitted. Thus, the overall number of samples is $K + N(2L + N_{cp} - 1)$. In contrast, if the frequency-domain data symbols at the K subcarriers are transmitted in practice using the traditional CP-OFDM waveform, it generally requires at least N_{ifft}-point IFFT, where $N_{ifft} = 2K$.

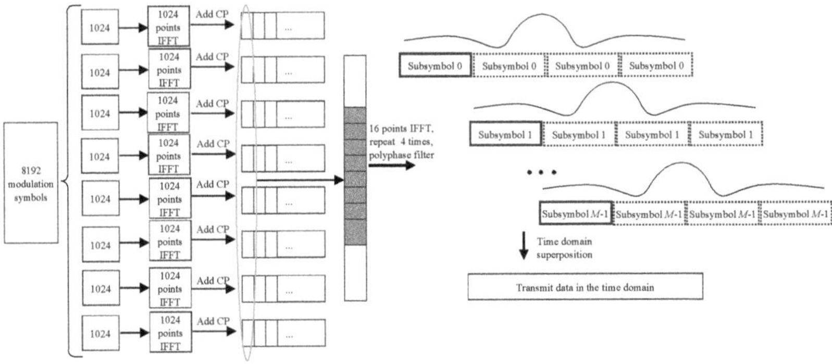

Figure 2. GFB-OFDM data-processing procedure with the number of modulated symbols $K = 8192$, the number of subbands $N = 8$ and the number of subcarriers in each subband $M = 1024$.

The aforementioned procedure demonstrates that GFB-OFDM can accomplish the IFFT of N_{ifft} points by utilizing N first-level IFFTs of M points and $N_{cp} + M$ second-level IFFTs of N_2-points, thereby enabling the realization of a large-point IFFT through two small-point IFFTs. Moreover, the number of subbands and the number of subcarrier in each subband can be chosen according to the required traffic on demand.

2.2. Polyphase Filter

The purpose of the polyphase filter is to filter each subband in order to eliminate the mirror interference between the subbands. The parameters of the filtering function need to be selected carefully, considering its impact on the overall performance of GFB-OFDM [23,24]. In this paper, we adopt a dual root-raised-cosine (DRRC) filtering function, which employs the time-domain representation of the RRC function to window and truncate frequency-domain RRC signals. The procedure of generating this filter is as follows.

The time-domain form of the frequency-domain RRC function can be expressed as

$$p(t) = \frac{(1-\alpha)T^2\text{sinc}(\frac{t}{T}(1-\alpha)) + \frac{4}{\pi}\alpha T^2\cos(\frac{\pi t}{T}(1+\alpha))}{T^2 - 16\alpha^2 t^2} \tag{8}$$

where α is the roll-off factor of the RRC filter and T is the length of the subsymbol mentioned earlier, which means that the half-value bandwidth of this frequency domain RRC function is $\frac{1}{T}$, where $\frac{1}{T}$ represents the subband frequency domain bandwidth.

The time-domain RRC function is employed to truncate the temporal extent of the filter function. The time-domain RRC function is expressed as

$$q(t) = \begin{cases} 1, & 0 \le |t| < T_b(1-\beta) \\ \sqrt{\frac{1}{2}(1+\cos(\frac{|t|-T_b(1-\beta)}{2T_b\beta}\pi))}, & T_b(1-\beta) \le |t| < T_b(1+\beta) \\ 0, & |t| \ge T_b(1+\beta) \end{cases} \tag{9}$$

where β is the roll-off factor of the RRC function and T_b is half of the time-domain bandwidth of the RRC function. Therefore, the length of this time-domain RRC function is $2T_b(1+\beta)$. Then, the time domain form $\mathbf{g}(t)$ of the DRRC filter is expressed as

$$\mathbf{g}(t) = q(t)p(t). \tag{10}$$

The time-domain length of the filtering function is equal to the length obtained by repeating each subsymbol in L times, i.e., $2T_b(1+\beta) = LT$. As L increases, which means the time-domain length of the filtering function becomes larger, the filtering function approaches an ideal frequency-domain RRC function. This enables an ideal elimination of interference between the subbands.

2.3. Receiver

The detection of CP-OFDM can be directly used as the receiver of GFB-OFDM, as illustrated in Figure 3, which is validated by Theorem 1.

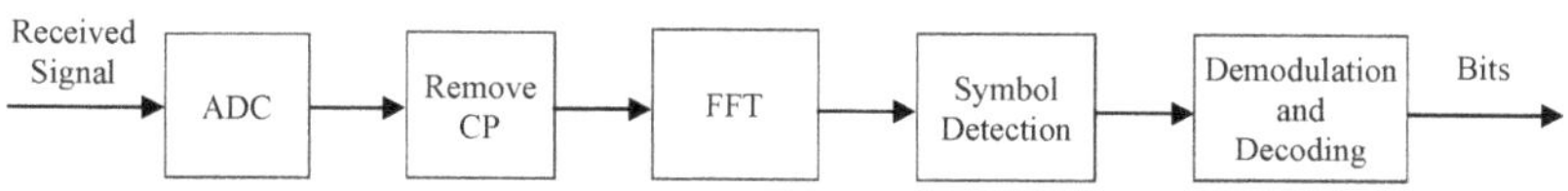

Figure 3. Processing at the receiver. The received time-domain signal is first converted into a digital signal via an ADC, followed by the removal of the cyclic prefix. After undergoing an FFT, the signal is converted to the frequency domain. In the frequency domain, it passes through the symbol detection module, as well as the demodulation and decoding module, resulting in the output information bits.

Theorem 1. *For a signal of length NM, the following two operations result in exactly the same output. The first operation: perform a NM-point IFFT; the second operation: perform N M-point IFFTs, M N-point IFFTs and polynomial filtering.*

Proof. Consider that $\mathbf{X}[l,k]$, consisting of NM information-bearing symbols, are transmitted on M subcarriers and N subbands. After performing first-level IFFT on N subbands, the data $x[m,k]$ are obtained as

$$x[m,k] = \frac{1}{M}\sum_{l=0}^{M-1} \mathbf{X}[l,k]e^{j2\pi\frac{ml}{M}} \tag{11}$$

where $j = \sqrt{-1}$, $m = 0,1,2,\ldots,M-1$. $l = 0,1,2,\ldots,M-1$ and $k = 0,1,2,\ldots,N-1$ denotes the index of subcarriers and subbands, respectively. Perform M times of second-level IFFTs on N points, resulting in $y[m,n]$, which can be expressed as

$$y[m,n] = \frac{1}{\sqrt{N}}\sum_{k=0}^{N-1} x[m,k]e^{j2\pi\frac{nk}{N}} \tag{12}$$

$$= \frac{1}{\sqrt{N}}\frac{1}{\sqrt{M}}\sum_{k=0}^{N-1}\sum_{l=0}^{M-1} \mathbf{X}[l,k]e^{j2\pi\frac{ml}{M}}e^{j2\pi\frac{nk}{N}} \tag{13}$$

$$= \frac{1}{\sqrt{NM}}\sum_{k=0}^{N-1}\sum_{l=0}^{M-1} \mathbf{X}[l,k]e^{j2\pi(\frac{ml}{M}+\frac{nk}{N})}. \tag{14}$$

For each subsymbol, repeat it L times to yield $y[m,n_{\text{ifft}}]$, where $LN = N_{\text{ifft}}$ and $n_{\text{ifft}} = 0,1,2,\ldots,N_{\text{ifft}}-1$.

$$y[m,n_{\text{ifft}}] = \frac{1}{\sqrt{N_{\text{ifft}}}}\sum_{k=0}^{N-1}\sum_{l=0}^{M-1} \mathbf{X}[l,k]e^{j2\pi(\frac{ml}{M}+\frac{n_{\text{ifft}}k}{N})}. \tag{15}$$

Here, the analysis should be conducted from a frequency domain perspective. Therefore, an FFT is performed on each symbol to obtain the frequency domain information $\mathbf{Y}[m,n_{\text{ifft}}]$, which is expressed as

$$\begin{aligned}
\boldsymbol{Y}[m, n_{\text{ifft}}] &= \frac{1}{\sqrt{N_{\text{ifft}}}} \sum_{n_i=0}^{N_{\text{ifft}}-1} \boldsymbol{y}[m, n_i] e^{-j2\pi \frac{n_i n_{\text{ifft}}}{N_{\text{ifft}}}} && (16)\\
&= \frac{1}{\sqrt{N_{\text{ifft}}}} \frac{1}{\sqrt{N_{\text{ifft}}}} \sum_{n_i=0}^{N_{\text{ifft}}-1} \sum_{k=0}^{N-1} \sum_{l=0}^{M-1} \boldsymbol{X}[l,k] e^{j2\pi(\frac{ml}{M} + \frac{n_i k}{N})} e^{-j2\pi \frac{n_i n_{\text{ifft}}}{N_{\text{ifft}}}} && (17)\\
&= \frac{1}{N_{\text{ifft}}} \sum_{k=0}^{N-1} \sum_{l=0}^{M-1} \boldsymbol{X}[l,k] e^{j2\pi \frac{ml}{M}} \sum_{n_i=0}^{N_{\text{ifft}}-1} e^{j2\pi(\frac{n_i k}{N} - \frac{n_i n_{\text{ifft}}}{N_{\text{ifft}}})} && (18)\\
&= \frac{1}{N_{\text{ifft}}} \sum_{k=0}^{N-1} \sum_{l=0}^{M-1} \boldsymbol{X}[l,k] e^{j2\pi \frac{ml}{M}} \sum_{n_i=0}^{N_{\text{ifft}}-1} e^{j2\pi n_i \frac{kM - n_{\text{ifft}}}{N_{\text{ifft}}}} && (19)\\
&= \sum_{k=0}^{N-1} \sum_{l=0}^{M-1} \boldsymbol{X}[l,k] e^{j2\pi \frac{ml}{M}} \delta[(kM - n_{\text{ifft}})_{N_{\text{ifft}}}]. && (20)
\end{aligned}$$

where "$[\cdot]_{N_{\text{ifft}}}$" denotes the modulo N_{ifft} operation and $\delta[\cdot]$ denotes the Dirac delta function. We can the obtain

$$\boldsymbol{Y}[m, n_{\text{ifft}}] = \begin{cases} \sum_{l=0}^{M-1} \boldsymbol{X}[l, \frac{n_{\text{ifft}}}{M}] e^{j2\pi \frac{ml}{M}}, & n_{\text{ifft}} = kM \\ 0, & n_{\text{ifft}} \neq kM. \end{cases} \quad (21)$$

Then, $\boldsymbol{Y}[m,p]$ is convolved with the rectangular filter $\text{FIL}[p]$. At the mean time, the temporal misalignment process is transformed into a phase rotation in the frequency domain. Then, we can obtain

$$\begin{aligned}
\boldsymbol{W}[m,b] &= (\boldsymbol{Y}[m,p] \otimes \text{FIL}[p]) e^{-j2\pi \frac{mN}{N_{\text{ifft}}} b} && (22)\\
&= \sum_{p=0}^{N_{\text{ifft}}-1} \boldsymbol{Y}[m,p] |\text{FIL}[b-p]| e^{-j2\pi \frac{mN}{N_{\text{ifft}}} b} && (23)\\
&= \sum_{\substack{p=0\\p=kM}}^{N_{\text{ifft}}-1} \sum_{l=0}^{M-1} \boldsymbol{X}[l, \frac{p}{M}] e^{j2\pi \frac{ml}{M}} |\text{FIL}[b-p]| e^{-j2\pi \frac{mN}{N_{\text{ifft}}} b} && (24)\\
&= \sum_{\substack{p=0\\p=kM}}^{N_{\text{ifft}}-1} \sum_{l=0}^{M-1} \boldsymbol{X}[l, \frac{p}{M}] |\text{FIL}[b-p]| e^{j2\pi m \frac{l-b}{M}} && (25)
\end{aligned}$$

where $b = 0,1,2,\ldots,N_{\text{ifft}} - 1$ and $p = 0,1,2,\ldots,N_{\text{ifft}} - 1$. "$\otimes$" and "$|\cdot|$" denotes the circular convolution symbol and the module value, respectively. $\text{FIL}[p] = 0$ when $p \geq M$, and $\text{FIL}[p] = \frac{1}{M}$ when $0 \leq p \leq M-1$. Sum the frequency domain data of different subsymbols to obtain

$$\begin{aligned}
\mathbf{Z}[b] &= \sum_{m=0}^{M-1} \boldsymbol{W}[m,b] && (26)\\
&= \sum_{\substack{p=0\\p=kM}}^{N_{\text{ifft}}-1} \sum_{l=0}^{M-1} \boldsymbol{X}[l, \frac{p}{M}] |\text{FIL}[b-p]| \sum_{m=0}^{M-1} e^{j2\pi m \frac{l-b}{M}} && (27)\\
&= M \sum_{\substack{p=0\\p=kM}}^{N_{\text{ifft}}-1} \sum_{l=0}^{M-1} \boldsymbol{X}[l, \frac{p}{M}] |\text{FIL}[b-p]| \delta[(l-b)_M] && (28)\\
&= M \sum_{\substack{p=0\\p=kM}}^{N_{\text{ifft}}-1} \boldsymbol{X}[(b)_M, \frac{p}{M}] |\text{FIL}[b-p]| && (29)\\
&= M \sum_{k=0}^{N-1} \boldsymbol{X}[(b)_M, k] |\text{FIL}[b-kM]| && (30)\\
&= \boldsymbol{X}[(b)_M, \lfloor \frac{b}{M} \rfloor] && (31)
\end{aligned}$$

where $\lfloor \cdot \rfloor$ denotes the rounding down operation. At this time, it means that the modulation symbols are completely recovered from the data $\mathbf{Z}[b]$. $\square$

The receiver first removes the cyclic prefix of each received OFDM symbol, followed by the FFT to recover the modulated symbols. Subsequently, the symbol detection, the demodulation, and the decoding are performed to ultimately retrieve the information elements. Note that the number of repetition L affects the inference between the subsymbols and the one between the subbands. That is, as L increases, the interference between the subsymbols becomes larger, the interference between the subbands becomes smaller, and the computational complexity of Equation (7) becomes higher. To reduce the complexity and the overhead of the cyclic prefix, a small L is preferred; however, it may introduce the interference between the subcarriers. The simulation results in Section 3 show that the interference between the subbands is tolerable when $L \geq 4$, and the interference between the subsymbols can be ignored. In this case, the receiver can receive in an almost lossless manner.

2.4. Receiver Analysis and Its Extension

In this section, we discuss a special case of Theorem 1 where oversampling and different filter lengths are considered. Similar to previous discussions, the total number of IFFT points is N_{ifft} and the data bandwidth is divided into N subbands as shown in Figure 4, each with M subcarriers. N_2 represents the second-level IFFT point number, where N_{ifft}, M, and N_2 are all positive integer power of two. The length of the filter is LN_2.

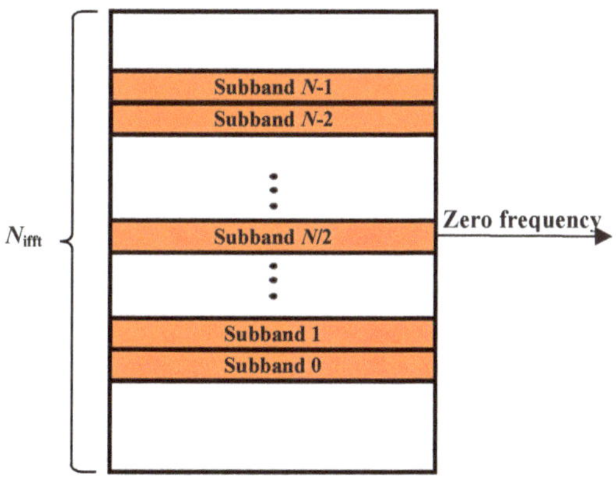

Figure 4. The locations of the N subbands in the frequency domain.

Firstly, let us analyze the case where $LN_2 = N_{\text{ifft}}$ for filter length selection. Due to the characteristic of orthogonal frequency domain subcarriers, we only need to track the process experienced by data on one subcarrier in the GFB-OFDM waveform at its transmission end in order to easily extend it to the transmission processes of all subcarriers. Assuming that one data d is transmitted on subcarrier P in subband Q, then we can analyze the transmission data processing of GFB-OFDM according to the following steps:

(1) The data $\boldsymbol{X}_{M\times N}$ across N subbands and M subcarriers can be expressed as

$$\mathbf{X}[m,n] = \begin{cases} d, & m = P \text{ and } n = Q \\ 0, & m \neq P \text{ or } n \neq Q. \end{cases} \tag{32}$$

(2) After performing first-level IFFT on N subbands, the data $\boldsymbol{x}_{M\times N}$ is obtained as

$$\boldsymbol{x}[m,n] = \text{IFFT}(\mathbf{X}[m,n]) = \begin{cases} \frac{d}{\sqrt{M}} e^{j2\pi\frac{Pm}{M}}, & n = Q \\ 0, & n \neq Q. \end{cases} \tag{33}$$

(3) Perform M times of second-level IFFTs on N_2 points, resulting in $y_{M\times N_2}$. The position of the subbands determines the mapping location for the data in the second-level IFFT, where $n_2 = 0,1,2,\dots,N_2-1$. Then, $y_{M\times N_2}$ can be expressed as

$$y[m,n_2] = \frac{d}{\sqrt{N_{\text{ifft}}}} e^{j2\pi(\frac{Pm}{M}+n_2\frac{N_s+Q}{N_2})}. \tag{34}$$

(4) For each subsymbol, repeat it L times to yield $y_{M\times LN_2}$. According to this assumption, $LN_2 = N_{\text{ifft}}$ can be written as $y_{M\times N_{\text{ifft}}}$, where $n_{\text{ifft}} = 0,1,2,\dots,N_{\text{ifft}}-1$.

$$y[m,n_{\text{ifft}}] = \frac{d}{\sqrt{N_{\text{ifft}}}} e^{j2\pi(\frac{Pm}{M}+n_{\text{ifft}}\frac{N_s+Q}{N_2})}. \tag{35}$$

(5) Here, the analysis should be conducted from a frequency domain perspective because it is believed that the frequency domain information after undergoing second-level IFFT should still contain the information from the original subcarriers. Therefore, an FFT is performed on each symbol to obtain the frequency domain information $\mathbf{Y}_{M\times N_{\text{ifft}}}$, which is expressed as

$$\mathbf{Y}[m,n_{\text{ifft}}] = \text{FFT}(y_{M\times N_{\text{ifft}}}) \tag{36}$$

$$= \frac{d}{\sqrt{N_{\text{ifft}}}} \sum_{n_i=0}^{N_{\text{ifft}}-1} e^{j2\pi(\frac{Pm}{M}+n_i\frac{N_s+Q}{N_2})} \frac{1}{\sqrt{N_{\text{ifft}}}} e^{-\frac{j2\pi n_i n_{\text{ifft}}}{N_{\text{ifft}}}} \tag{37}$$

$$= \frac{d}{N_{\text{ifft}}} \sum_{n_i=0}^{N_{\text{ifft}}-1} e^{j2\pi\frac{Pm}{M}} e^{j2\pi n_i(\frac{(N_s+Q)M}{N_2 M}-\frac{n_{\text{ifft}}}{N_{\text{ifft}}})}. \tag{38}$$

We can obtain

$$\mathbf{Y}[m,n_{\text{ifft}}] = \begin{cases} de^{j2\pi\frac{Pm}{M}}, & n_{\text{ifft}} = (N_s+Q)M \\ 0, & n_{\text{ifft}} \neq (N_s+Q)M. \end{cases} \tag{39}$$

(6) $\mathbf{Y}_{M\times N_{\text{ifft}}}$ is convolved with the rectangular filter FIL. At the mean time, the temporal misalignment process is transformed into a phase rotation in the frequency domain. Then, we can obtain

$$\mathbf{W}[m,b] = (\mathbf{Y}[m,(N_s+Q)M] \otimes \text{FIL})e^{-j2\pi\frac{mN_2}{N_{\text{ifft}}}b} \tag{40}$$

$$= d|\text{FIL}[b]|e^{j2\pi\frac{Pm}{M}} e^{-j2\pi\frac{mN_2}{N_{\text{ifft}}}b} \tag{41}$$

$$= d|\text{FIL}[b]|e^{j2\pi m\frac{P-b}{M}} \tag{42}$$

where $b = 0,1,2,\dots,N_{\text{ifft}}-1$ represents subcarriers in the frequency domain, "$\otimes$" is the circular convolution symbol, and "$|\cdot|$" denotes the module value.

(7) Sum the frequency domain data of different subsymbols to obtain $\mathbf{Z}_{N_{\text{ifft}}\times 1}$

$$\mathbf{Z}[b] = \sum_{m=0}^{M-1} \mathbf{W}[m,b] \tag{43}$$

$$= d|\text{FIL}[b]| \sum_{m=0}^{M-1} e^{j2\pi m\frac{P-b}{M}}. \tag{44}$$

Since the purpose of polyphase filtering is to filter data of each subband separately, it can be understood that the frequency domain carrier positions corresponding to the passband bandwidth of the filter are $b = \left[(N_s+Q)M-\frac{M}{2},\ (N_s+Q)M+\frac{M}{2}-1\right]$. It should be noted that the positions of the filters differ for different subbands. Additionally, because we have normalized the peak value of our polyphase filtering filters in the time domain, and because their frequency domain bandwidth corresponds to the subband width, we can obtain the frequency domain coefficients of these filters.

$$|\mathrm{FIL}[b]| = \frac{1}{M}. \tag{45}$$

According to (40), it can be similarly derived that, when $b = (N_s + Q)M - \frac{M}{2} + P$, we have

$$\mathbf{Z}\left[(N_s + Q)M - \frac{M}{2} + P\right] = d. \tag{46}$$

At this time, it means that the data on this subcarrier are represented as d, while the data on other subcarriers are zero. The time-domain information corresponding to $\mathbf{Z}_{N_{\mathrm{ifft}}\times 1}$, denoted as $z_{N_{\mathrm{ifft}}\times 1}$, represents the time-domain data after performing IFFT. At the receiver, CP-OFDM can be used to receive data in the frequency domain, and the data on each subcarrier satisfies orthogonality. This is consistent with the conclusion of Theorem 1.

More generally, when the length of the filter is $L(1 \le L \le \frac{N_{\mathrm{ifft}}}{N_2})$ and it is a DRRC filter with double roots, (35) becomes

$$\boldsymbol{y}[m, n_{\mathrm{ifft}}] = \begin{cases} \frac{d}{\sqrt{N_{\mathrm{ifft}}}} e^{\mathrm{j}2\pi(\frac{Pm}{M} + n_{\mathrm{ifft}}\frac{N_s+Q}{N_2})}, & n_{\mathrm{ifft}} < LN_2 \\ 0, & n_{\mathrm{ifft}} \ge LN_2. \end{cases} \tag{47}$$

Assuming the filter coefficients of LN_2 points, the total time-domain filter coefficients for one OFDM symbol length are given by fil $= [\mathrm{fil0}, \mathrm{zeros}(N_{\mathrm{ifft}} - LN_2, 1)]$. Then, the point-wise multiplication result of the symbol m is subjected to FFT and symbol shifting and we can obtain

$$\mathbf{W}[m, b] = \mathrm{FFT}(\boldsymbol{y}[m, n_{\mathrm{ifft}}] \times \mathrm{fil}(n_{\mathrm{ifft}}))e^{-\mathrm{j}2\pi\frac{mN_2}{N_{\mathrm{ifft}}}b} \tag{48}$$

$$= de^{\mathrm{j}2\pi\frac{(P-b)m}{M}}\mathrm{FFT}(e^{\mathrm{j}2\pi M n_{\mathrm{ifft}}\frac{N_s+Q}{MN_2}} \times \mathrm{fil}(n_{\mathrm{ifft}})) \tag{49}$$

$$= de^{\mathrm{j}2\pi\frac{(P-b)m}{M}}\mathrm{FIL}([b - M(N_s + Q)]_{N_{\mathrm{ifft}}}) \tag{50}$$

$$= de^{\mathrm{j}2\pi\frac{(P-b)m}{M}}|\mathrm{FIL}([b - M(N_s + Q)]_{N_{\mathrm{ifft}}})| \tag{51}$$

It should be noted that the peak value of the filter in the time domain occurs at $t = 0$ and that its frequency-domain coefficients are real numbers. The data of different subsymbols are then superimposed in the frequency domain.

$$\mathbf{Z}[b] = \sum_{m=0}^{M-1} \mathbf{W}[m, b] \tag{52}$$

$$= d|\mathrm{FIL}([b - M(N_s + Q)]_{N_{\mathrm{ifft}}})| \sum_{m=0}^{M-1} e^{\mathrm{j}2\pi m\frac{P-b}{M}}. \tag{53}$$

It is easy to determine that the passband of the filter is $b = [(N_s + Q)M - \frac{M}{2}, (N_s + Q)M + \frac{M}{2} - 1]$, so $\mathbf{Z}[b]$ only has a value when $b = (N_s + Q)M - \frac{M}{2} + P$, which is

$$\mathbf{Z}\left[(N_s + Q)M - \frac{M}{2} + P\right] = d|\mathrm{FIL}([P - \frac{M}{2}]_{N_{\mathrm{ifft}}})|. \tag{54}$$

It effectively filters the data on the original subcarrier while setting the information on other subcarriers to zero, achieving the effect of a two-level IFFT conversion with a first-level IFFT and filtering. Each subband is filtered individually, and then all different subbands are superimposed in the time domain.

2.5. *Advantages*

GFB-OFDM utilizes the two-level IFFTs to decompose the original large-size IFFT into two small-size IFFTs, enabling the operations of large-size IFFT dedicated to the ultra-wide bandwidth scenarios. Furthermore, the GFB-OFDM waveform possesses the following advantages:

(1) By alternating the number of subbands while fixing the subband bandwidth, it is possible to achieve data services in different channel bandwidths. Since the filtering is performed on each subbband, a single filter is sufficient. In contrast, each service with a different bandwidth required a different filter in CP-OFDM. This facilitates the flexible support for various channel bandwidth requirements in certain hardware scenarios.
(2) The polyphase filtering in GFB-OFDM efficiently employs an individual filter for each subband, thus mitigating the potential occurrence of the out-of-band leakage of the subbands. Therefore, it can accommodate various scenarios where different subcarrier spacings are used for different subbands, and these diverse subcarrier spacings can be uniformly processed using GFB-OFDM waveform processing.
(3) Different types of data can be assigned to different subbands within a single GFB-OFDM symbol, and they can be modulated either in a multicarrier manner or a single-carrier manner. The submodule for per subband processing in GFB-OFDM enables a simultaneous processing of different types of data-bearing subbands, achieving an integration of single-carrier processing and the multi-carrier processing.
(4) The receiver of the 4G/5G OFDM is the legacy of the proposed GFB-OFDM waveform without introducing any additional complexity.

3. Simulation Results

In this section, the performance evaluation and the analysis of the GFB-OFDM and the CP-OFDM of 5G NR are provided. In Table 1, it shows the simulation parameter setup. It is assumed that the CP-OFDM waveform can support an IFFT of 16,384-point (double bandwidth oversampled IFFT).

Table 1. Simulation Parameters.

Parameters	CP-OFDM	GFB-OFDM
Total number of subcarriers	8192	8192
Number of subbands	8	8
Number of subcarriers in each subband	1024	1024
Size of total IFFT	16,384	16,384
Size of first-level IFFT	/	2048
Size of second-level IFFT	/	16
Subcarrier spacing	480 kHz, 960 kHz	480 kHz, 960 kHz
Filter bandwidth	/	1 times the subband width
The length of the time domain filter	/	4 times the subsymbol length
Roll down factor α of frequency domain DRRC filter	/	0.15
Roll down factor β of the DRRC window filter	/	0.5
MCS	3/4 LDPC, 64 QAM	3/4 LDPC, 64 QAM
Length of CP	1/16 of the symbol length	1/16 of the symbol length
Channel	AWGN	AWGN

3.1. *Out-Of-Band Leakage*

The power spectral density (PSD) of CP-OFDM and GFB-OFDM are shown in Figure 5, with eight subbands and one subband chosen, respectively, as in Table 1. Due to the implementation of the polyphase filtering, the GFB-OFDM exhibits lower out-of-band leakage compared to CP-OFDM, as evident from the results in Figure 5a. This reduction in the out-of-band leakage allows for a decrease in the guard bandwidth and an improvement in the spectral efficiency. In Figure 5b, the out-of-band leakage of one subband is depicted, which reflects the inter-subband interference.

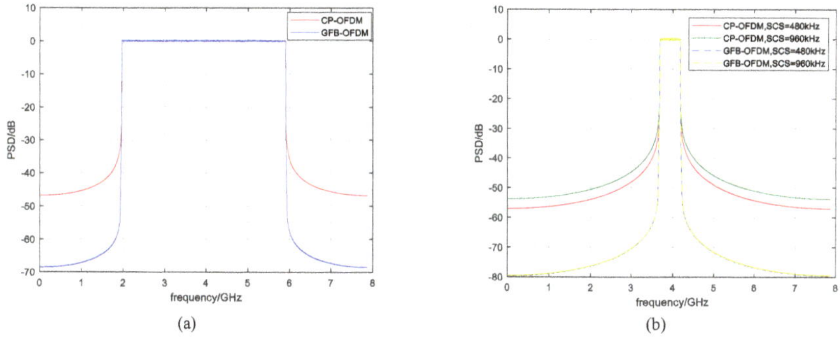

Figure 5. The out-of-band leakage of CP-OFDM and GFB-OFDM is evaluated. Specifically, the number of the subbands are 8 and 1 in the corresponding (**a**,**b**). The results demonstrate that the out-of-band leakage of GFB-OFDM is lower than the OFDM because of the polyphase filter operation.

It can be observed that the out-of-band leakage of one subband is significantly higher than that of GFB-OFDM. Specifically, the subband with a 960 kHz subcarrier spacing demonstrates heavier out-of-band leakage than that with a 480 kHz subcarrier spacing. Conversely, in GFB-OFDM, the out-of-band leakage across different subcarrier spacings remains consistently low.

3.2. BLER in Same Subcarrier Spacing

Let all subcarriers in the channel bandwidth have an identical subcarrier spacing of 480 kHz. In order to reduce the interference between subbands in GFB-OFDM, the first-level IFFT is oversampled with the double bandwidth, i.e., a 2048-point IFFT. Traditional CP-OFDM is adopted. Figure 6 presents the BLER of CP-OFDM and GFB-OFDM, showing that the BLER performance of the two waveforms are almost the same. This result confirms that the inter-subband interference can be ignored and that there is no degradation in the BLER of GFB-OFDM when $L \geq 4$.

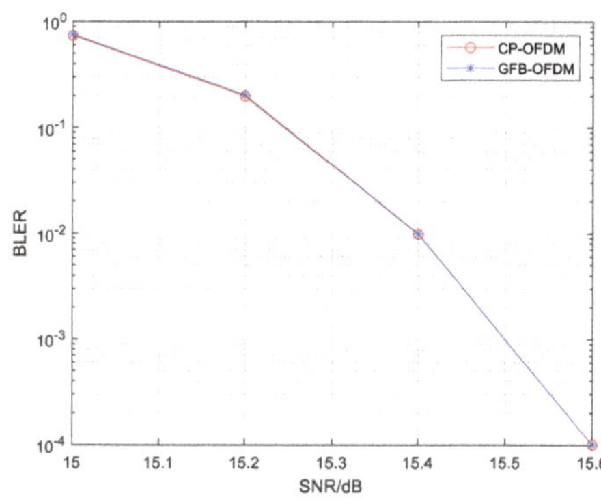

Figure 6. The The BLER of CP-OFDM and GFB-OFDM is evaluated for the same subcarrier spacing scenario. The rate-$\frac{3}{4}$ LDPC channel coding, the 64QAM modulation and the AWGN channel are employed. The result demonstrates that GFB-OFDM maintains a nearly identical BLER performance compared to CP-OFDM.

3.3. BLER in Different Subcarrier Spacing

When different subcarrier spacings exist within the channel bandwidth, a coexistence case of the subbands with different subcarrier spacings is shown in Figure 7a, where Subband 3 has a subcarrier spacing of 960 kHz while the rest have a spacing of 480 kHz. The traditional CP-OFDM cannot process the data from different subbands together during the IFFT. However, the proposed GFB-OFDM can handle all subbands together, demonstrating flexibility for various requirements. The specific procedure is illustrated in Figure 8. Specifically, after performing the first-level IFFT over the adjacent symbols of Subband 3 separately and then adding CP, they are concatenated and then serve as the input of the second-level IFFT along with the other subbands. In this simulation, the CP-OFDM waveform is obtained by the oversampled IFFT operations on the data with a subcarrier spacing of 480 kHz and 960 kHz, respectively, using 16,384 points. Subsequently, the two sets of the time-domain data are combined into one set.

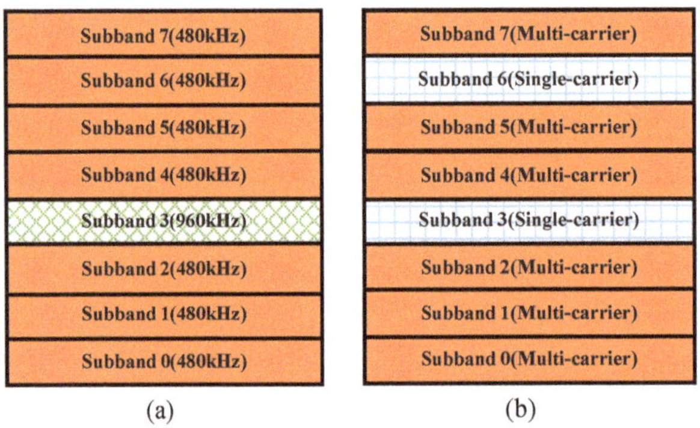

Figure 7. The locations of the eight subbands in the frequency domain under two cases. (**a**) represents the mixture scenarios of different subcarrier spacings, which were 480 kHz and 960 kHz, respectively. (**b**) represents single-carrier and multi-carrier mixture scenarios.

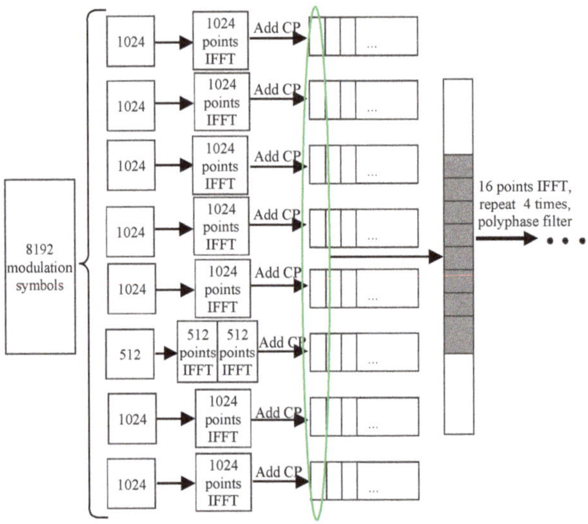

Figure 8. Subband data-processing procedure for different subcarrier spacing with the number of modulated symbols $K = 8192$, the number of subbands $N = 8$, and the number of subcarriers in each subband $M = 1024$ or $M = 512$.

Figure 9 presents a comparison of the BLER performance in demodulating data transmitted on subband 3. Due to the interference from the other subbands with different subcarrier spacings, the BLER performance of subband 3 degrades. It can be also observed that GFB-OFDM outperforms CP-OFDM since GFB-OFDM employs the polyphase filtering, resulting in low out-of-band leakage for each subband and thus smaller inter-subband interference compared to the CP-OFDM. Therefore, GFB-OFDM outperforms CP-OFDM in BLER in the different subcarrier spacing scenario, which facilitates a dynamic and flexible adjustment scheme for the subcarrier spacing between the subbands in regard to 6G application.

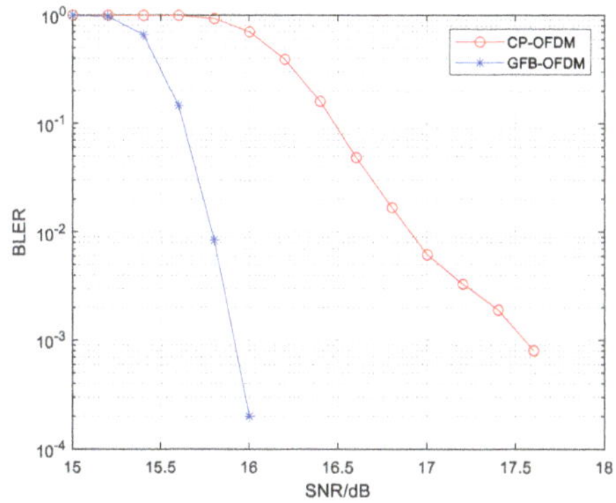

Figure 9. The BLER of CP-OFDM and GFB-OFDM is evaluated in different subcarrier spacing scenarios. The rate-$\frac{3}{4}$ LDPC channel coding, the 64 QAM modulation, and the AWGN channel are employed. Specifically, subbands with subcarrier spacings of 480 kHz and 960 kHz coexist. The result demonstrates that the BLER performance of GFB-OFDM is significantly superior to that of CP-OFDM.

3.4. BLER in Mixture Scenarios of Single-Carrier and Multi-Carrier

In the mixture scenario of the single-carrier and the multi-carrier modulation, as in Figure 7b, subband 3 and subband 6 are the single-carrier subbands; the remaining subbands are multi-carrier subbands with subcarrier spacings of 480 kHz which employ an oversampling IFFT operation with 16,384 points. CP-OFDM requires alternative processing for the single-carrier and the multi-carrier subbands, followed by time-domain superposition, with each single-carrier subband being sent individually. In contrast, the proposed GFB-OFDM scheme allows any subband to be used as either a single-carrier or a multi-carrier subband without increasing any complexity, enabling all single-carrier and multi-carrier subbands to be processed simultaneously. The specific process is illustrated in Figure 10.

Figure 11 presents a comparison of the BLER performance for demodulating the data transmitted on subband 3 at the receiver. Due to interference from other subbands, subband 3 experiences some degradation in terms of BLER performance. From the comparison results shown in the figure, it can be observed that GFB-OFDM outperforms CP-OFDM to some extent. This is mainly because GFB-OFDM has a polyphase filter effect at the transmitter, resulting in low out-of-band leakage for each subband and thus smaller inter-subband interference compared to CP-OFDM, leading to a better BLER performance. Therefore, in scenarios involving a mixture of single-carrier and multi-carrier transmissions, GFB-OFDM exhibits slightly better performance than CP-OFDM while maintaining lower processing complexity and greater flexibility.

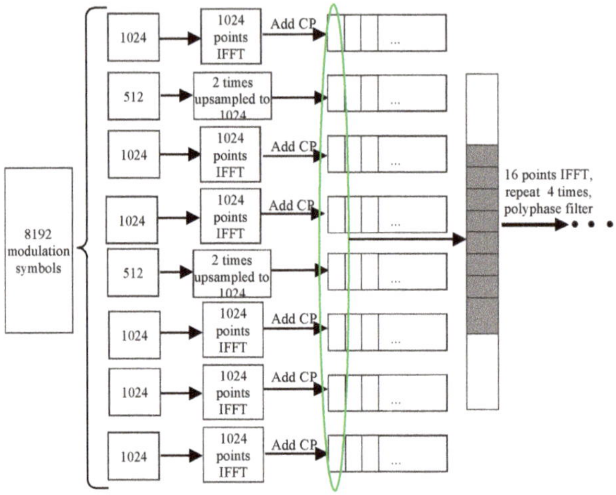

Figure 10. Subband data-processing procedure for a single-carrier and multi-carrier mixture scenario with the number of modulated symbols $K = 8192$, the number of subbands $N = 8$ and the number of subcarriers in each subband $M = 1024$ or $M = 512$.

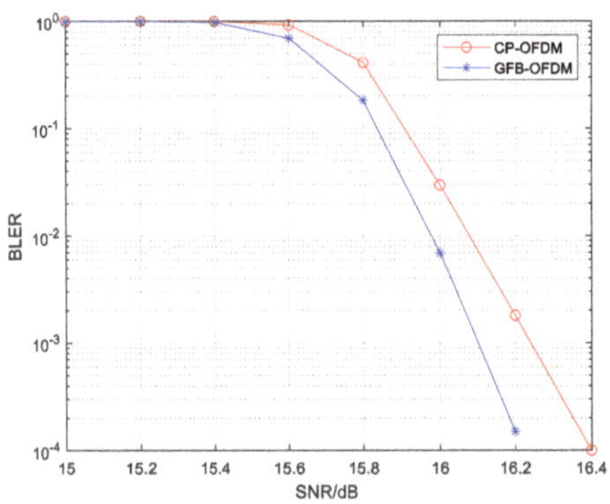

Figure 11. The BLER of CP-OFDM and GFB-OFDM is evaluated in single-carrier and multi-carrier mixture scenarios. The rate-$\frac{3}{4}$ LDPC channel coding, the 64 QAM modulation, and the AWGN channel are employed. The result demonstrates that the BLER performance of GFB-OFDM is superior to that of CP-OFDM.

3.5. Complexity Analysis

In general, the complexity of a multiplier is significantly higher than that of an adder. Hence, we focus on the comparison of the multiplication between CP-OFDM and GFB-OFDM. In addition, considering the flexibility in the subcarrier spacing adjustment, the CP-OFDM necessitates the inclusion of the filters for each subband. These filters are positioned differently to correspond to the frequency range of each subband respectively. Note that, since all subbands are the multi-carrier subbands, it illustrates the transmission process of CP-OFDM in Figure 12. Both GFB-OFDM and CP-OFDM require adding CP with a similar complexity. Thus, we can ignore the presence of CP during the comparison.

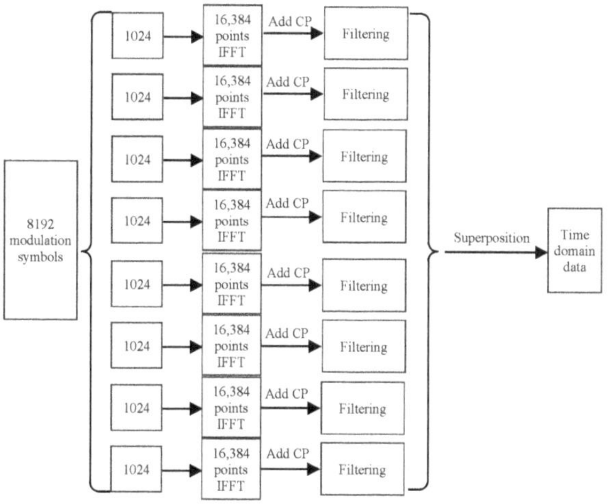

Figure 12. CP-OFDM data processing with filtering. The number of modulated symbols, subbands, and subcarriers are set as $K = 8192$, $N = 8$ and $M = 1024$.

First, we analyze the number of multiplications involved in the IFFT and the filtering of GFB-OFDM. The GFB-OFDM employs the two-level IFFTs, where it can be deduced from the processing procedure that the first-level IFFT is of M points. Hence, the overall number of multiplications in N subbands for first-level IFFT processing is $\frac{NM}{2}\log M$ [25]. The number of multiplications in M second-level IFFT points is $\frac{N_{\text{ifft}}}{M}$, so the total number of multiplications in M second-level IFFT is $\frac{N_{\text{ifft}}}{2}\log\frac{N_{\text{ifft}}}{M}$. In addition, the number of multiplications in the polyphase filtering process of GFB-OFDM can be determined using the following calculation. Considering that a second-level IFFT has $\frac{N_{\text{ifft}}}{M}$ points, the length of time-domain data after IFFT is $\frac{N_{\text{ifft}}}{M}$. Therefore, by repeating these data L times and performing point multiplication and windowing operations, the total number of multiplications becomes $\frac{LN_{\text{ifft}}}{M}$. Consequently, for one OFDM symbol, the overall number of multiplications required for point multiplication and windowing operations is $\frac{M \times LN_{\text{ifft}}}{M} = LN_{\text{ifft}}$.

Hence, the total number of multiplications involved in IFFT and filtering processing for GFB-OFDM is as follows:

$$\mathcal{O}(\text{GFB-OFDM}) = \frac{NM}{2}\log M + \frac{N_{\text{ifft}}}{2}\log\frac{N_{\text{ifft}}}{M} + LN_{\text{ifft}}. \tag{55}$$

We then analyze the number of multiplications involved in the IFFT and filtering process of CP-OFDM. The number of multiplications in the IFFT processing of CP-OFDM is $\frac{N_{\text{ifft}}N}{2}\log N_{\text{ifft}}$, which is determined based on N_{ifft}-point IFFT with N subbands.

The calculation process for the number of multiplications in the time-domain convolution filtering of CP-OFDM is as follows. The number of time-domain data points for one OFDM symbol is N_{ifft}, and the number of points included in the time-domain length of the filter function is $\frac{LN_{\text{ifft}}}{M}$. Therefore, the total number of multiplications for N subbands processed by time-domain convolution filtering is $\frac{NLN_{\text{ifft}}^2}{M}$. Hence, the total number of multiplications in the IFFT and filtering processing for CP-OFDM is

$$\mathcal{O}(\text{OFDM}) = \frac{N_{\text{ifft}}N}{2}\log N_{\text{ifft}} + \frac{NLN_{\text{ifft}}^2}{M}. \tag{56}$$

The calculation process above ignores cyclic prefix because its length has a minimal impact on the overall complexity. It is easy to observe that (36) is much smaller than (37). Figure 13 illustrates the relationship between multiplication complexity and the number of subbands for two waveforms, assuming $N_{\text{ifft}} = 16{,}384$, $L = 4$, the total number of subcarriers is 8192, and that it is divided into N subbands. It can be seen that GFB-OFDM has lower complexity compared to CP-OFDM and that the difference becomes more pronounced as the number of subbands N increases. For example, using the calculations in Section 3.2 as an example, the number of multiplications for transmitting one OFDM symbol in CP-OFDM and GFB-OFDM are approximately 9.0×10^6 and 1.1×10^5, respectively.

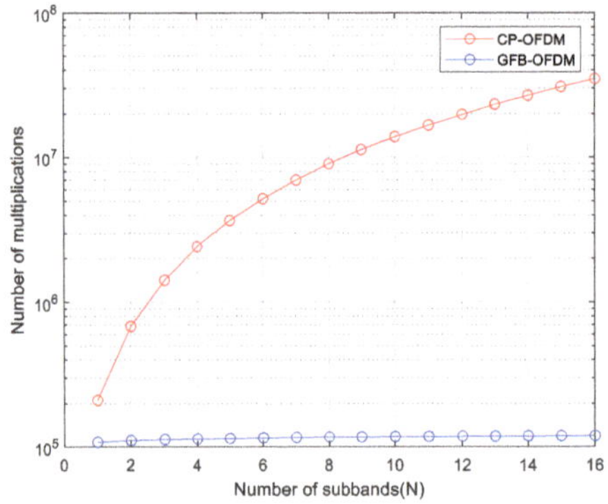

Figure 13. The complexity of CP-OFDM and GFB-OFDM is evaluated under a varying number of subbands. Under different numbers of subbands, the number of multiplications required by GFB-OFDM is less than that of CP-OFDM. In addition, as the number of subbands increases, the number of multiplications required by CP-OFDM increases rapidly. In contrast, the number of multiplications required by GFB-OFDM increases slowly.

As previously mentioned, GFB-OFDM exhibits lower complexity compared to CP-OFDM due to the following reasons:

(1) In CP-OFDM, subbands with different subcarrier spacings require independent processing and are then combined in the time domain using IFFT, leading to increased computational complexity. As the number of subbands requiring separate processing grows, the overall complexity rises substantially.

(2) Conversely, GFB-OFDM enables uniform processing of subbands with different subcarrier spacings, providing greater flexibility in their division. Modifying the size or increasing the number of subband divisions does not augment processing complexity while maintaining a constant bandwidth.

4. Conclusions

In this paper, we proposed the new waveform GFB-OFDM, which could be a candidate for 6G communication. It utilizes two levels of IFFTs, both in small sizes, and adopts the polyphase filtering to attain the same effect of the corresponding IFFT in the large size, which supports wideband communication. By adjusting the number of subbands for a fixed subband bandwidth, it enables data transmission in different channel bandwidths and thus facilitates a flexible support for various service requirements regarding hardware constraints. It can further adapt to the scenarios with various subcarrier spacings for each subband and accommodate both the single-carrier modulation and the multi-carrier modulation without increasing any computational complexity. We then provided the analysis and the simulation, and they verify that GFB-OFDM demonstrates the aforementioned advantages in the THz high-bandwidth communication and in the varying subcarrier spacing requirements. In future 6G communications that cover numerous complex application scenarios, different performance indicators could vary according to the specific applications. The proposed GFB-OFDM is capable of supporting the characteristic requirements of ultra-wide bandwidth, ultra-high speed, different data services, different subcarrier spacings, and different channel bandwidths in various scenarios and demands.

Author Contributions: Conceptualization, Y.X., J.H. and T.B.; Methodology, Y.X., J.H. and Y.H.; Software, J.H.; Validation, Y.X., T.B., J.H. and F.W.; Formal analysis, Y.H. and F.W.; Investigation, Z.X.; Resources, Y.X. and J.H.; Writing—original draft, J.H. and Y.H.; Writing – review and editing, Y.H. and F.W.; Visualization, X.N. All authors have read and agreed to the published version of the manuscript.

Funding: This work was supported by the ZTE Industry-University-Institute Cooperation Funds under Grant No. IA20240319006.

Institutional Review Board Statement: Not applicable.

Informed Consent Statement: Not applicable.

Data Availability Statement: Data are contained within the article.

Conflicts of Interest: The authors Yu Xin, Jian Hua and Tong Bao were employed by the ZTE Corporation, The remaining authors declare that the research was conducted in the absence of any commercial or financial relationships that could be construed as a potential conflict of interest. The authors declare that this study received funding from ZTE Industry-University-Institute Cooperation Funds. The funding sponsors had no role in the design of the study; in the collection, analyses, or interpretation of data; in the writing of the manuscript, and in the decision to publish the results.

References

1. Feng, W.; Wei, S.; Cao, J. 6G Technology Development Vision and Terahertz Communication. *Acta Phys. Sin.* **2021**, *70*, 244–303. [CrossRef]
2. Hosako, I.; Sekine, N.; Patrashin, M.; Saito, S.; Fukunaga, K.; Kasai, Y.; Baron, P.; Seta, T.; Mendrok, J.; Ochiai, S.; et al. At the Dawn of a New Era in Terahertz Technology. *Proc. IEEE* **2007**, *95*, 1611–1623. [CrossRef]
3. Yrjölä, S.; Matinmikko-Blue, M.; Ahokangas, P. Developing 6G Visions with Stakeholder Analysis of 6G Ecosystem. In Proceedings of the 2023 Joint European Conference on Networks and Communications & 6G Summit (EuCNC/6G Summit), Gothenburg, Sweden, 6–9 June 2023; pp. 705–710.
4. Inomata, M.; Yamada, W.; Kuno, N.; Sasaki, M.; Kitao, K.; Nakamura, M.; Ishikawa, H.; Oda, Y. Terahertz Propagation Characteristics for 6G Mobile Communication Systems. In Proceedings of the 2021 15th European Conference on Antennas and Propagation (EuCAP), Dusseldorf, Germany, 22–26 March 2021; pp. 1–5.
5. Sun, B.; Yu, X.; Chen, M.; Jiang, F.; Zhu, Z.; Zheng, Z. Research on Terahertz Communication System and the Prospect on Space Applications. In Proceedings of the 2023 IEEE 15th International Conference on Advanced Infocomm Technology (ICAIT), Hefei, China, 13–16 October 2023; pp. 67–72.
6. Rajeswari, L.M.; Manocha, S.K. Design of data adaptive IFFT/FFT block for OFDM system. In Proceedings of the 2011 Annual IEEE India Conference, Hyderabad, India, 15–17 December 2011; pp. 1–5.
7. Xin, Y.; Bao, T.; Hua, J.; Yu, G. A Novel Waveform Scheme for THz Communications. In Proceedings of the 2021 IEEE 94th Vehicular Technology Conference (VTC2021-Fall), Norman, OK, USA, 27–30 September 2021; pp. 1–6.
8. Yang, F.; Wang, Y.; Ding, L.; Qian, L. An improved equalization with real interference prediction scheme of the FBMC/OQAM system. *China Commun.* **2021**, *18*, 120–129. [CrossRef]
9. Gaspar, I.; Matthé, M.; Michailow, N.; Mendes, L.L.; Zhang, D.; Fettweis, G. Frequency-Shift Offset-QAM for GFDM. *IEEE Commun. Lett.* **2015**, *19*, 1454–1457. [CrossRef]
10. Datta, R.; Panaitopol, D.; Fettweis, G. Cyclostationary Detection of 5G GFDM Waveform in Cognitive Radio Transmission. In Proceedings of the 2014 IEEE International Conference on Ultra-WideBand (ICUWB), Paris, France, 1–3 September 2014; pp. 108–112.
11. Zhang, W.; Xin, Y.; Yu, G. FB-OFDM: A Novel Multicarrier Scheme for 5G. *ZTE Commun.* **2016**, *22*, 22–25.
12. Zhang, H.; Kong, D.; Xin, Y.; Xiao, L.; Jiang, T. Filter Bank Orthogonal Frequency Division Multiplexing With Index Modulation. *IEEE Commun. Lett.* **2021**, *25*, 3960–3964. [CrossRef]
13. Xin, Y.; Yu, G.; Yan, X.; Zhen, Y.; Jun, X.; Bo, G. FB-OFDM: A novel multicarrier scheme for 5G. In Proceedings of the 2016 European Conference on Networks and Communications (EuCNC), Athens, Greece, 27–30 June 2016; pp. 271–276.
14. Vakilian, V.; Wild, T.; Schaich, F.; ten Brink, S.; Frigon, J.F. Universal-filtered multi-carrier technique for wireless systems beyond LTE. In Proceedings of the 2013 IEEE Globecom Workshops (GC Wkshps), Atlanta, GA, USA, 9–13 December 2013; pp. 223–228.
15. Ferdous, M.J.; Hasan, M.; Salsabil, N.A.; Hossain, M.; Rahimi, M.J. Performance Analysis of OFDM and F-OFDM in Impulsive Noise Channel. In Proceedings of the 2022 IEEE International Women in Engineering (WIE) Conference on Electrical and Computer Engineering (WIECON-ECE), Naya Raipur, India, 30–31 December 2022; pp. 264–268.
16. Hadani, R.; Rakib, S.; Tsatsanis, M.; Monk, A.; Goldsmith, A.J.; Molisch, A.F.; Calderbank, R. Orthogonal Time Frequency Space. In Proceedings of the 2017 IEEE Wireless Communications and Networking Conference (WCNC), San Francisco, CA, USA, 19–22 March 2017; pp. 1–6.
17. Naikoti A. and Chockalingam A. Signal detection and channel estimation in OTFS. *ZTE Commun.* **2021**, *19*, 16–33.

18. Zhang, C.; Xing, W.; Yuan, J.H. Performance of LDPC coded OTFS systems over high mobility channels. *ZTE Commun.* **2021**, *19*, 45–53.
19. Liu, M.; Li, S.; Zhang, C.; Wang, B.; Bai, B. Coded Orthogonal Time Frequency Space Modulation. *ZTE Commun.* **2021**, *19*, 54–62.
20. Xiao, R.; Xin, Y.; Wei, Y.; Tian, K.; Zhang, X.; Hua, J.; Liu, X.; Yan, B.; Zhu, H. Performance Analysis of GFB-OFDM: A Waveform for Space-terrestrial Integrated Networks. *Mob. Commun.* **2023**, *47*, 71–78.
21. Xin, Y.; Hua, J.; Hu, L.; Yu, G. A Novel Waveform Scheme Suitable for Terahertz Communication: GFB-OFDM. *Radio Commun. Technol.* **2024**, *50*, 67–73.
22. Xin, Y.; Bao, T.; Hua, J.; Zhang, H. Generalized Filter Bank Orthogonal Frequency Division Multiplexing Systems for 6G. *Electronics* **2024**, *13*, 3006. [CrossRef]
23. Xin, Y.; Bao, T.; Hua, J. A Novel Waveform with 1+D Filter Scheme for THz Communications. In Proceedings of the 2022 International Wireless Communications and Mobile Computing (IWCMC), Dubrovnik, Croatia, 30 May–3 June 2022; pp. 1374–1379.
24. Rao, P.V.; Pransana, C.R.; Ravi, S. Design and ASIC Implementation of Root Raised Cosine Filter. *Eur. J. Sci. Res.* **2009**, *31*, 319–328.
25. Cooley, J.W.; Tukey, J.W. An algorithm for the machine calculation of complex Fourier series. *Math. Comput.* **1965**, *19*, 297–301. [CrossRef]

entropy

MDPI

Article

Random Frequency Division Multiplexing

Chanzi Liu, Jianjian Wu and Qingfeng Zhou *

The School of Electric Engineering and Intelligentization, Dongguan University of Technology, Dongguan 523808, China; liucz@dgut.edu.cn (C.L.); wjj9302@gmail.com (J.W.)
* Correspondence: enqfzhou@ieee.org

Abstract: In this paper, we propose a random frequency division multiplexing (RFDM) method for multicarrier modulation in mobile time-varying channels. Inspired by compressed sensing (CS) technology which use a sensing matrix (with far fewer rows than columns) to sample and compress the original sparse signal simultaneously, while there are many reconstruction algorithms that can recover the original high-dimensional signal from a small number of measurements at the receiver. The approach choose the classic sensing matrix of CS–Gaussian random matrix to compress the signal. However, the signal is not sparse which makes the reconstruction algorithms ineffective. We take full account of the great power of deep neural networks (DNN) to detect the signal as it is an underdetermined equation. The proposed RFDM establishes a novel signal modulation and detection method to target better transmission efficiency, and the simulation results show that the proposed method can achieve good BER, offering a new research paradigm to improve the spectrum efficiency of a multi-subcarrier, multi-antenna, multi-user system.

Keywords: RFDM; random matrix; DNN; MIMO

1. Introduction

Next-generation wireless systems and standards, beyond 5G and 6G, are expected to support a variety of services, such as vehicle-to-everything communications, autonomous driving, remote surgery, aerial vehicles, and operations in extremely high-frequency bands. However, the spectrum resource limitations, high propagation loss, complex network architectures, and energy consumption issues we face all require new modulation techniques and waveforms, which are able to cope with various challenging requirements and show robustness in high mobility scenarios.

Current systems are based on Orthogonal Frequency Division Multiplexing (OFDM) which can achieve near-optimal performance in time-invariant frequency selective channels. Nevertheless, due to its high PAPR (Peak-to-Average Power Ratio) and large Doppler frequency shifts, its performance drastically decreases by inter-carrier interference (ICI). To address this issue, Orthogonal Time Frequency Space (OTFS) is designed to multiplex information symbols in the delay-Doppler (DD) domain, effectively mitigating the effects of channel delays and Doppler shifts [1]. The affine frequency division multiplexing (AFDM) employs the inverse discrete affine Fourier transform to modulate symbols into a "warped" time-frequency domain to handle channel delays and Doppler shifts [2]. The interleave frequency division multiplexing (IFDM) proposed the interleave and Fourier transform to enhance the channels' statistical stability, and the low-complexity detectors are able to ensure bit error rate (BER) performance [3].

The core of OFDM, OTFS, AFDM, and IFDM lies in constructing sparse equivalent channel matrices that facilitate low-complexity signal detection algorithms, balancing optimal performance with implementation complexity. While the maximum likelihood

(ML) detector is optimal, its exhaustive search process makes it inefficient for multiple-input–multiple-output (MIMO) systems [4]. Although the zero-forcing (ZF) and minimum mean square error (MMSE) detectors which are widely used in MIMO systems provide suboptimal performance compared to ML; these detectors require the computation of the inverse of the Gram matrix of the channel. When the channel matrix has a large dimension, it will make these detectors impractical due to their high computational complexity, which poses challenges for the system's hardware efficiency and overall cost. Many other low-complexity detectors have been investigated. Paper [5] proposed a highly efficient detection technique achieved by dimensionality reduction in the channel matrix and the BER performance is comparable to the MMSE detector. Paper [6] presented a suboptimal ML detector which is suitable for the flexible implementation of OFDM-IM (index modulation) systems.

However, the new features of future communications [7], such as complex scenarios with unknown channel models, high speed, and accurate processing requirements, make traditional methods no longer suitable; embedding deep learning (DL) theories into communication systems has attracted a lot of attentions, and researchers believe that DL can achieve further performance improvements in complex scenarios for the following reasons: firstly, the deep network has been proven to be a universal function approximator with superior algorithmic learning ability. Secondly, handling large data is an essential feature of DL because of the instinctive nature of its distributed and parallel computing architectures. Finally, DL-based communication systems can break the artificial block structure to achieve global performance improvement because they are trained to optimize end-to-end performance.

There has been a lot of research on combining DL into MIMO communication, such as channel estimation, signal transmission, detection techniques, and so on. For example, Sun proposed a novel learn iterative search algorithm (LISA) for signal detection in a MIMO system with the DNN. Through training, optimal parameters of the neural networks are learned and thus near ML detection performance is obtained in both fixed and varying channel models under quadrature phase shift keying (QPSK) modulation than classical detectors [8]. A DL-aided Logarithmic Likelihood Ratio (LLR) correction method is proposed for improving the performance of MIMO receivers [9], where it is typical to adopt reduced-complexity algorithms for avoiding the excessive complexity of optimal full-search algorithms. A DNN is trained for detecting and correcting both over-confident and under-confident LLRs with a relatively low complexity compared to the popular reduced-complexity receiver detector techniques.

This paper investigates a RFDM algorithm for faster transmission efficiency. It compresses all symbols by a rectangular matrix. This matrix is the measurement matrix of CS, which can compress and measure signals simultaneously. Of course, the measurement matrix needs to satisfy certain conditions, and there are many matrices that meet these conditions; the Gaussian random matrix is one of the most commonly used measurement matrices which can ensure the reconstruction of the original high-dimensional signal with high probability. Meanwhile, since the signal is not a sparse signal, which is the premise of CS, the reconstruction algorithm cannot detect the signal from a small number of measured values. A DNN is leveraged to learn the implicit mapping function between the received compressed signals and original transmitted signal. The major contributions of this paper can be summarized as follows: (1) The investigated RFDM scheme compresses the original high-dimensional signal through a Gaussian random matrix, which can transmit more data, or the same amount of data, using fewer subcarriers or antennas. (2) A DNN detector is proposed to recover the original high-dimensional signal directly from the low-dimensional data by using the pre-trained DNN network. This novel signal multiplexing and detection method for MIMO communication systems is useful when the number of antennas or subcarriers is limited. That is to say, the main advantage is that the amount of data to be transmitted is reduced, or that the same subcarrier or transceiver antenna can transmit more data. The compression rate of our experimental setting is the fixed 0.75 which is related to the properties of measurement matrix and the accuracy of the reconstruction algorithm.

The remainder of this paper is organized as follows. The basic concept of OFDM and the origin of our idea are introduced in Section 2; this will help us to understand the feasibility of our solution. Section 3 describes the DNN-based RFDM algorithm which consists of two parts: the compression of the transmitted signal and the reconstruction of the signal based on DNN model. The schemes extending to multiple antennas are described in Section 4. The simulation results are analyzed in Section 5. Finally, the paper is concluded in Section 6.

2. Preliminaries

2.1. OFDM Modulation

A symbol vector $s \in \mathbb{C}^{N \times 1}$ in the frequency domain is modulated by the inverse fast fourier transform (IFFT) to generate a time-domain signal x,

$$x = \mathbf{F}_{\mathbf{N}}^{\mathbf{H}} s, \tag{1}$$

where $\mathbf{F}_{\mathbf{N}}$ denotes the N-point normalized FFT matrix and $\mathbf{F}_{\mathbf{N}}^{\mathbf{H}}$ is its conjugate transpose matrix. Then, a cyclic prefix (CP) is added to x and transmitted over mobile time-varying channels. After OFDM demodulation, the received signal is

$$y = \mathbf{F}_{\mathbf{N}}(\mathbf{H}x + v), \tag{2}$$

where $\mathbf{H}$ is a circulant channel matrix and $v \sim \mathcal{CN}(\mathbf{0}, \sigma^2\mathbf{I})$ is additive white Gaussian noise (AWGN).

2.2. Inspirations and Beginnings

CS is a signal processing technique that reconstructs a signal from a small number of measurements by exploiting its sparsity [10]. It relies on three principles: (1) the signal $s \in \mathbb{C}^{N \times 1}$ is sparse or can be represented with fewer non-zero coefficients in some specified domains, with at most K non-zero components. (2) This sparse or approximately sparse signal can be undersampled by a matrix $\mathbf{A}$ with specific properties. Since it projects the signal in a low-dimensional subspace spanned by the rows of the sensing matrix $\mathbf{A}$, $\mathbf{A}$ is either randomly generated or designed based on the restricted isometry property (RIP) or its incoherence with a representation basis [11]. Much effort has been made in the signal processing community to design $\mathbf{A}$ such that the structure of the data is preserved in the low-dimensional subspace. There are many matrices which satisfy the RIP, such as Gaussian random matrices, Bernoulli random matrices, and so on. (3) The reconstruction algorithm can uniquely solve the underdetermined equation by minimizing the following ℓ_0-norm:

$$\min ||s||_0 \ s.t. \ z = \mathbf{A}s,$$

where z is the received signal [12]. The nature of CS is to recover the original high-dimensional signal from a small number of measurements, inspired by CS principle; it is natural to apply this idea to the communication system [13]. Then, we propose the RFDM scheme. At the transmitter, the signal is compressed to lower dimensions and can be reconstructed at the receiver.

3. Random Frequency Division Multiplexing

3.1. RFDM

The proposed RFDM-based single-input–single-output (SISO) communication architecture is constructed as shown in Figure 1. A message bit sequence is digitally modulated as $\boldsymbol{s} \in \mathbb{C}^{N\times 1}$ with the power constraint $\frac{1}{N}\|s\|^2 = 1$; the elements $s_i, i = 1, 2, \ldots, N$ are individually taken from constellation set $\mathcal{Q}$, e.g., quadrature phase shift keying (QPSK) and quadrature amplitude modulation (QAM). Following serial-to-parallel (S/P) conversion and RFDM modulation, the obtained signal $\boldsymbol{x}$ is

$$x = \mathbf{A}\mathbf{F}_{\mathbf{N}}^{\mathbf{H}}\boldsymbol{s} = \mathbf{U}\boldsymbol{s}, \tag{3}$$

where $\mathbf{F_N}$ denotes the N-point normalized FFT matrix and $\mathbf{A} \in \mathbb{C}^{M\times N}(M < N)$ denotes the sensing matrix of CS, in which we use Gaussian random matrices here. And $\mathbf{U} = \mathbf{A}\mathbf{F}_{\mathbf{N}}^{\mathbf{H}} \in \mathbb{C}^{M\times N}$ denotes the random frequency (RF) transform. Then, a CP of a length at least equal to the maximum channel delay spread is added to $\boldsymbol{x}$. After the transmit filter, the signal $\boldsymbol{x}$ is sent out.

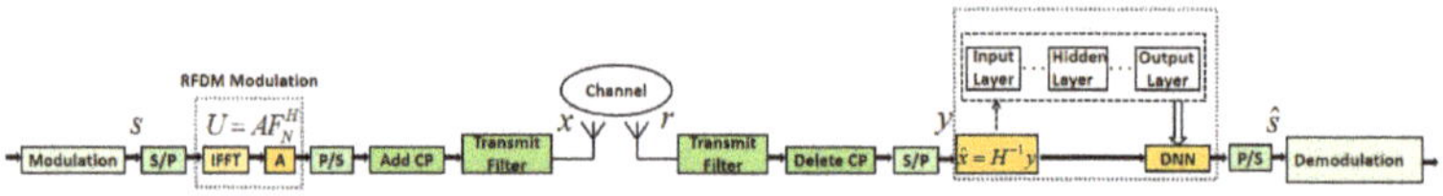

Figure 1. A RFDM-based multicarrier communication system.

The received signal $r[n]$ at the n-th slot is given by

$$r[n] = \sum_{p=0}^{P-1} x[n-p]g[n,p] + v[n], \tag{4}$$

where

$$g[n,p] = \sum_{i=1}^{L} h_i e^{j2\pi f_i (nT_s - pT_s)} P_{rc}(pT_s - \tau_i), \tag{5}$$

$p = 0, 1, \ldots, P-1$, P is the channel tap, $i = 1, 2, \ldots, L$, L is the number of multipaths, T_s is the system sampling interval, $\{h_i, f_i, \tau_i\}$ represent the complex channel gain, the Doppler shift, and the delay at the i-th path. P_{rc} is the raised-cosine roll-off filter which can reduce signal bandwidth and intersymbol interference.

3.2. DNN Detection Method

After the receiver filter and CP removal, (4) can be written as

$$y = \mathbf{H}x + v = \mathbf{H}\mathbf{A}\mathbf{F}_{\mathbf{N}}^{\mathbf{H}}\boldsymbol{s} + v. \tag{6}$$

Since the matrix $\mathbf{A}$ can compress the signal which will require fewer subcarriers or the number of antenna for transmitting the same data, and the original $\boldsymbol{s}$ is not sparse, these lead to the challenge of RFDM: developing a low-complexity detector with acceptable accuracy. This detector consists of two parts: (1) Estimate $\hat{x}$ according to the channel information as $\hat{x} = \mathbf{H}^{-1}y = x + \mathbf{H}^{-1}v$. (2) Reconstruct the original signal $\hat{\boldsymbol{s}}$ based on the DNN, as shown in Figure 1.

DNN, as one of the well-known algorithms of DL, can automatically learn and extract features from raw data, effectively model complex nonlinear relationships, and powerful computational capabilities to efficiently train on large datasets, achieving better generalization performance and enhancing the understanding and representation of data.

Generally, a DNN consists of C layers: an input layer, $C-2$ hidden layers, and an output layer. Given an input vector $\boldsymbol{d}$, the output of every layer can be written as follows:

$$\boldsymbol{d}_c = f(\mathbf{W}_c \cdot \boldsymbol{d}_{c-1} + \boldsymbol{b}_c), \tag{7}$$

where $\boldsymbol{d}_{c-1}$ represents the input of the c-th layer, $\boldsymbol{d}_c$ denotes its output, and $\mathbf{W}_c$ is the weight matrix associated with the $(c-1)$-th and c-th layers, which can be continuously updated during the training process using the backpropagation algorithm to minimize the loss function, allowing the network to better adapt to the training data. The weight matrices of different layers can learn various levels of features, assisting the network in more complex pattern recognition. $\boldsymbol{b}_c$ is the bias vector and $f(\cdot)$ is the activation function.

Consequently, the output of DNN can be mathematically expressed as:

$$\hat{\boldsymbol{s}} = F(\hat{\boldsymbol{x}};\Theta) = f^{(C)}(f^{(C-1)}(\cdots f^{(1)}(\hat{\boldsymbol{x}}))), \tag{8}$$

where Θ represents the learned parametric set and $\hat{\boldsymbol{x}}$ is the input of the network. The parameter set Θ is optimized by reducing the loss function defined as the distance between the prediction and the regression vector. We define T as the size of the dataset. The pair $\{\boldsymbol{x}^{(t)}, \boldsymbol{s}^{(t)}\}_{t=1}^{T}$ is utilized to train Θ to minimize the loss function. The mean squared error (MSE) is adopted as the loss function to express the distance between the transmitted vector $\boldsymbol{x}$ and the output of DNN $\boldsymbol{s}$ as:

$$\mathcal{L}(\boldsymbol{x}, \boldsymbol{s}) = \sum_{t=1}^{T} (\boldsymbol{x}^{(t)} - \mathbf{A}\mathbf{F}_{\mathbf{N}}^{\mathbf{H}} \boldsymbol{s}^{(t)})^2. \tag{9}$$

Although increasing the number of hidden layers, neurons, and samples increases the training time, this DNN model can be trained offline in advance. Once the parameters Θ of DNN model are obtained, the testing process is very fast. Then, the DNN-based network can predict $\hat{\boldsymbol{s}}$ from $\hat{\boldsymbol{x}}$.

It is important to note that the dataset collection process and DNN training are performed without affecting the classical communication system operation. Hence, it is feasible to collect a large dataset for capturing the dynamics in the environment because it does not interfere with the classical system operation.

4. Rfdm-Mimo

Let us consider the RFDM scheme in a multi-antenna scenario; Figure 2 shows the RFDM-MIMO scheme with the number of transmitting antennas N_t and receiving antennas N_r. Modulated signal $\boldsymbol{s}$ through S/P and RFDM modulation is divided into N_t segments, and CP is added to each segment. After the transmit filter, signal x_j is obtained and transmitted through the channel at the j-th antenna, $j = 1, \ldots, N_t$. After discarding the CP, the received signal $\overline{\boldsymbol{y}}$ can be rewritten as:

$$\overline{\boldsymbol{y}} = \overline{\mathbf{H}}\overline{\boldsymbol{x}} + \overline{\boldsymbol{v}}, \tag{10}$$

where $\overline{\boldsymbol{y}} = [\boldsymbol{y}_1^T, \ldots, \boldsymbol{y}_{N_r}^T]^T \in \mathbb{C}^{NN_r \times 1}$, $\overline{\boldsymbol{x}} = [\boldsymbol{x}_1^T, \ldots, \boldsymbol{x}_{N_t}^T]^T \in \mathbb{C}^{NN_t \times 1}$, $\overline{\mathbf{H}} = [\mathbf{H}_1^T, \ldots, \mathbf{H}_{N_r}^T]^T \in \mathbb{C}^{NN_r \times NN_t}$, $\overline{\boldsymbol{v}} = [\boldsymbol{v}_1^T, \ldots, \boldsymbol{v}_{N_r}^T]^T \in \mathbb{C}^{NN_r \times 1}$.

At the receiver, the complete RFDM-MIMO detector is like this: (1) Estimate $\hat{\overline{\boldsymbol{x}}}$ according to the channel information:

$$\hat{\overline{\boldsymbol{x}}} = \overline{\mathbf{H}}^{-1}\overline{\boldsymbol{y}} = \overline{\boldsymbol{x}} + \overline{\mathbf{H}}^{-1}\overline{\boldsymbol{v}}. \tag{11}$$

(2) Reconstruct the original signal $\hat{\boldsymbol{s}}$ based on the pre-trained DNN, as shown in Figure 2.

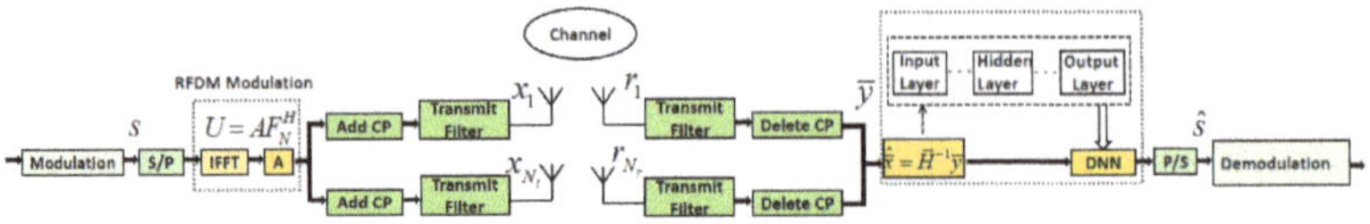

Figure 2. A RFDM-based multicarrier MIMO communication system.

5. Numerical Results

In this section, RFDM is compared to MMSE detectors in mobile time-varying channels. The multicarrier modulations use the same bandwidth, i.e., the system bandwidth is 1.4 MHz, the subcarrier spacing is Δf = 15,000 Hz, and the vehicle speed is 120 km/h with a maximum Doppler frequency shift 300 Hz. The number of subcarriers in the MMSE detector is $N_s = 128$ and the compression ratio $\rho = 0.75$, which means that the number of subcarriers for our RFDM is $N_s = 96$. The maximum number of multipath delay channel taps $L = 9$, and BPSK and QPSK are employed. Furthermore, the antenna number is considered with $N_t = N_r = 4$. Many simulation setups are from the 3GPP LTE-OFDM system [14]. We also assume that the channel estimation is perfect at the receiver and so we run 50,000 Monte Carlo simulations to estimate the average BER.

The sample of the training set is 50,000 to train the DNN model offline. The hidden layer produces an output after weighting (default function 'tansig') the input and summing with the deviation (default function 'learngdm'); the hidden layer produces an input to the output layer after activating the function (default function 'Sigmoid'); and finally, the hidden layer produces an output after weighting the input and summing with the deviation and passing through the corresponding function (default function 'purelin'). We set three layers—the input, hidden and output layers—in our simulations for simplicity. Once the *DNN* model is trained, then the receiving signal $\hat{x}$ is passed as the input to the *DNN* model to detect the original signal $\hat{s}$.

Figures 3 and 4 present the BER performance of the proposed RFDM and the MMSE scheme when using BPSK and QPSK modulation, respectively. It shows that our solution currently has a performance gap with MMSE, but when SNR exceeds 20, our bit error rate can achieve 10^{-3} which is still competitive, especially since MMSE uses 128 subcarriers while we only use 96 subcarriers.

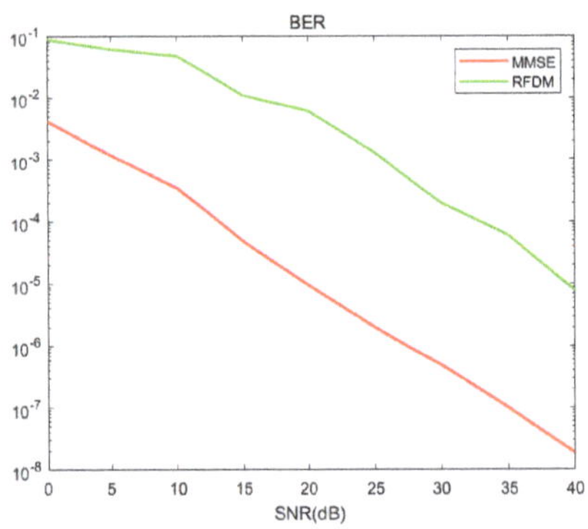

Figure 3. BER comparisons of RFDM with BPSK in 4 × 4 MIMO.

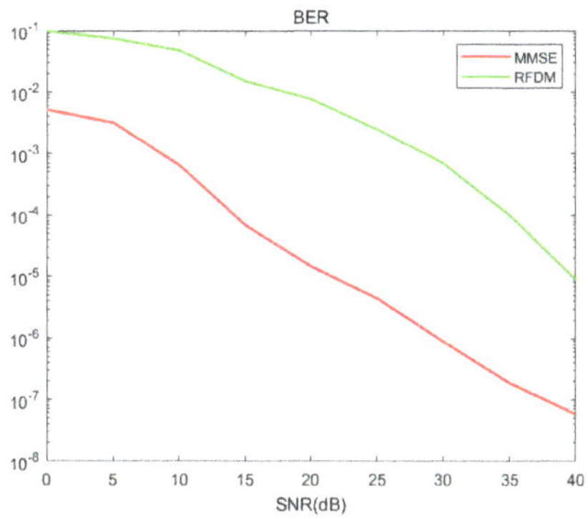

Figure 4. BER comparisons of RFDM with QPSK in 4 × 4 MIMO.

6. Conclusions

In this paper, we propose a novel RFDM scheme. Specifically, the signal is compressed to lower dimensions at the transmitter and can be reconstructed at the receiver. This means that we need fewer subcarriers or antennas to transmit the same amount of data. The DNN model is trained offline to learn the mapping function between an original signal. The BER performance of this proposed scheme is not better than the MMSE detection algorithm; however, we think it is a good attempt when the same quantity is transmitted, although only three quarters of the number of subcarriers of the MMSE scheme is used. It is still very important to improve the transmission speed. We will attempt to ascertain a better performance by avoiding the inversion of the channel matrix **H**, since its properties are directly considered and incorporated into DNN samples to train a good neural network model. There are too many training methods for neural network models, and ours used in this paper is too simple. We must optimize or find a better model to detect the signal. In addition, considering the properties of matrix **A**, the transmission scheme must be optimized comprehensively.

Author Contributions: Conceptualization and methodology, C.L., J.W. and Q.Z.; software, validation and writing—original draft preparation, C.L.; writing—review and editing, C.L., J.W. and Q.Z.; All authors have read and agreed to the published version of the manuscript.

Funding: This work was supported in part by NSFC under Grant 62171126 and in part by Guangdong Basic and Applied Basic Research Foundation 2022A1515140137.

Data Availability Statement: The data presented in this study are available on request from the corresponding author.

Conflicts of Interest: The authors declare no conflicts of interest.

References

1. Hadani, R.; Rakib, S.; Tsatsanis, M.; Monk, A.; Goldsmith, A.J.; Molisch, A.F.; Calderbank, R. Orthogonal time frequency space modulation. In Proceedings of the 2017 IEEE Wireless Communications and Networking Conference (WCNC), San Francisco, CA, USA, 19–22 March 2017; pp. 1–6.
2. Bemani, A.; Ksairi, N.; Kountouris, M. Affine frequency division multiplexing for next generation wireless communications. *IEEE Trans. Wirel. Commun.* **2023**, *22*, 8214–8229. [CrossRef]
3. Chi, Y.; Liu, L.; Ge, Y.; Chen, X.; Li, Y.; Zhang, Z. Interleave Frequency Division Multiplexing. *IEEE Wirel. Commun. Lett.* **2024**, *13*, 1963–1967. [CrossRef]
4. Datta, A.; Majhi, S. Restricted search space exploration with refinement for symbol detection in uplink massive mimo. *IEEE Trans. Veh. Technol.* **2023**, *72*, 13734–13739. [CrossRef]

5. Aafreen, R.; Khan, M.Z.A. Low-Complexity Detection Using Channel Dimensionality Reduction in 6G Uplink Systems. In Proceedings of the 2024 IEEE 99th Vehicular Technology Conference (VTC2024-Spring), Singapore, 24–27 June 2024; pp. 1–5.
6. Kim, K.-H. Low-Complexity Suboptimal ML Detection for OFDM-IM Systems. *IEEE Wirel. Commun. Lett.* **2023**, *12*, 416–420. [CrossRef]
7. Sun, Y.; Liu, J.; Wang, J.; Cao, Y.; Kato, N. When Machine Learning Meets Privacy in 6G: A Survey. *IEEE Commu. Surv. Tutor.* **2020**, *22*, 2694–2724. [CrossRef]
8. Sun, J.; Zhang, Y.; Xue, J.; Xu, Z. Learning to Search for MIMO Detection. *IEEE Trans. Wirel. Commun.* **2020**, *19*, 7571–7584. [CrossRef]
9. Chen, J.; Wang, T.Y.; Wu, J.Y.; Li, C.P.; Ng, S.X.; Maunder, R.G.; Hanzo, L. Deep Learning Aided LLR Correction Improves the Performance of Iterative MIMO Receivers. *IEEE Trans. Veh. Technol.* **2024**, *73*, 2045–2060. [CrossRef]
10. Candes, E.J.; Plan, Y. A Probabilistic and RIPless Theory of Compressed Sensing. *IEEE Trans. Inf. Theory* **2011**, *57*, 7235–7254. [CrossRef]
11. Monsalve, J.; Rueda-Chacon, H.; Arguello, H. Sensing Matrix Design for Compressive Spectral Imaging via Binary Principal Component Analysis. *IEEE Trans. Image Process.* **2020**, *29*, 4003–4012. [CrossRef]
12. Lal, B.; Gravina, R.; Spagnolo, F.; Corsonello, P. Compressed Sensing Approach for Physiological Signals: A Review. *IEEE Sens. J.* **2023**, *23*, 5513–5534. [CrossRef]
13. Liu, C.; Zhou, Q.; Wang, X.; Chen, K. MIMO Signal Multiplexing and Detection Based on Compressive Sensing and Deep Learning. *IEEE Access* **2019**, *7*, 127362–127372. [CrossRef]
14. 3GPP TS 36.211; Version 9.1.0. Release 9. LTE: April 2010; p. 87. Available online: https://www.etsi.org/deliver/etsi_ts/136200_136299/136211/09.01.00_60/ts_136211v090100p.pdf (accessed on 30 October 2024).

entropy

MDPI

Article

Optimal Decoding Order and Power Allocation for Sum Throughput Maximization in Downlink NOMA Systems

Zhuo Han [1], Wanming Hao [1,*], Zhiqing Tang [2] and Shouyi Yang [1]

[1] School of Electrical and Information Engineering, Zhengzhou University, Zhengzhou 450001, China; sincerezhuohan@sina.com (Z.H.); iesyyang@zzu.edu.cn (S.Y.)

[2] School of Cyber Science and Engineering, Zhengzhou University, Zhengzhou 450001, China; iezqtang@zzu.edu.cn

* Correspondence: iewmhao@zzu.edu.cn

Abstract: In this paper, we consider a downlink non-orthogonal multiple access (NOMA) system over Nakagami-*m* channels. The single-antenna base station serves two single-antenna NOMA users based on statistical channel state information (CSI). We derive the closed-form expression of the exact outage probability under a given decoding order, and we also deduce the asymptotic outage probability and diversity order in a high-SNR regime. Then, we analyze all the possible power allocation ranges and theoretically prove the optimal power allocation range under the corresponding decoding order. The demarcation points of the optimal power allocation ranges are affected by target data rates and total power, without an effect from the CSI. In particular, the values of the demarcation points are proportional to the total power. Furthermore, we formulate a joint decoding order and power allocation optimization problem to maximize the sum throughput, which is solved by efficiently searching in our obtained optimal power allocation ranges. Finally, Monte Carlo simulations are conducted to confirm the accuracy of our derived exact outage probability. Numerical results show the accuracy of our deduced demarcation points of the optimal power allocation ranges. And the optimal decoding order is not constant at different total transmit power levels.

Keywords: NOMA; decoding order; power allocation; outage probability; sum throughput

1. Introduction

Non-orthogonal multiple access (NOMA) is a potential candidate for a future multiple access technique. It was proposed by NTT DoCoMo in 2013, aiming at addressing the massive connectivity issue in wireless communication systems. NOMA has a good capacity in improving spectrum efficiency through sharing the same time/frequency/code resources, so it offers a better capacity performance compared with that of orthogonal multiple access (OMA). In power-domain NOMA, desired signals for different users with different power levels are combined using superposition coding (SC) at the transmitter, and each user decodes its own desired signal through successive interference cancellation (SIC). We refer to power-domain NOMA as NOMA in the rest of the paper.

1.1. Related Works

The performance of NOMA depends on the power allocation under the SC technique and the decoding order using the SIC technique. In earlier studies on NOMA systems, the SIC decoding order was prefixed using the ascending order of the users' channel gain [1], i.e., the user with the worse channel gain is allocated more power. The drawback of the prefixed SIC decoding order strategy is that it may bring performance degradation under fading channels, because the decoding order cannot vary with the variation in the instantaneous channel state information (CSI). To address this problem, some researchers have proposed a dynamic SIC decoding order strategies based on instantaneous CSI [2,3], instantaneous received power [4], and a ratio of the instantaneous received power to the target data rate [5]. The above-mentioned dynamic strategies can provide sensible decoding orders for fading channel scenarios and avoid exhaustive searching in all the possible decoding orders. However, they do not ensure the provision of the optimal decoding order to make full use of NOMA.

To find the optimal decoding order, some researchers have made efforts in theoretical analyses, given specific scenarios or conditions. For example, ref. [6] proved that the optimal decoding order for a sum rate maximization problem subjected to minimum user rate requirements under perfect CSI is determined using the ascending order of the channel gain. In [7], the optimal decoding order for the sum rate maximization problem under perfect CSI in a NOMA-CoMP system was proved to depend on the differentials of the channel gains. The authors of [8] showed that the optimal decoding order follows the descending order of the channel gain in the joint decoding order and computation resource allocation optimization problem under perfect CSI when the size of task data for two devices are equal. In [9], the authors proposed a necessary condition for a SIC decoding order to be optimal. In [10], it was demonstrated that decoding the far NOMA user first at the BS provides the best performance in two-user uplink (UL) cooperative non-orthogonal multiple access (C-NOMA) cellular networks. Ref. [11] derived a closed-form optimal decoding order for a total transmit power minimization problem subjected to outage constraints under statistical CSI. Ref. [12] derived the optimal decoding order with regard to the max–min fairness problem for a downlink NOMA system under statistical CSI. In studies based on theoretical analysis, refs. [6–10] assumed that perfect CSI is known to the transmitter, while refs. [11,12] considered the statistical CSI. For a majority of optimization problems involving an optimal decoding order, it is usually infeasible to find a closed-form solution through theoretical analysis. And experimental searching is used to find an optimal or suboptimal solution, such as in [13–15].

Similar to the optimal decoding order, the finding of optimal power allocation can also be solved through theoretical analysis [16–19] or experimental searching [20,21]. In [16], two closed-form optimal power allocation solutions were derived based on the Karush– Kuhn–Tucker (KKT) conditions by fixing the strong user's capacity and the weak user's capacity, respectively. In [17], the optimal power allocation was derived using KKT conditions because the energy-efficient (EE) maximization problem can be transformed into a convex optimization problem. Ref. [18] obtained the closed-form optimal power allocation for two paired users in an achievable sum secrecy rate maximization problem. Ref. [22] derived the optimal solutions of power allocation depending on which user is the D2D transmitter in the sum rate maximization problem subjected to minimum rate constraints. The authors of [19] derived the closed-form expression of power allocation based on KKT conditions to maximize the total throughput under the constraints of the data rate, sum transmit power, and atomicity of devices' tasks. It can be seen that a closed-form solution of optimal power allocation in certain optimization problems can only be obtained under special conditions, such as perfect CSI. Ref. [20] introduced a block-successive upper-bound minimization algorithm to find the globally optimal power allocation. Ref. [21] used deep reinforcement learning to solve the power allocation in a throughput maximization problem. By the way, some studies have used prefixed power allocation for its advantages in applications of the internet of things without full CSI [23].

1.2. Motivation and Contributions

Optimal solutions about decoding order and power allocation usually do not exist in non-convex optimization problems, or they cannot be found using current theoretical analysis methods. With such situations, experimental searching plays a part. A variety of searching algorithms have been designed to find the optimal decoding order and power allocation. However, these searching algorithms mainly conduct searching in the whole domain of the defined problem. When the number of users increases, the possible decoding orders and power allocation ranges also increase rapidly. The problems of algorithm inefficiency resulting from searching in the whole definition domain become more and more obvious. If the well-designed searching algorithms can search in a reduced power allocation range, then they will become more efficient in solving the joint decoding order and power allocation optimization problem by avoiding searching in useless ranges. To the best of our knowledge, the attainment of an optimal power allocation range through in-depth theoretical analysis to reduce the searching space in NOMA systems considering statistical CSI is still an open problem. This was the main motivation for us to carry out this study. We aimed to provided a reduced power allocation range to improve the searching efficiency of the well-designed searching algorithms, including machine learning algorithms, by avoiding searching in useless ranges. In this study, we considered a downlink NOMA system over Nakagami-m fading channels, where each node (one BS and multiple NOMA users) is equipped with a single antenna. We assumed that only statistical CSI is available to the BS. It is quite costly for the BS to achieve perfect CSI in practice, because wireless channels may vary randomly due to terminal mobility, weather variation, stochastic noise, and so on. So, the investigation on NOMA systems based on statistical CSI is of great necessity. A joint optimal decoding order and power allocation problem were formulated to maximize the sum throughput. The main contributions of this study are as follows:

- We derived a closed-form expression of the exact outage probability and corresponding asymptotic outage probability under given decoding orders. We also deduced the diversity order for insight of the characteristic of the outage probability in a high-signal-to-noise-ratio (high-SNR) regime.
- We analyzed all the possible power allocation ranges under each decoding order in detail, and we determined the optimal power allocation ranges to reduce the searching space. This is the main novelty of our work. The theoretical analysis results indicate that the demarcation points of the optimal power allocation ranges are affected by the target data rates and total power, without an effect from the CSI. In particular, the values of the demarcation points are proportional to the total power.
- We formulated a joint decoding order and power allocation optimization problem to maximize the sum throughput, which is solved by efficiently searching in the obtained optimal power allocation ranges.
- We conducted Monte Carlo simulations to confirm the accuracy of our derived exact outage probability. The numerical results show the accuracy of our deduced demarcation points of the optimal power allocation ranges. And we found that the optimal decoding order is not constant at different total transmit power levels.

The remainder of this paper is organized as follows. Section 2 describes the system model. Section 3 presents the outage probability analysis. Section 4 displays the power allocation analysis. Section 5 shows the formulation of the optimization problem. Section 6 shows the simulation results. Section 7 shows the discussion. Finally, Section 8 concludes this paper.

2. System Model

We consider a single-cell downlink NOMA system with a base station (BS) and K users, where each node is equipped with a single antenna. We assumed that the channel between the BS and each user experiences quasi-static Nakagami-m block fading. A Nakagami-m channel is a universal fading channel model, where parameter m defines the fading shape, and Ω is the average channel gain. In particular, Rayleigh channel is a special case of a Nakagami-m channel with $m = 1$. The CSI remains constant during one block but changes independently and randomly from one block to the next block. The BS only knows the long-term statistical CSI associated with each user, because it is costly to achieve perfect CSI via continual channel feedback. It is worth noting that the BS only conducts one-time power allocation over all the blocks, i.e., the power allocation factors remain unchanged as long as the statistical CSI of two users remains unchanged.

The signal received by user U_k can be expressed as

$$y_k = h_k(\sum_{k=1}^{K} \sqrt{a_k P}\, x_k) + n_k, \tag{1}$$

where $k = 1, 2, \ldots, K$, and $h_k \sim \text{Nak}(m_k, \Omega_k)$ denotes the channel coefficient of user U_k. x_k denotes the desired signal for user U_k, where $\mathbb{E}[x_k] = 0$, $\mathbb{E}[x_k^2] = 1$. a_k is the power allocation factor of user U_k, and $\sum_{k=1}^{K} a_k = 1$. P is the total transmission power for NOMA users. n_k is the additive white Gaussian noise (AWGN) with a zero mean and variance σ_k^2. The cumulative distribution function (CDF) of channel gain $g_k = |h_k^2|$ can be expressed as $F(g_k) = \gamma(m_k, \frac{m_k}{\Omega_k} g_k)/\Gamma(m_k)$. $\Gamma(m)$ is the gamma function, which can be calculated by `gamma(m)` in Matlab R2014a and more recent versions. $\gamma(m, x) = \int_0^x t^{m-1} e^{-t} dt$ denotes the lower incomplete gamma function, which can be calculated by `gammainc(x,m)*gamma(m)` in Matlab R2014a and more recent versions.

Let $\pi = (\pi(1), \pi(2), \ldots, \pi(K))$ denote a possible decoding order of the desired signals, where $\pi(j) = k$ means that signal x_k is the j-th one to be decoded, $j = 1, 2, \ldots, K$. According to the principle of the NOMA scheme, user $U_{\pi(i))}$ is able to subtract signal $x_{\pi(l))}$ using the SIC technique, $l \leq i$, while user $U_{\pi(i)}$ cannot subtract signal $x_{\pi(j)}$, $i < j$, which is treated as interference noise. Let $R_{\pi(l)}^{\pi(i)}$ denote the instantaneous data rate that user $U_{\pi(i)}$ can determine about the desired signal $x_{\pi(l)}$ when using the SIC technique, where $l \leq i$. Then, we have [1] ([Equation (3)])

$$R_{\pi(l)}^{\pi(i)} = \log_2 \left(1 + \frac{g_{\pi(i)} a_{\pi(l)} P}{g_{\pi(i)} \sum_{j=l+1}^{K} a_{\pi(j)} P + \sigma_{\pi(i)}^2}\right). \tag{2}$$

3. Outage Probability Analysis

In NOMA systems, a desired signal can be successfully decoded only when *the instantaneous data rate is no less than its target data rate* [1]. The outage probability reveals the probability that a user-desired signal cannot be successfully decoded at its own receiver.

3.1. Exact Outage Probability

Let $R_{\pi(i)}^{\text{F}}$ (in bits/s/Hz) denote the fixed target data rate of user $U_{\pi(i)}$, which is a predetermined parameter. We define event $\{R_{\pi(l)}^{\pi(i)} \geq R_{\pi(l)}^{\text{F}}\}$ as user $U_{\pi(i)}$ being able to decode signal $x_{\pi(l)}$, $l \leq i$. The outage probability of user $U_{\pi(i)}$ can be defined as

$$P_{\pi(i)}^{\text{out}} \triangleq 1 - \Pr[\bigcap_{l \leq k} \{R_{\pi(l)}^{\pi(i)} \geq R_{\pi(l)}^{\text{F}}\}]. \tag{3}$$

Let $\xi_{\pi(l)} = 2^{R_{\pi(l)}^{\text{F}}} - 1$. Then, $R_{\pi(l)}^{\pi(i)} \geq R_{\pi(l)}^{\text{F}}$ can be rewritten as

$$g_{\pi(i)} \geq (\xi_{\pi(l)} \sigma_{\pi(i)}^2 / P) / (a_{\pi(l)} - \xi_{\pi(l)} \sum_{j=l+1}^{K} a_{\pi(j)}) \tag{4}$$

Theorem 1. *Let $\bar{\gamma}_{\pi(i)} = P/\sigma_{\pi(i)}^2$ denote the SNR. The exact outage probability can be written as*

$$P_{\pi(i)}^{out} = \gamma(m_{\pi(i)}, \frac{m_{\pi(i)} \eta_{\pi(i)}}{\Omega_{\pi(i)} \bar{\gamma}_{\pi(i)}}) / \Gamma(m_{\pi(i)}), \tag{5}$$

where $\eta_{\pi(i)} = \max\{\frac{\xi_{\pi(l)}}{a_{\pi(l)} - \sum_{j=l+1}^{K} \xi_{\pi(l)} a_{\pi(j)}}\}$, $i = 1, 2, \ldots, K$, *and* $l = 1, 2, \ldots, i$.

Proof. $P_{\pi(i)}^{\text{out}}$ can be directly derived with the aid of $F(g_{\pi(i)})$. □

3.2. Asymptotic Outage Probability

The asymptotic outage probability can be used to calculate the exact outage probability in the high-SNR regime. The asymptotic outage probability also reveals the diversity order, which is the slope of the asymptotic line of exact outage probability.

Theorem 2. *We write the exact outage probability in (5) in a particular form as* $P^{out}_{\pi(i)} = (\Phi_{\pi(i)}\bar{\gamma}_{\pi(i)})^{-\delta_{\pi(k)}} + o(\bar{\gamma}_{\pi(i)}^{-\delta_{\pi(k)}})$*. When* $\bar{\gamma}_{\pi(i)} \to +\infty$*, the remainder term* $o(\bar{\gamma}_{\pi(i)}^{-\delta_{\pi(i)}})$ *can be omitted. The asymptotic outage probability can be obtained using*

$$ {}^{\infty}P^{out}_{\pi(i)} \approx \left(\Phi_{\pi(i)}\bar{\gamma}_{\pi(i)}\right)^{-\delta_{\pi(i)}}, \tag{6} $$

where $\Phi_{\pi(i)} = (\Gamma(m_{\pi(i)}+1))^{\frac{1}{m_{\pi(i)}}}(\frac{m_{\pi(i)}\eta_{\pi(i)}}{\Omega_{\pi(i)}})^{-1}$*, and* $\delta_{\pi(i)} = m_{\pi(i)}$ *is the diversity order.*

Proof. Let $C_{\pi(i)} = \frac{m_{\pi(i)}\eta_{\pi(i)}}{\Omega_{\pi(i)}}$. According to the series representation $\gamma(\alpha, x) = \sum_{n=0}^{\infty}\frac{(-1)^n x^{\alpha+n}}{n!(\alpha+n)}$, we can rewrite (5) as $P^{\text{out}}_{\pi(i)} = [(C_{\pi(i)}\bar{\gamma}_{\pi(i)}{}^{-1})^{m_{\pi(i)}}\sum_{n=0}^{\infty}\frac{(-1)^n(C_{\pi(i)}\bar{\gamma}_{\pi(i)}{}^{-1})^{m_{\pi(i)}+n}}{n!(m_{\pi(i)}+n)}]/\Gamma(m_{\pi(i)})$ $= [\frac{(C_{\pi(i)}\bar{\gamma}_{\pi(i)}{}^{-1})^{m_{\pi(i)}}}{m_{\pi(i)}} + \sum_{n=1}^{\infty}\frac{(-1)^n(C_{\pi(i)}\bar{\gamma}_{\pi(i)}{}^{-1})^{m_{\pi(i)}+n}}{n!(m_{\pi(i)}+n)}]/\Gamma(m_{\pi(i)})$ $= (C_{\pi(i)}\bar{\gamma}_{\pi(i)}{}^{-1})^{m_{\pi(i)}}/\Gamma(m_{\pi(i)}+1) + o(\bar{\gamma}_{\pi(i)}^{-m_{\pi(i)}}) = [\Gamma(m_{\pi(i)}+1)^{\frac{1}{m_{\pi(i)}}}C_{\pi(i)}{}^{-1}\bar{\gamma}_{\pi(i)}]^{-m_{\pi(i)}} + o(\bar{\gamma}_{\pi(i)}^{-m_{\pi(i)}}) = (\Phi_{\pi(i)}\bar{\gamma}_{\pi(i)})^{-m_{\pi(i)}} + o(\bar{\gamma}_{\pi(i)}^{-m_{\pi(i)}})$. □

3.3. Diversity Order

The diversity order is defined as $\delta_k \triangleq -\lim_{\bar{\gamma}\to+\infty}\frac{\lg P^{\text{out}}_k}{\lg\bar{\gamma}}$. In fact, the diversity order can be indicated by the slope of the asymptotic OP in the high-SNR regime.

Theorem 3. *The diversity order of a NOMA user can be written as*

$$ \delta_{\pi(i)} = m_{\pi(i)},\ i = 1, 2, \ldots, K. \tag{7} $$

Proof. With the aid of the asymptotic outage probability, we can use L'Hospital's Rule to obtain the diversity order: $\delta_{\pi(i)} = -\lim_{\bar{\gamma}_{\pi(i)}\to+\infty}\frac{\lg P^{\text{out}}_{\pi(i)}}{\lg\bar{\gamma}_{\pi(i)}} = -\lim_{\bar{\gamma}_{\pi(i)}\to+\infty}\frac{\lg^{\infty}P^{\text{out}}_{\pi(i)}}{\lg\bar{\gamma}_{\pi(i)}}$ $= -\lim_{\bar{\gamma}_{\pi(i)}\to+\infty}\frac{-m_{\pi(i)}\lg\bar{\gamma}_{\pi(i)} - m_{\pi(i)}\lg\Phi_{\pi(i)}}{\lg\bar{\gamma}_{\pi(i)}} = m_{\pi(i)}$. □

4. Power Allocation Analysis

Each decoding order has a corresponding power allocation range. The power allocation range originally generated by the decoding order can be further narrowed using our in-depth analysis. The reduced power allocation range is called the optimal power allocation range, which contributes to avoiding invalid searches in useless ranges and reducing the computation time. We take two-user scenario as examples first, and then we expand the analysis to multi-user scenarios.

4.1. Two-User NOMA Case

For a two-user scenario, the decoding order set is $\boldsymbol{\pi} = \{(1,2),(2,1)\}$, where $\pi = (1,2)$ and $\pi = (2,1)$ mean that U_1 and U_2 are the first users to be decoded, respectively.

4.1.1. Feasible Power Allocation Range I

We denote the power allocation range I as the feasible power allocation under decoding order $\pi = (1,2)$. The denominator of $\eta_{\pi(i)}$ in Theorem 1 can be written as $a_1 - \xi_1 a_2 > 0$,

$a_2 > 0$. Considering $a_1 + a_2 = 1$, we can obtain the condition for meeting the decoding order $\pi = (1,2)$ as

$$\{a_1 > \xi_1/(1+\xi_1)\} \cap \{a_2 < 1/(1+\xi_1)\} \tag{8}$$

With $\eta_{\pi(2)} = \max\{\frac{\xi_1}{a_1-\xi_1 a_2}, \frac{\xi_2}{a_2}\}$, we divide power allocation range I into two sub-ranges: power allocation range I-A and power allocation range I-B. In power allocation range I-A, $\eta_{\pi(2)} = \frac{\xi_2}{a_2}$, and the power allocation satisfies the inequalities as follows:

$$\{\frac{\xi_1(1+\xi_2)}{\xi_1+\xi_1\xi_2+\xi_2} \le a_1 < 1\} \cap \{0 < a_2 \le \frac{\xi_2}{\xi_1+\xi_1\xi_2+\xi_2}\} \tag{9}$$

In power allocation range I-B, $\eta_{\pi(2)} = \frac{\xi_1}{a_1-\xi_1 a_2}$, and the power allocation satisfies the inequalities as follows:

$$\{\frac{\xi_1}{1+\xi_1} < a_1 < \frac{\xi_1(1+\xi_2)}{\xi_1+\xi_1\xi_2+\xi_2}\} \cap \{\frac{\xi_2}{\xi_1+\xi_1\xi_2+\xi_2} < a_2 < \frac{1}{1+\xi_1}\} \tag{10}$$

4.1.2. Feasible Power Allocation Range II

We denote power allocation range II as the feasible power allocation under decoding order $\pi = (2,1)$. The denominator of $\eta_{\pi(i)}$ in Theorem 1 can be written as $a_2 - \xi_2 a_1 > 0$, $a_1 > 0$. Considering $a_1 + a_2 = 1$, we can obtain the condition meeting decoding order $\pi = (2,1)$ as follows:

$$\{a_2 > \xi_2/(1+\xi_2)\} \cap \{a_1 < 1/(1+\xi_2)\} \tag{11}$$

With $\eta_{\pi(1)} = \max\{\frac{\xi_2}{a_2-\xi_2 a_1}, \frac{\xi_1}{a_1}\}$, we divide power allocation range II into two sub-ranges: power allocation range II-A and power allocation range II-B. In power allocation range II-A, $\eta_{\pi(1)} = \frac{\xi_1}{a_1}$, and the power allocation satisfies the inequalities as follows:

$$\{\frac{\xi_2(1+\xi_1)}{\xi_1+\xi_1\xi_2+\xi_2} \le a_2 < 1\} \cap \{0 < a_1 \le \frac{\xi_1}{\xi_1+\xi_1\xi_2+\xi_2}\} \tag{12}$$

In power allocation range II-B, $\eta_{\pi(1)} = \frac{\xi_2}{a_2-\xi_2 a_1}$, the power allocation satisfies the inequalities as follows:

$$\{\frac{\xi_2}{1+\xi_2} < a_2 < \frac{\xi_2(1+\xi_1)}{\xi_1+\xi_1\xi_2+\xi_2}\} \cap \{\frac{\xi_1}{\xi_1+\xi_1\xi_2+\xi_2} < a_1 < \frac{1}{1+\xi_2}\} \tag{13}$$

Remark 1. *An arbitrary $\eta_{\pi(2)} = \frac{\xi_1}{a_1-\xi_1 a_2}$ in power allocation range I-A has an equal $\eta_{\pi(1)} = \frac{\xi_2}{a_2-\xi_2 a_1}$ in power allocation range II-A.*

4.1.3. Infeasible Power Allocation Range III

According to (8) and (11), there exists an infeasible power allocation range where $\xi_1\xi_2 \ge 1$ ($\frac{\xi_1}{1+\xi_1} \ge \frac{1}{1+\xi_2}$ or $\frac{1}{1+\xi_1} \le \frac{\xi_2}{1+\xi_2}$), making the denominator of $\eta_{\pi(i)}$ in Theorem 1 meaningless. We denote the infeasible power allocation range as the power allocation satisfying the inequality as follows:

$$\{\frac{1}{1+\xi_2} < a_1 < \frac{\xi_1}{1+\xi_1}\} \cap \{\frac{1}{1+\xi_1} < a_2 < \frac{\xi_2}{1+\xi_2}\} \tag{14}$$

Conversely, if $\xi_1\xi_2 < 1$ ($\frac{\xi_1}{1+\xi_1} < \frac{1}{1+\xi_2}$ or $\frac{1}{1+\xi_1} > \frac{\xi_2}{1+\xi_2}$), there is no infeasible power allocation range. Thus, feasible range I and feasible range II are stitched together seamlessly or even overlap partly.

Remark 2. *When $\xi_1\xi_2 \geq 1$, there exists an unfeasible power allocation range, which is just a part of all the possible power allocation ranges. It can be completely skipped when finding the optimal power allocation. The optimal power allocation exists in the remaining possible power allocation ranges.*

4.2. Multi-User NOMA Case

According to permutation theory, K users have $K!$ possible decoding orders, where $(\cdot)!$ is the factorial operator. We rewrite $\eta_{\pi(i)} = \max\{\frac{\xi_{\pi(l)}}{a_{\pi(l)} - \sum_{j=l+1}^{K} \xi_{\pi(l)} a_{\pi(j)}}\}$ as $\eta_{\pi(i)} = \max\{\eta_{\pi(i-1)}, \frac{\xi_{\pi(i)}}{a_{\pi(i)} - \sum_{j=i+1}^{K} \xi_{\pi(i)} a_{\pi(j)}}\}$, where $i = 1, 2, \ldots, K$, and $l = 1, 2, \ldots, i$. Then, the feasible power allocation range under each decoding order can be divided into 2^{K-1} sub-ranges.

Definition 1. *The sub-range r-A in feasible power allocation range r is defined as the sub-range satisfying*

$$\eta_{\pi(i)} = \frac{\xi_{\pi(i)}}{a_{\pi(i)} - \sum_{j=i+1}^{K} \xi_{\pi(i)} a_{\pi(j)}}. \tag{15}$$

It can be deduced from (15) that $\eta_{\pi(i-1)} \leq \eta_{\pi(i)}$. Similar with the analysis of the two-user NOMA case, we will show the power allocation range r-A first for the multi-user NOMA case, $r = \text{I}, \text{II}, \ldots, \text{K}!$. Denote $v_{\pi(i)}$ as the demarcation value of power allocation $a_{\pi(i)}$ in sub-range r-A. The demarcation values should satisfy $\eta_{\pi(1)} \leq \eta_{\pi(2)} \leq \cdots \leq \eta_{\pi(K)}$, which can be processed as

$$-\xi_{\pi(i+1)} v_{\pi(i)} + \xi_{\pi(i)}(1 + \xi_{\pi(i+1)}) v_{\pi(i+1)} \leq 0,\ 1 \leq i, i+1 \leq K. \tag{16}$$

Recall that $\sum_1^K a_{\pi(i)} = 1$. Then, we write an equation set as follows:

$$\begin{cases} -\xi_{\pi(i+1)} v_{\pi(i)} + \xi_{\pi(i)}(1 + \xi_{\pi(i+1)}) v_{\pi(i+1)} = 0,\ 1 \leq i, i+1 \leq K, \\ \sum_1^K v_{\pi(i)} = 1. \end{cases} \tag{17}$$

The demarcation values can be calculated using the equation set. In fact, the equation set can be written into a K-dimension matrix with K variables, which can be solved in Matlab 7.0 and more recent versions. Finally, we obtain sub-range r-A as follows:

$$\begin{cases} v_{\pi(1)} \leq a_{\pi(1)} < 1, \\ \cdots \\ \xi_{\pi(i)}(1 + \xi_{\pi(i)}) v_{\pi(i+1)} / \xi_{\pi(i+1)} \leq a_{\pi(i)} \leq v_{\pi(i)}, \\ \cdots \\ 0 < a_{\pi(K)} \leq v_{\pi(K)}. \end{cases} \tag{18}$$

It is clear that there are no CSI parameters in (17) or (18), i.e., the optimal power allocation ranges are only affected by the target data rates and total power. In a later section, we prove that sub-range r-A is the optimal power allocation range. For space reasons, we cannot display all of the analysis details of the useless $2^{K-1} - 1$ sub-ranges one by one in each power allocation range r.

5. Sum Throughput Maximization Problem

Sum throughput in NOMA systems is defined as the sum of each target data rate times its respective probability of a successful decoding [1]. Let $T = \sum_{k=1}^{K} R_k^{\text{F}}(1 - P_k^{\text{out}})$ denote the maximum sum throughput. Now, we formulate a sum throughput maximization problem as follows

$$\max_{\pi,\mathbf{a}} \; T \tag{19a}$$

$$\text{s.t. } a_{\pi(i)} \in \text{Range } r - A, r = 1, \ldots K! \tag{19b}$$

$$\sum_{i=1}^{K} a_{\pi(i)} = 1. \tag{19c}$$

Theorem 4. *For an arbitrary decoding order in a K-user case, the maximum sum throughput is always achieved in the optimal power allocation range r-A, where* $r = \mathrm{I}, \mathrm{II}, \ldots, \mathrm{K}!$, $\mathrm{K} \geq 2$.

Proof. Let $\eta^*_{\pi(i)} = \frac{\xi_{\pi(i)}}{a^*_{\pi(i)} - \sum_{j=i+1}^{K} \xi_{\pi(i)} a^*_{\pi(j)}}$ and $\eta^*_{\pi(i+1)} = \frac{\xi_{\pi(i+1)}}{a^*_{\pi(i+1)} - \sum_{j=i+1}^{K} \xi_{\pi(i+1)} a^*_{\pi(j)}}$. We set power allocation $\mathbf{a}^*$ from power allocation sub-range r-A to guarantee $\eta^*_{\pi(i+1)} \geq \eta^*_{\pi(i)}$ on the condition as shown in (18). Similarly, let $\eta'_{\pi(i)} = \frac{\xi_{\pi(i)}}{a'_{\pi(i)} - \sum_{j=i+1}^{K} \xi_{\pi(i)} a'_{\pi(j)}}$ and $\eta'_{\pi(i+1)} = \frac{\xi_{\pi(i+1)}}{a'_{\pi(i+1)} - \sum_{j=i+1}^{K} \xi_{\pi(i+1)} a'_{\pi(j)}}$. We set another power allocation $\mathbf{a}'$ from another power allocation sub-range r-B to meet $\eta'_{\pi(i+1)} \leq \eta'_{\pi(i)}$. Thus, for arbitrary $\eta'_{\pi(i+1)}$ from sub-range i-B, we can always find an equal $\eta^*_{\pi(i+1)}$ from sub-range i-A, i.e., $\eta^*_{\pi(i+1)} = \eta'_{\pi(i+1)}$. Then, we have the corresponding relationship $\eta^*_{\pi(i)} \leq \eta'_{\pi(i)}$. It is known that $P^{\text{out}}_{\pi(i)}(x)$ is a monotonically increasing function about variable x. If $\eta^*_{\pi(i)} \leq \eta^*_{\pi(i+1)} = \eta'_{\pi(i+1)} \leq \eta'_{\pi(i)}$, then $P^{\text{out}}_{\pi(1)}(\eta^*_{\pi(i)}) \leq P^{\text{out}}_{\pi(1)}(\eta'_{\pi(i)})$ and $P^{\text{out}}_{\pi(2)}(\eta^*_{\pi(2)}) = P^{\text{out}}_{\pi(2)}(\eta'_{\pi(2)})$ Now, we obtain the inequality $R^{\text{F}}_{\pi(i)} P^{\text{out}}_{\pi(i)}(\eta^*_{\pi(1)}) + R^{\text{F}}_{\pi(i+1)} P^{\text{out}}_{\pi(i+1)}(\eta^*_{\pi(i+1)}) \leq R^{\text{F}}_{\pi(i)} P^{\text{out}}_{\pi(i)}(\eta'_{\pi(i)}) + R^{\text{F}}_{\pi(i+1)} P^{\text{out}}_{\pi(i+1)}(\eta'_{\pi(i+1)})$. So, power allocation $\mathbf{a}^*$ from sub-range r-A can always provide greater sum throughput than that from other sub-ranges under the same decoding order. □

According to Theorem 4, we only need to search in optimal power allocation sub-range r-A under each decoding order to solve the optimization problem, where $r = \mathrm{I}, \mathrm{II}, \ldots, \mathrm{K}!$. The flow chat in Figure 1 below helps to express our method for solving the optimization problem. It is worth noting that both the optimal decoding order and optimal power allocation remain unchanged as long as the statistical CSI remains unchanged.

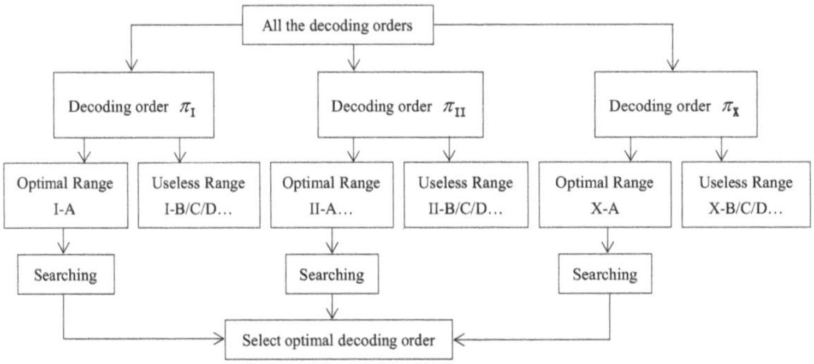

Figure 1. Method to solve the optimization problem. Each decoding order has an optimal power allocation sub-range, and the optimal decoding order is the one that produces the maximum sum throughput.

6. Simulations

We conducted Monte Carlo simulations (using 10^6 realizations) to verify the theoretical expression of our derived exact outage probability. Take a two-user case as an example. Set $\sigma_1^2 = \sigma_2^2 = 1$. Other related simulation parameters are given in the blank area on each figure for convenience, e.g., $\mathbf{R}^{\text{F}} = [R_1^{\text{F}} \; R_2^{\text{F}}]$ shows the target data rates of U_1 and U_2, respectively.

Figure 2 shows the outage probability versus the SNR at a fixed power allocation. Monte Carlo (MC) simulations well verified the accuracy of our theoretical exact outage probability. Our derived asymptotic (Asym) outage probability clearly expresses the characteristic of the exact outage probability in the high-SNR regime. The slope of the asymptotic outage probability represents the diversity order.

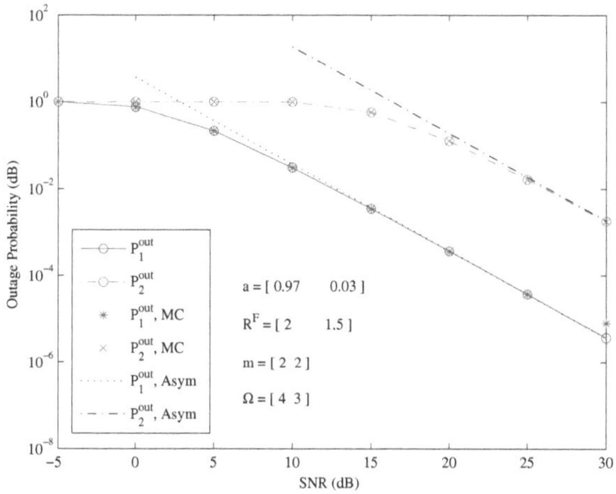

Figure 2. The outage probability of two users under a given decoding order and fixed power allocation.

Figure 3 depicts all the power allocation ranges for two users according to the analysis in Section 4.1, where $p_2 = a_2P$ and $p_1 = a_1P$ are the labels of the X-axis and Y-axis, respectively. We mark $M_1(0,P)$ and $V_1(\frac{P}{1+\xi_1}, \frac{\xi_1 P}{1+\xi_1})$ as the two demarcation points of power allocation range I, and $V_2(\frac{\xi_2 P}{\xi_1+\xi_1\xi_2+\xi_2}, \frac{\xi_1(1+\xi_2)P}{\xi_1+\xi_1\xi_2+\xi_2})$ is the demarcation point between power allocation range I-A and power allocation range I-B. Similarly, we mark $M_2(P,0)$ and $V_3(\frac{\xi_2 P}{1+\xi_2}, \frac{P}{1+\xi_2})$ as the two demarcation points of power allocation range II, and $V_4(\frac{\xi_2(1+\xi_1)P}{\xi_1+\xi_1\xi_2+\xi_2}, \frac{\xi_1 P}{\xi_1+\xi_1\xi_2+\xi_2})$ is the demarcation point between power allocation range II-A and power allocation range II-B. It can be found that the demarcation points are affected by the target data rates and total power, with no effect from the CSI. And the values of the demarcation points are proportional to the total power. In considering SNR $= P/\sigma^2$, this figure also shows the effect of the SNR on the power allocations when we fix σ^2.

Figure 4 sketches the exact outage probability of two users under different power allocation ranges. It shows that power allocation sub-range I-A outperforms sub-range I-B under the same decoding order $\pi = (1,2)$. Similarly, sub-range II-A outperforms sub-range II-B under the same decoding order $\pi = (2,1)$. Therefore, we just need to search in optimal power allocation sub-ranges I-A and II-A for the sum throughput maximization problem. Interestingly, the outage probability curve of range I-B perfectly overlaps with that of range II-B, which agrees well with our *Remark 1*. It is worth noting that the outage probabilities of the demarcation points V_2 and V_4 in Figure 3 are located identically at the same turning point marked by "o" in this figure. This indicates the accuracy of our theoretical analysis of demarcation points V_2 and V_4.

Figure 5 shows the maximum sum throughput values achieved by the optimal decoding order π^* versus the total transmit power. The maximum sum throughput curves under decoding order $\pi = (1,2)$ and $\pi = (2,1)$ intertwine together in different power regimes. Therefore, there does not exist a changeless optimal decoding order such that its corresponding maximum sum throughput is always greater than another one in all the power regimes. The method used to decide the optimal decoding order was to compare the maximum sum throughput values under different decoding orders. We selected the decoding order that produced the greater maximum sum throughput as the optimal decoding order.

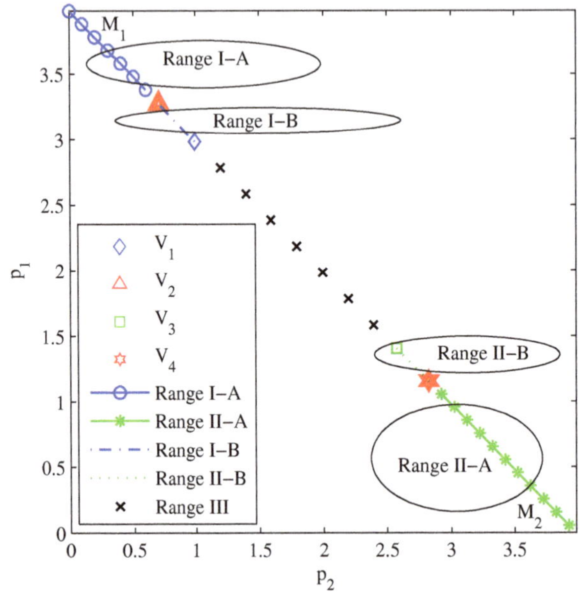

Figure 3. All the power allocation ranges for two-user NOMA.

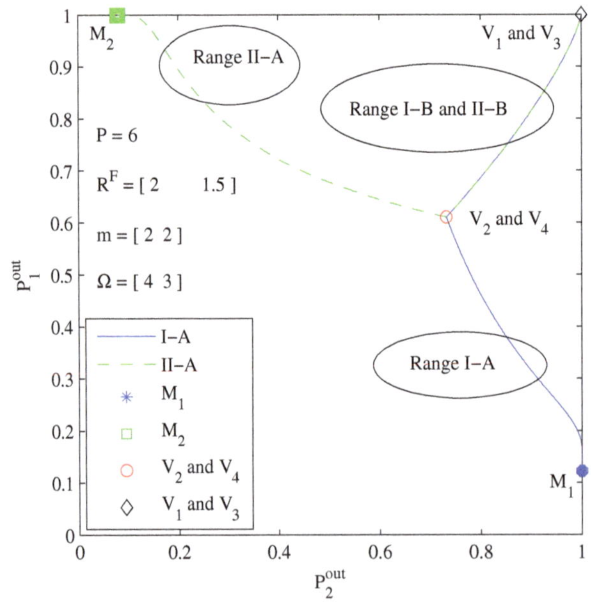

Figure 4. Outage probability scatter diagram of two users under different power allocation ranges.

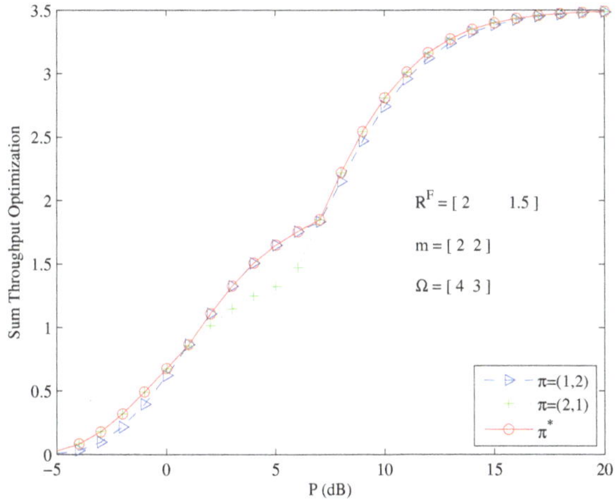

Figure 5. Optimal decoding order for sum throughput maximization problem at different total power levels.

7. Discussion

As we stated in the motivation part, well-designed searching algorithms, including machine learning algorithms, can be combined with our reduced power allocation range to improve the searching efficiency. This is the main advantage of our work. Without our theoretically optimal power allocation range, the searching algorithms previously proposed by researchers, such as those mentioned in the introduction part, need to search in the whole problem definition domain.

However, we just considered a single antenna at each node in this study. This is the obvious limitation of our work. We will try to conduct more meaningful work in future research by extending the proposed model to multiple antennas. For example, there exist some open problems about the optimal power allocation ranges considering beamforming or antenna selection. Our proposed theoretical analysis method used to achieve optimal power allocation and the joint optimization method in this paper were based on statistical CSI, whereas a majority of studies using beamforming usually consider perfect CSI. Thus, the extension to beamforming is quite challenging. In addition, it is also worth pondering how the increase in channel complexity affects the use of our proposed optimization method. When multiple antennas are used as nodes, the channel complexity will increase. Consider a downlink MISO-NOMA system as a simple example. If the antenna selection technique is adopted at the BS, the outage probability used in (5) will change, but the optimal power allocation range in Theorem 4 is still suitable.

8. Conclusions

We studied a single-cell downlink NOMA system considering statistical CSI. The closed-form expressions of the exact outage probability were derived. The asymptotic outage probability and diversity order were also deduced for insight on the characteristic of the outage probability in a high-SNR regime. In thoroughly analyzing the possible power allocation ranges, the optimal power allocation range under a certain decoding order was obtained in a theoretical form. The demarcation points of the optimal power allocation ranges are affected by the target data rates and total power, with no effect from the CSI. In particular, the values of the demarcation points are proportional to the total power. A sum throughput maximization problem was formulated, and it was solved by efficiently searching in our achieved optimal power allocation ranges. The numerical results show that our deduced demarcation points of the optimal power allocation ranges are accurate, and the optimal decoding order is not constant at different total transmit power levels.

Author Contributions: Conceptualization, Z.H.; Methodology, Z.H., W.H. and S.Y.; Software, Z.T.; Validation, W.H., Z.T. and S.Y.; Formal analysis, Z.H., Z.T. and S.Y.; Investigation, Z.H.; Resources, W.H., Z.T. and S.Y.; Data curation, W.H.; Writing—original draft, Z.H.; Writing—review & editing, W.H. and Z.T.; Visualization, Z.H.; Supervision, S.Y.; Project administration, S.Y. All authors have read and agreed to the published version of the manuscript.

Funding: This research was funded by [Songshan Laboratory Foundation] grant number [YYJC022022003] And The APC was funded by [Wangming Hao].

Institutional Review Board Statement: Not applicable.

Data Availability Statement: Data are contained within the article.

Conflicts of Interest: The authors declare no conflict of interest.

References

1. Wang, X.; Wang, J.; He, L.; Song, J. Outage Analysis for Downlink NOMA With Statistical Channel State Information. *IEEE Wirel. Commun. Lett.* **2018**, *7*, 142–145. [CrossRef]
2. Dao, V.L.; Hoang, L.N.; Girs, S. Outage Performance of Pairwise NOMA Allowing a Dynamic Decoding Order and Optimal Pairs of Power Levels. *IEEE Open J. Commun. Soc.* **2020**, *1*, 1886–1906. [CrossRef]
3. Zhang, Y.; Wang, H.M.; Zheng, T.X.; Yang, Q. Energy-efficient transmission design in non-orthogonal multiple access. *IEEE Trans. Veh. Technol.* **2017**, *66*, 2852–2857. [CrossRef]
4. Wang, J.; Xia, B.; Xiao, K. Outage Performance Analysis for Wireless Non-Orthogonal Multiple Access Systems. *IEEE Access* **2018**, *6*, 3611–3618. [CrossRef]
5. Chen, Z.; Liu, B. Switching Signal Decoding Order in Uplink NOMA System With Imperfect SIC. *IEEE Commun. Lett.* **2021**, *25*, 3528–3532. [CrossRef]
6. Yang, Z.; Xu, W.; Chen, M. On the optimality of power allocation for NOMA downlinks with individual QoS constraints. *IEEE Commun. Lett.* **2017**, *7*, 1649–1652. [CrossRef]
7. Dai, Y.; Lyu, L.; Cheng, N. Impact of Decoding Order on User Pairing in NOMA-CoMP Joint Transmission. *IEEE Trans. Veh. Technol.* **2023**, *72*, 9664–9669. [CrossRef]
8. Qian, L.P.; Feng, A.; Huang, Y.; Wu, Y.; Ji, B.; Shi, Z. Optimal SIC Ordering and Computation Resource Allocation in MEC-Aware NOMA NB-IoT Networks. *IEEE Internet Things J.* **2019**, *6*, 2806–2816. [CrossRef]
9. Zhang, X.; Chen, Y.; Zhang, J.; Lei, Y. Characterization of SINR Region for Multi-Cell Downlink NOMA Systems. In Proceedings of the ICC 2019–2019 IEEE International Conference on Communications (ICC), Shanghai, China, 20–24 May 2019; pp. 1–6. [CrossRef]
10. Elhattab, M.; Arfaoui, M.A.; Assi, C.; Ghrayeb, A.; Qaraqe, M. Power Allocation Optimization and Decoding Order Selection in Uplink C-NOMA Networks. *IEEE Commun. Lett.* **2023**, *27*, 352–356. [CrossRef]
11. Cui, J.; Ding, Z.; Fan, P. A Novel Power Allocation Scheme Under Outage Constraints in NOMA Systems. *IEEE Signal Process. Lett.* **2016**, *23*, 1226–1230. [CrossRef]
12. Shi, S.; Yang, L.; Zhu, H. Outage Balancing in Downlink Nonorthogonal Multiple Access With Statistical Channel State Information. *IEEE Trans. Wirel. Commun.* **2016**, *15*, 4718–4731. [CrossRef]
13. Hu, D.; Zhang, Q.; Li, Q. Joint Position, Decoding Order, and Power Allocation Optimization in UAV-Based NOMA Downlink Communications. *IEEE Syst. J.* **2020**, *14*, 2949–2960. [CrossRef]
14. Shi, Z.; Xie, X.; Lu, H.; Yang, H.; Cai, J. Deep Reinforcement Learning Based Dynamic User Access and Decode Order Selection for Uplink NOMA System With Imperfect SIC. *IEEE Wirel. Commun. Lett.* **2021**, *10*, 710–714. [CrossRef]
15. Rezvani, S.; Jorswieck, E.A.; Yamchi, N.M. Optimal SIC Ordering and Power Allocation in Downlink Multi-Cell NOMA Systems. *IEEE Trans. Wirel. Commun.* **2022**, *21*, 3553–3569. [CrossRef]
16. Wang, C.L.; Chen, J.Y.; Chen, Y.J. Power Allocation for a Downlink Non-Orthogonal Multiple Access System. *IEEE Wirel. Commun. Lett.* **2016**, *5*, 532–535. [CrossRef]
17. Hao, W.; Zeng, M.; Chu, Z.; Yang, S. Energy-efficient power allocation in millimeter wave massive MIMO with non-orthogonal multiple access. *IEEE Wirel. Commun. Lett.* **2017**, *6*, 782–785. [CrossRef]
18. Li, Y.; Chen, Y.; Ji, X. Secure User Pairing and Power Allocation for Downlink Non-Orthogonal Multiple Access against External Eavesdropping. *Entropy* **2024**, *26*, 64. [CrossRef] [PubMed]
19. Li, J.; Zhai, X.B.; Qian, H.; Zhang, R.; Liu, X. Joint Trajectory Design and Power Allocation in NOMA-Based UAV Networks. *IEEE Trans. Veh. Technol.* **2024**, *73*, 2345–2357. [CrossRef]
20. Zhou, X. An Efficient Block Successive Upper-Bound Minimization Algorithm for Caching a Reconfigurable Intelligent Surface-Assisted Downlink Non-Orthogonal Multiple Access System. *Electronics* **2024**, *13*, 791. [CrossRef]
21. Guo, S.; Zhao, X.; Zhang, W. Throughput Maximization for RF Powered Cognitive NOMA Networks with Backscatter Communication by Deep Reinforcement Learning. *IEEE Trans. Wirel. Commun.* **2023**, 1. [CrossRef]

22. Shen, K.Z.; So, D.K.C.; Tang, J.; Ding, Z. Power Allocation for NOMA With Cache-Aided D2D Communication. *IEEE Trans. Wirel. Commun.* **2024**, *23*, 529–542. [CrossRef]
23. Yang, M.; Chen, J.; Ding, Z.; Liu, Y.; Lv, L.; Yang, L. Joint Power Allocation and Decoding Order Selection for NOMA Systems: Outage-Optimal Strategies. *IEEE Trans. Wirel. Commun.* **2024**, *23*, 290–304. [CrossRef]

entropy

MDPI

Article

Joint User Association, Power Allocation and Beamforming for NOMA-Based Integrated Satellite–Terrestrial Networks

Peizhe Xin [1], Zihao Fu [1], Zhiyi Chen [1], Jing Jiang [1], Jing Zou [1], Yu Zhang [2,*] and Xinyue Hu [2]

[1] State Grid Economic and Technological Research Institute Co., Ltd., 102209 Beijing, China; xinpeizhe@chinasperi.sgcc.com.cn (P.X.); 18810555907@163.com (Z.F.); chenzhiyi@chinasperi.sgcc.com.cn (Z.C.); jiangjing@chinasperi.sgcc.com.cn (J.J.); zoujing@chinasperi.sgcc.com.cn (J.Z.)

[2] Information Materials and Intelligent Sensing Laboratory of Anhui Province, Anhui University, Hefei 230601, China; xinyuehu@ahu.edu.cn

* Correspondence: zhyu@ahu.edu.cn

Abstract: This paper investigated a non-orthogonal multiple access (NOMA)-based integrated satellite–terrestrial network (ISTN), where each user can select to access a terrestrial base station (BS) or the satellite according to the capacity of BS and their individual transmission requirements. A two-stage algorithm is proposed to solve the achievable sum rate maximizing resource optimization problem. In the first stage, user associations are determined based on individual preference lists and the backhaul capacities of the access points (APs). In the second stage, the power allocation, and the receiving beamforming vectors are optimized alternately. Within each iteration, the closed-form solution for the transmit power is derived. Simulation results show the effectiveness of the proposed algorithm and the benefits brought by NOMA. When the backhaul link capacity of terrestrial BSs is sufficient, users (UEs) prefer to access these BSs. Otherwise, the satellite can offer QoS guarantees to UEs. Furthermore, the overall system performance reaches its optimum when the number of UEs in the system matches the number of receive antennas at the APs.

Keywords: integrated satellite–terrestrial network (ISTN); non-orthogonal multiple access (NOMA); user association; power allocation; beamforming

1. Introduction

Although terrestrial cellular networks undergo several generations of evolution, half of the world still faces the challenges of achieving ubiquitous access, primarily due to the economic costs associated with base station (BS) deployment [1,2]. To address this challenge, the third-generation partnership project (3GPP) promotes a promising architecture, referred to as integrated satellite–terrestrial networks (ISTNs) [3,4], which integrate satellite communication and terrestrial networks and can provide wide-area coverage, universal multiaccess, and ubiquitous connectivity. ISTNs have a wide range of applications, such as personal mobile communication, transportation, aviation and navigation, Power Internet of Things (PIoT), emergency disaster relief, and so on.

Despite the advantages of ISTNs, scarce frequency resource and increased interference caused by heterogeneous networks limit their transmission performance. To overcome this, many researchers devote their efforts to studying resource allocations in ISTNs, mainly focusing on user scheduling, power allocation, and beamforming [5–8]. In [5], a distributed user association with a grouping mechanism was proposed to maximize the sum rate and balance the load of ISTN by jointly considering the backhaul capacity of BSs and mobility and delay of users (UEs). In [6], a power and frequency resource allocation scheme was proposed to minimize total energy and maximize spectral efficiency. A power control scheme for cognitive ISTNs was proposed in [7], in which the aim is to maximize the energy efficiency subject to interference power constraints imposed by terrestrial communications and outage constraints of satellite communications. In [8], the authors studied hybrid beamforming, user scheduling, and power allocation optimization in ISTN with the purpose of improving system sum rate and energy efficiency. The authors in [9] considered an uplink communication model consisting of a single satellite, a single terrestrial station, and multiple ground users. In this model, the BS serves as a satellite–terrestrial relay, bridging users and the satellite. The sum rate was maximized by optimizing user grouping and transmit power of users. Although elaborate resource allocation design can mitigate interference effectively, with the surge in user numbers, the congestion of spectrum resources has seriously degraded the communication rate in ISTNs.

Recently, non-orthogonal multiple access (NOMA) technology has presented an effective solution to promote spectrum efficiency. In a NOMA system, multiple UEs can share the same spectrum resource, with interference being effectively mitigated by the adoption of successive interference cancellation (SIC) technology at the receiver [10,11]. NOMA has been widely investigated in terrestrial networks. Ref. [12] investigated a sum rate maximization problem in a movable antenna-enabled uplink NOMA system. The author proposed a successive convex approximation-based algorithm to jointly optimize the movable antennas' positions, the decoding order, and the power control. The authors in [13] formulated a joint power allocation and channel assignment problem for uplink NOMA-based wireless communication networks. In order to solve the problem, a methodology that integrates convex optimization and machine learning techniques was proposed. In [14], a resource allocation problem for the coexistence of enhanced mobile broadband (eMBB) and ultra-reliable low-latency communication (URLLC) traffic scheduling was investigated; the sum rate of all eMBB users was maximized while maintaining the minimum data rate requirement of each eMBB user.

Research reveals that NOMA offers significant performance advantages compared with the traditional orthogonal multiple access (OMA) schemes. The integration of NOMA into ISTNs to enhance communication performance and resource utilization tends to be an inevitable trend. Considering applying NOMA technology to terrestrial networks, [1] and [15] investigated downlink ISTNs. The authors in [15] studied user association, user pairing, subchannel allocation, and power allocation algorithms. The work in [1] proposed a reconfigurable intelligent surface (RIS)-assisted downlink ISTN with NOMA, in which the energy efficiency was maximized by jointly optimizing the active beamforming vector at the BS and the passive beamforming matrix at the RIS. In [2], the authors investigated the joint optimization algorithm for the phase shift, transmit power, and receive beamforming of an uplink RIS-assisted NOMA-based ISTN. In the considered scenario, only one BS is deployed, and the direct UEs are served by space division multiple access technology. The authors of [16] investigated an uplink caching-based NOMA ISTN, where the satellite provides backhaul links for terrestrial BSs. In this work, a system utility function, which consists of the achieved terrestrial user rate and cross-tier interference caused by terrestrial BSs to satellite, was maximized by jointly optimizing user association, bandwidth assignment, and power allocation. In [17], a satellite–aerial–terrestrial uplink network was investigated. The authors considered a high-altitude platform as a grant-based entity serving multiple mobile terminals to realize satellite–terrestrial uplink communication. NOMA was also applied on the high-altitude platform to increase communication capacity.

Although many studies have focused on resource allocation in NOMA-based ISTNs, a majority of these papers have only implemented NOMA for a portion of UEs. And joint resource optimization is mainly divided into two categories: One considers fixed user association, optimizing power, and beamforming, which means UEs cannot flexibly choose to access BSs or satellites according to their individual requirements. The other category allows UEs to choose to access BSs or satellites, optimizing user association and power allocation. However, such research often assumes that all communication nodes are equipped with a single antenna, and therefore does not need to consider the impact of beamforming on user association. The joint optimization of user association, beamforming, and power allocation in NOMA-based ISTNs still poses challenges. Motivated by this, we investigated joint resource optimization in an uplink NOMA-based ISTN, where NOMA is applied for all UEs and each user can choose to access either the satellite or terrestrial BSs based on their communication rate requirements. Note that this work differs significantly from the investigations in [2,9,16,17], both in terms of the scenarios studied and the parameters optimized. The main contributions are summarized as follows:

- We propose an uplink NOMA-based ISTN, where each user can access either BSs or the satellite. With the goal of maximizing the achievable sum rate of all the UEs, an optimization problem of jointly optimizing the transmit power, receiving beamforming, and user association is established by considering the backhaul link capacity of BSs and the satellite, as well as individual quality of service (QoS) constraints.
- We develop a two-stage algorithm to solve the problem. At the first stage, each user is associated with a corresponding access point (AP), i.e., either a BS or the satellite, based on their preference list and the backhaul link capacity of the APs. At the second stage, the power allocation and the receiving beamforming vectors are optimized iteratively. Within each iteration, the closed-form solution for the transmit power is derived.
- The simulation results demonstrate the superiority of the proposed scheme compared with the random power allocation scheme and the traditional OMA scheme. In scenarios where the backhaul link capacity of terrestrial BSs is sufficient, UEs tend to access these BSs. However, when the backhaul link capacity of terrestrial BSs is insufficient, the satellite can offer QoS guarantees to UEs. Moreover, the system's overall performance achieves its peak when the number of UEs in the system aligns with the number of receive antennas at the APs.

2. System Model and Problem Formulation

2.1. System Model

As shown in Figure 1, an uplink communication scenario for the terrestrial–satellite network is investigated, in which M BSs and one satellite (The proposed algorithm can be easily extended to multiple satellite settings. The additional satellites can be viewed as additional APs.) are deployed to serve K terrestrial UEs. The UEs' data is uploaded to the BSs or satellite and then transmitted to the gateway of the core network via the backhaul links. Define the sets of UEs and BSs as $\mathcal{K} = \{1,2,\ldots,K\}$ and $\mathcal{M} = \{1,2,\ldots,M\}$, respectively. The satellite is denoted as $M+1$. And all the APs (i.e., BSs and satellite) constitute a set denoted by $\mathcal{R} = \mathcal{M} \cup M+1$. Assume that the m-th AP is equipped with N_m, $m = 1,\ldots,M+1$ antennas, while the terrestrial UEs are all deployed with single antennas. Denote $\mathcal{U}_m$ as the UEs associated to AP m, satisfying $\mathcal{K} = \{\mathcal{U}_1 \cup \mathcal{U}_2 \cdots \cup \mathcal{U}_{M+1}\}$, then the received signal of AP m from UE k is denoted by

$$\mathbf{y}_{m,k} = \mathbf{h}_{m,k}\sqrt{p_{m,k}}s_{m,k} + \sum_{j\in\mathcal{U}_m\backslash k} \mathbf{h}_{m,j}\sqrt{p_{m,j}}s_{m,j} + \sum_{i=1,i\neq m}^{M+1} \sum_{j\in\mathcal{U}_i} \mathbf{h}_{m,j}\sqrt{p_{i,j}}s_{i,j} + \mathbf{n}_m, \tag{1}$$

where $p_{m,k}$ and $s_{m,k}$ denote the transmit power and symbol of UE k served by AP m. $\mathbf{h}_{m,k} \in \mathbb{C}^{N_m\times 1}$ denotes the channel vector of UE k towards AP m. $\mathbf{n}_m \sim \mathcal{CN}(0,\sigma_m^2\mathbf{I})$ is the additive white Gaussian noise (AWGN) vector. Similar to most of the related works [4,18], we assume that all the wireless channels undergo slow fading, and the perfect channel state information of all channels are available at the gateway through feedback/training sent from the user terminals via the backhaul channel. This mechanism has already been adopted in DVB-S2 [19]. For the uplink NOMA communication scenario, the UEs with better channel conditions have the higher priority to be decoded through SIC [2]. Assume that channel coefficients of UEs towards AP m are sorted as

$$Q_m \triangleq \left|\mathbf{h}_{m,\pi_m(1)}\right|^2 \geq \left|\mathbf{h}_{m,\pi_m(2)}\right|^2 \geq \cdots \geq \left|\mathbf{h}_{m,\pi_m(K)}\right|^2, \forall m \in \mathcal{R}. \tag{2}$$

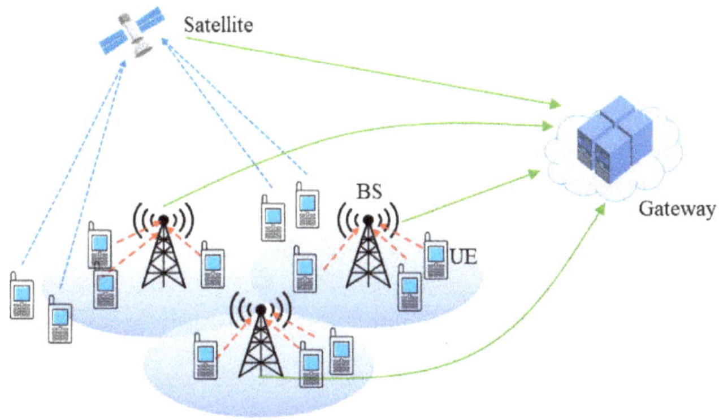

Figure 1. The NOMA-based terrestrial–satellite network architecture.

Let $\mathcal{S}_{m,k} = \{\pi_m(1), \pi_m(2), \cdots, \pi_m(k)\}$; thus, the received SINR of UE k at AP m is calculated by

$$SINR_{m,k} = \frac{\left|\mathbf{v}_{m,k}\mathbf{h}_{m,k}\right|^2 p_{m,k}}{\sum\limits_{j \in \mathcal{U}_m \setminus \mathcal{S}_{m,k}} \left|\mathbf{v}_{m,k}\mathbf{h}_{m,j}\right|^2 p_{m,j} + \sum\limits_{i=1, i \neq m}^{M+1} \sum\limits_{j \in \mathcal{U}_i} \left|\mathbf{v}_{m,k}\mathbf{h}_{m,j}\right|^2 p_{i,j} + \left\|\mathbf{v}_{m,k}\right\|^2 \sigma_m^2}, \quad (3)$$

where $\mathbf{v}_{m,k} \in \mathbb{C}^{1 \times N_m}$ is the receive beamformer for AP m to detect the signal from UE k. Thus, the achievable rate of UE k at AP m is given by

$$R_{m,k} = \log(1 + SINR_{m,k}) \quad (4)$$

Define $\alpha_{m,k}$ as the UE association indicator, where $\alpha_{m,k} = 1$ denotes that UE k is served by AP m, and $\alpha_{m,k} = 0$, otherwise. Therefore, the achievable transmission rate of UE k is written as

$$R_k = \sum_{m \in \mathcal{R}} \alpha_{m,k} R_{m,k} \quad (5)$$

And the achievable sum rate for AP m is given by

$$R_m = \sum_{k \in \mathcal{K}} \alpha_{m,k} R_{m,k} \quad (6)$$

2.2. Problem Formulation

In this work, we aim to maximize the achievable sum rate of all terrestrial UEs by optimizing the transmit power, receive beamformer, and UE association. The proposed problem is formulated as

$$\begin{aligned} \max_{\mathbf{v}_{m,k}, p_{m,k}, \alpha_{m,k}} \quad & R_{sum} = \sum_{k \in \mathcal{K}} R_k \\ \text{s.t.} \quad & \text{C1}: p_{m,k} \leq P_k^{max}, \quad \forall k \in \mathcal{K}, m \in \mathcal{R}, \\ & \text{C2}: \alpha_{m,k} \in \{0,1\}, \quad \forall k, m, \\ & \text{C3}: \sum_{m \in \mathcal{R}} \alpha_{m,k} = 1, \quad \forall k, \\ & \text{C4}: R_m \leq C_m, \quad \forall m \in \mathcal{M}, \\ & \text{C5}: R_k \geq r_k, \quad \forall k, \end{aligned} \quad (7)$$

where C1 limits the maximum transmit power of UE k, C2 constrains the UE user association indicators, and C3 denotes that a UE can only access one AP. Constraint C4 denotes the achieved sum rate of BS m should be less than its backhaul link capacity C_m. Constraint C5 guarantees the UE's QoS. Note that this work primarily considers communication scenarios where the access capacity of terrestrial BSs is limited, necessitating the introduction of satellites to provide service for terminals unable to connect to BSs. Consequently, satellites are specifically employed to guarantee normal communication in this region, and typically, satellite downlink communication rates are high. Thus, the backhaul link capacity constraint is not imposed on the satellite.

The optimization problem (7) is a mixed-integer non-linear programming problem, characterized by its pronounced non-convexity and NP complexity. Additionally, the update of the user association strategy triggers alterations in user channel conditions, making the direct transformation of the optimization problem (7) into a convex form exceedingly difficult. To address this complexity in a more efficient manner, the problem (7) has been strategically partitioned into three stages for resolution.

3. The Proposed Optimization Algorithm

3.1. User Association

With the fixed receive beamforming vector $\mathbf{v}_{m,k}$ and transmit power $p_{m,k}$, problem (7) is reduced to the following form

$$\begin{aligned} &\max_{\alpha_{m,k}} R_{sum} = \sum_{k \in \mathcal{K}} R_k \\ &\text{s.t.} \quad \text{C2}, \text{C3}, \text{C4}, \text{C5}, \end{aligned} \tag{8}$$

which is a non-convex integer programming that is difficult to solve. Here, we temporarily ignore the QoS constraint C5 and design an approach to determine user association.

Preparation: All the UEs establish their preference lists $\{Pre_k\}$ in accordance with the descending order of the channel gain towards the APs.

Judge and Join-in: Define $\mathcal{X}_1$ as the set of UEs that have already accessed an AP and $\mathcal{X}_2$ as the set of UEs that have not yet accessed any AP. Each UE in $\mathcal{X}_2$ sequentially sends the request signal to the APs in order of priority based on $\{Pre_k\}$. The AP determines whether the inclusion of the current user would cause the sum rate to exceed the backhaul link rate. If not, the AP allows the current user access. This judgment step is to ensure that the constraint C4 in (8) is satisfied. However, a challenging problem is that the subsequent UE association strategy, receive beamforming design, and power allocation can all affect the achievable sum rate of the current AP. To address this, an approximate upper bound on the achievable sum rate is derived as a criterion for making the decision.

Let $\boldsymbol{\alpha}_m = [\alpha_{m,1}, \alpha_{m,2}, \cdots, \alpha_{m,K}]^T$ and define $\mathbf{H}_m = [\mathbf{h}_{m,1}, \ldots, \mathbf{h}_{m,K}]\text{diag}(\boldsymbol{\alpha}_m)$ as the channel matrix from the UEs served by AP m to AP m, where $\text{diag}(\boldsymbol{\alpha})$ represents a diagonal matrix constructed from the vector $\boldsymbol{\alpha}$. Denote the interference channel matrix from the UEs served by AP i, $i \in \mathcal{R} \backslash m$ to AP m by

$$\overline{\mathbf{H}}_m = [\mathbf{H}_1\text{diag}(\boldsymbol{\alpha}_1), \cdots, \mathbf{H}_{m-1}\text{diag}(\boldsymbol{\alpha}_{m-1}), \mathbf{H}_{m+1}\text{diag}(\boldsymbol{\alpha}_{m+1}) \cdots, \mathbf{H}_{M+1}\text{diag}(\boldsymbol{\alpha}_{M+1})], \tag{9}$$

then, according to the minimum mean square error (MMSE) criterion, the upper bound of the achievable sum rate for AP m is

$$R_m^{up} = \log\det\left(\mathbf{I} + \frac{\mathbf{H}_m\mathbf{P}\mathbf{H}_m^H}{\overline{\mathbf{H}}_m\mathbf{P}\overline{\mathbf{H}}_m^H + \sigma_0^2\mathbf{I}}\right), \tag{10}$$

where $\mathbf{P} = \text{diag}(\mathbf{p}) = \text{diag}\left([p_1, p_2, \cdots, p_K]^T\right)$ and $p_k = \sum_{m \in \mathcal{R}} \alpha_{m,k} p_{m,k}$. The value of $p_{m,k}$ is set to the maximum, and $\alpha_{m,k}$ for $k \in \mathcal{X}_2$ is set to 0.

After the last AP reaches its saturation rate, if $\mathcal{X}_2 \neq \emptyset$, the remaining UEs reset their preference lists and repeat the above process.

3.2. Power Allocation

With the fixed receive beamforming vector $\mathbf{v}_{m,k}$ and UE association indicator $\alpha_{m,k}$, problem (7) is reduced to the following form:

$$\begin{aligned} \max_{p_{m,k}} \ & \sum_{m\in\mathcal{R}} \sum_{k\in\mathcal{U}_m} R_{m,k} \\ \text{s.t.} \ & \text{C1}' : p_{m,k} \le P_k^{max}, \quad \forall m \in \mathcal{R}, \forall k \in \mathcal{U}_m, \\ & \text{C4}' : \sum_{k\in\mathcal{U}_m} R_{m,k} \le C_m, \ \forall m \in \mathcal{M}, \\ & \text{C5}' : R_{m,k} \ge r_k, \ \forall m \in \mathcal{R}, \forall k \in \mathcal{U}_m, \end{aligned} \tag{11}$$

By introducing the Lagrange multiplier ρ_k associated with the constraint C5$'$, the partial Lagrange function of problem (11) can be derived as

$$\begin{aligned} \max_{p_{m,k}} \ & \sum_{m\in\mathcal{R}} \sum_{k\in\mathcal{U}_m} R_{m,k} + \sum_{m\in\mathcal{R}} \sum_{k\in\mathcal{U}_m} \rho_k (R_{m,k} - r_k) \\ \text{s.t.} \ & \text{C1}', \text{C4}' \end{aligned} \tag{12}$$

To tackle the non-convexity of (12), we first introduce the quadratic transform method in [20] to convert log-fractional rate functions in the objective function into convex quadratic expressions. By introducing auxiliary variables $\gamma_{m,k}$ and $\eta_{m,k}$, the log-fractional rate function can be transformed into

$$\begin{aligned} \tilde{R}_{m,k} \triangleq & \log(1+\gamma_{m,k}) - \gamma_{m,k} + 2\eta_{m,k}\sqrt{(1+\gamma_{m,k})\left|\mathbf{v}_{m,k}\mathbf{h}_{m,k}\right|^2 p_{m,k}} \\ & - \eta_{m,k}^2 \left(\sum_{j\in\{\mathcal{U}_m\setminus\mathcal{S}_{m,k}\}\cup k} \left|\mathbf{v}_{m,k}\mathbf{h}_{m,j}\right|^2 p_{m,j} + \sum_{i\in\mathcal{R}, i\ne m} \sum_{j\in\mathcal{U}_i} \left|\mathbf{v}_{m,k}\mathbf{h}_{i,j}\right|^2 p_{i,j} + \left\|\mathbf{v}_{m,k}\right\|^2 \sigma_m^2 \right) \end{aligned} \tag{13}$$

By substituting $\tilde{R}_{m,k}$ into (12), the problem is then transformed into the following:

$$\begin{aligned} \max_{p_{m,k}} \ & \sum_{m\in\mathcal{R}} \sum_{k\in\mathcal{U}_m} \tilde{R}_{m,k} + \sum_{m\in\mathcal{R}} \sum_{k\in\mathcal{U}_m} \rho_k (\tilde{R}_{m,k} - r_k) \\ \text{s.t.} \ & \text{C1}', \text{C4}' \end{aligned} \tag{14}$$

Given $p_{m,k}$, $\gamma_{m,k}$ and $\eta_{m,k}$ can be updated by

$$\gamma_{m,k} = \frac{\left|\mathbf{v}_{m,k}\mathbf{h}_{m,k}\right|^2 p_{m,k}}{\sum_{j\in\mathcal{U}_m\setminus\mathcal{S}_{m,k}} \left|\mathbf{v}_{m,k}\mathbf{h}_{m,j}\right|^2 p_{m,j} + \sum_{i\in\mathcal{R}, i\ne m} \sum_{j\in\mathcal{U}_i} \left|\mathbf{v}_{m,k}\mathbf{h}_{i,j}\right|^2 p_{i,j} + \left\|\mathbf{v}_{m,k}\right\|^2 \sigma_m^2} \tag{15}$$

$$\eta_{m,k} = \frac{\sqrt{(1+\gamma_{m,k})\left|\mathbf{v}_{m,k}\mathbf{h}_{m,k}\right|^2 p_{m,k}}}{\sum_{j\in\{\mathcal{U}_m\setminus\mathcal{S}_{m,k}\}\cup k} \left|\mathbf{v}_{m,k}\mathbf{h}_{m,j}\right|^2 p_{m,j} + \sum_{i\in\mathcal{R}, i\ne m} \sum_{j\in\mathcal{U}_i} \left|\mathbf{v}_{m,k}\mathbf{h}_{i,j}\right|^2 p_{i,j} + \left\|\mathbf{v}_{m,k}\right\|^2 \sigma_m^2} \tag{16}$$

With fixed $\gamma_{m,k}$ and $\eta_{m,k}$, the residual power allocation of (14) is convex without C4$'$. Accordingly, during the iterative process, we temporarily disregard constraint C4$'$. Following the completion of the iterations, adjustments will be made to the power allocation in order to satisfy constraint C4$'$. The corresponding Lagrangian dual function is derived as

$$\mathcal{L}(p_{m,k}) = \sum_{m\in\mathcal{R}} \sum_{k\in\mathcal{U}_m} \left(\tilde{R}_{m,k} + \rho_k(\tilde{R}_{m,k} - r_k)\right) \tag{17}$$

The Karush-Kuhn-Tucker (KKT) conditions are

$$\begin{aligned}\frac{\partial \mathcal{L}(p_{m,k})}{\partial p_{m,k}} =&(1+\rho_k)\eta_{m,k}\sqrt{\frac{(1+\gamma_{m,k})\left|\mathbf{v}_{m,k}\mathbf{h}_{m,k}\right|^2}{p_{m,k}}} - \sum_{k'\in\mathcal{S}_{m,k}\cup k}(1+\rho_{k'})\eta_{m,k'}^2\left|\mathbf{v}_{m,k'}\mathbf{h}_{m,k}\right|^2 \\ &- \sum_{m'\in\mathcal{R}\backslash m}\sum_{k'\in\mathcal{U}_i}(1+\rho_{k'})\eta_{m',k'}^2\left|\mathbf{v}_{m',k'}\mathbf{h}_{m,k}\right|^2 = 0,\ \forall m\in\mathcal{R}, k\in\mathcal{U}_m,\end{aligned} \tag{18}$$

Then, the optimal power allocation is derived by

$$p_{m,k}^* = \min\{\tilde{p}_{m,k}, P_k^{max}\} \tag{19}$$

where

$$\tilde{p}_{m,k} = \frac{((1+\rho_k)\eta_{m,k})^2(1+\gamma_{m,k})\left|\mathbf{v}_{m,k}\mathbf{h}_{m,k}\right|^2}{\left(\sum\limits_{k'\in\mathcal{S}_{m,k}\cup k}(1+\rho_{k'})\eta_{m,k'}^2\left|\mathbf{v}_{m,k'}\mathbf{h}_{m,k}\right|^2 + \sum\limits_{m'\in\mathcal{R}\backslash m}\sum\limits_{k'\in\mathcal{U}_i}(1+\rho_{k'})\eta_{m',k'}^2\left|\mathbf{v}_{m',k'}\mathbf{h}_{m,k}\right|^2\right)^2} \tag{20}$$

The Lagrangian multipliers ρ_k can be obtained by utilizing the gradient update method [21]. Specifically, in the t-th iteration, the Lagrangian multiplier is updated by the following expression

$$\rho_k^{(t)} = [\rho_k^{(t-1)} - \phi_k(\tilde{R}_{m,k} - r_k)]^+, \tag{21}$$

where the update step-size ϕ_k plays a crucial role in controlling the convergence of the Lagrangian multiplier.

After iterations, we need to deal with constraint C4′. Assume $\tilde{\mathcal{R}}$ representing the set of APs that cannot meet $\sum_{k\in\mathcal{U}_m} R_{m,k} \le C_m$, then we set $\Delta C_m = \sum_{k\in\mathcal{U}_m} R_{m,k} - C_m, \forall m \in \tilde{\mathcal{R}}$. The power allocation results can be adjusted as

$$p_{m,k} = \left(2^{\left(R_{m,k}-\Delta C_m\left(R_{m,k}/\sum\limits_{j\in\mathcal{U}_m}R_{m,j}\right)\right)} - 1\right)\frac{I_{m,k}}{\left|\mathbf{v}_{m,k}\mathbf{h}_{m,k}\right|^2},\ \forall m\in\tilde{\mathcal{R}} \tag{22}$$

where $I_{m,k} = \sum_{j\in\mathcal{U}_m\backslash\mathcal{S}_{m,k}}\left|\mathbf{v}_{m,k}\mathbf{h}_{m,j}\right|^2 p_{m,j} + \sum_{i=1,i\neq m}^{M+1}\sum_{j\in\mathcal{U}_i}\left|\mathbf{v}_{m,k}\mathbf{h}_{m,j}\right|^2 p_{i,j} + \left\|\mathbf{v}_{m,k}\right\|^2\sigma_m^2$.

3.3. Receive Beamforming

With the given transmit power $p_{m,k}$ and UE association indicator $\alpha_{m,k}$, the receive beamforming vectors $\mathbf{v}_{m,k}, \forall m,k$, can be obtained based on the optimal MMSE criteria to cope with the multiuser interference, which is obtained by

$$\mathbf{v}_{m,k}^* = \arg\max SINR_{m,k}, \tag{23}$$

which can be further represented as

$$\mathbf{v}_{m,k}^* = \arg\max\frac{\mathbf{v}_{m,k}\mathbf{Q}_1\mathbf{v}_{m,k}^H}{\mathbf{v}_{m,k}\mathbf{Q}_2\mathbf{v}_{m,k}^H}, \tag{24}$$

where $\mathbf{Q}_1 = \mathbf{h}_{m,k}p_{m,k}\mathbf{h}_{m,k}^H$, $\mathbf{Q}_2 = \sum_{j\in\mathcal{U}_m\backslash\mathcal{S}_{m,k}}\mathbf{h}_{m,j}p_{m,j}\mathbf{h}_{m,j}^H + \sum_{i=1,i\neq m}^{M+1}\sum_{j\in\mathcal{U}_i}\mathbf{h}_{m,j}p_{i,j}\mathbf{h}_{m,j}^H + \sigma_{m,k}^2\mathbf{I}$. It can be seen that (24) is a generalized Rayleigh quotient from [22]. Thus, the optimal solution is the eigenvector corresponding to the maximum generalized eigenvalue of matrix $\mathbf{Q}_2^{-1}\mathbf{Q}_1$, which is written as

$$\mathbf{v}_{m,k}^* = \lambda_{max}(\mathbf{Q}_2^{-1}\mathbf{Q}_1) \tag{25}$$

3.4. Complexity Analysis

The proposed two-stage algorithm to solve (7) is summarized in Algorithm 1. The first stage comprises steps 2 to 17, and the second stage comprises steps 18 to 25. In the first stage, the key operation is to calculate the achievable sum rate in (10), which needs a determinant operation with computational complexity $\mathcal{O}(N!)$. Thus, the complexity for the first stage in the worst case is $\mathcal{O}(N!K(M+1)T_1)$. Here, T_1 is a predefined maximum number of user association rounds. In the second stage, the key operation is to calculate $\mathbf{v}_{m,k}, \gamma_{m,k}, \eta_{m,k}$ and $p_{m,k}$. The complexity to calculate those parameters is $\mathcal{O}(N^3)$, $\mathcal{O}(N)$, $\mathcal{O}(N)$ and $\mathcal{O}(N)$, respectively. Thus, the complexity of the second stage is $\mathcal{O}\big((T_2^{in}(M+1)KN + (M+1)KN^3)T_2^{out}\big)$. Here, T_2^{out} and T_2^{in} represent the maximum iteration numbers for the outer and inner iterations, respectively.

Algorithm 1: The proposed two-stage algorithm to solve (7)

1 **Initialization:** $\alpha_{m,k} = 0$, $p_{m,k} = P_k^{max}$, ρ_k, $\gamma_{m,k}$, $\eta_{m,k}$, $\forall m \in \mathcal{R}, \forall k \in \mathcal{K}$, set $\mathcal{X}_1 = \emptyset$, $\mathcal{X}_2 = \mathcal{K}$; all the UEs set their preference lists $\{Pre_k\}$;
2 **for** *k=1 to card($\mathcal{X}_2$)* **do**
3   **while** $\{Pre_k\} \neq \emptyset$ **do**
4   UE k sends a request to the AP (denoted by AP i) with the highest priority in $\{Pre_k\}$;
5   Calculate R_i^{up} in (10);
6   **if** $R_i^{up} <= C_i$ **then**
7       $\alpha_{i,k} = 1$;
8       $\mathcal{X}_1 \leftarrow \mathcal{X}_1 \cup k, \mathcal{X}_2 \leftarrow \mathcal{X}_2 \setminus k$;
9   break;
10   **else**
11       $\{Pre_k\} \leftarrow \{Pre_k\} \setminus i$;
12   **end**
13   **end**
14 **end**
15 **if** $\mathcal{X}_2 \neq \emptyset$ **then**
16   All the UEs within $\mathcal{X}_2$ reset their preference lists and return to step 4;
17 **end**
18 **repeat**
19   Update $\mathbf{v}_{m,k}, \forall m,k$ in (25);
20   **repeat**
21   Update $\gamma_{m,k}$ and $\eta_{m,k}, \forall m,k$ in (15) and (16), respectively;
22   Compute $p_{m,k}, \forall m,k$ in (19);
23   **until** *Convergence*;
24 **until** *Convergence*;
25 Compute R_m, for all m to obtain $\tilde{R}$, and update $p_{m,k}$ for $m \in \tilde{R}$, $k \in \mathcal{U}_m$ in (22);
26 **Output:**$\{\alpha_{m,k}^*, p_{m,k}^*, \mathbf{v}_{m,k}^*\}$

According to the above analysis, the total computational complexity for our proposed algorithm is $\mathcal{O}\big((T_2^{in}(M+1)KN + (M+1)KN^3)T_2^{out} + N!K(M+1)T_1\big)$.

4. Simulation Results

This section presents simulation results to validate the performance of our proposed algorithms. We consider 4 BSs serving a square region on the ground with a size of 1 km $\times$ 1 km. The coordinates of the four vertices of the considered square region are (0 km, 0 km), (1 km, 0 km), (1 km, 1 km), and (0 km, 1 km), respectively. The four BSs are uniformly placed in the square region, with coordinates of (0.25 km, 0.25 km), (0.25 km, 0.75 km), (0.75 km, 0.75 km), and (0.75 km, 0.25 km), respectively. A low Earth orbit (LEO) satellite at an altitude of 300 km is deployed to serve the considered region. All the UEs are evenly and randomly distributed within the considered square region. The channels between UEs and BSs are modeled as Rayleigh channels [23], and the channels between UEs and the satellite are modeled as far-field line-of-sight channels [24]. The path loss exponents for the two types of links are set to 3 and 2, respectively. The maximum transmit powers P_k^{max} for all $k \in \mathcal{K}$ are set to be the same. Table 1 summarizes the simulation parameters. The simulations were conducted using Matlab 2021 B software, yielding numerical results from Monte Carlo simulations with 100 independent channel realizations.

Table 1. Simulation parameters.

Parameter	Setting
Path loss model	$\beta(\frac{d}{d_0})^{-\alpha}$ [15]
Parameters in path loss	$\beta= -30$ dB, $d_0 = 1$ m, $\alpha= 3, 2$ [15]
Noise power	-174 dBm/Hz [15]
Altitude of satellite	300 km [1]
Number of BSs	$M = 4$
Coordinates of BSs	(0.75 km, 0.75 km), (0.75 km, 0.25 km) (0.75 km, 0.75 km), (0.75 km, 0.25 km)
Backhaul link capacity for BS	$C_1 = C_2 = \cdots = C_M = C = 50$ bps/Hz
Number of receive antennas at BS	$N_1 = N_2 = \cdots = N_{M+1} = N = 10, 20, 40$
Number of users	$K = 10, 20, 30, 40$

Figure 2 gives the convergence of the proposed algorithm under different settings of K and N. The main iteration steps occur in the inner and outer iterations of stage 2 in Algorithm 1. During the inner iteration, parameters $\gamma_{m,k}, \eta_{m,k}$ are updated. The outer iteration involves the alternative optimization of $\mathbf{v}_{m,k}$ and $p_{m,k}$. Consequently, in the revised manuscript, we incorporated the convergence behavior for both the inner and outer iterations of stage 2 in Algorithm 1. It is seen that for both inner and outer iterations, the convergence is monotonic under all settings of K, N. For the inner iteration, it takes approximately 15 iterations to reach convergence. For the outer iteration, about four iterations are required to converge.

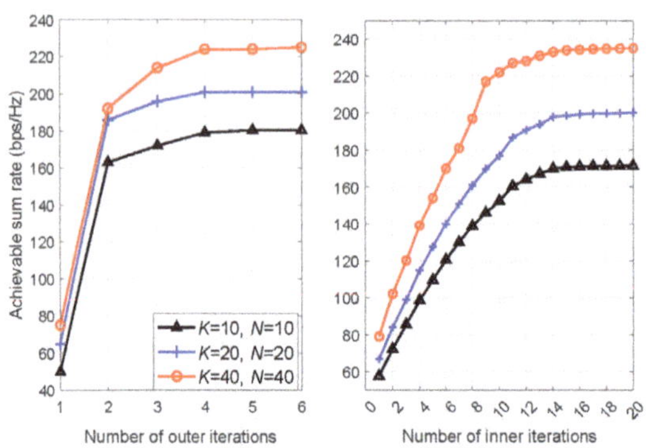

Figure 2. Convergence of the proposed algorithm under different settings of K and N.

Figure 3 presents the achievable sum rate versus the maximum transmit power. The total number of UEs is set to $K = 10$, $K = 20$, $K = 30$. The number of receive antennas is set to $N = 20$. It is seen that the achievable sum rate increases with the maximum transmit power. The NOMA scheme is obviously better than the OMA scheme, which indicates the benefits of adopting NOMA in uplink ISTNs. The performance degrades when the number of UEs is set to 30. This is because, when the number of receive antennas exceeds the number of UEs, the system has a sufficient degree of freedom to distinguish signals from different UEs and minimize interference among them. However, once the number of UEs exceeds the number of antennas, the system lacks the necessary degree of freedom for leveraging multiuser diversity, resulting in the degradation of system performance.

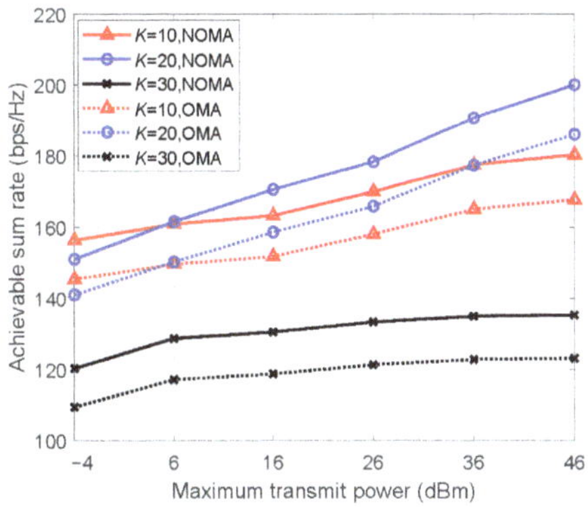

Figure 3. Achievable sum rate versus the maximum transmit power.

Figure 4 presents the achievable sum rate versus the number of UEs. The maximum transmit power is set as 26 dBm. The numbers of receive antennas for different APs are set to the same constant N. It is seen that the achievable sum rate increases with the number of UEs first, and then decreases sharply and finally becomes stable. The performance peak occurs when the total number of UEs is equal to the number of receive antennas. The reason is mentioned in the previous paragraph. Moreover, a comparison between the proposed algorithm and the random power allocation algorithm is conducted. For the random power allocation algorithm, the transmit power follows a uniform distribution in the range of $[0, 26]$ dBm. And we calculate the mean value of 20 transmit power sets. It is observed that our proposed algorithm performs better than the random power allocation algorithm.

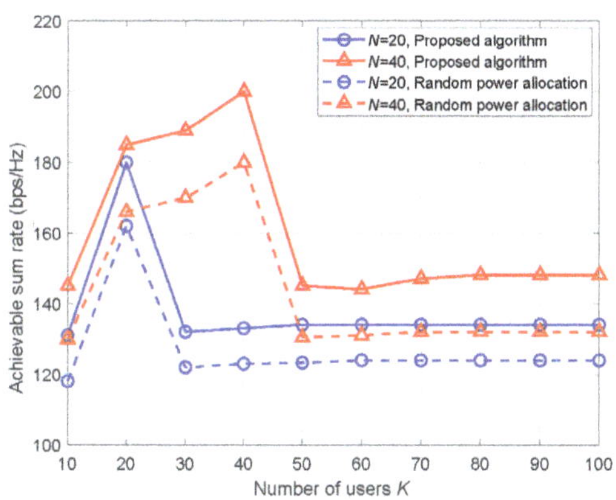

Figure 4. Achievable sum rate versus the number of UEs.

In Figure 5, the user association results are presented for various sets of backhaul link rate C. The maximum transmit power is set as 26 dBm. It is seen that the distribution of the number of UEs accessing different terrestrial BSs is relatively even. As the backhaul link capacity increases, fewer UEs will access the satellite. This is because UEs will prioritize accessing terrestrial BSs with better channel conditions. However, in situations where the capacity of terrestrial BSs is constrained, UEs will turn to accessing the satellite in order to guarantee the fulfillment of their QoS requirements.

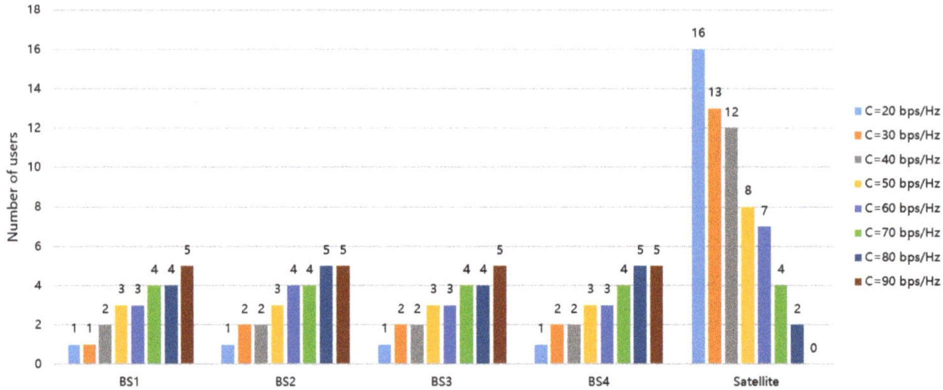

Figure 5. User association results.

Figure 6 presents the achievable sum rate versus the backhaul link capacity. The total number of UEs K and receive antennas N are set as follows: $K = 10$, $N = 10$, $K = 20$, $N = 20$, and $K = 40$, $N = 40$. The maximum transmit power is set as 26 dBm. It is seen that the performance increases with the increase in backhaul link capacity. When the backhaul link capacity is relatively low, the sum rate for 10 UEs is higher than the sum rates for 20 UEs and 40 UEs. This is because when the backhaul link capacity is small, the number of UEs that terrestrial BSs can accommodate is relatively limited, and excess UEs will access the satellite. This results in severe interference among the UE signals received by the satellite, thereby impairing the overall performance. When the number of UEs is relatively low, terrestrial BSs are sufficient to provide service to all UEs even with limited backhaul link capacity. As the backhaul link capacity of these BSs continues to rise, the achievable sum rate approaches saturation, leading to a flattened trend in the sum rate for 10 UEs thereafter.

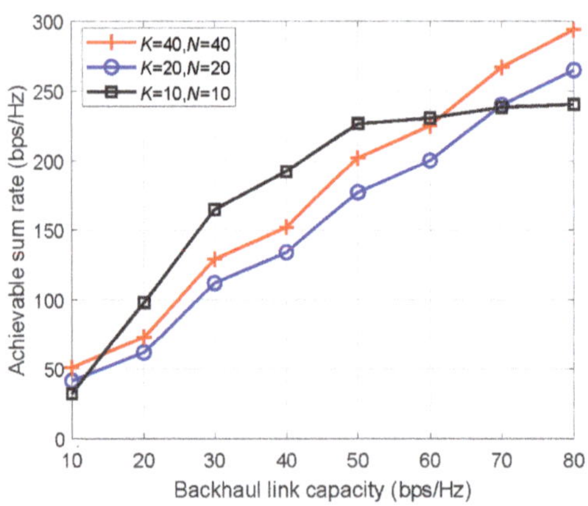

Figure 6. Achievable sum rate versus the backhaul link capacity.

5. Conclusions

This paper investigates a NOMA-based uplink ISTN, in which each user can access the satellite or terrestrial BSs according to their individual QoS requirements. In order to maximize the achievable sum rate, an optimization problem is established to jointly optimize the transmit power, user association, and receive beamforming. A two-stage algorithm is developed to solve the non-convex problem. At the first stage, user associations are determined based on individual preference lists and the backhaul capacities of the APs. At the second stage, the power allocation and the receiving beamforming vectors are alternately optimized in an iterative process. Within each iteration, the closed-form solution for the transmit power is derived. Simulation results validate the effectiveness of the proposed algorithm and the benefits brought by NOMA. UEs prefer to access terrestrial BSs when their backhaul link capacity is sufficient; otherwise, they can rely on the satellite for QoS guarantees. Furthermore, the achievable sum rate achieves its peak when the number of UEs aligns with the number of receive antennas at the APs.

Author Contributions: Conceptualization, P.X. and Z.F.; methodology, Z.C. and J.J.; validation, Y.Z. and X.H.; writing—review and editing, J.Z.; funding acquisition, P.X. All authors have read and agreed to the published version of the manuscript.

Funding: This research was funded by the science and technology project of the state grid corporation of China: Research and application on critical technology of integrated communication, sensing, and computation in power wireless network under hierarchical cooperative control mode of the distribution network (No. 5700-202356629A-3-2-ZN).

Institutional Review Board Statement: Not applicable.

Data Availability Statement: Data are contained within the article. Further inquiries can be directed to the corresponding author.

Conflicts of Interest: Authors Peizhe Xin, Zihao Fu, Zhiyi Chen, Jing Jiang, Jing Zou were employed by the company State Grid Economic and Technological Research Institute Co., Ltd. The remaining authors declare that the research was conducted in the absence of any commercial or financial relationships that could be construed as a potential conflict of interest.

References

1. Lv, Z.; Gong, F.; Chen, G.; Li, G.; Hui, T.; Xu, S. Energy Efficiency Design in RIS-Assisted Satellite–Terrestrial Integrated Networks with NOMA. *IEEE Wirel. Commun. Lett.* **2024**, *13*, 1948–1952. [CrossRef]
2. Liu, X.; Zhao, B.; Lin, M.; Ouyang, J.; Wang, J.-B.; Wang, J. IRS-Aided Uplink Transmission Scheme in Integrated Satellite-Terrestrial Networks. *IEEE Trans. Veh. Technol.* **2023**, *72*, 1847–1861. [CrossRef]
3. Zhu, X.; Jiang, C. Integrated Satellite-Terrestrial Networks Toward 6G: Architectures, Applications, and Challenges. *IEEE Internet Things J.* **2022**, *9*, 437–461. [CrossRef]
4. Lin, M.; Lin, Z.; Zhu, W.-P.; Wang, J.-B. Joint beamforming for secure communication in cognitive satellite terrestrial networks. *IEEE J. Sel. Areas Commun.* **2018**, *36*, 1017–1029. [CrossRef]
5. Dai, C.-Q.; Li, S.; Wu, J.; Chen, Q. Distributed User Association with Grouping in Satellite–Terrestrial Integrated Networks. *IEEE Internet Things J.* **2022**, *9*, 10244–10256. [CrossRef]
6. Chan, S.; Lee, H.; Kim, S.; Oh, D. Intelligent Low Complexity Resource Allocation Method for Integrated Satellite-Terrestrial Systems. *IEEE Wirel. Commun. Lett.* **2022**, *11*, 1087–1091. [CrossRef]
7. Ruan, Y.; Li, Y.; Zhang, R.; Jiang, L. Energy Efficient Power Control for Cognitive Multibeam-Satellite Terrestrial Networks with Poisson Distributed Users. *IEEE Trans. Cogn. Commun. Netw.* **2022**, *8*, 964–974. [CrossRef]
8. Peng, D.; Bandi, A.; Li, Y.; Chatzinotas, S.; Ottersten, B. Hybrid Beamforming, User Scheduling, and Resource Allocation for Integrated Terrestrial-Satellite Communication. *IEEE Trans. Veh. Technol.* **2021**, *70*, 8868–8882.
9. Zhu, L.; Bai, L.; Zhou, L.; Choi, J. Efficient User Scheduling for Uplink Hybrid Satellite-Terrestrial Communication. *IEEE Trans. Wirel. Commun.* **2023**, *22*, 1885–1899. [CrossRef]
10. Li, X.; Zhao, M.; Zeng, M.; Mumtaz, S.; Menon, V.G.; Ding, Z.; Dobre, O.A. Hardware Impaired Ambient Backscatter NOMA Systems: Reliability and Security. *IEEE Trans. Commun.* **2021**, *69*, 2723–2736. [CrossRef]
11. Zhang, Y.; Zhao, X.; Zhou, Z.; Qin, P.; Geng, S.; Xu, C. Robust Resource Allocation for Lightweight Secure Transmission in Multicarrier NOMA-Assisted Full Duplex IoT Networks. *IEEE Internet Things J.* **2022**, *9*, 6443–6457. [CrossRef]
12. Li, N.; Wu, P.; Ning, B.; Zhu, L. Sum Rate Maximization for Movable Antenna Enabled Uplink NOMA. *IEEE Wirel. Commun. Lett.* **2024**, *13*, 2140–2144. [CrossRef]
13. Ghanbarzadeh, V.; Zahabi, M.; Amiriara, H.; Jafari, F.; Kaddoum, G. Resource Allocation in NOMA Networks: Convex Optimization and Stacking Ensemble Machine Learning. *IEEE Open J. Commun. Soc.* **2024**, *5*, 5276–5288. [CrossRef]
14. Esmaeily, A.; Mendis, H.K.; Mahmoodi, T.; Kralevska, K. Beyond 5G Resource Slicing with Mixed-Numerologies for Mission Critical URLLC and eMBB Coexistence. *IEEE Open J. Commun. Soc.* **2023**, *4*, 727–747. [CrossRef]

15. Wang, L.; Wu, Y.; Zhang, H.; Choi, S.; Leung, V.C.M. Resource Allocation for NOMA Based Space-Terrestrial Satellite Networks. *IEEE Trans. Wirel. Commun.* **2021**, *20*, 1065–1075. [CrossRef]
16. Zhang, Y.; Zhang, H.; Zhou, H.; Long, K.; Karagiannidis, G.K. Resource Allocation in Terrestrial-Satellite-Based Next Generation Multiple Access Networks with Interference Cooperation. *IEEE J. Sel. Areas Commun.* **2022**, *40*, 1210–1221. [CrossRef]
17. Kong, H.; Lin, M.; Han, L.; Zhu, W.-P.; Ding, Z.; Alouini, M.-S. Uplink Multiple Access with Semi-Grant-Free Transmission in Integrated Satellite-Aerial-Terrestrial Networks. *IEEE J. Sel. Areas Commun.* **2023**, *41*, 1723–1736. [CrossRef]
18. Lin, Z.; Lin, M.; Wang, J.-B.; de Cola, T.; Wang, J. Joint Beamforming and Power Allocation for Satellite-Terrestrial Integrated Networks with Non-Orthogonal Multiple Access. *IEEE J. Sel. Top. Signal Process.* **2019**, *13*, 657–670. [CrossRef]
19. Morello, A.; Mignone, V. DVB-S2: The second generation standard for satellite broad-band services. *Proc. IEEE* **2006**, *94*, 210–227. [CrossRef]
20. Shen, K.; Yu, W. Fractional programming for communication systems—Part I: Power control and beamforming. *IEEE Trans. Signal Process.* **2018**, *66*, 2616–2630. [CrossRef]
21. Cheung, K.T.K.; Yang, S.; Hanzo, L. Achieving maximum energy-efficiency in multi-relay OFDMA cellular networks: A fractional programming approach. *IEEE Trans. Commun.* **2013**, *16*, 2746–2757. [CrossRef]
22. Iimori, H.; de Abreu, G.T.F.; Alexandropoulos, G.C. MIMO Beamforming Schemes for Hybrid SIC FD Radios with Imperfect Hardware and CSI. *IEEE Trans. Wirel. Commun.* **2019**, *18*, 4816–4830. [CrossRef]
23. Di, B.; Zhang, H.; Song, L.; Li, Y.; Li, G.Y. Ultra-Dense LEO: Integrating Terrestrial-Satellite Networks Into 5G and Beyond for Data Offloading. *IEEE Trans. Wirel. Commun.* **2019**, *18*, 47–62. [CrossRef]
24. Toka M.; Lee, B.; Seong, J.; Kaushik, A.; Lee, J.; Lee, J.; Lee, N.; Shin, W.; Poor, H.V. RIS-Empowered LEO Satellite Networks for 6G: Promising Usage Scenarios and Future Directions. *IEEE Commun. Mag.* **2024**, *62*, 128–135. [CrossRef]

entropy

MDPI

Article

Simultaneously Transmitting and Reflecting Reconfigurable Intelligent Surfaces Empowered Cooperative Rate Splitting with User Relaying †

Kangchun Zhao, Yijie Mao * and Yuanming Shi

School of Information Science and Technology, Shanghai Tech University, Shanghai 201210, China; zhaokch12022@shanghaitech.edu.cn (K.Z.); shiym@shanghaitech.edu.cn (Y.S.)
* Correspondence: maoyj@shanghaitech.edu.cn

† This paper is an extended version of our paper published in the IEEE 98th Vehicular Technology Conference, VTC2023-Fall, Hong Kong, China, 10–13 October 2023.

Abstract: In this work, we unveil the advantages of synergizing cooperative rate splitting (CRS) with user relaying and simultaneously transmitting and reflecting reconfigurable intelligent surface (STAR RIS). Specifically, we propose a novel STAR RIS-assisted CRS transmission framework, featuring six unique transmission modes that leverage various combinations of the relaying protocols (including full duplex-FD and half duplex-HD) and the STAR RIS configuration protocols (including energy splitting-ES, mode switching-MS, and time splitting-TS). With the objective of maximizing the minimum user rate, we then propose a unified successive convex approximation (SCA)-based alternative optimization (AO) algorithm to jointly optimize the transmit active beamforming, common rate allocation, STAR RIS passive beamforming, as well as time allocation (for HD or TS protocols) subject to the transmit power constraint at the base station (BS) and the law of energy conservation at the STAR RIS. To alleviate the computational burden, we further propose a low-complexity algorithm that incorporates a closed-form passive beamforming design. Numerical results show that our proposed framework significantly enhances user fairness compared with conventional CRS schemes without STAR RIS or other STAR RIS-empowered multiple access schemes. Moreover, the proposed low-complexity algorithm dramatically reduces the computational complexity while achieving very close performance to the AO method.

Keywords: cooperative rate splitting (CRS); simultaneously transmitting reconfigurable intelligent surface (STAR RIS); rate splitting multiple access (RSMA); max-min fairness (MMF)

1. Introduction

In sixth-generation (6G) and beyond mobile communications, an increasing number of devices connect to the wireless network, causing the following problems: (1) Multi-device connections introduce significant multi-user interference, severely degrading system performance. (2) Ensuring a favorable communication environment for each user becomes highly challenging due to the vast number of devices accessing the transmission network. To deal with these problems, a novel non-orthogonal interference management strategy in the physical (PHY) layer named rate-splitting multiple access (RSMA) comes into our sight. It emerges as a promising technique for enhancing spectral efficiency and user fairness [1]. RSMA follows the design principle where user messages are divided into common and private parts at the transmitter. The common parts are encoded into common streams and decoded by multiple users. Meanwhile, the private parts are independently encoded into private streams and decoded only by the corresponding users, following the decoding of the common streams and their subsequent removal through successive interference cancellation (SIC). This empowers RSMA to partially treat interference as noise and partially decode interference. It encompasses various multiple access (MA) schemes like space division MA (SDMA), which fully treats interference as noise, and non-orthogonal MA (NOMA), which fully decodes interference, treating them as special cases [2].

However, as the common streams of RSMA are required to be decoded by multiple users, the achievable rate will be limited by the user with the weakest channel strength. To overcome this limitation, a novel RSMA scheme named cooperative rate splitting (CRS) is proposed in [3,4]. This scheme empowers users with stronger channel conditions, acting as relaying users, to transmit the decoded common stream to users with weaker channel conditions. CRS thereby boosts the received signal strengths at the weaker users and improves the achievable common rate. It not only enhances spectral efficiency but also extends radio coverage, promotes user fairness, and reduces energy consumption [4,5]. There are two transmission phases in CRS, one is the direct transmission phase in which the base station (BS) transmits signals to all users, and the other is the cooperative transmission phase in which the relaying users transmit the decoded common streams to the destination users. According to whether these two phases are executed at the same time, there are two protocols in CRS: *half duplex (HD)* [4] and *full duplex (FD)* [6]. In the *HD* protocol, the direct and cooperative transmission phases are executed in different time slots, while in the *FD* protocol, these two transmission phases are executed at the same time. Both protocols demonstrate significant potential in realizing the aforementioned advantages of CRS.

In parallel, reconfigurable intelligent surface (RIS), gaining increasing attention recently, has been viewed as a promising technique in future wireless networks [7–9]. An RIS is a two-dimensional (2D) meta-surface containing numerous passive and low-cost reflecting elements that enable tuning the phase shift of the incident signal. By employing the RIS, the line-of-sight (LoS) channel between the BS and users can be reconfigured, effectively improving the spectral efficiency, and energy efficiency and extending the communication coverage. Nevertheless, conventional RIS can solely reflect signals, limiting its coverage to a 180° area. This has motivated the emergence of a novel RIS that enables reflection and transmits the incident signal at the same time, which is called simultaneously transmitting and reflecting RIS (STAR RIS) [10] or intelligent omni-surface (IOS) [11]. According to different operating methods, STAR RIS has three protocols: *energy splitting (ES)*, *mode switching (MS)*, and *time switching (TS)*. In the *ES* protocol, all elements can reflect and transmit the incident signal at the same time. In the *MS* protocol, each element either reflects or transmits the incident signal. In the *TS* protocol, the transmission time is split into two time slots. One is the reflection time slot, where all elements reflect the incident signal, and the other is the transmission time slot, where all elements transmit the incident signal. By adjusting the amplitude and phase shift of the reflected and transmitted signals, all three protocols of STAR RIS offer full spatial coverage and additional degrees of freedom (DoF) for the system.

1.1. Related Works

In view of the advantages of STAR RIS, many existing works have explored its integration with various MA schemes [12–17]. In [12], the authors investigated the coverage range of a two-user STAR RIS-aided orthogonal MA (OMA) or NOMA system. In [13], the beamforming optimization was studied in an uplink STAR RIS-aided NOMA system for outage probability minimization and secrecy capacity maximization. In [14], the sum rate maximization was investigated in a downlink STAR RIS-aided NOMA system, wherein the decoding order and transmit beamforming are jointly optimized. These studies have demonstrated that STAR RIS offers benefits such as extending communication coverage and enhancing the sum rate. Besides, the integration of STAR RIS and RSMA has recently been investigated in some existing works [15–17]. In [15], the primary focus was on maximizing the sum secrecy rate in a STAR RIS-aided simultaneous wireless information and power transfer (SWIPT) system using RSMA. In [16], the sum-rate maximization problem for a STAR RIS-aided uplink RSMA system is studied. In [17], efforts were directed toward minimizing the outage probability while considering the spatial correlation among the

STAR RIS channels. The aforementioned STAR RIS-aided RSMA works [15–17] emphasize the advantages of STAR RIS in enabling full-space wireless networks and enhancing interference management, leading to improvements in both spectral and energy efficiency. However, so far, the integration of STAR RIS and CRS has not been investigated yet. And the performance of max-min fairness for STAR RIS-aided RSMA remains unexplored.

1.2. Contributions

Considering the aforementioned advantages of both CRS and STAR RIS, along with the identified research gaps in the current literature, there is a compelling incentive to integrate these two techniques. On one hand, the CRS transmission scheme provides robust and flexible interference management to enhance the performance of STAR RIS. On the other hand, STAR RIS reconfigures and improves wireless channels, thereby further enhancing the performance of CRS. Their integration can lead to a mutually beneficial solution. In this work, we delve into the STAR RIS-aided CRS system, and the main contributions of this paper are summarized as follows:

- We propose a novel downlink STAR RIS-aided CRS transmission framework, empowering a STAR RIS to assist both the direct and cooperative transmission phases of CRS. Within this framework, we investigate six different transmission modes including various combinations of CRS relaying protocols (*HD* and *FD*) and STAR RIS operating protocols (*ES*, *MS*, and *TS*).
- We formulate a new resource allocation problem with the objective of maximizing the minimum user rate. To solve this problem, the STAR RIS passive beamforming, the BS active beamforming, common rate allocation, and time slot allocation (in *HD* or *TS* protocols) are jointly optimized under the transmit power constraint at the BS and the energy conservation constraints at the STAR RIS. Due to the non-convexity of the formulated problem, we propose an alternative optimization (AO) algorithm to solve the problem. This approach involves decomposing the original problem into two subproblems: the STAR RIS passive beamforming optimization and transmit active beamforming optimization. Each subproblem is then solved using a successive convex approximation (SCA)-based method. Through iterative solving of the two subproblems, we attain a near-optimal solution until convergence.
- We further propose a low-complexity algorithm to solve the formulated problem. For the passive beamforming design, we derive a closed-form solution for the STAR RIS passive beamforming based on the gradient descent approach. To ensure the derived solution meets the STAR RIS constraints, we further use the symmetric unitary projection based on singular value decomposition (SVD) to project the solution into the feasible set of constraints. To transmit active beamforming, we use the zero-forcing (ZF) approach to fix the beamforming direction. Subsequently, we simply optimize the power allocation using SCA. Apart from perfect channel state information at the transmitter (CSIT), we also extend the two proposed algorithms to address the ergodic rates (ERs) optimization problems under imperfect CSIT.
- We evaluate the performance of the proposed STAR RIS-aided CRS framework and show the effectiveness of our proposed algorithms by numerical results. Our analysis reveals that the STAR RIS-aided CRS scheme outperforms other STAR RIS-aided MA schemes. We also offer insights into the preferred regions for the six proposed transmission protocols. Moreover, the results demonstrate that our proposed low-complexity algorithm achieves comparable performance while significantly reducing CPU time compared to the AO algorithm.

1.3. Organization

The subsequent sections of the paper are organized as follows. Section 2 delineates the system model and formulates the max-min fairness problem. Section 3 details the AO optimization framework for the problem and the proposed low-complexity algorithm. Section 4 presents numerical results. Section 5 concludes the paper.

2. System Model and Problem Formulation

As illustrated in Figure 1, we consider a STAR RIS-assisted multi-user multiple-input single-output (MISO) downlink transmission network, with CRS supported during the transmission. There is a BS equipped with L transmit antennas, a STAR RIS containing N elements indexed by the set $\mathcal{N} = \{1, \cdots, N\}$, and K users indexed by the set $\mathcal{K} = \{1, 2, \cdots, K\}$. Each user is assumed to be low-mobility and equipped with a single transmit antenna and a receive antenna. Given the ability of the STAR RIS to transmit and reflect the incident signal simultaneously, we divide the transmission space into two distinct subspaces, namely, the reflection space and the transmission space. Users in the reflection space receive signals reflected by the STAR RIS and are indexed by the set $\mathcal{K}_r$. Conversely, users in the transmission space receive signals transmitted by the STAR RIS and are indexed by the set $\mathcal{K}_t$. The two user sets satisfy $\mathcal{K}_r \cup \mathcal{K}_t = \mathcal{K}$ and $\mathcal{K}_r \cap \mathcal{K}_t = \emptyset$. Both subspaces allow user relaying, which further divides the users in each subspace into two user groups: the relaying user group $\mathcal{K}_{i1}$ and the destination user group $\mathcal{K}_{i2}, i \in \{r, t\}$. The user groups satisfy $\mathcal{K}_{r1} \cap \mathcal{K}_{r2} = \emptyset, \mathcal{K}_{t1} \cap \mathcal{K}_{t2} = \emptyset, \mathcal{K}_{r1} \cup \mathcal{K}_{r2} = \mathcal{K}_r$, and $\mathcal{K}_{t1} \cup \mathcal{K}_{t2} = \mathcal{K}_t$. For simplicity, we denote the relaying user set and the destination user set in the full space as $\mathcal{K}_1 = \mathcal{K}_{t1} \cup \mathcal{K}_{r1}$ and $\mathcal{K}_2 = \mathcal{K}_{t2} \cup \mathcal{K}_{r2}$, respectively. Without loss of generality, we assume users in $\mathcal{K}_1$ have better channel conditions than users in $\mathcal{K}_2$.

In this work, we consider a block-fading channel model with T transmission blocks indexed by the set $\mathcal{T} = \{1, \cdots, T\}$. The channels between the BS and STAR RIS, the BS and user-k, STAR RIS and user-k are, respectively, denoted by $\mathbf{E}(t) \in \mathbb{C}^{N \times L}$, $\mathbf{g}_k(t) \in \mathbb{C}^{L \times 1}$, $\mathbf{h}_k(t) \in \mathbb{C}^{N \times 1}$, where t denotes the t-th block, $t \in \mathcal{T}$. Besides, the channel between user-i to user-j is denoted by $h_{i,j}$. As the pioneer study on STAR RIS-empowered CRS, we employ a simplified CSI model by assuming perfect knowledge of CSI at the BS for all communication links, i.e., perfect CSI at the transmitter (CSIT). The imperfect CSIT setting can be easily extended based on the existing works on RSMA with imperfect CSIT. It is important to highlight that our primary emphasis is on exploring the diverse performance achieved by various transmission modes of STAR RIS-empowered CRS.

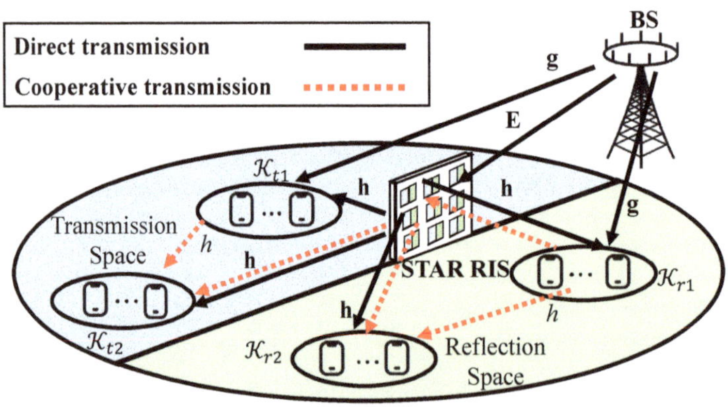

Figure 1. The transmission architecture of the proposed STAR RIS-assisted CRS.

2.1. Transmission Modes of STAR RIS-Empowered CRS

CRS involves two transmission phases: the direct transmission phase and the cooperative transmission phase. During the direct transmission phase, the BS sends signals to all users. In the cooperative transmission phase, users in $\mathcal{K}_1$ act as relaying users to transmit the decoded common stream to users in $\mathcal{K}_2$, utilizing either the *HD* or *FD* protocols. The *FD* protocol allows the simultaneous transmission and reception for relaying users. However, this approach introduces self-interference at the relaying users. In the *HD* protocol, each time block is divided into two consecutive parts, respectively, for the two transmission phases. Let λ $(0 < \lambda \leq 1)$ denote the fraction of time allocated to the direct transmission phase, and $(1 - \lambda)$ is the fraction allocated to the cooperative transmission phase.

In this paper, both transmission phases of CRS are assisted by the STAR RIS. According to the principle of single-connected STAR RIS in [10], each element of STAR RIS can operate in the reflection mode and/or the transmission mode according to different operating protocols [10].

The reflection and transmission matrices are all diagonal matrices, denoted as $\mathbf{\Theta}_i = \text{diag}(\sqrt{\beta_{1,i}}e^{j\theta_{1,i}}, \cdots, \sqrt{\beta_{N,i}}e^{j\theta_{N,i}})$, $i \in \{r, t\}$. In the *ES* protocol, each element of STAR RIS simultaneously reflects and transmits the incident signal. Hence, $\beta_{n,r}$ and $\beta_{n,t}$ should satisfy $\beta_{n,r} + \beta_{n,t} = 1$ and $\beta_{n,r}, \beta_{n,t} \in [0, 1]$. The *MS* protocol is a special case of *ES* where each element operates in either reflection or transmission mode, resulting in binary values for the amplitude coefficients, i.e., $\beta_{n,r}, \beta_{n,t} \in \{0, 1\}$. In the *TS* protocol, each element of STAR RIS switches between the transmission and reflection modes in each transmission block. Define $0 \leq \lambda_r \leq 1$ as the percentage of time allocated to the reflection mode in each transmission block, and $(1 - \lambda_r)$ as the percentage of time allocated to the transmission mode. The *TS* protocol focuses on designing λ_r and the phase shifts, where $\beta_{n,r} = 1, \beta_{n,t} = 0$ during the λ_r percentage of time and $\beta_{n,r} = 0, \beta_{n,t} = 1$ for the remaining $(1 - \lambda_r)$.

In the proposed STAR RIS-assisted CRS transmission framework, we pair the aforementioned relaying protocols (*HD* and *FD*) with the three STAR RIS operating protocols (*ES*, *MS*, and *TS*), yielding six distinct transmission modes for STAR RIS-assisted CRS. These modes are summarized in Figure 2 and explained as follows:

- *FD-ES (FE)* mode: This refers to the use of the *FD* protocol for CRS and the *ES* protocol for the STAR RIS. As illustrated in Figure 2a, *FE* enables each element of the STAR RIS to operate in both transmission and reflection modes simultaneously in the non-orthogonal direct and cooperative transmission phases.
- *HD-ES (HE)* mode: This refers to the use of the *HD* protocol for CRS and the *ES* protocol for STAR RIS. As illustrated in Figure 2b, *HE* enables each element of the STAR RIS to operate in both transmission and reflection modes simultaneously in the orthogonal direct and cooperative transmission phases.
- *FD-MS (FM)* mode: This refers to the use of the *FD* protocol for CRS and the *MS* protocol for STAR RIS. As illustrated in Figure 2c, *FM* enables each element of the STAR RIS to operate in either transmission or reflection mode in the non-orthogonal direct and cooperative transmission phases.
- *HD-MS (HM)* mode: This refers to the use of the *HD* protocol for CRS and the *MS* protocol for the STAR RIS. As illustrated in Figure 2d, *HM* enables each element of the STAR RIS to operate in either transmission or reflection mode in the orthogonal direct and cooperative transmission phases.
- *FD-TS (FT)* mode: This refers to the use of the *FD* protocol for CRS and the *TS* protocol for the STAR RIS. As illustrated in Figure 2e, each transmission block is split into two time slots based on λ_r $(0 \leq \lambda_r \leq 1)$. *FT* facilitates all elements of the STAR RIS to function in the reflection mode during the initial portion λ_r of the time, while they operate in the transmission mode during the subsequent portion $1 - \lambda_r$ of the time, across the non-orthogonal direct and cooperative transmission phases.
- *HD-TS (HT)* mode: This refers to the use of the *HD* protocol for CRS and the *TS* protocol for the STAR RIS. As illustrated in Figure 2f, each transmission block is split into two time slots based on λ_r: one for the reflection mode and the other for the transmission mode of STAR RIS. Then, each time slot is further split into two orthogonal parts for the direct and cooperative transmission phases of CRS, resulting in a total of four orthogonal transmission slots. Here, we introduce λ_1 $(0 < \lambda_1 \leq \lambda_r)$ and λ_2 $(0 < \lambda_2 \leq 1 - \lambda_r)$ to represent the time allocation between the two transmission phases in the reflection and transmission modes, respectively.

Hereinafter, we use superscript j to represent the selected transmission mode, where $j \in \{FE, FM, FT, HE, HM, HT\}$.

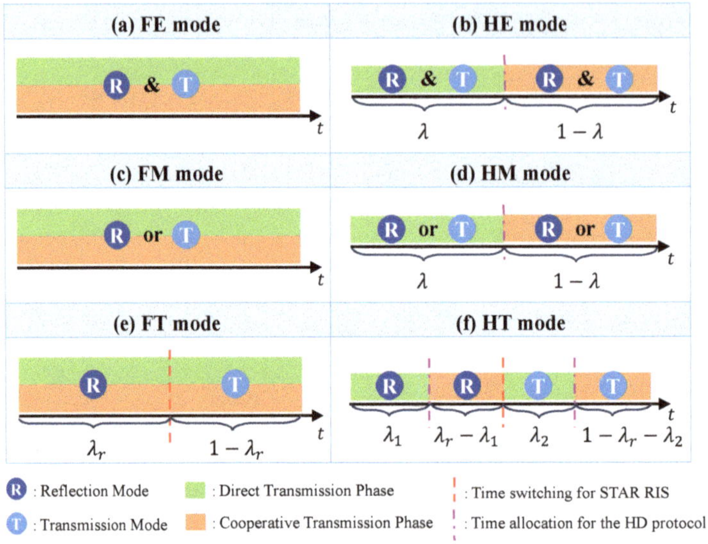

Figure 2. Six transmission modes of the proposed STAR RIS-assisted CRS.

2.2. Transmit Signal Model

For all six transmission modes described in the previous subsection, the primary differences among them lie in the received signals and the approaches employed to process the received signals. The transmit signals in both the direct and cooperative transmission phases remain consistent, as specified in this subsection.

2.2.1. Direct Transmission Phase

We denote W_k as the message intended for user-k. At the BS, according to the principle of one-layer RSMA [2,18], W_k is split into a common sub-message $W_{c,k}$ and a private sub-message $W_{p,k}$. All common sub-messages $W_{c,1}, \cdots, W_{c,K}$ are collected and then encoded into a common stream s_0, which is required to be decoded by all users. The individual private sub-message $W_{p,k}$ is independently encoded into a private stream s_k, which is required to be decoded by user-k only. Denote $\mathbf{s} = [s_0, s_1, \ldots, s_K]^T$ and $\mathbf{P} = [\mathbf{p}_0, \mathbf{p}_1, \ldots, \mathbf{p}_K] \in \mathbb{C}^{L\times(K+1)}$, respectively, as the stream vector and the transmit beamforming matrix, where $\mathbf{p}_k \in \mathbb{C}^{L\times 1}$. The transmit signal at the BS in the direct transmission phase is given by

$$\mathbf{x}^{[1]} = \mathbf{P}\mathbf{s} = \sum_{k=0}^{K} \mathbf{p}_k s_k, \tag{1}$$

where the superscript [1] denotes the direct transmission phase. Here, assuming $\mathbb{E}\{\mathbf{s}\mathbf{s}^H\} = \mathbf{I}$, the transmit power of the BS is limited by $\mathrm{tr}(\mathbf{P}\mathbf{P}^H) \leq P_t$, where P_t is the maximum transmit power at the BS.

2.2.2. Cooperative Transmission Phase

Users in $\mathcal{K}_1$ employ the decode-and-forward (DF) protocol to decode, re-encode, and subsequently transmit the decoded s_0 to users in $\mathcal{K}_2$. The transmit signal at the relaying user-k in the cooperative transmission phase is

$$x_k^{[2]} = \sqrt{P_k} s_0, \tag{2}$$

where the superscript [2] denotes the cooperative transmission phase and P_k is the relaying transmit power at user-k.

2.3. Received Signal Model for FD-Based Transmission Modes

In the FD-based transmission modes (*FE*, *FM*, and *FT*), both the direct and cooperative transmission phases share the same radio resources. Self-interference is inevitably among the relaying users in $\mathcal{K}_1$ as they simultaneously receive signals from the BS and transmit the common stream s_0 to users in $\mathcal{K}_2$, [19,20]. We denote the self-interference channel as $I_k(t)$, which follows the same distribution $I_k(t) \sim \mathcal{CN}(0, \Omega_I^2)$ for all $k \in \mathcal{K}_1$. The signal received by users in $\mathcal{K}_2$ from $\mathcal{K}_1$, denoted as $\hat{s}_0(t)$, is a delayed version of s_0, where $\hat{s}_0(t) = s_0(t - t_d)$. Here, t_d represents the processing time for users in $\mathcal{K}_1$ to decode and forward s_0. To ensure that the relaying users transmit and receive $\hat{s}_0(t)$ within the same block, we assume that t_d is much smaller than one block period [20,21]. In this work, we focus on optimizing $\mathbf{P}$, $\boldsymbol{\Theta}_r$ and $\boldsymbol{\Theta}_t$ in each transmission block. Hence, we omit t in the following. Next, we will specify the received signals and achievable rates for *FE*, *FM*, and *FT*, respectively.

2.3.1. FE/FM

When $j = FE$ or $j = FM$, the signal received at each relaying user-k in $\mathcal{K}_1$ is given by

$$y_k^j = \Big(\underbrace{\mathbf{g}_k^H + \mathbf{h}_k^H \boldsymbol{\Theta}_i \mathbf{E}}_{\tilde{\mathbf{g}}_{k,i}^H}\Big)\mathbf{x}^{[1]} + I_k\sqrt{P_k}\hat{s}_0 + n_k, i \in \{r, t\}, \tag{3}$$

where n_k is the additive white Gaussian noise (AWGN) which follows $n_k \sim \mathcal{CN}(0, \sigma^2)$ and $I_k\sqrt{P_k}\hat{s}_0$ is the self-interference suffered at user-k. Each user-k in $\mathcal{K}_1$ sequentially decodes the common stream s_0 and the private stream s_k with the assistance of SIC. Let $\tilde{\mathbf{g}}_{k,i}$ denote the effective channel between the BS and user-k, the SINRs of decoding s_0 and s_k at user-k in $\mathcal{K}_1$ are, respectively, given as

$$\gamma_{c,k}^{FD} = \gamma_{c,k}^{FD,[1]} = \frac{\left|\tilde{\mathbf{g}}_{k,i}^H \mathbf{p}_0\right|^2}{\sum_{m\in\mathcal{K}}\left|\tilde{\mathbf{g}}_{k,i}^H \mathbf{p}_m\right|^2 + |I_k|^2 P_k + \sigma^2}, \tag{4}$$

$$\gamma_k^{FD} = \frac{\left|\tilde{\mathbf{g}}_{k,i}^H \mathbf{p}_k\right|^2}{\sum_{m\in\mathcal{K},m\neq k}\left|\tilde{\mathbf{g}}_{k,i}^H \mathbf{p}_m\right|^2 + |I_k|^2 P_k + \sigma^2}. \tag{5}$$

Each user-k in $\mathcal{K}_2$ receives signals from both the BS and the relaying users in $\mathcal{K}_1$, which are given as

$$y_k^j = \tilde{\mathbf{g}}_{k,i}^H \mathbf{x}^{[1]} + \sum_{m\in\mathcal{K}_1}\Big(\underbrace{h_{m,k} + \mathbf{h}_k^H \boldsymbol{\Theta}_i^H \mathbf{h}_m}_{\tilde{h}_{m,k,i}}\Big) x_m^{[2]} + n_k,\ i \in \{r, t\}. \tag{6}$$

Each user in $\mathcal{K}_2$ then employs the maximal ratio combining (MRC) approach to ensure the received signals are properly co-phased and merged [22]. Denote the effective channel between user-m and user-n as $\tilde{h}_{m,n,i}$, the SINRs of decoding the s_0 and s_k at user-k in $\mathcal{K}_2$ are given as

$$\gamma_{c,k}^{FD} = \underbrace{\frac{\left|\tilde{\mathbf{g}}_{k,i}^H \mathbf{p}_0\right|^2}{\sum_{m\in\mathcal{K}}\left|\tilde{\mathbf{g}}_{k,i}^H \mathbf{p}_m\right|^2 + \sigma^2}}_{\gamma_{c,k}^{FD,[1]}} + \underbrace{\frac{\sum_{m\in\mathcal{K}_1}\left|\tilde{h}_{m,k,i}\right|^2 P_m}{\sigma^2}}_{\gamma_{c,k}^{FD,[2]}}, \tag{7}$$

$$\gamma_k^{FD} = \frac{\left|\tilde{\mathbf{g}}_{k,i}^H \mathbf{p}_k\right|^2}{\sum_{m\in\mathcal{K},m\neq k}\left|\tilde{\mathbf{g}}_{k,i}^H \mathbf{p}_m\right|^2 + \sigma^2}. \tag{8}$$

With the SINRs of s_0 and s_k in (4), (5), (7) and (8), the achievable common and private rates are obtained as $R^j_{c,k} = \log_2(1+\gamma^{FD}_{c,k})$ and $R^j_k = \log_2(1+\gamma^{FD}_k)$, $j \in \{FE, FM\}$. To guarantee all users decode the common stream, the following condition should satisfy [2]:

$$R^j_c = \min\{R^j_{c,k} | k \in \mathcal{K}\}, \tag{9}$$

R^j_c is shared by all users, we denote C_k as the part of R^j_c that is allocated to user-k, it satisfies $\sum_{k\in\mathcal{K}} C_k \leq R^j_c$. Hence, the total achievable rate of user-k is

$$R^j_{tot,k} = R^j_k + C_k, \forall k \in \mathcal{K}. \tag{10}$$

Note that Equations (9) and (10) remain applicable for other transmission modes by setting j to $FT/HE/HM/HT$. In the following discussion of these transmission modes, the equations will not be repeated.

2.3.2. FT

In FT mode, there are two time slots: all elements of STAR RIS operate in the reflection mode during the first time slot (λ_r) while all elements of STAR RIS operate in the transmission mode during the second time slot $(1-\lambda_r)$. We denote the SINRs for decoding s_0 and s_k in the reflection and transmission time slots, respectively, as $\gamma^{FD,i}_{c,k}$ and $\gamma^{FD,i}_k$, where $i \in \{r,t\}$. In the reflection time slot, users in $\mathcal{K}_r$ receive signals from the BS and the STAR RIS while users in $\mathcal{K}_t$ receive signals from the BS only. The corresponding SINRs are given as

$$\begin{aligned} \gamma^{FD,r}_{c,k} &= \gamma^{FD}_{c,k}\,|_{\boldsymbol{\Theta}_i=\boldsymbol{\Theta}_r},\ \gamma^{FD,r}_k = \gamma^{FD}_k\,|_{\boldsymbol{\Theta}_i=\boldsymbol{\Theta}_r}, k \in \mathcal{K}_r,\\ \gamma^{FD,r}_{c,k} &= \gamma^{FD}_{c,k}\,|_{\boldsymbol{\Theta}_i=\mathbf{0}},\ \gamma^{FD,r}_k = \gamma^{FD}_k\,|_{\boldsymbol{\Theta}_i=\mathbf{0}}, k \in \mathcal{K}_t, \end{aligned} \tag{11}$$

where $\gamma^{FD}_{c,k}$ and γ^{FD}_k are defined in (4) and (5), respectively. In the transmission time slot, users in $\mathcal{K}_r$ receive signals from the BS only while users in $\mathcal{K}_t$ receive signals from both BS and STAR RIS. The corresponding SINRs are given as

$$\begin{aligned} \gamma^{FD,t}_{c,k} &= \gamma^{FD}_{c,k}\,|_{\boldsymbol{\Theta}_i=\mathbf{0}},\ \gamma^{FD,t}_k = \gamma^{FD}_k\,|_{\boldsymbol{\Theta}_i=\mathbf{0}}, k \in \mathcal{K}_r,\\ \gamma^{FD,t}_{c,k} &= \gamma^{FD}_{c,k}\,|_{\boldsymbol{\Theta}_i=\boldsymbol{\Theta}_t},\ \gamma^{FD,t}_k = \gamma^{FD}_k\,|_{\boldsymbol{\Theta}_i=\boldsymbol{\Theta}_t}, k \in \mathcal{K}_t. \end{aligned} \tag{12}$$

The achievable common and private rates for user-k are given as

$$\begin{aligned} R^{FT}_{c,k} &= \lambda_r \log_2(1+\gamma^{FD,r}_{c,k}) + (1-\lambda_r)\log_2(1+\gamma^{FD,t}_{c,k}),\\ R^{FT}_k &= \lambda_r \log_2(1+\gamma^{FD,r}_k) + (1-\lambda_r)\log_2(1+\gamma^{FD,t}_k). \end{aligned} \tag{13}$$

By substituting $R^{FT}_{c,k}$ and R^{FT}_k into (9) and (10), we obtain the corresponding achievable rate of each user in the FT mode.

2.4. *Received Signal Model for HD-Based Transmission Modes*

In HD-based transmission modes, the direct and cooperative transmission phases are orthogonal in each transmission block.

In contrast to FD-based modes, there is no self-interference at relaying users, but the transmission rates for the direct and transmission phases are scaled by the pre-log factors $\lambda(0 < \lambda \leq 1)$ and $(1-\lambda)$, respectively. In the following, we specify the received signals and achievable rates for HE, HM, and HT.

2.4.1. HE/HM

In the direct transmission phase, the signal received at user-k is

$$y^{HD,[1]}_k = \tilde{\mathbf{g}}^H_{k,i}\mathbf{x}^{[1]} + n_k, \forall k \in \mathcal{K}, j \in \{HE, HM\}. \tag{14}$$

The SINRs of decoding s_0 and s_k in the direct transmission phase are given as

$$\gamma_{c,k}^{HD,[1]} = \frac{\left|\tilde{\mathbf{g}}_{k,i}^{H}\mathbf{p}_0\right|^2}{\sum_{m\in\mathcal{K}}\left|\tilde{\mathbf{g}}_{k,i}^{H}\mathbf{p}_m\right|^2+\sigma^2}, \tag{15}$$

$$\gamma_{k}^{HD} = \frac{\left|\tilde{\mathbf{g}}_{k,i}^{H}\mathbf{p}_0\right|^2}{\sum_{m\in\mathcal{K},m\neq k}\left|\tilde{\mathbf{g}}_{k,i}^{H}\mathbf{p}_m\right|^2+\sigma^2}. \tag{16}$$

In the cooperative transmission phase, users in $\mathcal{K}_1$ forward the decoded s_0 to users in $\mathcal{K}_2$. The signal received by user-k in $\mathcal{K}_2$ in the cooperative transmission phase is

$$y_k^{HD,[2]} = \sum_{m\in\mathcal{K}_1} \tilde{h}_{m,k,i}\sqrt{P_m}s_0 + n_k, \forall k\in\mathcal{K}_2. \tag{17}$$

The SINR of decoding s_0 at user-k in $\mathcal{K}_2$ in the cooperative transmission phase is given as

$$\gamma_{c,k}^{HD,[2]} = \frac{\sum_{m\in\mathcal{K}_1}\left|\tilde{h}_{m,k,i}\right|^2 P_m}{\sigma^2}. \tag{18}$$

With the SINR in (15), (16) and (18), we obtain the achievable common and private rates in *HE* and *HM* modes ($j\in HE/HM$) as

$$\begin{aligned} R_{c,k}^{j} &= \lambda\log_2\left(1+\gamma_{c,k}^{HD,[1]}\right), \forall k\in\mathcal{K}_1 \\ R_{c,k}^{j} &= \lambda\log_2\left(1+\gamma_{c,k}^{HD,[1]}\right)+(1-\lambda)\log_2\left(1+\gamma_{c,k}^{HD,[2]}\right), \forall k\in\mathcal{K}_2 \end{aligned} \tag{19}$$

and

$$R_k^j = \lambda\log_2\left(1+\gamma_k^{HD}\right). \tag{20}$$

By substituting (19) and (20) into (9) and (10), we obtain the corresponding achievable rate of each user in the *HE* and *HM* modes.

2.4.2. HT

In the *HT* mode, there are four time slots: two for direct transmission with STAR RIS reflection or transmission, and two for cooperative transmission with STAR RIS reflection or transmission. We use $\gamma_{c,k}^{HD,[s],i}$ and $\gamma_k^{HD,i}$, $i\in\{r,t\}$, $s\in\{1,2\}$ to denote the SINRs of decoding s_0 and s_k in the direct or cooperative transmission phase. The superscript i indicates whether it is in the reflection or transmission time slot. Specifically, in the two reflection time slots, respectively, for cooperative and transmission phases, we have

$$\begin{aligned} \gamma_{c,k}^{HD,[s],r} &= \gamma_{c,k}^{HD,[s]}\ |_{\mathbf{\Theta}_i=\mathbf{\Theta}_r},\ \gamma_k^{HD,r} = \gamma_k^{HD}\ |_{\mathbf{\Theta}_i=\mathbf{\Theta}_r}, k\in\mathcal{K}_r, \\ \gamma_{c,k}^{HD,[s],r} &= \gamma_{c,k}^{HD,[s]}\ |_{\mathbf{\Theta}_i=\mathbf{0}},\ \gamma_k^{HD,r} = \gamma_k^{HD}\ |_{\mathbf{\Theta}_i=\mathbf{0}}, k\in\mathcal{K}_t, \end{aligned} \tag{21}$$

where $\gamma_{c,k}^{HD,[s]}, s\in\{1,2\}$ and γ_k^{HD} are defined in (15), (18) and (16), respectively. In the two transmission time slots, respectively, for cooperative and transmission phases, we have

$$\begin{aligned} \gamma_{c,k}^{HD,[s],t} &= \gamma_{c,k}^{HD,[s]}\ |_{\mathbf{\Theta}_i=\mathbf{0}},\ \gamma_k^{HD,t} = \gamma_k^{HD}\ |_{\mathbf{\Theta}_i=\mathbf{0}}, k\in\mathcal{K}_r, \\ \gamma_{c,k}^{HD,[s],t} &= \gamma_{c,k}^{HD,[s]}\ |_{\mathbf{\Theta}_i=\mathbf{\Theta}_t},\ \gamma_k^{HD,t} = \gamma_k^{HD}\ |_{\mathbf{\Theta}_i=\mathbf{\Theta}_t}, k\in\mathcal{K}_t. \end{aligned} \tag{22}$$

Hence, the achievable common and private rates for user-k are given as

$$\begin{aligned}
R_{c,k}^{HT} &= \lambda_1 \log_2\left(1+\gamma_{c,k}^{HD,[1],r}\right) + \lambda_2 \log_2\left(1+\gamma_{c,k}^{HD,[1],t}\right), \forall k \in \mathcal{K}_1, \\
R_{c,k}^{HT} &= \lambda_1 \log_2\left(1+\gamma_{c,k}^{HD,[1],r}\right) + (\lambda_r - \lambda_1) \log_2\left(1+\gamma_{c,k}^{HD,[2],r}\right) \\
&\quad + \lambda_2 \log_2\left(1+\gamma_{c,k}^{HD,[1],t}\right) + (1-\lambda_r-\lambda_2) \log_2\left(1+\gamma_{c,k}^{HD,[2],t}\right), \quad \forall k \in \mathcal{K}_2.
\end{aligned} \tag{23}$$

$$R_k^{HT} = \lambda_1 \log_2\left(1+\gamma_k^{HD,r}\right) + \lambda_2 \log_2\left(1+\gamma_k^{HD,t}\right). \tag{24}$$

2.5. Problem Formulation

In this work, we aim to jointly optimize the active beamforming $\mathbf{P}$ at the BS, the common rate allocation $\mathbf{c} = [C_1, \cdots, C_K]$, as well as the passive beamforming $\mathbf{\Theta}_r$, $\mathbf{\Theta}_t$ at the STAR RIS. For the *HD*-based transmission modes, the corresponding time allocation coefficients are also jointly optimized. Our objective is to maximize the minimum user rate (max-min rate) among all users subject to the transmit power constraint. The formulated problems for the six transmission modes of the proposed STAR RIS-aided CRS model are illustrated as follows:

2.5.1. FE

$j = FE$, the optimization problem is formulated as

$$\begin{aligned}
\mathcal{P}_{FE}: \quad & \max_{\mathbf{P},\mathbf{c},\mathbf{\Theta}_r,\mathbf{\Theta}_t} \min_{k\in\mathcal{K}} R_{tot,k}^{FE} && \text{(25a)}\\
\text{s.t.} \quad & \sum_{k\in\mathcal{K}} C_k \le R_c^j, C_k \ge 0, \forall k \in \mathcal{K} && \text{(25b)}\\
& \theta_{n,i} \in [0, 2\pi), \forall n \in \mathcal{N}, i \in \{r,t\}, && \text{(25c)}\\
& \mathrm{tr}\left(\mathbf{P}\mathbf{P}^H\right) \le P_t && \text{(25d)}\\
& \beta_{n,r}, \beta_{n,t} \in [0,1], \beta_{n,r}+\beta_{n,t} = 1, \forall n \in \mathcal{N}, && \text{(25e)}
\end{aligned}$$

where constraint (25b) guarantees the common stream and is successfully decoded by all users. Constraints (25c) and (25e) are the constraints for the phase shift and amplitude ranges for each element of the STAR RIS, respectively. (25d) limits the transmit power at the BS.

2.5.2. FM

$j = FM$, the formulated problem is given as

$$\begin{aligned}
\mathcal{P}_{FM}: \quad & \max_{\mathbf{P},\mathbf{c},\mathbf{\Theta}_r,\mathbf{\Theta}_t} \min_{k\in\mathcal{K}} R_{tot,k}^{FM} && \text{(26a)}\\
\text{s.t.} \quad & \text{(25b)–(25d)} \\
& \beta_{n,r}, \beta_{n,t} \in \{0,1\}, \beta_{n,r}+\beta_{n,t} = 1, \forall n \in \mathcal{N}. && \text{(26b)}
\end{aligned}$$

2.5.3. FT

$j = FT$, the time allocation variable λ_r is jointly optimized with the beamforming matrices and the common rate allocation. The optimization problem is

$$\begin{aligned}
\mathcal{P}_{FT}: \quad & \max_{\mathbf{P},\mathbf{c},\mathbf{\Theta}_r,\mathbf{\Theta}_t,\lambda_r} \min_{k\in\mathcal{K}} R_{tot,k}^{FT} && \text{(27a)}\\
\text{s.t.} \quad & \text{(25b)–(25d)}, \\
& \beta_{n,r}, \beta_{n,t} \in [0,1], \forall n \in \mathcal{N}, && \text{(27b)}\\
& 0 \le \lambda_r \le 1. && \text{(27c)}
\end{aligned}$$

2.5.4. HE

$j = HE$, the time allocation variable λ is jointly optimized with the beamforming matrices and the common rate allocation. The formulated problem is

$$\mathcal{P}_{HE}: \max_{\mathbf{P},\mathbf{c},\mathbf{\Theta}_r,\mathbf{\Theta}_t,\lambda} \min_{k\in\mathcal{K}} R_{tot,k}^{HE} \tag{28a}$$
$$\text{s.t.} \quad \text{(25b)–(25e)}$$
$$0 < \lambda \leq 1. \tag{28b}$$

2.5.5. HM

$j = HM$, the corresponding optimization problem is formulated as

$$\begin{aligned}\mathcal{P}_{HM}: &\max_{\mathbf{P},\mathbf{c},\mathbf{\Theta}_r,\mathbf{\Theta}_t,\lambda} \min_{k\in\mathcal{K}} R_{tot,k}^{HM} \\ \text{s.t.} \quad &\text{(25b)–(25d), (26b), (28b).}\end{aligned} \tag{29}$$

2.5.6. HT

$j = HT, \lambda_r, \lambda_1, \lambda_2$ are jointly optimized. The corresponding optimization problem is formulated as

$$\mathcal{P}_{HT}: \max_{\mathbf{P},\mathbf{c},\mathbf{\Theta}_r,\mathbf{\Theta}_t,\lambda_r,\lambda_1,\lambda_2} \min_{k\in\mathcal{K}} R_{tot,k}^{HT} \tag{30a}$$
$$\text{s.t.} \quad \text{(25b)–(25d), (27b), (27c)}$$
$$0 < \lambda_1 \leq \lambda_r, 0 < \lambda_2 \leq 1 - \lambda_r. \tag{30b}$$

From problems (25)–(30) for the six transmission modes, we observe that $\mathbf{P}$, $\mathbf{\Theta}_r$, $\mathbf{\Theta}_t$ are highly coupled in the non-convex and fractional expressions for the achievable common and private rates. Besides, problem (26) and (29) involve integer programming due to (25b). All problems are highly non-convex and difficult to solve. In the next section, we will propose two algorithms to solve the problems.

3. Proposed Optimization Frameworks

To solve the aforementioned six non-convex optimization problems, in this section, we first propose an AO framework that alternatively optimizes the STAR RIS passive beamforming matrices $\mathbf{\Theta}_r$ and $\mathbf{\Theta}_t$, and the remaining variables. To ease the computational complexity, we further propose a low-complexity algorithm to address the optimization problems. In the end, we extend the proposed optimization framework to the imperfect CSIT scenario.

3.1. Proposed AO Algorithm

We divide the original non-convex problems into two subproblems, one for the STAR RIS passive beamforming design, and the other for the joint BS active beamforming, common rate, and time allocation design. The approaches for solving the two subproblems are detailed in the following:

3.1.1. STAR RIS Passive Beamforming Optimization

With the transmit beamforming $\mathbf{P}$, common rate allocation $\mathbf{c}$ and time allocation variables λ_t, λ_1 and λ_2 fixed, we focus on optimizing $\mathbf{\Theta}_r$ and $\mathbf{\Theta}_t$. First, we introduce an auxiliary variable t^j to represent the objective function for the transmission mode j, leading to the following constraint:

$$R_{p,k}^{j} + C_k \geq t^j, \forall k \in \mathcal{K}, j \in \{FE, FM, FT, HE, HM, HT\}. \tag{31}$$

Although the achievable rate expressions may vary across different transmission modes, our main focus is the SINR expressions as the optimization variables $\mathbf{\Theta}_r$ and $\mathbf{\Theta}_t$ are embedded solely within the SINR. Note that the non-convexity of the SINR expressions remains consistent across all six transmission modes. The key distinction in optimizing the STAR RIS across these modes lies in the constraints imposed by the *ES*, *MS*, and *TS* protocols.

In the following, we begin by illustrating the SCA-based method employed to address the non-convexity of the SINR expressions, using the *FE* mode as an example. Following this, we elaborate on the penalty methods proposed to manage the STAR RIS constraints for the *ES*, *MS*, and *TS* protocols, respectively.

- **Step 1.1: The SCA method for non-convex SINRs:**

Let $\psi_{n,i} = \sqrt{\beta_{n,i}} e^{j\theta_{n,i}}, n \in \mathcal{N}, i \in \{r,t\}$ denote each element of $\mathbf{\Theta}_r$ and $\mathbf{\Theta}_t$. With $\boldsymbol{\psi}_i = [\psi_{1,i}, \ldots, \psi_{N,i}]^T$, we have $\mathbf{g}_k^H \mathbf{p}_m + \mathbf{h}_k^H \mathbf{\Theta}_i \mathbf{E} \mathbf{p}_m = \bar{g}_{km} + \mathbf{s}_{km}^H \boldsymbol{\psi}_i$, $\mathbf{h}_n^H \mathbf{\Theta}_i \mathbf{h}_m = \bar{\mathbf{h}}_{m,n}^H \boldsymbol{\psi}_i$, where $\bar{g}_{km} = \mathbf{g}_k^H \mathbf{p}_m$, $\mathbf{s}_{km} = (\mathrm{diag}(\mathbf{h}_k^H)\mathbf{E}\mathbf{p}_m)^*$ and $\bar{\mathbf{h}}_{m,n} = (\mathrm{diag}(\mathbf{h}_n^H)\mathbf{h}_m)^* \in \mathbb{C}^{N\times 1}$. In this way, the optimization variables are represented by $\boldsymbol{\psi}_r$ and $\boldsymbol{\psi}_t$.

Taking the *FE* problem (25) as an example, we first introduce slack variables $\boldsymbol{\delta} = [\delta_1, \ldots, \delta_K]^T$ and $\boldsymbol{\delta}_c = [\delta_{c,1}, \ldots, \delta_{c,K}]^T$ to, respectively, denote the SINRs of s_k and s_0 in the direct transmission phase. Furthermore, we introduce $\boldsymbol{\xi} = [\xi_1, \ldots, \xi_{K_2}]$ to denote the SINRs of the destination users in the cooperative transmission phase, where K_2 is the number of users in $\mathcal{K}_2$. The optimization problem (25) for updating the STAR RIS passive beamforming matrices is then equivalently transformed into

$$
\begin{aligned}
\mathcal{P}_{FE-1}: \quad & \max_{t^{FE}, \boldsymbol{\psi}_r, \boldsymbol{\psi}_t, \boldsymbol{\delta}, \boldsymbol{\delta}_c, \boldsymbol{\xi}} \quad t^{FE} && \text{(32a)} \\
\text{s.t.} \quad & \log_2(1+\delta_k) + C_k \geq t^{FE}, \forall k \in \mathcal{K}, && \text{(32b)} \\
& \log_2(1+\delta_{c,k}) \geq \textstyle\sum_{m\in\mathcal{K}} C_m, k \in \mathcal{K}_1, && \text{(32c)} \\
& \log_2(1+\delta_{c,k}+\xi_k) \geq \textstyle\sum_{m\in\mathcal{K}} C_m, k \in \mathcal{K}_2, && \text{(32d)} \\
& \delta_{c,k} \leq \gamma_{c,k}^{FD,[1]}, \forall k \in \mathcal{K}, && \text{(32e)} \\
& \delta_k \leq \gamma_k^{FD}, \forall k \in \mathcal{K}, && \text{(32f)} \\
& \xi_k \leq \gamma_{c,k}^{FD,[2]}, \forall k \in \mathcal{K}_2, && \text{(32g)} \\
& \text{(25e)}.
\end{aligned}
$$

For constraints (32e) and (32f), we further introduce slack variables $\boldsymbol{\eta} = [\eta_1, \ldots, \eta_K]^T$ and $\boldsymbol{\eta}_c = [\eta_{c,1}, \ldots, \eta_{c,K}]^T$ to, respectively, denote the denominators of the SINRs for s_0 and s_k. (32e) and (32f) are equivalently transformed to

$$
\begin{aligned}
& \eta_{c,k} \geq \mathrm{de}(\gamma_{c,k}^{FD,[1]}), \forall k \in \mathcal{K} && \text{(33a)} \\
& \eta_k \geq \mathrm{de}(\gamma_k^{FD}), \forall k \in \mathcal{K} && \text{(33b)} \\
& \delta_{c,k}\eta_{c,k} \leq \left| \bar{g}_{k0} + \mathbf{s}_{k0}^H \boldsymbol{\psi}_i \right|^2, && \text{(33c)} \\
& \delta_k \eta_k \leq \left| \bar{g}_{kk} + \mathbf{s}_{kk}^H \boldsymbol{\psi}_i \right|^2, && \text{(33d)}
\end{aligned}
$$

where $\mathrm{de}(\cdot)$ is an operator defined to extract the denominator of a fraction, i.e., $\mathrm{de}(\frac{a}{b}) = b$. For constraint (33c), the left-hand side (LHS) equals to $\delta_{c,k}\eta_{c,k} = \frac{1}{4}(\delta_{c,k}+\eta_{c,k})^2 - \frac{1}{4}(\delta_{c,k}-\eta_{c,k})^2$. By further approximating $\delta_{c,k}\eta_{c,k}$ at the point $(\delta_{c,k}^{[z]}, \eta_{c,k}^{[z]})$ in iteration $[z]$ using the first-order Taylor approximation of $(\delta_{c,k}-\eta_{c,k})^2$, we have

$$
\delta_{c,k}\eta_{c,k} \leq \frac{1}{4}(\delta_{c,k}+\eta_{c,k})^2 - \frac{1}{2}(\delta_{c,k}^{[z]} - \eta_{c,k}^{[z]})(\delta_{c,k}-\eta_{c,k}) + \frac{1}{4}(\delta_{c,k}^{[z]} - \eta_{c,k}^{[z]})^2 \triangleq \nu(\delta_{c,k}^{[z]}, \eta_{c,k}^{[z]}). \quad \text{(34)}
$$

We also approximate the right-hand side (RHS) of (33c) at $\boldsymbol{\psi}_i^{[z]}$ by the first-order Taylor approximation of $\left|\bar{g}_{k0} + \mathbf{s}_{k0}^H \boldsymbol{\psi}_i\right|^2$ as

$$\left|\bar{g}_{k0} + \mathbf{s}_{k0}^H \boldsymbol{\psi}_i\right|^2 \geq 2\Re\{(\mathbf{s}_{k0}^H \boldsymbol{\psi}_i^{[z]} + \bar{g}_{k0})^H \mathbf{s}_{k0}^H \boldsymbol{\psi}_i\} - |\mathbf{s}_{k0}^H \boldsymbol{\psi}_i|^2 + |\bar{g}_{k0}|^2 \triangleq \varpi(\boldsymbol{\psi}_i^{[z]}, \boldsymbol{\psi}_i, \bar{g}_{k0}, \mathbf{s}_{k0}). \quad (35)$$

Based on this approach, constraints (33c) and (33d) are transformed to

$$\nu(\delta_{c,k}^{[z]}, \eta_{c,k}^{[z]}) \leq \varpi(\boldsymbol{\psi}_i^{[z]}, \boldsymbol{\psi}_i, \bar{g}_{k0}, \mathbf{s}_{k0}), \forall k \in \mathcal{K}_1, \quad (36a)$$

$$\nu(\delta_k^{[z]}, \eta_k^{[z]}) \leq \varpi(\boldsymbol{\psi}_i^{[z]}, \boldsymbol{\psi}_i, \bar{g}_{kk}, \mathbf{s}_{kk}), \forall k \in \mathcal{K}. \quad (36b)$$

And constraint (32g) is approximated as

$$\xi_k \sigma^2 \leq P_m \varpi(\boldsymbol{\psi}_i^{[z]}, \boldsymbol{\psi}_i, h_{m,k}, \mathbf{h}_{m,k}), \forall k \in \mathcal{K}. \quad (37)$$

Therefore, utilizing the SCA method, the non-convex SINR expressions (32e)–(32g) are replaced by (36) and (37), which are convex. This approach can be applied directly to the other five transmission modes, so we simplify by omitting their discussion here.

- **Step 1.2: The penalty method for non-convex STAR RIS constraints**:

After Step 1.1, the optimization problems for passive beamforming in all six modes become convex except for the constraints related to the phase and amplitude of the STAR RIS. In the following, we specify the penalty methods proposed to handle these constraints for the *ES*, *MS*, and *TS* protocols.

ES: For constraint (25e), we adopt the penalty method proposed in [23]. By introducing a large positive constant C, the objective function is transformed into $t^j + C\sum_{n=1}^N(|\psi_{n,r}|^2 + |\psi_{n,t}|^2 - 1), j \in \{FE, HE\}$ with an additional constraint

$$|\psi_{n,r}|^2 + |\psi_{n,t}|^2 \leq 1. \quad (38)$$

Using the first-order Taylor approximation of $(|\psi_{n,r}|^2 + |\psi_{n,t}|^2)$ at iteration $[z]$, we obtain the approximated objective function as

$$t^j + C\sum_{n=1}^{N} \Re\{2(\psi_{n,r}^{[z]})^* \psi_{n,r} - |\psi_{n,r}^{[z]}|^2 + 2(\psi_{n,t}^{[z]})^* \psi_{n,t} - |\psi_{n,t}^{[z]}|^2\}. \quad (39)$$

MS: For constraint (26b), it is obvious that the amplitude coefficient of each STAR RIS element in the *FM* and *HM* modes is an integer chosen from 0 to 1. To transform constraints (26b), it is worth noting that no matter whether each element is 0 or 1, it always satisfies (38) and

$$-|\psi_{n,i}|^2 + |\psi_{n,i}| = 0, i \in \{r, t\}. \quad (40)$$

(40) forces the amplitudes of $\psi_{n,r}$ and $\psi_{n,t}$ to 0 or 1. Due to constraint (38), $0 \leq |\psi_{n,i}| \leq 1$, we obtain that

$$-|\psi_{n,i}|^2 + |\psi_{n,i}| \geq 0, i \in \{r, t\}. \quad (41)$$

Hence, constraint (26b) can be replaced by adding a penalty term to the objective function as $t^j - C\sum_{n=1}^N(-|\psi_{n,r}|^2 + |\psi_{n,r}| - |\psi_{n,t}|^2 + |\psi_{n,t}|), j \in \{FM, HM\}$. Applying the first-order Taylor approximation to the penalty term $-|\psi_{n,r}|^2 - |\psi_{n,t}|^2$ at iteration $[z]$, the objective function is transformed to

$$t^j - C\sum_{n=1}^{N}\left(A^{[z]} - 2\Re\{(\psi_{n,r}^{[z]})^* \psi_{n,r}\} - 2\Re\{(\psi_{n,t}^{[z]})^* \psi_{n,t}\}\right), \quad (42)$$

where $A^{[z]} = |\psi_{n,r}^{[z]}|^2 + |\psi_{n,t}^{[z]}|^2 + |\psi_{n,r}| + |\psi_{n,t}|$.

TS: For *TS* modes, we also adopt the penalty method proposed in [23] to address constraints (27b), which transforms the objective function into $t^j + C\sum_{n=1}^{N}(|\psi_{n,r}|^2 - 1 + |\psi_{n,t}|^2 - 1), j \in \{FT, HT\}$ with an additional constraint

$$|\psi_{n,i}|^2 \leq 1, n \in \mathcal{N}, i \in \{r,t\}. \tag{43}$$

The objective function is then approximated in the same way as (39).

After Steps 1.1 and 1.2, the STAR RIS optimization subproblems for all six transmission modes become convex quadratically constrained quadratic program (QCQP) and can be solved using the CVX optimization toolbox. The detailed process to solve the STAR RIS passive beamforming optimization problem is summarized in Algorithm 1.

Algorithm 1: STAR RIS passive beamforming optimization algorithm for six transmission modes

Initialize: $j, z = 0, t^{j,[z]}, \mathbf{\Theta}_r^{[z]}, \mathbf{\Theta}_t^{[z]}$, and slack variables for mode j;
repeat
  $z \leftarrow z + 1$;
  Updating $\mathbf{\Theta}_r^{[z]}, \mathbf{\Theta}_t^{[z]}$ by employing Steps 1.1 and 1.2 to approximate the problem and then solving the approximated convex problem using CVX ;
until *convergence*;

3.1.2. Joint Optimization of the Active Beamforming, Common Rate and Time Allocation

Given $\mathbf{\Theta}_r$ and $\mathbf{\Theta}_t$, the effective channels of all users are fixed. To jointly optimize the remaining variables, we first introduce an auxiliary variable x^j to denote the objective function for the transmission mode j, leading to the following constraints:

$$R_{p,k}^{j} + C_k \geq x^j, \forall k \in \mathcal{K}, j \in \{FE, FM, FT, HE, HM, HT\}. \tag{44}$$

In this subproblem, non-convexity arises from both the fractional SINR expressions and the coupling of time allocation variables within the rate expressions. Both of these challenges are tackled using SCA in the following steps.

- **Step 2.1: The SCA method for time allocation variables:**

In this step, we first specified the SCA methods employed to address the coupled time allocation variables within the rate expressions. Note that in the *FE* and *FM* modes, there are no time allocation variables. The non-convexity of the corresponding optimization problems arises solely from the SINR expressions, which are directly addressed in Step 2.2.

FT: To optimize λ_r in problem (27), we first introduce the slack variables $\boldsymbol{\alpha}^i = [\alpha_1^i, \ldots, \alpha_K^i]^T$, and $\boldsymbol{\alpha}_c^i = [\alpha_{c,1}^i, \ldots, \alpha_{c,K}^i]^T$, $i \in \{r,t\}$ to denote the private and common rates. This allows us to transform (44) and (25b) to

$$\lambda_r\alpha_k^r + (1-\lambda_r)\alpha_k^t + C_k \geq x^{FT}, \tag{45a}$$
$$\lambda_r\alpha_{c,k}^r + (1-\lambda_r)\alpha_{c,k}^t \geq \textstyle\sum_{m\in\mathcal{K}} C_m, \tag{45b}$$
$$\alpha_k^i \leq \log_2(1+\gamma_k^{FD,i}), \tag{45c}$$
$$\alpha_{c,k}^i \leq \log_2(1+\gamma_{c,k}^{FD,i}). \tag{45d}$$

For the bilinear function $\lambda_r\alpha_k^r$, it can be approximated by the first-order Taylor approximation at iteration $[z]$ based on $\nu(\cdot)$ in (34). Hence, constraints (45a) and (45b) are approximated as

$$-\nu(\alpha_k^{r,[z]}, -\lambda_r) + \alpha_k^t - \nu(\alpha_k^{t,[z]}, -\lambda_r) + C_k \geq x^{FT}, \tag{46a}$$
$$-\nu(\alpha_{c,k}^{r,[z]}, -\lambda_r) + \alpha_{c,k}^t - \nu(\alpha_{c,k}^{t,[z]}, -\lambda_r) \geq \textstyle\sum_{m\in\mathcal{K}} C_m. \tag{46b}$$

HE/HM: To optimize λ in problems (28) and (29), akin to the *FT* mode, we introduce the slack variables $\boldsymbol{\alpha}$, and $\boldsymbol{\alpha}_c^i$ to represent the private and common rates, thereby facilitating the transformation of (44) and (25b) to

$$\lambda\alpha_k + C_k \geq x^j, \tag{47a}$$
$$\lambda\alpha_{c,k} \geq \sum_{m\in\mathcal{K}} C_m, k \in \mathcal{K}_1 \tag{47b}$$
$$\lambda\alpha_{c,k} + (1-\lambda)\log(1+\gamma_{c,k}^{HD,[2]}) \geq \sum_{m\in\mathcal{K}} C_m, k \in \mathcal{K}_2 \tag{47c}$$
$$\alpha_k \leq \log_2(1+\gamma_k^{HD}), \tag{47d}$$
$$\alpha_{c,k} \leq \log_2(1+\gamma_{c,k}^{HD,[1]}). \tag{47e}$$

Using the same first-order Taylor approximation method in (46), constraints (47a)–(47c) are approximated to

$$-\nu(\alpha_k^{[z]}, -\lambda) + C_k \geq x^j, \tag{48}$$
$$-\nu(\alpha_{c,k}^{[z]}, -\lambda) \geq \sum_{m\in\mathcal{K}} C_m, \forall k \in \mathcal{K}_1, \tag{49}$$
$$-\nu(\alpha_{c,k}^{[z]}, -\lambda) + (1-\lambda)\log_2(1+\gamma_{c,k}^{HD,[2]}) \geq \sum_{m\in\mathcal{K}} C_m, \forall k \in \mathcal{K}_2. \tag{50}$$

HT: To jointly optimize λ_r, λ_1, and λ_2 in problems (28), we also introduce the slack variables $\boldsymbol{\alpha}^i$, and $\boldsymbol{\alpha}_c^i$, $i \in \{r, t\}$ to transform the problem. By applying the approximation method described in (46), we approximate (44) and (25b) to

$$-\nu(\alpha_k^{r,[z]}, -\lambda_1) - \nu(\alpha_k^{t,[z]}, -\lambda_2) + C_k \geq x^{HT}, \tag{51a}$$
$$-\nu(\alpha_{c,k}^{r,[z]}, -\lambda_1) - \nu(\alpha_{c,k}^{t,[z]}, -\lambda_2) \geq \sum_{m\in\mathcal{K}} C_m, \forall k \in \mathcal{K}_1, \tag{51b}$$
$$\begin{aligned}&-\nu(\alpha_{c,k}^{r,[z]}, -\lambda_1) + (\lambda_r - \lambda_1)\log_2(1+\gamma_{c,k}^{HD,[2],r}) - \\ &\nu(\alpha_{c,k}^{r,[z]}, -\lambda_2) + (1-\lambda_r-\lambda_2)\log_2(1+\gamma_{c,k}^{HD,[2],t}) \geq \sum_{m\in\mathcal{K}} C_m, \forall k \in \mathcal{K}_2,\end{aligned} \tag{51c}$$
$$\alpha_{c,k}^i \geq \log(1+\gamma_{c,k}^{HD,[1],i}), \tag{51d}$$
$$\alpha_k^i \geq \log(1+\gamma_k^{HD,i}). \tag{51e}$$

Following Step 2.1, the time allocation variables are decoupled from the non-convex SINR expressions for all transmission modes. Next, we proceed to address the classical non-convex SINR expressions.

- **Step 2.2: The SCA method for non-convex SINRs**:

Similar to Step 1.1, the SCA method for SINRs is applicable to all transmission modes. Therefore, we illustrate this approach using the *FE* problem (25) as an example, without detailing other modes.

We first introduce slack variables $\boldsymbol{\iota} = [\iota_1, \cdots, \iota_K]^T$, $\boldsymbol{\iota}_c = [\iota_{c,1}, \cdots, \iota_{c,K}]^T$ to denote the SINRs of s_k and s_0. Constraints (44) and (25b) are equivalently transformed to

$$\log_2(1+\iota_k) + C_k \geq x^{FE}, \forall k \in \mathcal{K}, \tag{52a}$$
$$\log_2(1+\iota_{c,k}) \geq \sum_{m\in\mathcal{K}} C_m, k \in \mathcal{K}, \tag{52b}$$
$$\iota_{c,k} \leq \gamma_{c,k}^{FD,[1]}, \forall k \in \mathcal{K}, \tag{52c}$$
$$\iota_k \leq \gamma_k^{FD}, \forall k \in \mathcal{K}. \tag{52d}$$

Constraints (52c) and (52d) remain non-convex, we then introduce variables $\boldsymbol{\zeta} = [\zeta_1, \cdots, \zeta_K]^T$, $\boldsymbol{\zeta}_c = [\zeta_{c,1}, \cdots, \zeta_{c,K}]^T$ to, respectively, denote the denominators of SINRs for s_k and s_0. (52c) and (52d) are equivalently transformed to

$$\zeta_{c,k} \geq \mathrm{de}(\gamma_{c,k}^{FD,[1]}), \forall k \in \mathcal{K}, \tag{53a}$$

$$\zeta_k \geq \mathrm{de}(\gamma_k^{FD}), \forall k \in \mathcal{K}, \tag{53b}$$

$$\iota_{c,k}\zeta_{c,k} \leq \left|\tilde{\mathbf{g}}_{i,k}^H \mathbf{p}_0\right|^2, \tag{53c}$$

$$\iota_k\zeta_k \leq \left|\tilde{\mathbf{g}}_{i,k}^H \mathbf{p}_k\right|^2. \tag{53d}$$

Following the approaches specified in Step 1.1 to address the non-convexity in (33), we approximate constraints (53c) and (53d) at iteration $[z]$ to

$$\nu(\iota_{c,k}^{[z]}, \zeta_{c,k}^{[z]}) \leq \varpi(\mathbf{p}_0^{[z]}, \mathbf{p}_0, 0, \tilde{\mathbf{g}}_{i,k}), \forall k \in \mathcal{K}, \tag{54a}$$

$$\nu(\iota_k^{[z]}, \zeta_k^{[z]}) \leq \varpi(\mathbf{p}_k^{[z]}, \mathbf{p}_k, 0, \tilde{\mathbf{g}}_{i,k}), \forall k \in \mathcal{K}. \tag{54b}$$

Based on the approximation methods specified in Steps 2.1 and 2.2, the joint active beamforming, common rate, and time allocation optimization problems for all six transmission modes become QCQP and can be solved using CVX. The detailed process of the SCA methods to solve this subproblem is illustrated in Algorithm 2.

Algorithm 2: Joint transmit active beamforming and resource allocation algorithm

---

Initialize: $j, z = 0, x^{j,[z]}, \mathbf{P}^{[z]}, \mathbf{c}^{[z]}, \lambda_r^{[z]}, \lambda^{[z]}, \lambda_1^{[z]}, \lambda_2^{[z]}$ and slack variables according to the corresponding mode j;

repeat

  $z \leftarrow z + 1$;

  Updating $\mathbf{P}^{[z]}, \mathbf{c}^{[z]}, \lambda_r^{[z]}, \lambda^{[z]}, \lambda_1^{[z]}, \lambda_2^{[z]}$ uses the SCA methods in Steps 2.1 and 2.2;

until *convergence*;

3.1.3. Alternative Optimization

Based on Algorithms 1 and 2, the proposed AO algorithm that solves the two subproblems iteratively is shown in Algorithm 3. Starting from a feasible point $(\mathbf{P}^{[0]}, \boldsymbol{\Theta}_r^{[0]}, \boldsymbol{\Theta}_t^{[0]}, \mathbf{c}^{[0]}, \lambda_r^{[0]}, \lambda^{[0]}, \lambda_1^{[0]}, \lambda_2^{[0]})$, at each iteration $[z]$, we first update $\boldsymbol{\Theta}_r^{[z]}$ and $\boldsymbol{\Theta}_t^{[z]}$ by Algorithm 1. Given $\boldsymbol{\Theta}_r^{[z]}$ and $\boldsymbol{\Theta}_t^{[z]}$, we then update $\mathbf{P}^{[z]}$, the common rate and time allocation based on Algorithm 2. By iteratively solving these two subproblems, the objective function is updated until convergence.

Algorithm 3: Proposed AO algorithm

Initialize: $j, z = 0, \boldsymbol{\Theta}_r^{[z]}, \boldsymbol{\Theta}_t^{[z]}, \mathbf{P}^{[z]}, \mathbf{c}^{[z]}, \lambda_r^{[z]}, \lambda^{[z]}, \lambda_1^{[z]}, \lambda_2^{[z]}$;

repeat

  $z \longleftarrow z + 1$

  Given $\mathbf{P}^{[z-1]}, \mathbf{c}^{[z]}, \lambda_r^{[z]}, \lambda^{[z]}, \lambda_1^{[z]}, \lambda_2^{[z]}$, use Algorithm 1 to obtain $\boldsymbol{\Theta}_t^{[z]}$ and $\boldsymbol{\Theta}_r^{[z]}$.

  Given $\boldsymbol{\Theta}_t^{[z]}$ and $\boldsymbol{\Theta}_r^{[z]}$, use Algorithm 2 to obtain $\mathbf{P}^{[z]}, \mathbf{c}^{[z]}, \lambda_r^{[z]}, \lambda^{[z]}, \lambda_1^{[z]}, \lambda_2^{[z]}$.

until *convergence*;

3.2. Proposed Low-Complexity Algorithm

In this subsection, we propose a low-complexity algorithm for the STAR RIS passive beamforming and transmit beamforming optimization subproblems.

3.2.1. STAR RIS Passive Beamforming Optimization

We begin by specifying the proposed low-complexity algorithm tailored for the *ES* protocol. Subsequently, we extend this algorithm to the *MS* and *TS* protocols.

ES: To develop a low-complexity algorithm for designing $\mathbf{\Theta}_r$ and $\mathbf{\Theta}_t$ in the ES protocol, we consider the following problem aimed at maximizing the sum of channel gains:

$$\begin{aligned} \max_{\mathbf{\Theta}_r,\mathbf{\Theta}_t} \ & \sum_{k=1}^{K} \|\mathbf{g}_k^H + \mathbf{h}_k^H \mathbf{\Theta}_i \mathbf{E}\|^2, i \in \{r,t\} \\ \text{s.t.} \ & \text{(25c), (25e).} \end{aligned} \tag{55}$$

Constraints (25c) and (25e) are non-convex, which make the problem (55) difficult to solve. To tackle such a non-convex problem, we first relax constraints (25c) and (25e) to $\|\mathbf{\Theta}_r\|_F^2 + \|\mathbf{\Theta}_t\|_F^2 \leq N$, where $\|\cdot\|_F$ denotes the Frobenius norm. This approximation relaxes the feasible solutions of $\mathbf{\Theta}_r$ and $\mathbf{\Theta}_t$ to a convex set $\mathcal{S} = \{(\mathbf{\Theta}_r, \mathbf{\Theta}_r) | \|\mathbf{\Theta}_r\|_F^2 + \|\mathbf{\Theta}_t\|_F^2 \leq N\}$. Therefore, problem (55) is relaxed to

$$\max_{\mathbf{\Theta}} f(\mathbf{\Theta}) \triangleq \|\mathbf{G}^H + \mathbf{H}^H \mathbf{\Theta} \mathbf{E}_x\|_F^2 \tag{56a}$$

$$\text{s.t. } \mathbf{\Theta} \in \mathcal{S}, \tag{56b}$$

where $\mathbf{\Theta} = \begin{bmatrix} \mathbf{\Theta}_r & 0 \\ 0 & \mathbf{\Theta}_t \end{bmatrix}$, $\mathbf{G} = [\mathbf{G}_r; \mathbf{G}_t], \mathbf{G}_r = \{\mathbf{g}_k | k \in \mathcal{K}_r\}, \mathbf{G}_t = \{\mathbf{g}_k | k \in \mathcal{K}_t\}$, $\mathbf{H} = [\mathbf{H}_r; \mathbf{H}_t], \mathbf{H}_r = \{\mathbf{h}_k | k \in \mathcal{K}_r\}, \mathbf{H}_t = \{\mathbf{h}_k | k \in \mathcal{K}_t\}$ and $\mathbf{E}_x = [\mathbf{E}; \mathbf{E}]$.

Inspired by the method proposed in [24], we employ the gradient decent method at the point $\mathbf{\Theta}_0 = \mathbf{0}$ to derive a closed-form solution of (56). The corresponding gradient of $f(\mathbf{\Theta})$ at $\mathbf{\Theta}_0 = \mathbf{0}$ is

$$\nabla_{\mathbf{\Theta}} f(\mathbf{\Theta}_0) = \mathbf{H}\mathbf{G}^H \mathbf{E}_x^H, \tag{57}$$

By using (57) as the decent direction, i.e., $\mathbf{D}_0 = \nabla_{\mathbf{\Theta}} f(\mathbf{\Theta}_0)$, we then take the Armijo rule to determine the step size α as

$$f(\mathbf{\Theta}_0 + \alpha \mathbf{D}_0) \geq f(\mathbf{\Theta}_0) + \phi \cdot \alpha \cdot \text{tr}(\nabla_{\mathbf{\Theta}} f(\mathbf{\Theta}_0)^H \mathbf{D}_0), \tag{58}$$

where $\phi \in (0, 0.5)$ is a constant shrinkage factor. Substituting (57) into (58), we obtain

$$\alpha^2 \|\mathbf{H}^H \mathbf{H} \mathbf{G}^H \mathbf{E}_x^H \mathbf{E}_x\|_F^2 + \alpha \cdot (2 - \phi) \|\mathbf{H}\mathbf{G}^H \mathbf{E}_x^H\|_F^2 \geq 0. \tag{59}$$

To guarantee the solution is on the boundary of the convex set $\mathcal{S}$, we set $\alpha = \frac{\sqrt{N}}{\|\mathbf{H}\mathbf{G}^H \mathbf{E}_x^H\|_F}$. Then, we obtain the following solution of problem (56) as

$$\mathbf{\Theta} = \frac{\sqrt{N}}{\|\mathbf{H}\mathbf{G}^H \mathbf{E}_x^H\|_F} \mathbf{H}\mathbf{G}^H \mathbf{E}_x^H. \tag{60}$$

However, the solution (60) for the relaxed problem (56) may not satisfy the original constraints (25c), (25e). Therefore, we propose to project $\mathbf{\Theta}$ into the feasible region. Following [24], we first define the operator sym$(\cdot)$ that projects any square matrix $\mathbf{X}$ to a symmetric matrix, i.e.,

$$\text{sym}(\mathbf{X}) = \frac{1}{2}\left(\mathbf{X}^T + \mathbf{X}\right), \tag{61}$$

and the operator uni$(\cdot)$ which projects an arbitrary matrix $\mathbf{X}$ to a unitary matrix as

$$\text{uni}(\mathbf{X}) = \mathbf{U}\mathbf{V}^H, \tag{62}$$

where $\mathbf{U}$ and $\mathbf{V}$ are unitary matrices obtained by the singular value decomposition (SVD) of $\mathbf{X}$, i.e., $\mathbf{X} = \mathbf{U}\mathbf{S}\mathbf{V}^H$. Consider $\mathbf{X}$ with dimension N and rank R. Then, the unitary matrices $\mathbf{U}$

and $\mathbf{V}$ can be partitioned as $\mathbf{U} = [\mathbf{U}_R, \mathbf{U}_{N-R}]$ and $\mathbf{V} = [\mathbf{V}_R, \mathbf{V}_{N-R}]$, respectively. Based on the operators in (61) and (62), we could then define the symmetric unitary projection as [24]

$$\text{symuni}(\mathbf{X}) = \text{uni}(\text{sym}(\mathbf{X})) = \hat{\mathbf{U}}\mathbf{V}^H, \tag{63}$$

where $\hat{\mathbf{U}} = [\mathbf{U}_R, \mathbf{V}_{N-R}^*]$.

Utilizing the symmetric unitary projection, we then derive a feasible STAR RIS passive beamforming solution $\boldsymbol{\Theta}$. It is noteworthy that both $\boldsymbol{\Theta}_r$ and $\boldsymbol{\Theta}_t$ are diagonal matrices, implying that $\boldsymbol{\Theta}$ is also diagonal. The diagonal entries of $\boldsymbol{\Theta}$ are determined by

$$\text{diag}\{\boldsymbol{\Theta}\} = \frac{\sqrt{2}}{2}\text{symuni}\left\{\text{diag}\left(\mathbf{H}\mathbf{G}^H\mathbf{E}_x^H\right)\right\}. \tag{64}$$

Since the unitary matrix $\text{symuni}\{\text{diag}(\mathbf{H}\mathbf{G}^H\mathbf{E}_x^H)\} \in \mathbb{C}^{2N\times 2N}$ whose dimension is $2N$, we have $\|\text{symuni}\{\text{diag}(\mathbf{H}\mathbf{G}^H\mathbf{E}_x^H)\}\|_F^2 = 2N$. To ensure constraint (25e), we scale each entry with $\frac{\sqrt{2}}{2}$.

MS: For the *MS* mode, we first utilize (64) to derive $\boldsymbol{\Theta}_r$ and $\boldsymbol{\Theta}_t$. As each element can only operate in either reflection or transmission mode, we then compare the reflection and transmission amplitude coefficients of each STAR RIS element. The element with the higher coefficient is designated as 1, while the one with the lower coefficient is designated as 0. Based on (64), the reflection and transmission coefficients of each STAR RIS element are designed as

$$\begin{aligned} &\beta_{r,n} = \begin{cases} 1, \text{ if } |\boldsymbol{\Theta}_{n,n}| - |\boldsymbol{\Theta}_{N+n,N+n}| \geq 0 \\ 0, \text{ if } |\boldsymbol{\Theta}_{n,n}| - |\boldsymbol{\Theta}_{N+n,N+n}| < 0 \end{cases}, \forall n \in \mathcal{N} \\ &\beta_{t,n} = 1 - \beta_{r,n}. \end{aligned} \tag{65}$$

TS: For the *TS* mode, in the reflection time slots, $\boldsymbol{\Theta}_t = \mathbf{0}$, the amplitudes of all elements in $\boldsymbol{\Theta}_r$ are 1. The solution of the reflection matrix is given as

$$\boldsymbol{\Theta}_r = \text{symuni}\{\text{diag}\left(\mathbf{H}_r\mathbf{G}_r^H\mathbf{E}^H\right)\}. \tag{66}$$

In the transmission time slots, the transmission matrix is given as

$$\boldsymbol{\Theta}_t = \text{symuni}\{\text{diag}\left(\mathbf{H}_t\mathbf{G}_t^H\mathbf{E}^H\right)\}. \tag{67}$$

3.2.2. Joint Optimization of the Active Beamforming, Common Rate and Time Allocation

With fixed $\boldsymbol{\Theta}_r$ and $\boldsymbol{\Theta}_t$, we aim to simplify the design of the remaining variables. To achieve this, we reduce the optimization dimension by fixing the direction of the precoders. Subsequently, we focus on jointly optimizing the power, common rate, and time allocation variables.

Let $\bar{\mathbf{g}}_k = \mathbf{g}_k/||\mathbf{g}_k||$, $\bar{\mathbf{G}} = [\bar{\mathbf{g}}_1, \ldots, \bar{\mathbf{g}}_K]$, $\bar{\mathbf{p}}_k = \mathbf{p}_k/||\mathbf{p}_k||$, and $\bar{\mathbf{P}} = [\bar{\mathbf{p}}_1, \ldots, \bar{\mathbf{p}}_K]$. We transform the active beamformings in the form of $\mathbf{p}_c = \sqrt{\rho_c P_t}\bar{\mathbf{p}}_c, \mathbf{p}_k = \sqrt{\rho_k P_t}\bar{\mathbf{p}}_k$, where ρ_c and ρ_k are the power allocation coefficients. For the beamforming direction $\bar{\mathbf{p}}_k$ of the private streams, we adopt zero-forcing (ZF) beamforming, i.e., $\bar{\mathbf{P}} = \bar{\mathbf{G}}(\bar{\mathbf{G}}^H\bar{\mathbf{G}})^{-1}$ [25]. For the beamforming direction $\bar{\mathbf{p}}_c$ of the common stream, we utilize maximum ratio transmission (MRT) and SVD methods as $\bar{\mathbf{p}}_c = \mathbf{u}_c$, where $\mathbf{u}_c = \mathbf{U}_c(:,1), \mathbf{G} = \mathbf{U}_c\mathbf{S}_c\mathbf{V}_c$ [2]. Based on the aforementioned beamforming direction design, the optimization problems for all six transmission modes are simplified. In the following, we illustrate this simplification by considering the *FD* mode as a representative example:

$$\max_{\rho,c} \min_{k\in\mathcal{K}} C_k + R^{\mathrm{FE}}_{p,k} \tag{68a}$$
$$\text{s.t.} \quad \textstyle\sum_{m\in\mathcal{K}} C_m \leq R^{\mathrm{FE}}_{c,k}, \forall k \in \mathcal{K}, \tag{68b}$$
$$\rho_c + \textstyle\sum_{k\in\mathcal{K}} \rho_k \leq 1, \tag{68c}$$

where $\rho = [\rho_c, \rho_1, \ldots, \rho_K]$ are the power allocation coefficients of the BS, we have $R^{\mathrm{FE}}_{p,k} = \log_2(1 + \rho_k P_t |\tilde{\mathbf{g}}_k^H \bar{\mathbf{p}}_k|^2/\sigma^2)$, $R^{\mathrm{FE}}_{c,k} = \log_2(1 + \rho_c P_t |\tilde{\mathbf{g}}_k^H \bar{\mathbf{p}}_c|^2/(\rho_k P_t |\tilde{\mathbf{g}}_k^H \bar{\mathbf{p}}_k|^2 + \sigma^2))$. The optimization problem can be directly solved using the SCA-based approach specified in Algorithm 2. The optimization approach for other transmission modes aligns with that of the FE mode. For *FT/HE/HM/HT* modes, the additional time allocation variables are jointly optimized using the SCA-based approach specified in Step 2.1.

The detailed process of the proposed low-complexity algorithm to solve problems (25)–(30) is outlined in Algorithm 4. Contrasted with Algorithm 3, the primary difference lies in the closed-form STAR RIS passive beamforming design and the closed-form active beamforming direction design. Additionally, there is no alternating optimization between the two subproblems. Each subproblem is solved once before the output.

Algorithm 4: Proposed low-complexity algorithm

1 **Initialize:** $j, \mathbf{P}^{[0]}, \mathbf{c}^{[0]}, \lambda_r^{[0]}, \lambda^{[0]}, \lambda_1^{[0]}, \lambda_2^{[0]}$;

2 Update $\mathbf{\Theta}_t$ and $\mathbf{\Theta}_r$ through (64) if $j = FE/HE$, (65) if $j = FM/HM$, (66) and (67) if $j = FT/HT$.

3 Given $\mathbf{\Theta}_t$ and $\mathbf{\Theta}_r$, update $\mathbf{P}$, $\mathbf{c}$, λ_r, λ, λ_1, λ_2 by solving the joint power, common rate, and time allocation problem via Algorithm 2.

3.3. *Convergence and Computational Complexity Analysis*

3.3.1. Convergence Analysis

In Algorithm 1, the SCA method guarantees the objective t^j monotonically increases, meaning that $t^{j,[z-1]} \leq t^{j,[z]}$. This arises from the fact that the solution of $\mathbf{\Theta}_r$ and $\mathbf{\Theta}_t$ at iteration $[z-1]$ is a feasible point of the STAR RIS problem at iteration $[z]$. Due to STAR RIS power constraint (25e), there is an upper bound for $t^{j,[z]}$. Hence, the convergence of Algorithm 1 is guaranteed. Similarly, in Algorithm 2, the SCA method guarantees that $x^{j,[z-1]} \leq x^{j,[z]}$ since the solution at iteration $[z-1]$ is an also feasible point of the joint optimization problem at iteration $[z]$. Due to the power constraint (25d) at the BS, there is an upper bound for the objective value. Hence, the convergence of the Algorithm 2 is guaranteed. The convergence of Algorithms 1 and 2 implies that the objective of the AO algorithm increases monotonically, which guarantees the convergence of Algorithm 3.

3.3.2. Computational Complexity of Algorithm 3

The computational complexity of Algorithm 1 is $\mathcal{O}(N^{3.5}\log_2(1/\epsilon_1))$, where ϵ_1 is the convergence tolerance. The complexity of Algorithm 2 is $\mathcal{O}((KN_t)^{3.5}\log_2(1/\epsilon_2))$, where ϵ_2 is the corresponding convergence tolerance. Therefore, the computational complexity of Algorithm 3 is $\mathcal{O}(T(N^{3.5}\log_2(1/\epsilon_1) + (KN_t)^{3.5}\log_2(1/\epsilon_2)))$, where T denotes the number of iterations of the AO algorithm.

3.3.3. Computational Complexity of Algorithm 4

For the proposed low-complexity algorithm, the computational complexity of the STAR RIS passive beamforming design is mainly occupied by the unitary projection with the SVD method, which is $\mathcal{O}(N^3)$. For the active beamforming optimization subproblem, the computational complexity is $\mathcal{O}(K^{3.5}\log_2(1/\epsilon_2))$. Therefore, the overall computational complexity of Algorithm 4 is $\mathcal{O}(N^3 + K^{3.5}\log_2(1/\epsilon_2))$, which is significantly lower than that of Algorithm 3.

3.4. Imperfect CSIT

Due to quantization errors in CSI feedback, feedback delays, and many other issues, it is difficult for the BS to obtain the perfect CSIT. Hence, in this subsection, we focus on an imperfect CSIT scenario where the direct channel between the BS and users, the RIS and users is imperfect. Since the locations of the BS and STAR RIS are typically fixed, perfect CSIT between the BS and the STAR RIS is considered. In order to simplify the notations, we introduce a block diagonal matrix $\mathbf{Q}$ including all perfect channel coefficients, which is given as

$$\mathbf{Q} = diag([\mathbf{g}_1, \ldots, \mathbf{g}_K], [\mathbf{h}_1, \ldots, \mathbf{h}_K], \mathbf{E}). \tag{69}$$

We denote the estimated CSI between the BS and user-k as $\widehat{\mathbf{g}}_k$, and the estimated CSI between the STAR RIS and user-k as $\widehat{\mathbf{h}}_k$. Besides, we denote $\widetilde{\mathbf{g}}_k$ and $\widetilde{\mathbf{h}}_k$ as the corresponding estimation errors, both following independent and identically distributed (i.i.d) zero-mean complex Gaussian distributions. The corresponding error covariance matrices are denoted as $\mathbb{E}_{\mathbf{g}_k|\widehat{\mathbf{g}}_k}\{\mathbf{g}_k\mathbf{g}_k^H|\widehat{\mathbf{g}}_k\} = \mathbf{R}_{g,k}$ and $\mathbb{E}_{\mathbf{h}_k|\widehat{\mathbf{h}}_k}\{\mathbf{h}_k\mathbf{h}_k^H|\widehat{\mathbf{h}}_k\} = \mathbf{R}_{h,k}$, respectively. The relationship between the real channel matrix $\mathbf{Q}$ and the estimated channel matrix $\widehat{\mathbf{Q}}$ is modeled as

$$\mathbf{Q} = \widehat{\mathbf{Q}} + \widetilde{\mathbf{Q}}, \tag{70}$$

where $\widehat{\mathbf{Q}} = diag\Big([\widehat{\mathbf{g}}_1, \ldots, \widehat{\mathbf{g}}_K], \Big[\widehat{\mathbf{h}}_1, \ldots, \widehat{\mathbf{h}}_K\Big], \mathbf{E}\Big)$ and $\widetilde{\mathbf{Q}} = diag\Big([\widetilde{\mathbf{g}}_1, \ldots, \widetilde{\mathbf{g}}_K], \Big[\widetilde{\mathbf{h}}_1, \ldots, \widetilde{\mathbf{h}}_K\Big], \mathbf{0}\Big)$.

With partial CSIT, the instantaneous rate cannot guarantee decodability for the users. A more robust approach is to optimize the ER which characterizes the long-term rate performance across all channel states. The ERs for the common and private streams at user-k are defined as

$$\overline{R}^j_{c,k} = \mathbb{E}_{\{\widehat{\mathbf{Q}},\mathbf{Q}\}}\Big\{R^j_{c,k}\Big\}, \tag{71a}$$

$$\overline{R}^j_{p,k} = \mathbb{E}_{\{\widehat{\mathbf{Q}},\mathbf{Q}\}}\Big\{R^j_{p,k}\Big\}. \tag{71b}$$

Optimizing ERs directly is very challenging. Instead, we can leverage average rates (ARs) to obtain ERs [26]. With a given channel estimation $\widehat{\mathbf{Q}}$, ARs for the common and private streams at user-k are defined as

$$\widehat{R}^j_{c,k} = \mathbb{E}_{\{\mathbf{Q}|\widehat{\mathbf{Q}}\}}\Big\{R^j_{c,k}|\widehat{\mathbf{Q}}\Big\}, \tag{72a}$$

$$\widehat{R}^j_{p,k} = \mathbb{E}_{\{\mathbf{Q}|\widehat{\mathbf{Q}}\}}\Big\{R^j_{p,k}|\widehat{\mathbf{Q}}\Big\}. \tag{72b}$$

The relation between ERs and ARs is as follows [26]

$$\overline{R}^j_{c,k} = \mathbb{E}_{\widehat{\mathbf{Q}}}\Big\{\widehat{R}^j_{c,k}\Big\}, \tag{73a}$$

$$\overline{R}^j_{p,k} = \mathbb{E}_{\widehat{\mathbf{Q}}}\Big\{\widehat{R}^j_{p,k}\Big\}. \tag{73b}$$

To ensure that the common message s_0 is successfully decoded by all users, the ER of the common stream s_0 should satisfy

$$\overline{R}^j_c = \min_{k\in\mathcal{K}}\Big\{\overline{R}^j_{c,k}\Big\}. \tag{74}$$

By replacing the instantaneous rates in (25)–(30) with the average rates in (72), we obtain the stochastic AR optimization problem for a given channel estimation $\widehat{\mathbf{Q}}$. A conventional approach to address such stochastic problems is the sampled average approximation (SAA) approach [2], which transforms the problems into deterministic ones. However, such an approach requires a large number of channel samples and significantly increases the computational complexity. To avoid these issues, we follow [27] and obtain the lower bounds for the ARs. For example, the AR $\widehat{R}^j_{c,k}$ in (72) is bounded by

$$\begin{aligned}\widehat{R}^{j}_{c,k} &= \mathbb{E}_{\mathbf{Q}|\widehat{\mathbf{Q}}}\left\{\log_2\left(1+\frac{|\mathbf{g}_k^H\mathbf{p}_c+\mathbf{h}_k^H\mathbf{\Theta}_i\mathbf{E}\mathbf{p}_c|^2}{\sum_{j=1}^{K}|\mathbf{g}_k^H\mathbf{p}_j+\mathbf{h}_k^H\mathbf{\Theta}_i\mathbf{E}\mathbf{p}_j|^2+\sigma_k^2}\right)\right\}\\ &\geq \log_2\left(1+\frac{|\widehat{\mathbf{g}}_k^H\mathbf{p}_c+\widehat{\mathbf{h}}_k^H\mathbf{\Theta}_i\mathbf{E}\mathbf{p}_c|^2}{\sum_{j=1}^{K}\left(|\widehat{\mathbf{g}}_k^H\mathbf{p}_j|^2+\mathbf{p}_j^H\mathbf{R}_{g,k}\mathbf{p}_j+\mathbf{q}_{i,j}^H\mathbf{R}_{h,k}\mathbf{q}_{i,j}\right)+\mathbf{p}_c^H\mathbf{R}_{h,k}\mathbf{p}_c+\mathbf{q}_{i,c}^H\mathbf{R}_{h,k}\mathbf{q}_{i,c}+\sigma_k^2}\right),\\ &\triangleq \widehat{R}^{j,lb}_{c,k},\end{aligned} \tag{75}$$

where $\mathbf{q}_{i,j}=\mathbf{\Theta}_i\mathbf{E}\mathbf{p}_j$. The inequality is obtained by treating $\mathbf{e}_{g,k}^H\mathbf{p}_c s_c+\mathbf{e}_{h,k}^H\mathbf{q}_{i,c}s_c+\sum_{j=1}^{K}(\mathbf{e}_k^H\mathbf{p}_j+\mathbf{e}_{h,k}^H\mathbf{q}_{i,j})s_j$ as independent Gaussian noise and applying Jensen's inequality [28]. Similarly, the AR $\widehat{R}^{j}_{p,k}$ is bounded by

$$\begin{aligned}\widehat{R}^{j}_{p,k} &= \mathbb{E}_{\mathbf{Q}|\widehat{\mathbf{Q}}}\left\{\log_2\left(1+\frac{|\mathbf{g}_k^H\mathbf{p}_k+\mathbf{h}_k^H\mathbf{\Theta}_i\mathbf{E}\mathbf{p}_k|^2}{\sum_{j\neq k}^{K}|\mathbf{g}_k^H\mathbf{p}_j+\mathbf{h}_k^H\mathbf{\Theta}_i\mathbf{E}\mathbf{p}_j|^2+\sigma_k^2}\right)\right\}\\ &\geq \log_2\left(1+\frac{|\widehat{\mathbf{g}}_k^H\mathbf{p}_k+\widehat{\mathbf{h}}_k^H\mathbf{\Theta}_i\mathbf{E}\mathbf{p}_k|^2}{\sum_{j\neq k}^{K}|\widehat{\mathbf{g}}_k^H\mathbf{p}_j|^2+\sum_{j=1}^{K}\mathbf{p}_j^H\mathbf{R}_{g,k}\mathbf{p}_j+\sum_{j=1}^{K}\mathbf{q}_{i,j}^H\mathbf{R}_{h,k}\mathbf{q}_{i,j}+\sigma_k^2}\right)\\ &\triangleq \widehat{R}^{j,lb}_{p,k}.\end{aligned} \tag{76}$$

With the lower bound (75), the ER of the common stream is, therefore, bounded by

$$\overline{R}_c=\min_{k\in\mathcal{K}}\left\{\mathbb{E}_{\widehat{\mathbf{Q}}}\left\{\widehat{R}^{j}_{c,k}\right\}\right\}\geq\min_{k\in\mathcal{K}}\left\{\mathbb{E}_{\widehat{\mathbf{Q}}}\left\{\widehat{R}^{j,lb}_{c,k}\right\}\right\}\geq\mathbb{E}_{\widehat{\mathbf{Q}}}\left\{\min_{k\in\mathcal{K}}\left\{\widehat{R}^{j,lb}_{c,k}\right\}\right\}. \tag{77}$$

Similarly, the ER of the private stream is bounded by

$$\overline{R}_p=\mathbb{E}_{\widehat{\mathbf{Q}}}\left\{\widehat{R}^{j}_{p,k}\right\}\geq\mathbb{E}_{\widehat{\mathbf{Q}}}\left\{\widehat{R}^{j,lb}_{p,k}\right\}. \tag{78}$$

By replacing the instantaneous rates R_c^j and $R^j_{p,k}$ in (25)–(30) with the lower bounds $\min_{k\in\mathcal{K}}\widehat{R}^{lb}_{c,k}$ and $\widehat{R}^{lb}_{p,k}$ in (75) and (76), we construct six long-term deterministic imperfect CSIT optimization problems. As the non-convexity in these problems remains the same as the perfect CSIT ones, our proposed two algorithms can be employed to address the imperfect CSIT optimization problems directly. Therefore, we omit the algorithm framework here.

4. Numerical Results

In this section, we evaluate the performance of our proposed STAR RIS-assisted CRS system framework along with the two proposed algorithms.

4.1. Simulation Setting

The setting of the simulation follows [29]. Consider a three-dimensional (3D) space, the BS is located at $(0,0,0)$m, the STAR RIS is located at $(0,50,0)$m. Users are randomly distributed within a circle, with the center being the location of the STAR RIS and a radius of $r=5$ m. We follow the method in [4] to separate the users into group $\mathcal{K}_1$ and $\mathcal{K}_2$. We only choose one user with the largest channel gain as the relaying user, other users are destination users. The channels between BS and users, as well as between user-m and user-n are assumed to be Rayleigh fading, and are denoted as $\mathbf{g}_k=L_{\mathrm{BU},k}\bar{\mathbf{g}}_k, h_{m,n}=L_{\mathrm{UU}}\bar{h}_{m,n}$, where $L_{\mathrm{BU},k}$ and L_{UU} denote the path loss between the BS and users as well as the path loss between users. $\bar{\mathbf{g}}_k$ and $\bar{h}_{m,n}$ follow the complex Gaussian distribution with a certain variance, i.e., $\bar{\mathbf{g}}_k\sim\mathcal{CN}(0,\sigma_k^2), \bar{h}_{m,n}\sim\mathcal{CN}(0,\sigma_{m,n}^2)$. The channels between the BS and STAR RIS, as well as between STAR RIS and users are assumed to follow the Rician fading model with the Rician factor being 3 dB. The path loss for all the aforementioned channels is defined as $L_x = L_0(d_x/d_0)^{-\alpha_x}, x \in \{\text{BU}, \text{UU}, \text{BR}, \text{RU}\}$, where $d_0 = 1$ m is the reference distance, $L_0 = -30$ dB is the path loss at reference distance, d_x is the distance of different channels, and α_x denotes different channel path exponents. The channel path loss exponents between the BS and STAR RIS, as well as between STAR RIS and users are 2.2 [14]. The channel path loss exponent between the BS and users is 3.76 [23]. Besides, the relaying power is $P_k = 0.5P_t$, the convergence tolerance is $\epsilon = 10^{-3}$, and the noise power is $\sigma = -90$ dBm. All results are obtained by averaging over 100 random channel realizations. The variances of user channels are set to $\sigma_k^2 = 1$ for the relaying user, $\sigma_k^2 = 0.3$ for the destination users, and $\sigma_{m,n}^2 = 1$.

The following schemes are compared in this section:

- **CRS-FE/FM/FT/HE/HM/HT**: This refers to the STAR RIS-assisted CRS for six different transmission modes, as we proposed in Section 2.
- **CRS-HD/FD**: This refers to the existing CRS transmission schemes based on the *HD* or *FD* protocols without the assistance of STAR RIS [4,6].
- **RSMA-ES**: This refers to the existing STAR RIS-assisted RSMA transmission scheme without user relaying [16]. Here we only consider the *ES* protocol since it is more general than the *MS* and *TS* protocols.
- **SDMA-ES**: This refers to the existing STAR RIS-assisted SDMA transmission scheme. Again, we only consider the *ES* protocol.
- **NOMA-ES**: This refers to the existing STAR RIS assisted NOMA transmission scheme [29] with the *ES* protocol.
- **RSMA**: This refers to the conventional one-layer RSMA transmission scheme, which does not involve user relaying or assistance from STAR RIS [2].

4.2. Simulation Results

4.2.1. Comparison Among STAR RIS Operating Protocols

From Figure 3, we observe that the *FE/HE* mode consistently achieves a higher max-min rate than the *FM/HM* and *FT/HT* modes. This is because the *ES* protocol is more general than the *MS* and *TS* protocol, and it ensures that each STAR RIS element has a larger tunable range of the amplitude coefficient. As the number of STAR RIS elements increases, the max-min rate performance gain between *FM/HM* and *FT/HT* modes decreases. Surprisingly, when N is larger than 110, the *FT/HT* mode achieves a higher max-min rate than the *FM/HM* mode. This indicates that the *TS* protocol outperforms the *MS* protocol when the STAR RIS element is large. With an increasing number of STAR RIS elements, the performance improvement gained from having all elements operating within the same transmission/reflection space is outweighed by the performance degradation caused by partial time service.

Due to the superior max-min rate performance of *ES* compared to *MS* and *TS*, in the following simulation results, we only illustrate the proposed STAR RIS-assisted CRS in *FE* and *HE* modes for clarity.

4.2.2. Comparison Between FD and HD Relaying Protocols

Figure 4 illustrates the max-min rate performance versus transmit SNR. We compare our proposed CRS-FE/HE with the conventional RSMA scheme and the conventional CRS schemes with *HD* and *FD* protocols. Algorithm 3 is employed to optimize the corresponding problems of the proposed schemes. We set $N = 50, K = 4, N_t = 4$. The max-min rate increases with SNR for all transmission schemes. It is noteworthy that when SNR is below 25 dB, the CRS-FE scheme achieves a higher max-min rate than the CRS-HE scheme. Conversely, when the SNR exceeds 25 dB, the CRS-HE scheme outperforms the CRS-FE scheme. This divergence stems from the introduction of self-interference in *FD* relaying. As the SNR increases, the level of self-interference also rises. This indicates that the *HD* protocol is preferred in the high SNR regime while the *FD* protocol is preferred in the low and moderate SNR regimes. For consistently ensuring the optimal max-min rate performance gain, we can employ *FD* relaying at low and moderate SNR and *HD* relaying at high SNR.

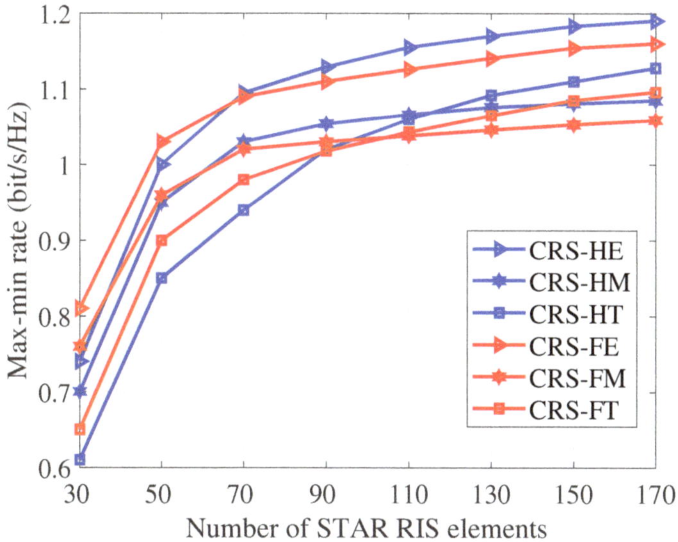

Figure 3. Max-min rate versus the number of STAR RIS elements N, SNR = 20 dB, $K = 4$, $N_t = 4$.

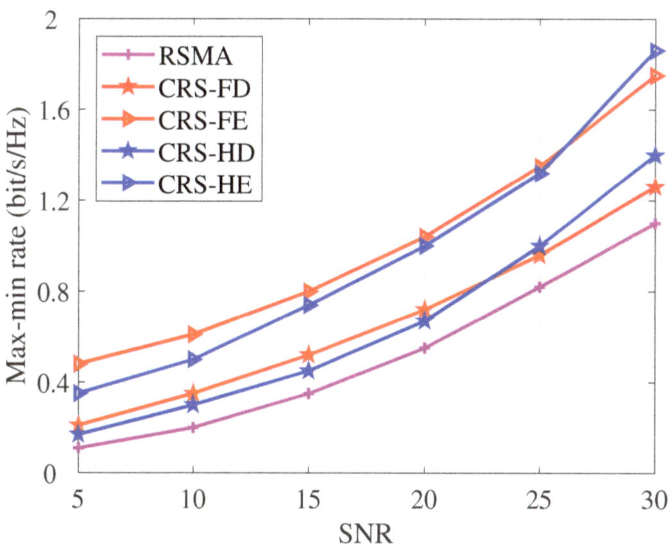

Figure 4. Max-min rate versus SNR, $N = 50$, $K = 4$, $N_t = 4$.

From Figure 4, it is obvious that with the assistance of STAR RIS and CRS, CRS-FE/HE achieves explicit max-min rate gain over conventional schemes. Specifically, CRS-FE demonstrates an average relative performance gain of 117.1% and 52.4% over conventional RSMA and CRS-FD, respectively. This demonstrates the significant performance enhancement achieved by integrating STAR RIS and CRS.

4.2.3. Comparison Among Different MA Schemes

In Figure 5, we compare various STAR RIS-assisted transmission schemes with the *ES* protocol when $N = 50$, $K = 4$, $N_t = 4$. The results show that our proposed STAR RIS-aided CRS scheme outperforms other MA schemes. It achieves an average max-min rate gain of 42.6% over the RSMA-ES scheme, 90.9% over the SDMA-ES scheme, and 77.2% over the NOMA-ES scheme. Besides, the RSMA-ES scheme outperforms NOMA-ES and SDMA-ES. By splitting the user message into common and private parts, RSMA facilitates more flexible interference management in all SNR regimes.

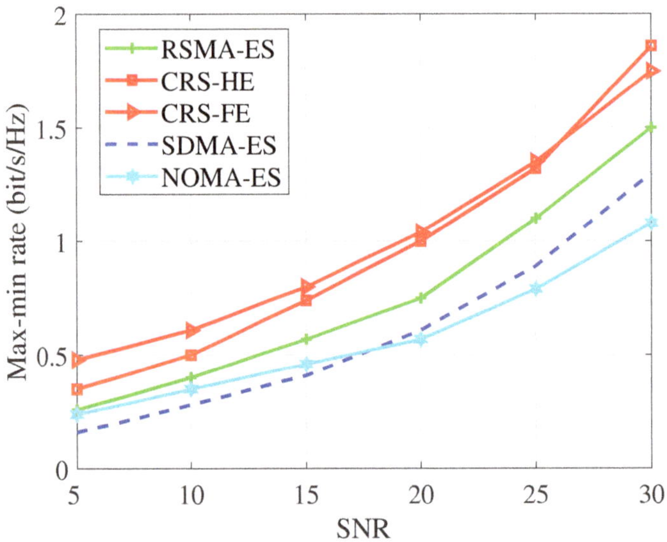

Figure 5. Max-minrate versus SNR, $N = 50$, $K = 4$, $N_t = 4$.

4.2.4. Comparison Between the Two Proposed Algorithms

In Figure 6, we compare the two proposed optimization algorithms for the proposed STAR RIS-aided CRS in the *ES* mode. We set SNR = 20 dB, $K = 4$, $N_t = 4$. In comparison to the proposed AO algorithm, the low-complexity algorithm reduces the average CPU time by 99.2% while incurring only a 9.1% performance loss at $N = 64$. This confirms the effectiveness of the proposed low-complexity algorithm. This efficiency is primarily due to the low-complexity algorithm eliminating the need for CVX operations to update $\mathbf{\Theta}_r$ and $\mathbf{\Theta}_t$, along with the alternative optimization between two subproblems. Instead, CVX is exclusively used for joint power, common rate, and time allocation, leading to a significant reduction in the optimization dimension. The results suggest that if the system can accommodate high time complexity, Algorithm 3 should be employed to achieve near-optimal max-min rate performance. Conversely, if the system is time-sensitive, the low-complexity Algorithm is recommended.

4.2.5. Imperfect CSIT

Figure 7 illustrates the ergodic max-min rate versus the number of STAR RIS elements, where SNR = 20 dB, $K = 4$, $N_t = 4$. The estimation errors $\mathbf{e}_{g,k}$ and $\mathbf{e}_{h,k}$ follow $\mathcal{CN}(\mathbf{0}, \kappa\sigma_k^2\mathbf{I})$, where σ_k^2 represents the channel variance and κ represents the CSIT error variance ratio. In this work, we set $\kappa = 0.2$. As the number of STAR RIS elements increases, the ergodic max-min rate increases accordingly. Our proposed low-complexity algorithm achieves performance close to that of the AO algorithm. This observation is consistent with our findings under perfect CSIT.

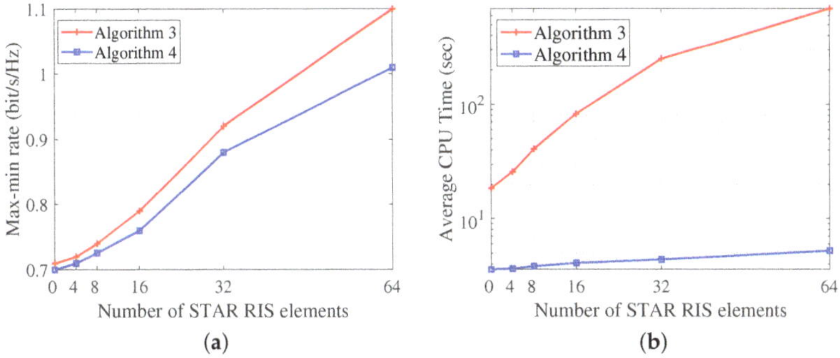

Figure 6. The performance of Algorithm 3 and 4. (**a**) Max-min rate versus N. (**b**) Average CPU time versus N.

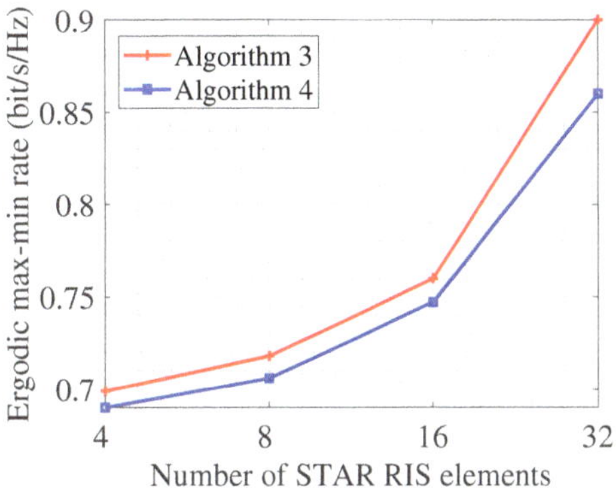

Figure 7. Ergodic max-min rate versus the number of STAR RIS elements N, SNR $= 20$ dB, $K = 4$, $N_t = 4$, $\kappa = 0.2$.

5. Conclusions

In this paper, we propose a novel STAR RIS-assisted CRS transmission with six different transmission modes, namely, *HE*, *HM*, *HT*, *FE*, *FM*, and *FT*. With the objective of maximizing the minimum user rate, we then propose a unified SCA-based AO algorithm to optimize the BS active beamforming and the STAR RIS passive beamforming iteratively under the transmit power constraint at the BS and the law of energy conservation at the STAR RIS. Meanwhile, we propose a novel low-complexity resource allocation algorithm that designs the BS active beamforming and the STAR RIS passive beamforming in closed form. Numerical results show that our proposed STAR RIS aid-CRS system achieves superior max-min rate performance gain over the existing CRS schemes and the STAR RIS-aided MA schemes. Furthermore, we show that the *FD* relaying has better max-min rate performance at low and moderate SNR while the *HD* relaying is preferred at high SNR. The *ES* protocol offers superior max-min rate performance compared to the *MS* and *TS* protocols, albeit with increased hardware complexity. With a large number of STAR RIS elements, the *TS* protocol demonstrates a significantly higher max-min rate gain compared to the *MS* protocol. Our proposed AO algorithm is better suited for systems that can accommodate higher time complexity, while the low-complexity algorithm is preferable for systems that are more time-sensitive.

Author Contributions: Conceptualization, Y.M.; Methodology, K.Z. and Y.M.; software, K.Z.; investigation, K.Z.; resources, Y.M.; writing—original draft preparation, K.Z.; writing—review and editing, Y.M. and Y.S. All authors have read and agreed to the published version of the manuscript.

Funding: This research was funded by the National Nature Science Foundation of China under Grant 62201347, Shanghai Sailing Program under Grant 22YF1428400.

Institutional Review Board Statement: Not applicable.

Data Availability Statement: The data represented in this study are available on request from the corresponding author.

Acknowledgments: The authors thank the anonymous reviewers and the Editor for their helpful comments and review of this article.

Conflicts of Interest: The authors declare no conflicts of interest.

References

1. Clerckx, B.; Mao, Y.; Jorswieck, E.A.; Yuan, J.; Love, D.J.; Erkip, E.; Niyato, D. A primer on rate-splitting multiple access: Tutorial, myths, and frequently asked questions. *IEEE J. Sel. Areas Commun.* **2023**, 1265–1308. [CrossRef]
2. Mao, Y.; Dizdar, O.; Clerckx, B.; Schober, R.; Popovski, P.; Poor, H.V. Rate-splitting multiple access: Fundamentals, survey, and future research trends. *IEEE Commun. Surv. Tut.* **2022**, *24*, 2073–2126. [CrossRef]
3. Zhang, J.; Clerckx, B.; Ge, J.; Mao, Y. Cooperative Rate Splitting for MISO Broadcast Channel with User Relaying, and Performance Benefits Over Cooperative NOMA. *IEEE Signal Process. Lett.* **2019**, *26*, 1678–1682. [CrossRef]
4. Mao, Y.; Clerckx, B.; Zhang, J.; Li, V.O.K.; Arafh, M.A. Max-Min Fairness of *K*-User Cooperative Rate-Splitting in MISO Broadcast Channel With User Relaying. *IEEE Trans. Wirel. Commun.* **2020**, *19*, 6362–6376. [CrossRef]
5. Khisa, S.; Almekhlafi, M.; Elhattab, M.; Assi, C. Full duplex cooperative rate splitting multiple access for a MISO broadcast channel with two users. *IEEE Commun. Lett.* **2022**, *26*, 1913–1917. [CrossRef]
6. Li, T.; Zhang, H.; Zhou, X.; Yuan, D. Full-duplex cooperative rate-splitting for multigroup multicast with SWIPT. *IEEE Trans. Wirel. Commun.* **2021**, *21*, 4379–4393. [CrossRef]
7. Pan, Q.; Wu, J.; Zheng, X.; Yang, W.; Li, J. Differential privacy and IRS empowered intelligent energy harvesting for 6G Internet of Things. *IEEE Internet Things J.* **2021**, *9*, 22109–22122. [CrossRef]
8. Basar, E.; Di Renzo, M.; De Rosny, J.; Debbah, M.; Alouini, M.S.; Zhang, R. Wireless communications through reconfigurable intelligent surfaces. *IEEE Access* **2019**, *7*, 116753–116773. [CrossRef]
9. Li, Z.; Yuan, W.; Li, B.; Wu, J.; You, C.; Meng, F. Reconfigurable Intelligent Surface Aided OTFS: Transmission Scheme and Channel Estimation. *IEEE Internet Things J.* **2023**, *10*, 19518–19532. [CrossRef]
10. Liu, Y.; Mu, X.; Xu, J.; Schober, R.; Hao, Y.; Poor, H.V.; Hanzo, L. STAR: Simultaneous transmission and reflection for 360° coverage by intelligent surfaces. *IEEE Wirel. Commun.* **2021**, *28*, 102–109. [CrossRef]
11. Zhang, H.; Di, B. Intelligent omni-surfaces: Simultaneous refraction and reflection for full-dimensional wireless communications. *IEEE Commun. Surv. Tut.* **2022**, *24*, 1997–2028. [CrossRef]
12. Wu, C.; Liu, Y.; Mu, X.; Gu, X.; Dobre, O.A. Coverage characterization of STAR-RIS networks: NOMA and OMA. *IEEE Commun. Lett.* **2021**, *25*, 3036–3040. [CrossRef]
13. Zhang, Z.; Chen, J.; Liu, Y.; Wu, Q.; He, B.; Yang, L. On the secrecy design of STAR-RIS assisted uplink NOMA networks. *IEEE Trans. Wirel. Commun.* **2022**, *21*, 11207–11221. [CrossRef]
14. Zuo, J.; Liu, Y.; Ding, Z.; Song, L.; Poor, H.V. Joint design for simultaneously transmitting and reflecting (STAR) RIS assisted NOMA systems. *IEEE Trans. Wirel. Commun.* **2022**, *22*, 611–626. [CrossRef]
15. Hashempour, H.R.; Bastami, H.; Moradikia, M.; Zekavat, S.A.; Behroozi, H.; Swindlehurst, A.L. Secure SWIPT in STAR-RIS Aided Downlink MISO Rate-Splitting Multiple Access Networks. *arXiv* **2022**, arXiv:2211.09081.
16. Katwe, M.; Singh, K.; Clerckx, B.; Li, C.P. Improved Spectral Efficiency in STAR-RIS Aided Uplink Communication using Rate Splitting Multiple Access. *IEEE Trans. Wirel. Commun.* **2023**, 5365–5382. [CrossRef]
17. Dhok, S.; Sharma, P.K. Rate-splitting multiple access with STAR RIS over spatially-correlated channels. *IEEE Trans. Commun.* **2022**, *70*, 6410–6424. [CrossRef]
18. Clerckx, B.; Joudeh, H.; Hao, C.; Dai, M.; Rassouli, B. Rate splitting for MIMO wireless networks: A promising PHY-layer strategy for LTE evolution. *IEEE Commun. Mag.* **2016**, *54*, 98–105. [CrossRef]
19. Riihonen, T.; Werner, S.; Wichman, R. Mitigation of Loopback Self-Interference in Full-Duplex MIMO Relays. *IEEE Trans. Signal Process.* **2011**, *59*, 5983–5993. [CrossRef]
20. Sabharwal, A.; Schniter, P.; Guo, D.; Bliss, D.W.; Rangarajan, S.; Wichman, R. In-Band Full-Duplex Wireless: Challenges and Opportunities. *IEEE J. Sel. Areas Commun.* **2014**, *32*, 1637–1652. [CrossRef]
21. Zhong, C.; Suraweera, H.A.; Zheng, G.; Krikidis, I.; Zhang, Z. Wireless Information and Power Transfer with Full Duplex Relaying. *IEEE Trans. Commun.* **2014**, *62*, 3447–3461. [CrossRef]

22. Hu, J.; Beaulieu, N.C. Performance Analysis of Decode-and-Forward Relaying with Selection Combining. *IEEE Commun. Lett.* **2007**, *11*, 489–491. [CrossRef]
23. Yang, Z.; Chen, M.; Saad, W.; Xu, W.; Shikh-Bahaei, M.; Poor, H.V.; Cui, S. Energy-efficient wireless communications with distributed reconfigurable intelligent surfaces. *IEEE Trans. Wirel. Commun.* **2021**, *21*, 665–679. [CrossRef]
24. Fang, T.; Mao, Y. A Low-Complexity Beamforming Design for Beyond-Diagonal RIS aided Multi-User Networks. *IEEE Commun. Lett.* **2024**, *28*, 203–207. [CrossRef]
25. Oestges, C.; Clerckx, B. *MIMO Wireless Communications: From Real-World Propagation to Space-Time Code Design*; Academic Press: Cambridge, MA, USA, 2010.
26. Joudeh, H.; Clerckx, B. Sum-Rate Maximization for Linearly Precoded Downlink Multiuser MISO Systems with Partial CSIT: A Rate-Splitting Approach. *IEEE Trans. Commun.* **2016**, *64*, 4847–4861. [CrossRef]
27. Kim, D.; Choi, J.; Park, J.; Kim, D.K. Max–Min fairness beamforming with rate-splitting multiple access: Optimization without a toolbox. *IEEE Wirel. Commun. Lett.* **2022**, *12*, 232–236. [CrossRef]
28. Boyd, S.; Boyd, S.P.; Vandenberghe, L. *Convex Optimization*; Cambridge University Press: Cambridge, UK, 2004.
29. Mu, X.; Liu, Y.; Guo, L.; Lin, J.; Schober, R. Simultaneously transmitting and reflecting (STAR) RIS aided wireless communications. *IEEE Trans. Wirel. Commun.* **2021**, *21*, 3083–3098. [CrossRef]

entropy

Article

Rate Optimization of Intelligent Reflecting Surface-Assisted Coal Mine Wireless Communication Systems

Yang Liu *, Zhao Yang, Bin Wang and Yanhong Xu

School of Communication and Information Engineering , Xi'an University of Science and Technology, Xi'an 710054, China; yangz@stu.xust.edu.cn (Z.Y.); wangbin@mail.xidian.edu.cn (B.W.); yhxu@xust.edu.cn (Y.X.)
* Correspondence: lyang@xust.edu.cn

Abstract: This paper proposes a three-step joint rate optimization method for intelligent reflecting surface (IRS)-assisted coal mine wireless communication systems. Different from terrestrial IRS-assisted communication scenarios, in coal mines, IRSs can be installed flexibly on the tops of rectangular tunnels to address the issues of signals being blocked and interfered with by mining equipment. Therefore, it is necessary to optimize the IRS deployment position, the transmit power and IRS phase shifts to achieve the maximum effective achievable rate at user stations equipped with the proposed system. However, due to the complex channel models of coal mines, the optimization problem of IRS deployment position is non-convex. To solve this problem, two auxiliary variables along with logarithmic operations and Taylor approximation are introduced. On this basis, a three-step joint rate optimization involving the transmit power, IRS phase shifts and IRS deployment position is proposed to maximize the effective achievable rates at the user station. The simulation results show that compared with other rate optimization schemes, the effective achievable rates at the user station using the proposed joint rate optimization scheme can be improved by approximately 12.32% to 54.17% for different parameter configurations. It is also pointed out that the deployment position of the IRS can converge to the same optimal position independent of the initial deployment position. Moreover, we investigate the effects of the roughness of the tunnel walls in a coal mine on the effective achievable rates at the user station, and the simulation results indicate that the proposed three-step joint rate optimization scheme performs better in the coal mine scenario regardless of the roughness.

Keywords: intelligent reflecting surface; achievable rates; rate optimization; coal mine wireless communication system

1. Introduction

Coal mines, as a vertical industrial application of 5th/6th-Generation (5G/6G) wireless communication networks, face critical challenges due to poor wireless propagation environments, such as uneven and rough tunnel walls, coal dust and high humidity, obstructions from mining equipment (ME) and so on [1]. These adverse factors result in complex and special wireless transmission characteristics in coal mines, including severe transmission loss, signal attenuation and distortion, and multipath interference. In particular, ME in coal mines will block direct communication links and degrade efficient signal transmission, which will create potential safety risks [2–4]. Traditional relay solutions have the issues of higher costs and power consumption, more complicated maintenance and lower efficiency. Therefore, there is an urgent need for new wireless communication technology with the attributes of low power consumption, low costs and high reliability to overcome these challenges and improve the quality of coal mine wireless communication systems.

As a candidate technology of 6G, intelligent reflecting surfaces (IRSs) have emerged due to their intelligent manipulation of the wireless propagation environment, flexible reconfiguration, low power consumption and cost and high spectral and energy efficiency, providing an innovative and effective means to realize excellent communication by directly reshaping wireless propagation channels in favor of signal transmission [5–9]. Since IRSs are economical and easy to deploy in the tunnels of coal mines and can address signals that are blocked and interfered with by obstructions, they have promise for applications in coal mines to enhance communication metrics [10]. Among them, the effective achievable rate, as a crucial metric of coal mine communication systems, plays a vital role in ensuring safety in a coal mine.

Much work has focused on the rate optimization of various IRS-assisted communication scenarios. In [11], the joint optimization of active transmit and passive reflective beamforming is investigated. High-quality suboptimal solutions are obtained by alternating optimization, semi-definite relaxation (SDR) and Gaussian randomization to maximize the target rate. To reduce the computational complexity of Gaussian randomized SDR, a simple neural network is proposed to replace SDR at the cost of a reduced secrecy rate [12]. Furthermore, the block coordinate ascent method (BCAM) is employed to find stationary solutions to the formulated non-convex optimization problem, which can reduce computational complexity to linear time with the number of IRS elements without compromising the secrecy rate too much [13]. These works only performed the joint optimization of transmit power and IRS phase shifts, which are insufficient, as IRS deployments need to be involved in the optimization problem. In [14], it was proven that if the distance between the access point (AP) and user equipment (UE) is fixed, the IRS should be deployed near the AP or UE to achieve the maximum transmission rate in a single UE system. For an unfixed distance between the AP and UE, the strategy of IRS deployment positions in IRS-assisted wireless communication systems is theoretically analyzed and experimentally validated in [15], which shows that the optimal IRS deployment is related to the relationship between the distance from the AP to the UE and the distance from the IRS to the AP-UE line. Different from terrestrial communication scenarios, in underground coal mines, IRSs can usually be installed on the tops of tunnels to reduce the absorption and scattering of signals by various rock materials and address the issues of signals being blocked and interfered with by ME. Therefore, it is necessary to optimize the IRS deployment position to improve the effective achievable rate for IRS-assisted coal mine wireless communication systems.

In this paper, we propose a joint rate optimization involving transmit power, IRS phase shifts, and the IRS deployment position for an IRS-assisted coal mine wireless communication system with interference from ME to maximize the effective achievable rate at the user station. Specifically, we first develop an IRS-assisted coal mine communication channel model by considering the roughness of tunnel walls and scatters in the space based on a geometry-based stochastic model (GBSM), a 3rd-generation partnership project (3GPP) standardized modeling framework [16], to characterize a rectangular coal mine tunnel communication channel. Based on this, a three-step joint rate optimization scheme is proposed. Due to the confined tunnel construction in a coal mine, the IRS deployment position optimization becomes complex and non-convex. To address this issue, two auxiliary variables along with logarithmic operations and Taylor approximation are introduced into the optimization problem. In the simulations, we first validate the convergence of the proposed three-step joint rate optimization scheme and the analysis results show that the IRS deployment position can converge to the same optimal value independent of the initial IRS deployment position. Next, we also compare the effective achievable rates at the user station under the proposed three-step joint rate optimization scheme and other rate optimization schemes. The comparison results show that the effective achievable rates can be improved by approximately 12.32% and 54.17% by varying the total transmit power under different parameter configurations. Finally, we investigate the effects of the roughness of the tunnel walls of a coal mine on the effective achievable rates at the user station, and the simulation results indicate that the proposed three-step joint rate optimization scheme performs better in a coal mine scenario regardless of the roughness.

This paper is organized as follows. In Section 2, we establish an IRS-assisted wireless communication system model with the channel model based on a three-dimensional (3D) GBSM in a coal mine and formulate the joint rate optimization problem. In Section 3, we propose an optimization algorithm for the IRS deployment position and a three-step joint optimization scheme to maximize the effective achievable rate at the user station. In Section 4, we simulate and compare the results under the proposed rate optimization scheme and other schemes. In Section 5, we conclude our work.

2. System Model and Problem Formulation

2.1. Channel Modeling and System Model

In this paper, we consider a rectangular coal mine tunnel and establish a coal mine channel model based on a 3D multiple input–multiple output (MIMO) GBSM under the 3GPP standard framework [17]. A three-dimensional description of an IRS-assisted coal mine wireless communication system is illustrated in Figure 1, which consists of a base station (BS) equipped with K antennas, an IRS with N low-cost passive reflecting elements, a single-antenna User and one single-antenna ME. In Figure 1, the gray lines represent the direct links between antenna pairs, and the red blank and blue solid circles represent the collections of multipath components generated by the reflection and refraction of the signal due to the roughness of the tunnel walls and scatters in the space. In this system model, the IRS is installed on the tops of tunnels and can be flexibly deployed along the horizontal axis, which forms a plane together with the BS, User and ME. We established a three-dimensional Cartesian coordinate system with the BS as the origin, denoting the coordinates of the BS, IRS, ME and User as $\mathbf{b} = [0, 0, 0]$, $\mathbf{I} = [I_x, I_y, I_z]$, $\mathbf{m} = [m_x, m_y, m_z]$ and $\mathbf{u} = [u_x, u_y, u_z]$, respectively. We assume that $I_y = m_y = u_y = 0$. Moreover, the direct link between the BS and the User is blocked by the mining equipment, which also interferes with the reflection link.

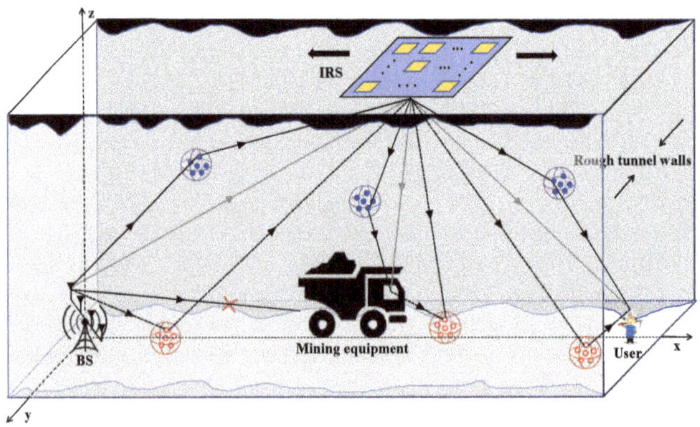

Figure 1. Illustration of 3D IRS-assisted coal mine wireless communication system model.

Next, we give a representation of the IRS-cascaded channel with ideal phase modulation combing using IRS technology. Here, it is assumed that the IRS phase modulation is ideal and independent of the angles of arrival in the BS-IRS and IRS–User channels. The IRS-cascaded channel between the BS and User can be represented as the products of the channel matrix of the BS-IRS link, the IRS phase shift matrix and the channel matrix of the IRS–User link, where both channel matrices can be modeled as the 3D MIMO GBSM for the coal mine [18].

Denote the channel matrix between the BS and IRS as $\mathbf{H}^{BI}(f) \in \mathbb{C}^{K \times N}$ and the entry of the kth row and nth column as $h_{kn}^{BI}(f)$ corresponding to the channel impulse response between the kth antenna of the BS and the nth reflecting element of the IRS. Here, $\mathbb{C}$ represents the complex-valued field. As shown in [18], the channel impulse response can be defined as the weighted sum of the Line-of-Sight (LoS) component and the Non-Line-of-Sight (NLoS) component. Therefore,

$$h_{kn}^{BI}(f) = \sqrt{\frac{\kappa}{\kappa+1}} h_{kn}^{BI,L}(f) + \sqrt{\frac{1}{\kappa+1}} h_{kn}^{BI,NL}(f), \quad (1)$$

where κ is the Rician factor and f is the working frequency constrained by the carrier frequency f_c and the bandwidth B within a range $[f_c - B/2, f_c + B/2]$, $k = 1, 2, \cdots, K$ and $n = 1, 2, \cdots, N$.

The LoS component $h_{kn}^{BI,L}(f)$ is given by

$$h_{kn}^{BI,L}(f) = \sqrt{\frac{C_0 (D_{kn}^{BI,L})^{-2}}{\sigma}} \cdot e^{j2\pi\tau_{kn}^{BI,L}(f_c - f)}, \quad (2)$$

where C_0 represents the path loss for a propagation distance of 1 meter, $D_{kn}^{BI,L}$ is the distance of the LoS path between the kth antenna of BS and the nth reflecting element of IRS, σ is the noise standard deviation, $\tau_{kn}^{BI,L}$ is the propagation delay of the LoS component computed by $\tau_{kn}^{BI,L} = D_{kn}^{BI,L}/c$, c is the speed of light, and f_c is the carrier frequency.

The NLoS component $h_{kn}^{BI,NL}(f)$ is given by

$$h_{kn}^{BI,NL}(f) = \sum_{m=1}^{M} \sqrt{\frac{C_0 (D_{kn,m}^{BI,NL})^{-2}}{\sigma}} \cdot \sqrt{P_{kn,m}^{BI,NL}} \cdot e^{j2\pi\tau_{kn,m}^{BI,NL}(f_c - f)} \cdot e^{-j\Delta\phi_m} \cdot \rho_s^{r_m}, \quad (3)$$

where M denotes the total number of NLoS paths, $D_{kn,m}^{BI,NL}$ and $P_{kn,m}^{BI,NL}$ indicate the propagation distance and the normalized power of the mth NLoS path between the kth antenna of the BS and the nth reflecting element of the IRS, respectively. $\tau_{kn,m}^{BI,NL}$ is calculated as $D_{kn,m}^{BI,NL}/c$. $\Delta\phi_m$ represents the phase difference caused by the rough tunnel walls of the mth NLoS path and ρ_s is the roughness attenuation factor. The reflection phase can be changed by the roughness of the coal mine tunnel wall surface as shown in Figure 2.

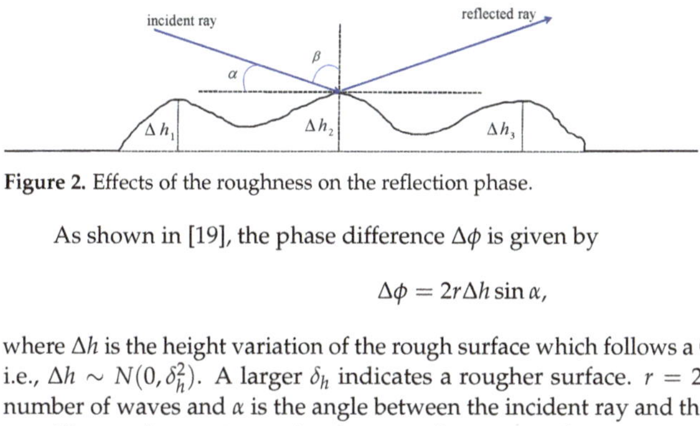

Figure 2. Effects of the roughness on the reflection phase.

As shown in [19], the phase difference $\Delta\phi$ is given by

$$\Delta\phi = 2r\Delta h \sin\alpha, \quad (4)$$

where Δh is the height variation of the rough surface which follows a Gaussian distribution, i.e., $\Delta h \sim N(0, \delta_h^2)$. A larger δ_h indicates a rougher surface. $r = 2\pi f_c/c$ represents the number of waves and α is the angle between the incident ray and the horizontal plane.

The roughness attenuation factor can be expressed as

$$\rho_s = \exp\left\{-8\left(\frac{\pi\delta_h \cos\beta}{\lambda}\right)^2\right\}, \quad (5)$$

where β represents the angle between the incident ray and the vertical plane.

Similarly, the channel matrix between the IRS and User is denoted as $\mathbf{H}^{IU}(f) \in \mathbb{C}^{N\times 1}$ and the channel matrix between the IRS and ME as $\mathbf{H}^{IM}(f) \in \mathbb{C}^{N\times 1}$. The corresponding channel impulse responses $h_n^{IU}(f)$ between the nth reflecting element to the User and $h_n^{IM}(f)$ between the nth reflecting element to the ME can be given in analogy with Equation (1).

$$h_n^{IU}(f) = \sqrt{\frac{\kappa}{\kappa+1}} h_n^{IU,L}(f) + \sqrt{\frac{1}{\kappa+1}} h_n^{IU,NL}(f), \tag{6}$$

where

$$h_n^{IU,L}(f) = \sqrt{\frac{C_0 (D_n^{IU,L})^{-2}}{\sigma}} \cdot e^{j2\pi\tau_n^{IU,L}(f_c - f)}. \tag{7}$$

$$h_n^{IU,NL}(f) = \sum_{m=1}^{M} \sqrt{\frac{C_0 (D_{n,m}^{IU,NL})^{-2}}{\sigma}} \cdot \sqrt{P_{n,m}^{IU,NL}} \cdot e^{j2\pi\tau_{n,m}^{IU,NL}(f_c - f)} \cdot e^{-j\Delta\phi_m} \cdot \rho_s^{r_m}. \tag{8}$$

$$h_n^{IM}(f) = \sqrt{\frac{\kappa}{\kappa+1}} h_n^{IM,L}(f) + \sqrt{\frac{1}{\kappa+1}} h_n^{IM,NL}(f), \tag{9}$$

where

$$h_n^{IM,L}(f) = \sqrt{\frac{C_0 (D_n^{IM,L})^{-2}}{\sigma}} \cdot e^{j2\pi\tau_n^{IM,L}(f_c - f)}. \tag{10}$$

$$h_n^{IM,NL}(f) = \sum_{m=1}^{M} \sqrt{\frac{C_0 (D_{n,m}^{IM,NL})^{-2}}{\sigma}} \cdot \sqrt{P_{n,m}^{IM,NL}} \cdot e^{j2\pi\tau_{n,m}^{IM,NL}(f_c - f)} \cdot e^{-j\Delta\phi_m} \cdot \rho_s^{r_m}. \tag{11}$$

2.2. Problem Formulation

For simplicity, we consider that the distances for the different antenna pairs of the LoS path are almost the same, i.e., $D_{kn}^{BI,L} = D^{BI,L}$ for $k = 1, \cdots, K$ and $n = 1, \cdots, N$ in Equation (2). The same operations are made for other links, i.e., $D_n^{IU,L} = D^{IU,L}$ in Equation (7) and $D_n^{IM,L} = D^{IM,L}$ in Equation (10). Similarly, the distances for different NLoS paths are also almost equal and are linear to the distance of the LoS path, i.e., $D_{kn,m}^{BI,NL} = \epsilon D^{BI,L}$, $D_{n,m}^{IU,NL} = \epsilon D^{IU,L}$ and $D_{n,m}^{IM,NL} = \epsilon D^{IM,L}$ for $m = 1, 2, \cdots, M$. For fixed coal mine tunnels, ϵ is generally set as $(1, 1.7]$. Therefore, based on Equations (1)–(3), the channel matrix $\mathbf{H}^{BI}(f)$ between BS and IRS can be expressed as

$$\mathbf{H}^{BI}(f) = \sqrt{\frac{C_0 (D^{BI,L})^{-2}}{\sigma}} \cdot \left(\sqrt{\frac{\kappa}{\kappa+1}} \tilde{\mathbf{H}}^{BI,L} + \frac{1}{\epsilon} \sqrt{\frac{1}{\kappa+1}} \tilde{\mathbf{H}}^{BI,NL} \right), \tag{12}$$

where $\tilde{\mathbf{H}}^{BI,L}$ and $\tilde{\mathbf{H}}^{BI,NL}$ represent the channel matrices constituted by the channel impulse responses of the LoS and NLoS components between the BS and IRS, excluding the constant coefficients. Here, $D^{BI,L}$ can be computed by

$$D^{BI,L} = \sqrt{\|\mathbf{b} - \mathbf{I}\|^2}. \tag{13}$$

Similarly, the channel matrix $\mathbf{H}^{IU}(f)$ between the IRS and User can be expressed as

$$\mathbf{H}^{IU}(f) = \sqrt{\frac{C_0 (D^{IU,L})^{-2}}{\sigma}} \cdot \left(\sqrt{\frac{\kappa}{\kappa+1}} \tilde{\mathbf{H}}^{IU,L} + \frac{1}{\epsilon} \sqrt{\frac{1}{\kappa+1}} \tilde{\mathbf{H}}^{IU,NL} \right) \tag{14}$$

with

$$D^{IU,L} = \sqrt{\|\mathbf{I} - \mathbf{u}\|^2}. \tag{15}$$

The channel matrix $\mathbf{H}^{IM}(f)$ between the IRS and ME can be expressed as

$$\mathbf{H}^{IM}(f) = \sqrt{\frac{C_0 (D^{IM,L})^{-2}}{\sigma}} \cdot \left(\sqrt{\frac{\kappa}{\kappa+1}} \tilde{\mathbf{H}}^{IM,L} + \frac{1}{\epsilon} \sqrt{\frac{1}{\kappa+1}} \tilde{\mathbf{H}}^{IM,NL} \right) \tag{16}$$

with

$$D^{IM,L} = \sqrt{||\mathbf{I} - \mathbf{m}||^2}. \tag{17}$$

To maximize the effective achievable rate at the user station for IRS-assisted coal mine wireless communications, it is assumed that the channel state information (CSI) of the links between the BS and IRS, IRS and User, and IRS and ME is fully known. The BS transmits an effective message s with a mean of 0 and unit variance to the User employing beamforming. The transmit beamforming vector for s is denoted by $\mathbf{w} \in \mathbb{C}^{K\times 1}$ with the constraint $tr(\mathbf{w}) \leq P$, where P represents the total transmit power of the BS. The IRS reflection is modeled as $\boldsymbol{q} \triangleq [q_1, ..., q_N]^T$, where $q_n = \psi_n e^{j\theta_n}$, $\psi_n \in [0,1]$, and $\theta_n \in [0, 2\pi)$ denote the amplitude reflection coefficient and phase shift, respectively, $n = 1, ..., N$. For simplicity, we assume that $\psi_n = 1$ and thus $|q_n| = 1$, $n = 1, ..., N$. Due to severe path loss, the signals reflected by the IRS two or more times can be neglected. Based on the established system model, the receiving signals at the User and ME are given by

$$y_U(f) = (\mathbf{H}^{IU}(f))^T\mathbf{Q}(\mathbf{H}^{BI}(f))^T\mathbf{w}s + n_U, \tag{18}$$

$$y_M(f) = (\mathbf{H}^{IM}(f))^T\mathbf{Q}(\mathbf{H}^{BI}(f))^T\mathbf{w}s + n_M, \tag{19}$$

where $\mathbf{Q} \triangleq diag(\mathbf{q}) \in \mathbb{C}^{N\times N}$ represents a diagonal phase shift matrix, n_U and n_M denote Gaussian noises at the User and ME with means of zero and variances of σ_U^2 and σ_M^2, respectively. Therefore, the achievable rates at the User and ME are given by

$$R_U(f) = \log_2(1 + \frac{|(\mathbf{H}^{IU}(f))^T\mathbf{Q}(\mathbf{H}^{BI}(f))^T\mathbf{w}|^2}{\sigma_U^2}), \tag{20}$$

$$R_M(f) = \log_2(1 + \frac{|(\mathbf{H}^{IM}(f))^T\mathbf{Q}(\mathbf{H}^{BI}(f))^T\mathbf{w}|^2}{\sigma_M^2}). \tag{21}$$

Due to the interference from the ME on the IRS reflection link, the effective achievable rate at the user station is given by

$$R_{eff}(f) = R_U(f) - R_M(f). \tag{22}$$

Our objective is thus to maximize the effective achievable rate $R_{eff}(f)$ at the user station by jointly optimizing the transmit beamforming vector $\mathbf{w}$, the IRS phase shift vector $\mathbf{q}$ and the horizontal coordinate I_x of the IRS. The optimization problem can be formulated as follows.

$$\begin{aligned} &\underset{\mathbf{w},\mathbf{q},I_x}{maximize} \quad R_{eff}(f) \\ s.t.\ &C1 : tr(\mathbf{w}) \leq P, \\ &C2 : |q_n| = 1, \forall n, \\ &C3 : \theta_n \in [0, 2\pi), \\ &C4 : I_x \in [0, u_x], \\ &C5 : f \in [f_c - B/2, f_c + B/2]. \end{aligned} \tag{23}$$

Since the objective function $R_{eff}(f)$ is non-convex with respect to $\mathbf{w}$, $\mathbf{q}$ and I_x, it is not feasible to solve this optimization problem directly with these three variables. To address this problem, we perform alternative joint optimization for $R_{eff}(f)$ by fixing any two variables of $\mathbf{w}$, $\mathbf{q}$ or I_x and optimizing the remaining one variable. Jointly optimizing $\mathbf{w}$ and $\mathbf{q}$ is similar to [13]. Here, we focused on optimizing the horizontal coordinate I_x of IRS based on the channel model of a coal mine as described in Section 2.1 and embedded it into the three-step joint optimization to maximize $R_{eff}(f)$.

3. A Three-Step Joint Optimization Algorithm

For simplicity, it is assumed that the working frequency f is fixed and the noise powers at the User and ME are the same, i.e., $\sigma_U^2 = \sigma_M^2 = \sigma^2$. Compared to the size of the tunnels in the coal mine, the effects of the heights of the BS, ME and User on the signals can be ignored. Based on the system model in Section 2, the horizontal coordinate I_x of the IRS can range from 0 to u_x, which represents the horizontal range of deploying IRSs on the tops of tunnels. According to the optimization problem in Equation (23), we first optimize I_x given $\mathbf{w}$, $\mathbf{q}$ and then perform three-step joint optimization to maximize $R_{eff}(f)$.

3.1. Optimizing I_x Given $\mathbf{w}$ and $\mathbf{q}$

Given $\mathbf{w}$ and $\mathbf{q}$, the optimization problem in Equation (23) can be transformed into

$$\begin{aligned} &\underset{I_x}{maximize} \quad R_{eff} \\ &s.t. \;\; C4: I_x \in [0, u_x]. \end{aligned} \tag{24}$$

By substituting Equations (12)–(16) into the effective achievable rate at the user station in Equations (20)–(22), we can obtain

$$\begin{aligned} R_{eff} &= R_U - R_M \\ &= \log_2\left((\frac{C_0(D^{IU,L})^{-2}C_0(D^{BI,L})^{-2}}{\sigma^2})|((\tilde{\mathbf{H}}^{IU})^T\mathbf{Q}(\tilde{\mathbf{H}}^{BI})^T)\mathbf{w}|^2 + \sigma^2 \right) \\ &\quad - \log_2\left((\frac{C_0(D^{IM,L})^{-2}C_0(D^{BI,L})^{-2}}{\sigma^2})|((\tilde{\mathbf{H}}^{IM})^T\mathbf{Q}(\tilde{\mathbf{H}}^{BI})^T)\mathbf{w}|^2 + \sigma^2 \right), \end{aligned} \tag{25}$$

where $\tilde{\mathbf{H}}^{BI} = \sqrt{\frac{\kappa}{\kappa+1}}\tilde{\mathbf{H}}^{BI,L} + \frac{1}{\epsilon}\sqrt{\frac{1}{\kappa+1}}\tilde{\mathbf{H}}^{BI,NL}$, $\tilde{\mathbf{H}}^{IU} = \sqrt{\frac{\kappa}{\kappa+1}}\tilde{\mathbf{H}}^{IU,L} + \frac{1}{\epsilon}\sqrt{\frac{1}{\kappa+1}}\tilde{\mathbf{H}}^{IU,NL}$, $\tilde{\mathbf{H}}^{IM} = \sqrt{\frac{\kappa}{\kappa+1}}\tilde{\mathbf{H}}^{IM,L} + \frac{1}{\epsilon}\sqrt{\frac{1}{\kappa+1}}\tilde{\mathbf{H}}^{IM,NL}$.

Since σ^2 is much smaller than the left term of " + " in the above equation, it can be ignored. Based on Equations (13)–(17), Equation (25) can be further transformed into

$$\begin{aligned} R_{eff} =& R_U - R_M \\ =& \underbrace{\log_2\left(\frac{(C_0^2/\sigma^2)|((\tilde{\mathbf{H}}^{IU})^T\mathbf{Q}(\tilde{\mathbf{H}}^{BI})^T)\mathbf{w}|^2}{(I_z^2 + (u_x - I_x)^2)(I_z^2 + I_x^2)} \right)}_{\gamma_1} \\ & \underbrace{- \log_2\left(\frac{(C_0^2/\sigma^2)|((\tilde{\mathbf{H}}^{IM})^T\mathbf{Q}(\tilde{\mathbf{H}}^{BI})^T)\mathbf{w}|^2}{(I_z^2 + (m_x - I_x)^2)(I_z^2 + I_x^2)} \right)}_{\gamma_2}. \end{aligned} \tag{26}$$

According to the fact that γ_1 is convex with respect to $(u_x - I_x)^2$ and I_x^2 in Equation (26), the bound of γ_1 can be derived by its first-order Taylor expansion at $(u_x - I_x^t)^2$ and $(I_x^t)^2$. Therefore, γ_1 can be expressed as

$$
\begin{aligned}
\gamma_1 &= \log_2\left(\frac{(C_0^2/\sigma^2)\left|((\tilde{\mathbf{H}}^{IU})^T\mathbf{Q}(\tilde{\mathbf{H}}^{BI})^T)\mathbf{w}\right|^2}{(I_z^2+(u_x-I_x)^2)(I_z^2+I_x^2)}\right) \\
&= \log_2\left(\frac{C_0^2}{\sigma^2}\left|((\tilde{\mathbf{H}}^{IU})^T\mathbf{Q}(\tilde{\mathbf{H}}^{BI})^T)\mathbf{w}\right|^2\right)+\log_2\left(\frac{1}{I_z^2+(u_x-I_x)^2}\right)+\log_2\left(\frac{1}{I_z^2+I_x^2}\right) \\
&\geq \log_2\left(\frac{C_0^2}{\sigma^2}\left|((\tilde{\mathbf{H}}^{IU})^T\mathbf{Q}(\tilde{\mathbf{H}}^{BI})^T)\mathbf{w}\right|^2\right) \\
&\quad+\log_2\left(\frac{1}{I_z^2+(u_x-I_x^t)^2}\right)+\frac{I_z^2+(u_x-I_x^t)^2}{\ln 2}\cdot\left(\frac{-((u_x-I_x)^2-(u_x-I_x^t)^2)}{(I_z^2+(u_x-I_x^t)^2)^2}\right) \\
&\quad+\log_2\left(\frac{1}{I_z^2+(I_x^t)^2}\right)+\frac{I_z^2+(I_x^t)^2}{\ln 2}\cdot\left(\frac{-(I_x^2-(I_x^t)^2)}{(I_z^2+(I_x^t)^2)^2}\right) \\
&= \gamma_1^{lowerbd}.
\end{aligned}
\tag{27}
$$

Next, we derive the bound of γ_2 by introducing two auxiliary variables μ_1 and μ_2 constrained by

$$\mu_1 \leq (m_x - I_x)^2, \tag{28}$$

$$\mu_2 \leq I_x^2. \tag{29}$$

Therefore, γ_2 can be expressed as

$$
\begin{aligned}
\gamma_2 &= \log_2\left(\frac{(C_0^2/\sigma^2)\left|((\tilde{\mathbf{H}}^{IM})^T\mathbf{Q}(\tilde{\mathbf{H}}^{BI})^T)\mathbf{w}\right|^2}{(I_z^2+(m_x-I_x)^2)(I_z^2+I_x^2)}\right) \\
&= \log_2\left(\frac{C_0^2}{\sigma^2}\left|((\tilde{\mathbf{H}}^{IM})^T\mathbf{Q}(\tilde{\mathbf{H}}^{BI})^T)\mathbf{w}\right|^2\right)+\log_2\left(\frac{1}{I_z^2+(m_x-I_x)^2}\right)+\log_2\left(\frac{1}{I_z^2+I_x^2}\right) \\
&\leq \log_2\left(\frac{C_0^2}{\sigma^2}\left|((\tilde{\mathbf{H}}^{IM})^T\mathbf{Q}(\tilde{\mathbf{H}}^{BI})^T)\mathbf{w}\right|^2\right)+\log_2\left(\frac{1}{I_z^2+\mu_1}\right)+\log_2\left(\frac{1}{I_z^2+\mu_2}\right) \\
&= \gamma_2^{upperbd}.
\end{aligned}
\tag{30}
$$

Since the constraints on μ_1 and μ_2 in Equations (28) and (29) are non-convex, we use their first-order Taylor expansions to transform them into convex constraints.

$$\mu_1 \leq (m_x - I_x^t)^2 + 2\cdot(m_x - I_x^t)\cdot(I_x - I_x^t), \tag{31}$$

$$\mu_2 \leq (I_x^t)^2 + 2\cdot I_x^t \cdot (I_x - I_x^t). \tag{32}$$

Based on these above transformations, the optimization problem for the horizontal deployment position of the IRS in Equation (24) can be transformed into the following convex optimization problem.

$$
\begin{aligned}
\underset{I_x,\mu_1,\mu_2}{maximize} \quad & \gamma = \gamma_1^{lowerbd} - \gamma_2^{upperbd} \\
s.t. \quad & C1: I_x \in [0, u_x], \\
& C2: \mu_1, \mu_2.
\end{aligned}
\tag{33}
$$

The complete solution with an iterative loop is organized in Algorithm 1, where I_x^t denotes the horizontal coordinate of the IRS in the t-th iteration.

Algorithm 1 Optimization Scheme for the Horizontal Position of IRS

1: Initialize the optimized transmit beamforming vector $\mathbf{w}^{opt}$ of BS and the phase shift $\mathbf{q}^{opt}$ of IRS.
2: Initialize the horizontal coordinate of IRS as I_x^0 and set $t = 0$. Denote the convergence threshold as ξ_1.
3: **repeat**
4: The solution of the horizontal coordinate I_x is obtained by solving

$$\begin{aligned} \max_{I_x,\mu_1,\mu_2} \ & \gamma = \gamma_1^{lowerbd} - \gamma_2^{upperbd} \\ s.t. \ & C1 : \|\mathbf{w}\|^2 \leq P, \\ & C2 : |q_n| = 1, \ \forall n, \\ & C3 : I_x \in [0, u_x], \\ & C4 : \mu_1, \mu_2. \end{aligned}$$

5: $t = t + 1$.
6: $I_x^t = I_x$.
7: **until** $|R_{eff}^t - R_{eff}^{t-1}| \leq \xi_1$.

3.2. Three-Step Joint Optimization with $\mathbf{w}$*,* $\mathbf{q}$ *and* I_x

The solutions to the optimization problem in Equation (23) can be divided into three steps as shown in Algorithm 2, which optimize the transmit beamforming vector $\mathbf{w}$, the phase shift $\mathbf{q}$ and the horizontal coordinate I_x of the IRS jointly and alternately. The transmit beamforming vector $\mathbf{w}$ of the BS can be optimized using the normalized eigenvector corresponding to the largest eigenvalue and the optimal phase shift $\mathbf{q}$ can be obtained by using the BCAM algorithm proposed in [13]. The horizontal deployment position I_x of the IRS can be optimized by applying the proposed Algorithm 1. The alternate joint optimization is performed by fixing two variables of $\mathbf{w}$, $\mathbf{q}$ or I_x and optimizing the remaining one.

Algorithm 2 Alternate Joint Optimization Scheme

1: Initialize the transmit beamforming vector $\mathbf{w}$ of BS, the phase shift $\mathbf{q}$ of IRS, the horizontal coordinate I_x^0 of IRS and set $t = 0$. Denote the convergence threshold as ξ_2.
2: **repeat**
3: Fix $\mathbf{q}$ and I_x^t, and optimize $\mathbf{w} \to \mathbf{w}^{opt}$,
4: Fix $\mathbf{w}^{opt}$ and I_x^t, and optimize $\mathbf{q} \to \mathbf{q}^{opt}$,
5: Fix $\mathbf{w}^{opt}$ and $\mathbf{q}^{opt}$, and optimize $I_x \to I_x^{opt}$.
6: $t = t + 1$.
7: $\mathbf{q} = \mathbf{q}^{opt}$.
8: $I_x^t = I_x^{opt}$.
9: **until** $|R_{eff}^t - R_{eff}^{t-1}| \leq \xi_2$.

4. Simulation Results

In this section, we perform different simulations to demonstrate the effectiveness of the proposed three-step optimization scheme based on the channel model of a coal mine. The parameter configurations are listed in Table 1 and here we set the working frequency to be 915 MHz, a commonly used commercial working frequency in coal mine wireless communication systems, as pointed out in [17,20] .

We first simulated the effects of different initial horizontal deployment positions on the convergence to the optimal horizontal deployment position using the horizontal coordinate optimization algorithm proposed in Section 3.1 with the total transmit power $P = 15$ dBm and the roughness $\delta_h = 0.02$ as shown in Figure 3. The three initial horizontal coordinate I_x^0 were set to be 110 m, 130 m, and 145 m, respectively. From Figure 3, it can be seen that when the number of iterations was increased to about 7, these three curves all convergd to the same optimal horizontal coordinate $I_x^{opt} = 148.16$ m, which indicates that the optimized horizontal coordinate obtained by the proposed optimization algorithm is independent of the initial deployment positions of the IRS.

Table 1. Parameterconfigurations for simulations.

System Parameters	Value
Number of BS antennas, K	4
Number of IRS reflection elements, $N_x \times N_y$	8×8
ME coordinate, $[m_x, m_y, m_z]$	[145 m, 0, 0]
User coordinate, $[u_x, u_y, u_z]$	[150 m, 0, 0]
IRS vertical coordinate, I_z	5 m
Normalized path loss, C_0	−30 dB
Noise standard deviation, σ	−80 dBm
Carrier frequency, f_c	1.2 GHz
Working frequency, f	915 MHz

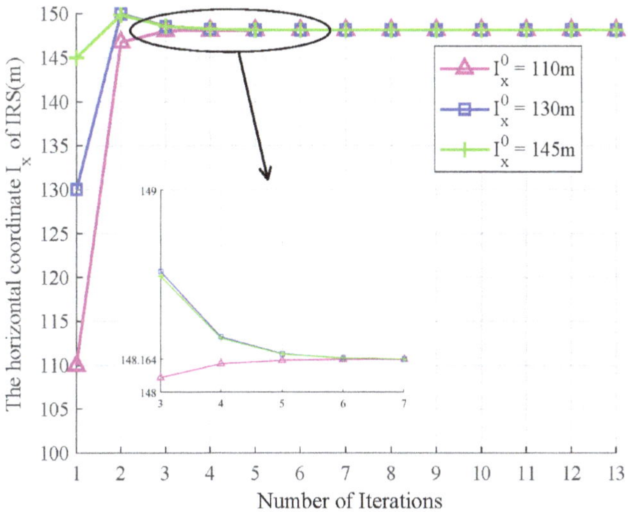

Figure 3. Convergence analysis under different initial horizontal coordinates of IRS with $P = 15$ dBm and $\delta_h = 0.02$.

Next, we compared the effective achievable rates at the user using different rate optimization schemes, including the proposed three-step joint rate optimization algorithm, the SDR algorithm in [11] and the BCAM algorithm in [13] under the coal mine channel model, as shown in Figure 4. It can be observed that the effective achievable rates obtained by the proposed three-step joint rate optimization scheme are the highest. Compared to the SDR algorithm and the BCAM algorithm, which optimized **w** and **q** alternately, the effective achievable rates could be improved by approximately 14.39% and 18.64% when varying the total transmit power P from −5 dBm to 25 dBm, respectively. In addition, 25.09% and 36.69% improvements were obtained in comparison with these two algorithms which only optimized **q** without beamforming at BS. The simulation results verify the necessities of optimizing the deployment position of the IRS in coal mines and the effectiveness of the proposed three-step joint rate optimization scheme.

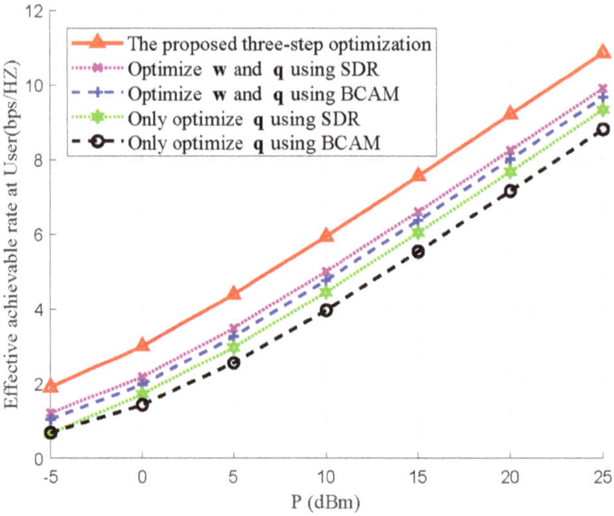

Figure 4. Comparisons of the effective achievable rates at the user station between different rate optimization schemes with varying P and fixed $\delta_h = 0.02$.

In the following section, we investigated the effects of the roughness of the tunnels in a coal mine on the effective achievable rates for users benefiting from the proposed three-step joint rate optimization scheme and the proposed alternate optimization of **w** and **q** in [13], as shown in Figure 5. Here, we chose three typical values of roughness, 0, 0.05 and 0.1, which means that the tunnels were completely smooth, moderately rough and very rough, respectively. The simulation results show that as the roughness δ_h increases, the effective achievable rates at the user station, as expected, decrease for both rate optimization schemes. However, regardless of the roughness, the effective achievable rates obtained by using the proposed three-step rate optimization scheme are always higher than those obtained by the scheme in [13], which indicates that the proposed three-step joint rate optimization scheme which includes the optimization of the deployment position of the IRS performs better in the coal mine scenario.

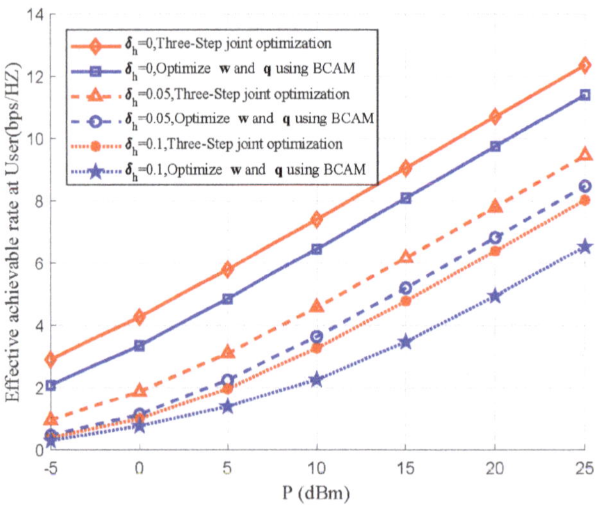

Figure 5. Effects of the roughness of the tunnel walls in the coal mine on the effective achievable rates at the user station using different rate optimization schemes.

Finally, in order to further verify the effectiveness and generality of the proposed three-step joint optimization algorithm, we conducted supplementary simulations for another working frequency in coal mine wireless communication systems. We set the working frequency and carrier frequency to be $f = 2.45$ GHz and $f_c = 3$ GHz, and increased the distance between the User and BS to 160 m. The convergence to the optimal IRS deployment horizontal position is shown in Figure 6. We can obtain the similar conclusion that, by increasing the number of iterations to about 9, these three curves all converge to the same optimal horizontal coordinate $I_x^{opt} = 158.77$ m, regardless of the initial deployment positions of the IRS. Moreover, the effective achievable rates at the user station using these three different rate optimization schemes under the coal mine channel model are also plotted in Figure 7. The simulation results show that when varying the total transmit power P from -5 dBm to 25 dBm, 12.32% and 18.62% improvements could be obtained compared with those that alternately optimized **w** and **q**, while 31.16% and 54.17% improvements could be obtained compared with those that only optimized **q**.

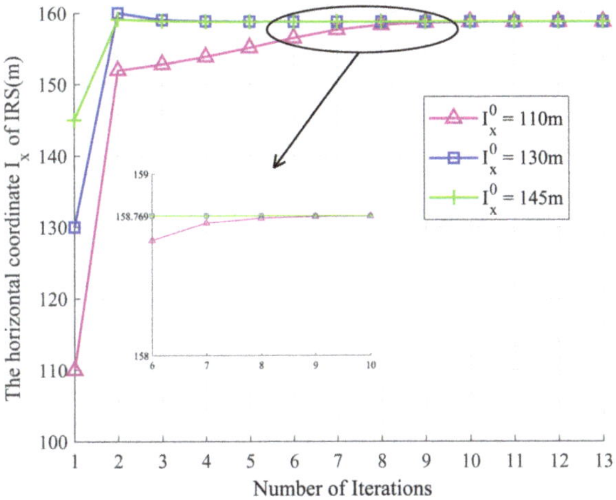

Figure 6. Comparisons of the effective achievable rates at the user station between different rate optimization schemes with varied P and fixed $\delta_h = 0.02$.

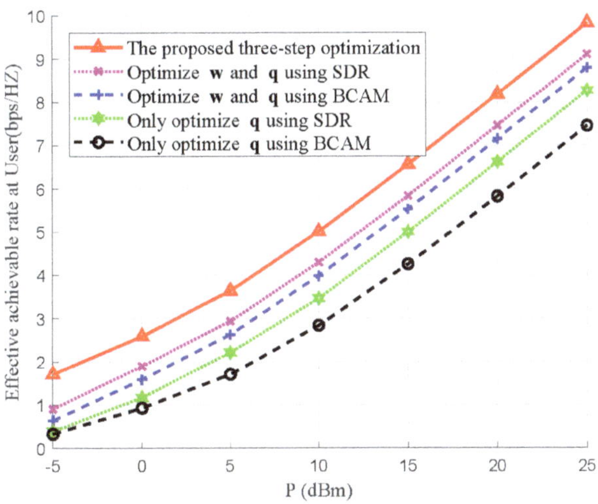

Figure 7. Comparisons of the effective achievable rates at the user station between different rate optimization schemes with varied P and fixed $\delta_h = 0.02$.

5. Conclusions

This paper established an IRS-assisted coal mine communication system modeled by a 3D MIMO GBSM, for which a three-step joint optimization involving the transmit power, the IRS phase shifts and the IRS deployment position was proposed to maximize the effective achievable rate at the user station. To solve the non-convexity of the optimization problem for the IRS deployment position, logarithmic operations, Taylor relaxation and two auxiliary variables were introduced to transform it into a convex problem. The simulation results show that the proposed three-step joint rate optimization scheme performed better in the coal mine scenario regardless of the roughness of the tunnels, and the convergence to the optimal deployment position of the IRS was independent of the initial deployment position. By comparison, approximately 12.32% to 54.17% improvements in the effective achievable rates at the user station could be obtained by using the proposed rate optimization scheme under different parameter configurations. Our work provided the feasibility of deploying an IRS to enhance the performance of coal mine wireless communication systems and verified the necessity to optimize the IRS deployment position to improve the effective achievable rates at the user station.

Author Contributions: Conceptualization, Y.L. and Y.X.; Methodology, Y.L.; Validation, Z.Y.; Formal analysis, Z.Y.; Writing—original draft, Z.Y.; Writing—review & editing, Y.L., Z.Y. and Y.X.; Funding acquisition, B.W. All authors have read and agreed to the published version of the manuscript.

Funding: This research was funded by National Natural Science Foundation of China grant number 62271386, 62471385, 61801371.

Data Availability Statement: Data are contained within the article.

Conflicts of Interest: The authors declare no conflict of interest.

References

1. Ranjany, A.; Misraz, P.; Dwivediz, B.; Sahuy, H.B. Channel modeling of wireless communication in underground coal mines. In Proceedings of the 8th International Conference on Communication Systems and Networks (COMSNETS), Bangalore, India, 5–10 January 2016; pp. 1–2.
2. Ben Mabrouk, I.; Talbi, L.; Nedil, M.; Hettak, K. MIMO-UWB Channel Characterization Within an Underground Mine Gallery. *IEEE Trans. Antennas Propag.* **2012**, *60*, 4866–4874. [CrossRef]
3. de Oliveira Gomes, P.H.; Guieiro, G.; de Almeida, E.P.L.; Garcia, L.G.U. Evaluation of Shadowing Caused by Mining Machinery in V2I Communications. In Proceedings of the IEEE 29th Annual International Symposium on Personal, Indoor and Mobile Radio Communications (PIMRC), Bologna, Italy, 9–12 September 2018; pp. 63–67.
4. Játiva, P.P.; Azurdia-Meza, C.A.; Sánchez, I.; Seguel, F.; Zabala-Blanco, D.; Firoozabadi, A.D.; Gutiérrez, C.A.; Soto, I. A VLC Channel Model for Underground Mining Environments With Scattering and Shadowing. *IEEE Access* **2020**, *8*, 185445–185464. [CrossRef]
5. ElMossallamy, M.A.; Zhang, H.; Song, L.; Seddik, K.G.; Han, Z.; Li, G.Y. Reconfigurable Intelligent Surfaces for Wireless Communications: Principles, Challenges, and Opportunities. *IEEE Trans. Cogn. Commun. Netw.* **2020**, *6*, 990–1002. [CrossRef]
6. Liu, Y.; Liu, X.; Mu, X.; Hou, T.; Xu, J.; Di Renzo, M.; Al-Dhahir, N. Reconfigurable Intelligent Surfaces: Principles and Opportunities. *IEEE Commun. Surveys Tut.* **2021**, *23*, 1546–1577. [CrossRef]
7. Zhang, Z.; Dai, L.; Chen, X.; Liu, C.; Yang, F.; Schober, R.; Poor, H.V. Active RIS vs. Passive RIS: Which Will Prevail in 6G? *IEEE Trans. Commun.* **2023**, *71*, 1707–1725. [CrossRef]
8. Wang, B.; Yuan, Z.; Lu, J.; Zhang, X. Multitask Collaborative Learning Neural Network for Radio Signal Classification. *IEEE Trans. Commun.* **2024**, 1. [CrossRef]
9. Yuan, X.; Zhang, Y.-J.A.; Shi, Y.; Yan, W.; Liu, H. Reconfigurable-Intelligent-Surface Empowered Wireless Communications: Challenges and Opportunities. *IEEE Wirel. Commun.* **2021**, *28*, 136–143. [CrossRef]
10. Kisseleff, S.; Chatzinotas, S.; Ottersten, B. Reconfigurable Intelligent Surfaces in Challenging Environments: Underwater, Underground, Industrial and Disaster. *IEEE Access* **2021**, *9*, 150214–150233. [CrossRef]
11. Cui, M.; Zhang, G.; Zhang, R. Secure Wireless Communication via Intelligent Reflecting Surface. *IEEE Wireless Commun. Lett.* **2019**, *8*, 1410–1414. [CrossRef]
12. Song, Y.; Khandaker, M.R.A.; Tariq, F.; Wong, K.-K.; Toding, A. Truly Intelligent Reflecting Surface-Aided Secure Communication Using Deep Learning. In Proceedings of the IEEE 93rd Vehicular Technology Conference (VTC2021-Spring), Helsinki, Finland, 25–28 April 2021; pp. 1–6.

13. Kumar, V.; Flanagan, M.F.; Kwan Ng, D.W.; Tran, L.-N. On the Secrecy Rate under Statistical QoS Provisioning for RIS-assisted MISO Wiretap Channel. In Proceedings of the IEEE Global Communications Conference (GLOBECOM), Madrid, Spain, 7–11 December 2021; pp. 1–6.
14. Wu, Q.; Zhang, S.; Zheng, B.; You, C.; Zhang, R. Intelligent Reflecting Surface-Aided Wireless Communications: A Tutorial. *IEEE Trans. Commun.* **2021**, *69*, 3313–3351. [CrossRef]
15. Ren, Y.; Zhou, R.; Teng, X.; Meng, S.; Zhou, M.; Tang, W.; Li, X.; Li, C.; Jin, S. On Deployment Position of RIS in Wireless Communication Systems: Analysis and Experimental Results. *IEEE Wirel. Commun. Lett.* **2023**, *12*, 1756–1760. [CrossRef]
16. Liu, Y.; Zhang, J.; Zhang, Y.; Gong, H.; Jiang, T.; Liu, G. How to Extend 3-D GBSM to Integrated Sensing and Communication Channel With Sharing Feature? *IEEE Wirel. Commun. Lett.* **2024**, *13*, 2045–2049. [CrossRef]
17. Ji, X.; Wang, C.-X.; Chang, H. A Novel MIMO Channel Model for Underground Mine Communications. In Proceedings of the IEEE/CIC International Conference on Communications in China (ICCC), Dalian, China, 10–12 August 2023; pp. 1–5.
18. Gong, H.; Zhang, J.; Zhang, Y.; Zhou, Z.; Liu, G. How to Extend 3-D GBSM to RIS Cascade Channel With Non-Ideal Phase Modulation? *IEEE Wirel. Commun. Lett.* **2024**, *13*, 555–559. [CrossRef]
19. Zhou, C. Ray Tracing and Modal Methods for Modeling Radio Propagation in Tunnels With Rough Walls. *IEEE Trans. Antennas Propag.* **2017**, *65*, 2624–2634. [CrossRef] [PubMed]
20. Gangwar, K.; Chen, G.C.Y.; Chan, K.K.M.; Gangwar, R.K.; Rambabu, K. Antenna System for Communication in Underground Mining Environment to Ensure Miners Safety. *IEEE Access* **2021**, *9*, 150162–150171. [CrossRef]

entropy

MDPI

Review

A Roadmap for NF-ISAC in 6G: A Comprehensive Overview and Tutorial

Azar Hakimi, Diluka Galappaththige and Chintha Tellambura *

Department of Electrical and Computer Engineering, University of Alberta, Edmonton, AB T6G 1H9, Canada; hakimina@ualberta.ca (A.H.); diluka.lg@ualberta.ca (D.G.)
* Correspondence: ct4@ualberta.ca

Abstract: Near-field (NF) integrated sensing and communication (ISAC) has the potential to revolutionize future wireless networks. It enables simultaneous communication and sensing operations on the same radio frequency (RF) resources using a shared hardware platform, maximizing resource utilization. NF-ISAC systems can improve communication and sensing performance compared to traditional far-field (FF) ISAC systems by exploiting the unique propagation characteristics of NF spherical waves with an additional distance dimension. Despite its potential, NF-ISAC research is still in its early stages, and a comprehensive survey of the technology is lacking. This paper systematically explores NF-ISAC technology, providing an in-depth analysis of both NF and FF systems, their applicability in various scenarios, and different channel models. It highlights the advantages and philosophies of ISAC, examining both narrow-band and wide-band NF-ISAC systems. Case studies and simulations offer deeper insights into NF-ISAC design philosophies. Additionally, the paper reviews the existing NF-ISAC literature, methodologies, potentials, and conclusions, and discusses future research areas, challenges, and applications.

Keywords: near-field communications; integrated sensing and communication; channel modeling; optimization problem

1. Introduction

The sixth generation (6G) of cellular networks is anticipated to be available in early 2030, and the research journey has already commenced [1]. The advent of 6G is poised to contribute to the establishment of a human-friendly intelligent society. Consequently, the targets set for 6G excel those of previous generations. Therefore, it is set to enable new applications, such as extended reality (XR), holographic communication, pervasive intelligence, Internet of Things (IoT), autonomous driving in vehicle-to-everything (V2X), smart traffic control, smart homes, unmanned aerial vehicles (UAVs), and factory automation [2]. Meanwhile, the International Telecommunications Union (ITU) has announced its aspirations for 6G, including a timeline, future technology trends, recommended frameworks, and so on [3]. This marks the official start of the journey towards 6G standardization. They defined six major 6G use cases, including three extended fifth generation (5G) application scenarios, namely, immersive enhanced mobile broadband (eMBB+), massive machine type communication (mMTC+), hyper ultra-reliable and low-latency communication (URLLC+), and three new items that will thrive in the new era of 6G, namely, integrated sensing and communication (ISAC), integrated artificial intelligence (AI) and communication, and ubiquitous connectivity.

To facilitate these application requirements, the data rate and connectivity of 6G must surpass that of the 5G by a factor of 100 and 10 times, respectively. This results in a minimum peak rate of 1 Tbps (terabit per second) and connectivity of 10^7 devices per square kilometer [4]. Nevertheless, applications requiring such high data rates and connectivity cannot be adequately supported within a saturated sub-6 GHz bandwidth. On the other hand, these applications also necessitate ultra-reliable, low-latency, and high-precision sensing in addition to communication [5–7]. In this regard, numerous prospective 6G candidate technologies have emerged, including extremely large-scale multiple-input multiple-output (XL-MIMO) or extremely large-scale antenna arrays (ELAAs), ISAC, and millimeter wave (mmWave) and terahertz (THz) communication [8–11].

Integrating emerging ISAC systems with ELAAs operating in higher frequency bands leads to a wireless paradigm known as near-field (NF)-ISAC [12–14]. In particular, large antenna arrays typically provide a high spatial resolution, needed for angle-of-arrival (AoA) estimation and angular beamforming [12–14]. In addition, when the aperture of antenna arrays is extremely large and is combined with extremely high operating frequencies, the inevitable NF effect occurs. The NF and far-field (FF) are two intrinsic electromagnetic (EM) zones associated with antenna arrays (Section 2). Current wireless networks primarily operate in the FF region due to small antenna array apertures and lower operating frequencies. However, with the emergence of ELAAs and high-frequency bands, the NF zone extends to tens or even hundreds of meters, gaining prominence in sensing and communication applications [2]. While the NF effect introduces complexity to communication and sensing channels, it also presents new opportunities for NF-ISAC through the use of ELAAs [2,12–14]. Despite this potential, NF-ISAC remains unexplored, which has motivated a thorough investigation of its capabilities.

Motivation and Contribution

Although several surveys and tutorials have addressed various aspects of ISAC and NF communication separately [2,6,9,10,15,16], there are only a few works specifically focused on NF- ISAC [13,14,17].

In particular, ref. [13] discusses the opportunities and challenges of NF communication and sensing compared to FF counterparts. It highlights the advantages of NF, NF-ISAC designs, associated challenges, and existing solutions while suggesting future directions. Reference [14] compares NF and FF characteristics, emphasizing NF's advantages in communication and sensing. The included simulations demonstrate the potential for power savings in NF-ISAC systems. In [17], the fundamentals, challenges, and opportunities of NF wide-band ISAC systems are discussed. The study emphasizes the implications of significant angular-delay correlation in NF wide-band systems on communication and sensing functions, such as signal multiplexing and range sensing.

These studies have inspired this work, recognizing the need for more comprehensive research. The contributions of this paper are summarized as follows:

1. In contrast to previous survey papers [13,14,17], this study offers a comprehensive examination covering all facets of NF-ISAC systems. Specifically, it thoroughly analyzes NF and FF systems, exploring their respective applications in communication and sensing scenarios. Various channel model scenarios for NF and FF are presented. Additionally, the merits and philosophies of ISAC are explored, paving the way to investigate both narrow-band and wide-band systems within NF-ISAC.
2. To gain deeper insights into this domain, an NF-ISAC integrated with a non-orthogonal multiple access (NOMA) communication system and a target sensing system is proposed. Three case studies are defined. Case 1 prioritizes the sensing system, aiming to solve a sensing signal-to-noise ratio (SNR) maximization problem while ensuring quality of service (QoS) for the communication system. Conversely, Case 2 focuses on optimizing the communication system by maximizing its sum rate while maintaining QoS for the sensing system. Case 3 introduces the sum-weighted rate of users' communication and sensing rates.
3. An extensive simulation is conducted to evaluate the performance of the proposed case studies. The proposed design is compared with three other benchmarks for deeper insights: far-field (FF), communication-only, and sensing-only schemes. Numerical results indicate that the proposed NF design outperforms the FF counterpart in communication and sensing, demonstrating superior efficiency and effectiveness.

4. A thorough literature review of the existing works is conducted, exploring various studies, methodologies, and findings in NF-ISAC. Insightful conclusions are drawn from the review, focusing on potential future research directions and the challenges. By examining current trends and knowledge gaps, valuable insights are provided to guide future research endeavors.

This survey study distinguishes itself from earlier ones by focusing on all aspects of NF-ISAC systems. It covers the most up-to-date research progress not reported in previous surveys. By exploring recent breakthroughs in NF-ISAC research, the study aims to expand the knowledge base and encourage further investigation in this rapidly developing field. Ultimately, this survey is a valuable resource for scholars, practitioners, and enthusiasts of NF-ISAC wireless research.

The structure of this paper is outlined in Figure 1.

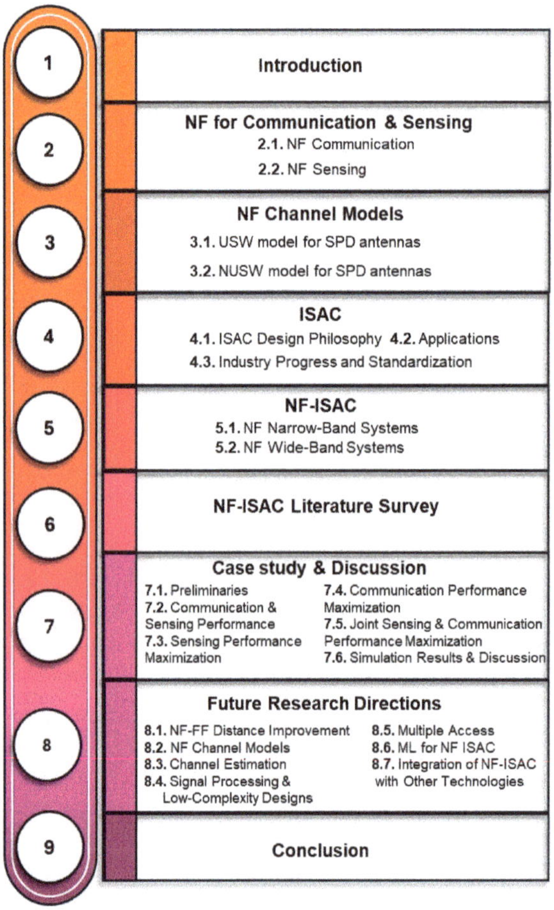

Figure 1. Outlook of this paper.

Notation: Vectors (lower-case) and matrices (upper-case) are bold. $\mathbf{A}^{\mathrm{T}}$, $\mathbf{A}^{\mathrm{H}}$, $\mathrm{Tr}(\mathbf{A})$, $\mathrm{rank}(\mathbf{A})$, and $\mathrm{diag}(\mathbf{A})$ denote transpose, Hermitian transpose, trace, rank, and diagonal operator. $\mathbf{A} \succeq \mathbf{0}$ denotes a positive semidefinite matrix. $\mathbf{I}_M$ is the $M \times M$ identity matrix. The Euclidean norm, expectation, and absolute value operators are denoted by $\|\cdot\|$, $\mathbb{E}[\cdot]$, and $|\cdot|$. $\mathcal{CN}(\boldsymbol{\mu}, \mathbf{R})$ is a complex Gaussian vector with mean $\boldsymbol{\mu}$ and co-variance $\mathbf{R}$. The operators $(\cdot)^{-1}$ and $\otimes$ denote the inversion of a matrix and the Kronecker product.

$\mathbb{C}^{M\times N}$ represents $M \times N$ complex matrices. Finally, $\mathcal{M} \triangleq \{1,\ldots,M\}$, $\mathcal{K} \triangleq \{1,\ldots,K\}$, $\mathcal{L} \triangleq \{1,\ldots,L\}$, $\mathcal{K}_k \triangleq \mathcal{K} \setminus \{k\}$, and $\mathcal{L}_l \triangleq \mathcal{L} \setminus \{l\}$.

2. Near-Field for Communication and Sensing

The EM propagation environment of a radio frequency (RF) source, also known as a radiation source, can be separated into three zones (Figure 2): reactive NF, radiating NF, and radiating FF [2]. It is important to note that the relative distance between the transceivers distinguishes these regions rather than the precise location of the transmitter.

1. *Reactive NF:* The boundary separating the reactive near field (NF) from the radiating NF is called the Fresnel distance, calculated as $0.5\sqrt{D^3/\lambda}$, where D is the antenna aperture size and λ is the electromagnetic wavelength. Thus, the region within this distance (i.e., $r < 0.5\sqrt{D^3/\lambda}$) from the transmitter is the reactive NF. In this region, the electric and magnetic field components are out of phase, causing the EM field's energy to oscillate rather than being fully radiated. Evanescent waves, or non-propagating fields, dominate, rapidly diminishing with distance. Significant amplitude and phase nonlinearities occur across the transmitter antenna array [2].
2. *Radiating NF:* The boundary between the NF and FF is calculated as $2D^2/\lambda$ and is referred to as the Rayleigh distance. Indeed, the radiating NF is situated in the region between the Fresnel and Rayleigh distances (e.g., $0.5\sqrt{D^3/\lambda} \leq r \leq 2D^2/\lambda$). The electric and magnetic fields are perpendicular and in phase, generating propagating waves. However, the fields have not yet developed into ordinary planar waves, and the angular field distribution is influenced by the distance between receivers, resulting in spherical wavefronts. Spherical waves, which have nonlinear phase changes across the antenna aperture and varying amplitude depending on transceiver distance, thus dominate radio propagation [2].
3. *Radiating FF:* This region is beyond the Rayleigh distance, which surrounds the radiating NF region. In this region, the signal paths between each point on the transmitter and the receiver can be treated as parallel, i.e., the angular field distribution is almost independent of the distance between the receiver and transmitter. This generates planar wavefronts with linear phase fluctuations and radio propagation, with the planar waves having the slowest decay rate of all [2].

As the reactive NF region or Fresnel region is often small, i.e., a few wavelengths and evanescent waves decay exponentially with distance, this work focuses only on the radiating NF region. For the sake of brevity, it is called the NF region throughout the rest of the paper.

While NF propagation has always existed, previous generations of wireless technologies have primarily relied on FF propagation due to the small antenna array sizes and sub-6 GHz operating frequencies. However, advancements in ELAAs with hundreds or even thousands of antennas (typically more than 100) and the use of higher frequency bands, such as mmWave (10 GHz to 100 GHz) and THz (100 GHz to 10 THz), have fundamentally shifted the EM propagation environment from FF to NF. For example, an ELAA with a length of 7.4 m operating at 2.6 GHz has a Rayleigh distance of 950 m.

There is no strict distance threshold or boundary between the NF and FF zones; instead, the transition occurs gradually [2,12,14]. Thus, the boundary between NF and FF depends on the unique application scenarios. As a result, various metrics have been developed to describe the field boundary. The two most typically utilized criteria are the phase error and channel gain error perspectives [2].

- *Phase error perspective:* If the phase difference is less than 22.5°, the wavefronts have minimal curvature and can be approximated as plane waves; otherwise, they retain the spherical wave [14,18]. Some relevant metrics are the Rayleigh distance, Fraunhofer condition, and extended Rayleigh distance for MIMO transceivers and reconfigurable intelligent surfaces (RIS). Among these, the Rayleigh distance, the most widely used metric, is proportional to the product of the carrier frequency ($1/\lambda$) and the square of the array aperture size (D^2), i.e., $2D^2/\lambda$. The Fraunhofer condition, which satisfies the Fraunhofer diffraction equation, simulates wave diffraction. For antennas bigger than a half-wavelength, the NF and FF are defined in terms of the Fraunhofer distance, i.e., $d_F = 2D^2/\lambda$. The extended Rayleigh distance based on the phase difference defines the NF and FF between the transmitter and receiver with large antenna arrays. The NF region is defined as $r < 4D_t D_r/\lambda$, where r is the distance of the first antenna at the receiver from the first antenna at the transmitter, and D_t and D_r are the array aperture of the transmitter and the receiver, respectively. These distances primarily pertain to the field boundary near the main axis of the antenna aperture.

- *Channel gain error perspective:* This provides a more precise definition of the field boundary for off-axis locations. The Friis formula, which states that channel gain diminishes with the inverse of distance squared [19], applies in the FF region but not the NF region. Therefore, the FF region is defined as where the Friis formula can approximate the actual channel gain with tolerable error; otherwise, it is the NF region. In this case, the field boundary depends not only on aperture size and wavelength, but also on the angle of departure, angle of arrival, and the shape of the transmit antenna aperture.

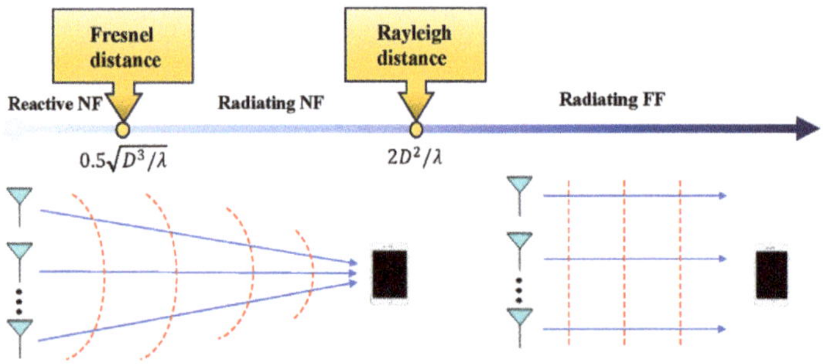

Figure 2. FF planar wavefront versus NF spherical wavefront.

Unlike FF planar wavefronts, which are predominantly influenced by the propagation angle, NF spherical waves introduce an additional distance dimension, which substantially impacts the shape and behavior of the wavefront. In particular, the FF steering vector points to a specific direction, whereas the NF focusing vector focuses on a specific location [14]. This emphasizes a critical distinction between FF beam steering and NF beam-focusing, resulting in substantial theoretical and technological shifts in NF communication and sensing compared to conventional FF counterparts.

Figure 2 illustrates waveform propagation in the NF and FF, showing the spherical wave in the NF and the planar wave in the FF, distinguishable by the Rayleigh distance.

2.1. NF Communication

The properties of NF spherical waves can increase channel degrees of freedom (DoFs) and boost system capacity [2,14]. In FF communication, the phase difference between signals received or transmitted by different antennas is linear, which is mainly governed by the wavefront's angle of arrival or departure, as well as the antenna spacing [2,14]. Since the FF wavefront is planar, the phase difference between antennas varies linearly with their spatial position along the propagation direction, i.e., the angles between each transmitter antenna element and the receiver are almost identical in the FF region due to the large distance between them. This leads to a uniform phase shift across all antennas, and each link's propagation distance grows linearly with the antenna index. This results in a single DoF in the product of the line of sight (LoS) and steering vector [2,14]. In contrast, the phase difference between the NF transceiver antennas is nonlinear and dependent on the distance between each antenna [2]. In NF spherical wavefronts, the variation in distances from the transmitter to each receiver antenna leads to varying phase shifts. Thus, each antenna has its propagation link distance information, resulting in enhanced DoF [2]. Moreover, traditional FF communication cannot distinguish users with the same angle, resulting in significant inter-user interference. However, NF communication can distinguish users with the same angle based on different distances and effectively mitigate inter-user interference owing to the nonlinear phases that relate to the distance of each antenna [2].

The additional distance dimension of spherical waves can mitigate multi-user interference. In particular, it helps to de-correlate the multi-user channel, pushing it closer to the optimal propagation condition. In contrast to FF communication beamforming, i.e., guiding beam energy in a specific direction, NF communication beamforming based on spherical wavefronts achieves a new function of beam-focusing, which focuses beam energy in a particular location/region [2,12,14]. This increases the received signal power at the intended user and removes interference from undesired users, allowing for massive connectivity. As the rank of the NF LoS MIMO channel exceeds one, the spatial multiplexing gain can potentially increase.

2.2. NF Sensing

Sensing requires both signal transmission and reception. In signal transmission, the focused and tightly confined beams in NF beam-focusing allow the signal energy to be focused on the target point with less signal spreading, resulting in increased signal power at the target. On the other hand, as NF spherical waves convey both distance (range) and angle information, NF reception provides selective signal capturing, allowing for targeted reception from specific transmitters while minimizing interference. It also reduces the need for distributed arrays and their synchronization [12,14]. NF beam-focusing improves the sensing SNR of echo signals, allowing for more precise estimation. Combining NF reception with advanced estimate techniques and spectrum search algorithms yields higher position estimation accuracy than the FF counterpart. For example, multiple antenna arrays with known spatial configurations allow for the indirect inference of distance information based on signal intensity, time of arrival, or phase discrepancies between antennas [14]. The additional feature of spherical wavefronts in the range provides improved recognition accuracy, facilitating applications such as human-activity recognition.

However, the distinct advantages of both NF communication and NF sensing are highly dependent on proper transmission architecture [14]. Conventional FF transmission designs, in particular, cannot be easily transferred to NF due to model mismatches, such as the channel model, resulting in significant performance loss. For instance, at close ranges, FF beams become divergent and broader, increasing user interference and angle estimation errors [14]. Consequently, existing FF beamforming strategies drastically reduce NF communication rates and sensing accuracy. Therefore, to fully realize the benefits of NF, specialized transmission designs and improved beamforming techniques that account for the features of NF channels are necessary [14].

3. Near-Field Channel Models

This section covers the foundations of the NF channel model, as illustrated in Figure 3. Conventional FF systems, with signal propagation distances greater than the Rayleigh distance, primarily use planner wave-based channel models. This leads to a uniform planar wave (UPW) with a linear phase and uniform power fluctuation of signals. However, NF systems must utilize spherical wave-based channel models to accurately capture NF characteristics, such as nonlinear phase variations [2]. In addition, the NF zone can be separated into uniform and non-uniform amplitude regions using the uniform power distance to account for signal amplitude variations. NF channel models may be generally classified into two categories based on Rayleigh and uniform power distances: (i) uniform spherical wave (USW) models, and (ii) non-uniform spherical wave (NUSW) models. These NF channel models vary in accuracy and attributes according to the involved assumptions.

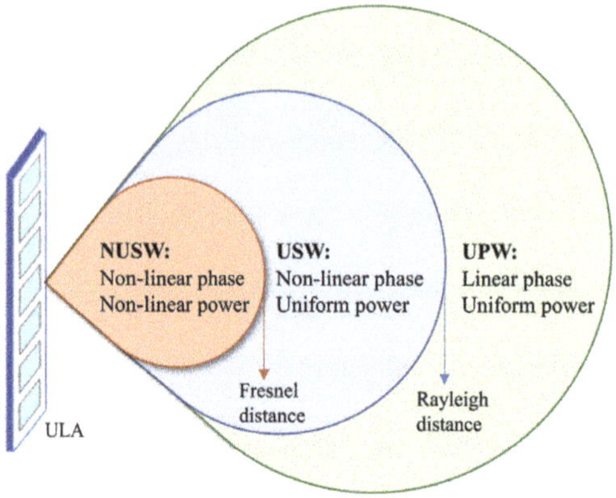

Figure 3. NF and FF separation.

NF channel models are further sub-classified as deterministic or statistical models that take scattering and small-scale fading into account [2]. In particular, deterministic models employ ray tracing, geometric optics, or EM wave propagation theories to precisely determine channel gain, mainly for LoS or channels with a finite number of paths. In contrast, statistical models incorporate the channel's average behavior, fading effects, and time-varying properties, making them appropriate for defining rich-scattering environments [2]. On the other hand, transceiver antenna types such as spatially discrete (SPD) and continuous-aperture (CAP) antennas significantly influence NF channel models [2]. Several widely used NF channel models and their characteristics are reviewed next.

3.1. USW Model for SPD Antennas

Without loss of any generality, assume $M = 2\hat{M} + 1$ antenna transmitter and single-antenna receiver, where the transmitter antenna index is given by $m \in \{-\hat{M}, \ldots, \hat{M}\}$. The coordinates of the m-th element of the transmitter antenna array and the receiver are denoted as $\mathbf{s}_m = [s_{m,x}, s_{m,y}, s_{m,z}]^{\mathrm{T}}$ for $m \in \{-\hat{M}, \ldots, \hat{M}\}$ and $\mathbf{r} = [r_x, r_y, r_z]^{\mathrm{T}}$, respectively. It is further assumed that the central element of an antenna array is located in the origin of the coordinate system, i.e., $\mathbf{s}_0 = [0, 0, 0]^{\mathrm{T}}$. In addition, θ and ϕ are the azimuth and elevation angles of the receiver, respectively, relative to the $x - z$ plane. The propagation distance between the m-th antenna element and the receiver is $r_m = \|\mathbf{r} - \mathbf{s}_m\|$ [2]. Therefore, the channel coefficient between the m-th transmit antenna element and the receiver is given by [2,9]:

$$h_{m,\mathrm{NF}} = \beta_m e^{-j\frac{2\pi}{\lambda} r_m}, \tag{1}$$

where β_m represents the channel gain (amplitude) for the m-th link. Let $r = \|\mathbf{r} - \mathbf{s}_0\|$ be the propagation distance between the receiver and the central element of the transmit antenna array. Assuming r is greater than the uniform power distance, resulting in $\beta_{-\hat{M}} \approx \beta_{-\hat{M}+1} \approx \cdots \approx \beta_{\hat{M}} = \beta$, the NF LoS channel between the transmitter and receiver can be represented as:

$$\mathbf{h}_{\mathrm{NF}}^{\mathrm{LoS}} = \beta e^{-j\frac{2\pi}{\lambda} r} \mathbf{a}(\mathbf{r}), \tag{2}$$

where $\mathbf{a}(\mathbf{r})$ is the NF array response vector, given as:

$$\mathbf{a}(\mathbf{r}) = \left[e^{-j\frac{2\pi}{\lambda}\left(\|\mathbf{r}-\mathbf{s}_{-\hat{M}}\|-r\right)}, \ldots, e^{-j\frac{2\pi}{\lambda}\left(\|\mathbf{r}-\mathbf{s}_{\hat{M}}\|-r\right)} \right]^{\mathrm{T}}. \tag{3}$$

As a result, the phase of the m-th entry of (3) is not a linear function of $\mathbf{s}_m$.

In rich-scattering environments, the receiver receives signals that scatterers reflect through non-LoS (NLoS) channels. The randomness of these multi-path NLoS components also causes channels to be random. Thus, a statistical channel model is necessary. Let L be the total number of scatterers, $\mathbf{r}_l$ the coordinate of the l-th scatterer, and $\tilde{\beta}_l$ the channel gain, including the random reflection coefficient of the l-th scatterer. Then, the NF multi-path channel can be modeled as:

$$\mathbf{h}_{\mathrm{NF}} = \mathbf{h}_{\mathrm{NF}}^{\mathrm{LoS}} + \sum_{l \in \mathcal{L}} \underbrace{\tilde{\beta}_l \mathbf{a}(\mathbf{r}_l)}_{\mathrm{NLoS}}, \tag{4}$$

where the random phase of $\tilde{\beta}_l$ is assumed to be independent and identically distributed (i.i.d.) and uniformly distributed in $(-\pi, \pi]$. The NF array response vectors of two different antenna array geometries, uniform linear array (ULA) and uniform planar array (UPA), are discussed next to gain further insight.

- *Uniform linear array:* A ULA is a one-dimensional linear antenna array with equal antenna spacing of d. The antenna array is placed in the $x-y$ plane, with the origin of the coordinate system at the center of the ULA, resulting in $\phi = 90°$, i.e., the z-axis can be disregarded. The coordinates of the receiver and m-th ULA element are $\mathbf{r} = [r\cos(\theta), r\sin(\theta)]^{\mathrm{T}}$ and $\mathbf{s}_m = [md, 0]^{\mathrm{T}}$. The propagation distance can be approximated as:

$$\begin{aligned} \|\mathbf{r} - \mathbf{s}_m\| &= \sqrt{r^2 + m^2 d^2 - 2rmd\cos(\theta)} \\ &\overset{(a)}{\approx} r - md\cos(\theta) + \frac{m^2 d^2 \sin^2(\theta)}{2r}, \end{aligned} \tag{5}$$

  where the step (a) is obtained by using Fresnel approximation [18]. The m-th entry of the antenna array response for a ULA is thus given as:

$$[\mathbf{a}_{\mathrm{ULA}}(r,\theta)]_m = e^{-j\frac{2\pi}{\lambda}\left(-md\cos(\theta) + \frac{m^2 d^2 \sin^2(\theta)}{2r}\right)}. \tag{6}$$

- *Uniform planar array:* A UPA is a two-dimensional array of antennas uniformly arranged in a rectangular grid. The UPA is placed in the $x-z$ plane, with the origin of the coordinate system at the center of the UPA. Assume $M = M_x \times M_z$ antenna elements, with $M_x = 2\hat{M}_x + 1$ and $M_z = 2\hat{M}_z + 1$, and d_x and d_z antenna spacings in the x and z directions, respectively. The coordinates of the receiver and (m,n)-th ULA element are $\mathbf{r} = [r\cos(\theta)\sin(\phi), r\sin(\theta)\sin(\phi), r\cos(\phi)]^{\mathrm{T}}$ and $\mathbf{s}_{m,n} = [md_x, 0, nd_z]^{\mathrm{T}}$, respectively, where $m \in \{-\hat{M}_x, \ldots, \hat{M}_x\}$ and $n \in \{-\hat{M}_z, \ldots, \hat{M}_z\}$. The propagation distance is then approximated as:

$$\begin{aligned} \|\mathbf{r} - \mathbf{s}_m\| &= \sqrt{r^2 + m^2 d_x^2 + n^2 d_z^2 - 2rmd_x\cos\theta\sin\phi - 2rnd_z\cos\phi} \\ &\overset{(b)}{\approx} r - md_x\cos\theta\sin\phi + \frac{m^2 d_x^2 (1 - \cos^2\theta\sin^2\phi)}{2r} \\ &\quad - nd_z\cos\phi + \frac{n^2 d_z^2 \sin\phi}{2r}, \end{aligned} \tag{7}$$

  where the step (b) is computed via Fresnel approximation, assuming $d_x/r \ll 1$ and $d_z/r \ll 1$, and omitting the bi-linear term [18]. This approximation is adequate for the USW model [2]. By eliminating the constant phase, i.e., $e^{-j\frac{2\pi}{\lambda}r}$, the array response vector's phase can be separated into two components: (i) $-md_x\cos\theta\sin\phi + \frac{m^2 d_x^2(1-\cos^2\theta\sin^2\phi)}{2r}$ and (ii) $-nd_z\cos\phi + \frac{n^2 d_z^2\sin\phi}{2r}$, which only depend on m and n. Thereby, the NF array response vector for a UPA can be given as:

$$\mathbf{a}_{\mathrm{UPA}}(r,\theta,\phi) = \mathbf{a}_x(r,\theta,\phi) \otimes \mathbf{a}_z(r,\theta,\phi), \tag{8}$$

where the m-th entry of $\mathbf{a}_x(r,\theta,\phi)$ and the n-th entry of $\mathbf{a}_z(r,\theta,\phi)$ are given as:

$$[\mathbf{a}_x(r,\theta,\phi)]_m = e^{-j\frac{2\pi}{\lambda}\left(-md_x \cos\theta \sin\phi + \frac{m^2 d_x^2 (1-\cos^2\theta \sin^2\phi)}{2r}\right)}, \tag{9a}$$

$$[\mathbf{a}_z(r,\theta,\phi)]_n = e^{-j\frac{2\pi}{\lambda}\left(-nd_z \cos\phi + \frac{n^2 d_z^2 \sin\phi}{2r}\right)}. \tag{9b}$$

3.2. NUSW Model for SPD Antennas

When the propagation distance r is less than the uniform power distance, the channel gain changes are not negligible. The channel gains of various links are thus not uniform and must be estimated independently. In particular, the channel gain between the m-th transmit and the n-th receiver antenna elements can be calculated using the free-space path loss as $\beta_{mn} = 1/\sqrt{4\pi||\mathbf{r}_n - \mathbf{s}_m||^2}$. The respective channel coefficient of the NUSW model is thus given as [2]:

$$h_{\mathrm{NF}}(\mathbf{s}_m, \mathbf{r}_n) = \frac{1}{\sqrt{4\pi||\mathbf{r}_n - \mathbf{s}_m||^2}} e^{-j\frac{2\pi}{\lambda}||\mathbf{r}_n - \mathbf{s}_m||^2}. \tag{10}$$

3.3. Channel Characteristics

Modeling the NF propagation channel presents several unique challenges. For instance, NF systems operating in high-frequency bands encounter electromagnetic wavefronts with pronounced curvature and complex interactions with objects, leading to scattering, reflection, and diffraction behaviors that differ significantly from those in lower-frequency bands [20]. As a result, traditional models like Friis' free space path loss, which are effective for conventional channel (e.g., FF) conditions, may not accurately capture the intricacies of NF propagation.

Theoretical modeling of the NF channel requires accounting for wavefront curvature, elevated path loss exponents, and the impact of small-scale phenomena, such as scattering from particles or surface roughness. Additionally, these models must consider heightened sensitivity to environmental factors, such as humidity and material properties, which become more significant at higher frequencies like mmWave and THz bands [20].

While advancements have been made in understanding and modeling NF propagation, existing models—particularly those developed for lower frequency bands, like the 3GPP TR36.873 for below 6 GHz—may not fully capture the NF behavior at higher frequencies. For example, in LoS conditions observed in urban micro (UMi) and urban macro (UMa) environments, the path loss at higher frequencies follows Friis' free space path loss model. However, in NLoS conditions, a higher path loss slope (or path loss exponent) is observed, indicating that shadow fading, path loss, and delay spreads in NF high-frequency scenarios can differ significantly from conventional channel characteristics [21].

4. Integrated Sensing and Communications

This section briefly reviews ISAC fundamentals and recent developments before proceeding to the NF-ISAC networks.

The fundamental premise of ISAC revolves around the fusion of sensing and communication operations, leveraging the same hardware platform, shared spectrum, collaborative signal processing techniques, and even a unified control framework [15]. When conventional communication systems strive to achieve channel capacity, the goal is to introduce as much randomness into the signaling as possible [22]. Conversely, a certain degree of determinism in the wireless signal becomes imperative to enhance sensing performance. Consequently, the endeavor to concurrently implement both sensing and communication functions gives rise to a pivotal trade-off, succinctly encapsulating the essence of the random-deterministic dilemma, which is but one facet of the multifaceted conflicts of interest arising in the amalgamation of sensing and communication within a single system. It is precisely these intricate challenges and trade-offs that render ISAC research not only compelling but also imperative in advancing our understanding of integrated wireless systems.

Sensing plays a crucial role in 6G, particularly in applications such as localization [16], smart transportation [23], smart cities [24], and other location-based services. Currently, communication and sensing systems utilize separate spectrum resources. However, given the rapid growth of wireless services and mobile devices, using separate frequency bands may not be spectrum-friendly [25]. Nevertheless, the transition to higher frequency bands like mmWave and THz and super-massive MIMO technology results in sensing and communication sharing similar system architectures, channel characteristics, and signal processing methods [25]. Consequently, both sensing and communication can leverage the same equipment and waveform simultaneously, marking a new era for dual-functional systems, i.e., ISAC [5]. This convergence creates possibilities for enhanced spectral efficiency, beamforming efficiency, and improved cost and size efficiency in integrated functionalities within 6G networks [7].

ISAC is emerging as a pivotal technology across various application scenarios. For instance, ISAC can aid in autonomous driving by significantly impacting traffic congestion management, enhancing safety measures, and improving overall reliability, primarily through high-resolution obstacle detection [26]. Moreover, the convergence of ISAC with UAVs opens up opportunities for offloading communication and sensing tasks onto ISAC, leading to a mutual enhancement in capacity and flexibility [27]. Notably, there is a growing enthusiasm reflected in recent publications concentrating on various aspects of ISAC, including waveform design [28–30], coding design [31–33], experimental testing [34–36], and signal processing [37].

4.1. ISAC Design Philosophy

ISAC aims to integrate sensing and communication and achieve direct trade-offs and mutual benefits. To illustrate these, let us consider a general ISAC system having an M ULA antenna base station (BS), K single-antenna users, and N targets. The BS transmitted signal at the l-th time slot, $\mathbf{x}(l) \in \mathbb{C}^{M\times 1}$, can be given as:

$$\mathbf{x}(l) = \sum_{k\in\mathcal{K}} \sqrt{\rho}\mathbf{w}_k s_k(l) + \sqrt{(1-\rho)}\mathbf{s}_t(l), \tag{11}$$

where $s_k(l)$ and $\mathbf{w}_k \in \mathbb{C}^{M\times 1}$ are the intended data and the communication beamforming for the k-th user, respectively. In (11), $\mathbf{s}_t(l) \sim \mathcal{CN}(0, \mathbf{R}_t)$ is the sensing signal with the covariance matrix of $\mathbf{R}_t = \mathbb{E}[\mathbf{s}_t\mathbf{s}_t^{\mathrm{H}}] \succeq 0$, which is designed to extend the degrees of freedom of the BS transmit signal to achieve enhanced sensing performance [15]. Finally, $0 \leq \rho \leq 1$ is the power allocation factor between communication and sensing, which determines the priority and the level of integration between communication and sensing [15].

Depending on the level of integration, ISAC design falls into three different categories: (i) Communication-centric design with high ρ ($\rho \to 1$), (ii) sensing-centric design with low ρ ($\rho \to 0$), and (iii) joint design with moderate ρ [38].

1. *Communication-centric design:* This term pertains to the design of a communication signal that can serve a dual purpose for sensing as well [39]. It uses the minimum amount of modification to incorporate wireless sensing. Let us consider a downlink communication and sensing system. A straightforward approach involves utilizing the communication signal and extracting target information from the echoes.
2. *Sensing-centric design:* Sensing has a higher priority than communication, i.e., using the radar (sensing) for communication as a secondary function [40]. For example, wireless communication capacity can be added to a radar sensor by embedding communication symbols in the output waveform. In practice, the information contained within the sensing signal should not compromise the integrity of the sensing function [40].
3. *Joint design:* Within this classification, the signal is collaboratively designed with equal/well-designed priorities for sensing and communication to achieve improved

trade-offs between two functionalities, such as a more flexible resource allocation framework between sensing and communication functions. Consequently, the joint signal design offers greater flexibility and higher DoF to effectively balance the requirements of both sensing and communication [41].

With these diverse design principles, ISAC elevates traditional wireless communication networks to a new dimension and profoundly impacts the current information society [15]. In particular, sensing capabilities might become a standard feature in next-generation wireless networks, serving as both an auxiliary approach and a fundamental service for many users and applications. On the other hand, sensory data can improve communication performance, e.g., sensing-aided vehicle beamforming and resource management. Furthermore, with sensing capabilities, future mobile networks can observe their surroundings and become more intuitive [15]. These networks continuously perceive their surroundings and offer services like monitoring traffic, weather, and human activity recognition. These collected/sensed data are the foundation for developing intelligence for the ISAC network and related applications, including smart homes, transportation, and cities.

4.2. *ISAC Applications*

This subsection overviews some potential application scenarios facilitated by ISAC's communication and sensing capabilities (Figure 4).

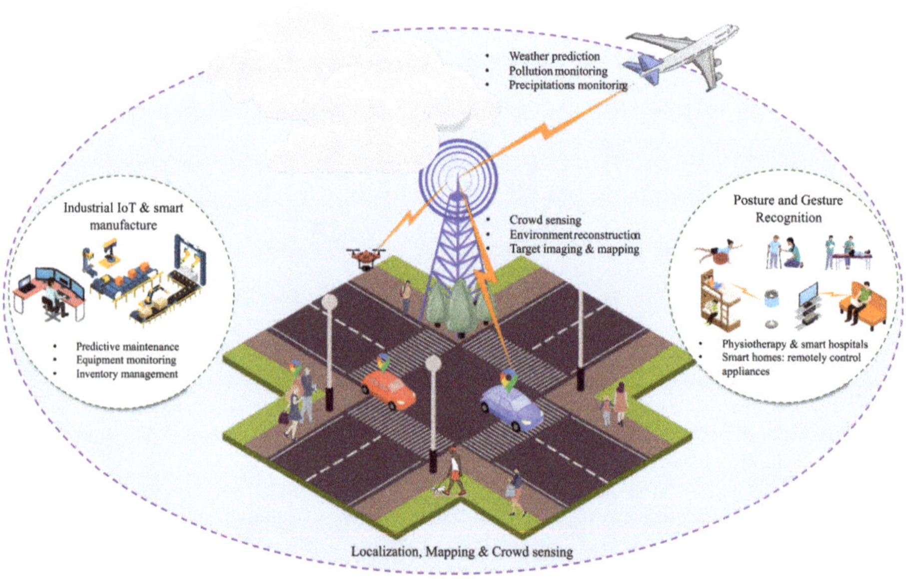

Figure 4. ISAC technology for future wireless networks.

4.2.1. Human Activity Recognition

Activity recognition enables computing systems to monitor, analyze, and support individuals in their daily lives by recording their behaviors. These data have several potential applications, particularly in the healthcare industry [42]. As over-the-air signals are influenced by both static and moving objects, as well as dynamic human activities, amplitude/phase variations in wireless signals could be used to detect or recognize human presence, proximity, falls, sleep, breathing, and daily activities by extracting range, Doppler, or micro-Doppler features. Recognizing or detecting a driver's blink rate with a high sensing resolution can help detect fatigued driving to enhance road safety. In addition, integrating sensing capability into current commercial wireless equipment, such as Wi-

Fi devices, can detect and identify residents' behaviors, resulting in an innovative and human-centric living environment [42].

4.2.2. Localization and Tracking

Localization is essential for standardizing, implementing, and utilizing all generations of networks. Due to bandwidth and antenna restrictions, resulting in low range and angle resolutions, existing cellular networks (e.g., fourth generation (4G) and 5G) can only give measurement data with meter-level accuracy to aid with global navigation satellite systems [10,43]. According to the 5G New Radio (NR) Release 17 key parameter indications, the highest needed horizontal/vertical localization accuracies in industrial IoT applications are 0.2 m/1 m. However, the current systems fail to meet this requirement and, hence, cannot facilitate future applications. In particular, indoor human activity identification, autonomous robotics, and manufacturing demand improved location precision to pinpoint user placements. Current wireless localization technologies rely on device-based implementation, in which wireless equipment (e.g., a smartphone) determines the location of an object through signal interactions and geometrical relationships with other Wi-Fi access points or BSs. However, this method restricts the number of objects that may be located and does not apply to diverse scenarios [10,43].

Nevertheless, ISAC-enabled cellular networks can achieve greater localization accuracy than existing localization systems by using extra Doppler processing and essential information from multi-path components. Additionally, a cellular network with sensing capabilities is useful for more than just detecting the position of a specific object using a smartphone; it can also collect spectroscopic and geometric information from the surrounding environment.

4.2.3. V2X

Autonomous vehicles have the potential to revolutionize the transportation business by improving highway capacity and traffic flow, reducing fuel consumption and pollution, and lowering accident rates [44,45]. To accomplish this, automobiles are equipped with communication transceivers and various sensors that extract ambient information while exchanging data with roadside units (RSUs), other vehicles, and even pedestrians [44,45]. ISAC also addresses electromagnetic compatibility and spectrum congestion issues as a feasible solution. For instance, ISAC-aided V2X can provide environmental information for quick vehicle platooning, secure and seamless access, and simultaneous localization and mapping (SLAM) [44,45]. In addition, RSU networks can expand the sensing range of a passing vehicle beyond its own LoS and field of view.

4.2.4. Smart Manufacturing and Industrial IoT

Wireless networks in sectors such as construction, automobile manufacturing, and product lines have sparked the industrial IoT revolution, resulting in orders of magnitude gains in automation and production efficiency [46]. Such cases include network nodes and robots collaborating to complete complicated and often sensitive tasks that need large-scale connection and impose severe latency constraints. Nevertheless, ISAC can provide significant benefits in such smart factory situations, in addition to the ultra-fast, low-latency connections. By integrating sensing functions, industrial nodes and robots can navigate, coordinate, and map their surroundings more efficiently, potentially reducing signaling overhead [46].

4.2.5. Environmental Monitoring

Another compelling application of ISAC is environmental monitoring, including areas such as weather prediction, pollution detection, and precipitation monitoring [47]. Communication signals across different frequencies can be leveraged to monitor environmental changes [10]. For instance, mmWave signals are close to the water vapor absorption bands, making them sensitive to humidity [10]. Analyzing data from mmWave links in conjunction with other factors like path loss makes it possible to monitor atmospheric conditions effectively. This capability allows ISAC systems to provide real-time environmental insights, making them valuable tools for comprehensive environmental monitoring.

4.2.6. NF-ISAC Use Cases

While NF-ISAC and conventional ISAC may share some common use cases, there are also unique use cases with features that are specific to NF-ISAC. NF-ISAC systems can deliver high-precision positioning and navigation by exploiting the fine spatial resolution available in the NF region. This capability is precious in environments where GPS signals are weak or unavailable, such as inside buildings or warehouses [20]. Additionally, NF-ISAC systems can accurately detect and classify human movements using radar and other sensing techniques, making them ideal for monitoring daily activities and health-related behaviors, such as remote patient monitoring and fitness tracking [15].

Furthermore, NF-ISAC systems can be optimized for efficient wireless power transfer (WPT) and energy harvesting (EH) by leveraging the focused electromagnetic fields in the NF region, which is particularly beneficial for powering small IoT devices or sensors in remote or hard-to-reach locations. Moreover, NF-ISAC can eliminate the need for explicit synchronization between reference stations, thereby reducing system overhead [20].

4.3. Industry Progress and Standardization

As the groundwork for 6G research gains momentum, ISAC has emerged as a focal point, garnering substantial attention from key industry players. Major companies like Ericsson, NTT DOCOMO, ZTE, China Mobile, China Unicom, Intel, and Huawei have all underscored the pivotal role of sensing in their 6G white papers and Wi-Fi 7 visions, emphasizing its significance in shaping the future of wireless technology [48–50]. In particular, ref. [51] define harmonized sensing and communication as one of the three emerging opportunities in 5.5G, i.e., beyond 5G, with the primary goal of using the sensing capabilities of existing massive MIMO BSs and enabling future UAVs and automotive vehicles. In [50], Huawei envisions the 6G new air interface supporting simultaneous wireless communication and sensor signals. This allows ISAC-enabled cellular networks to "see" the actual/physical environment, which is one of 6G's distinguishing features. In [52], Nokia also introduces a unified mmWave system as a model for future indoor ISAC technologies.

Furthermore, the Institute of Electrical and Electronics Engineers (IEEE) standardization association and the 3rd Generation Partnership Project (3GPP) have substantially contributed to the evolution of ISAC-related specifications. Notably, IEEE 802.11 established the wireless local-area network (WLAN) Sensing Topic Interest Group and Study Group in 2019, followed by the creation of the official Task Group IEEE 802.11bf in 2020. These initiatives aim to define the necessary modifications to existing Wi-Fi standards, aligning them with 802.11-compliant waveforms to enhance sensing capabilities.

In parallel, the NR Release 16 specification has redefined the positional reference signal, endowing it with a more regular signal structure and a significantly wider bandwidth. This redesign facilitates improved signal correlation and parameter estimation, including precise time-of-arrival estimation, demonstrating the concerted efforts across industry and standards bodies to advance ISAC technology [53].

5. Near-Field Integrated Sensing and Communications

NF-ISAC must optimize communication and sensing performance on a shared system architecture and hardware platform to meet future network requirements. To accomplish this, the impacts of NF channels and transmission must be considered when balancing communication and sensing operations and their trade-offs for narrow-band and wide-band systems.

5.1. Near-Field Narrow-Band Systems

For effective joint communication and sensing, NF beam-focusing designs must alleviate severe path loss in high-frequency bands. Compared to conventional time-division-based beamforming, which alters communication and sensing beams at separate times, it is more promising to generate multiple beams concurrently to enhance communication and sensing performance [2]. In particular, communication beams should be focused on users at specific locations/regions to improve NF communication performance. Conversely, the detection/estimation goal primarily determines NR sensing beam control. Sensing beams should dynamically scan the region of interest in the angular and range domains to estimate target parameters (e.g., angle, range, and Doppler). The multi-beam design based on array partition, which divides the ELAA into many sub-arrays, each responsible for controlling a sub-beam, is one simple yet effective existing approach. However, it limits the number of antennas allocated per sub-array; thus, communication users or sensing targets may be located in the FF of the sub-array, reducing the beam-focusing effect. To strike the communication and sensing performance trade-off, it is required to precisely determine the number of sub-arrays and optimize the antenna allocation for these sub-arrays.

Conversely, target tracking necessitates adaptive beam designs to track moving targets. Both communication and sensing beams must be directed toward communication users and/or sensing targets to boost communication and sensing SNR. Hence, although there are weak channel correlations, multiple beams must be generated to ensure high communication and sensing SNR when the targets and communication users are at the same angle but at different ranges [54]. This contrasts significantly with the FF region, where only one beam is required when they exist at the same angle.

Narrow-band ISAC is suitable for applications requiring low-power, low-latency sensing and communication in NF settings [2,54]. Examples include smart homes, IoT devices, human–machine interaction, automotive applications, and security and surveillance. Narrow-band ISAC, used in smart homes and IoT devices, allows for efficient home automation, wearable technology, and human–machine interaction via proximity detection and gesture recognition. For instance, smart speakers may recognize user motions or identify people nearby while interacting with other devices. It is also useful in automotive applications that need in-cabin sensing, such as monitoring driver alertness or operating infotainment systems by gesture. Furthermore, NF-narrow-band ISAC is critical in security and surveillance as it can provide short-range proximity detection to prevent unauthorized access while maintaining continuous communication with central security systems [2,54].

5.2. Near-Field Wide-Band Systems

Wide-band NF-ISAC is more complex due to the prominent beam-split effect. In particular, as the operating frequency affects the NF beam-focusing location/region, beams at different frequencies in wide-band systems concentrate at different locations, affecting communication and sensing performance [55].

For communication users, the beam-split effect degrades received SNR as the beam energy distributes in different places [55]. One feasible option for addressing this issue is integrating true-time-delay (TTD) devices with ELAA to generate frequency-dependent beams. TTDs can introduce adjustable time delays into signals, resulting in frequency-dependent phase changes. Conversely, the NF beam-split effect can efficiently improve sensing performance. Specifically, the beam-split effect can generate numerous beams simultaneously, each focused at the same range but at different angles [56]. There is a fundamental trade-off between TTD allocation for communication and sensing. Furthermore, various design challenges, such as how to improve radar sensing performance (e.g., Cramér-Rao bound (CRB) minimization) under communication performance constraints and vice versa, must be addressed.

Wide-band ISAC is ideal for high-resolution sensing and data-rich communication in NF environments [55,56]. It can be used in imaging radar systems for environmental mapping, detecting objects, and identification, which is especially useful in robotics and medical health monitoring. Wide-band ISAC improves the user experience in augmented and virtual reality systems by allowing high-fidelity interaction, such as precise hand-tracking and object recognition, while maintaining high-speed communication with processors. Automotive applications also benefit from wide-band ISAC, such as advanced driver assistance systems, which use high-precision sensing for parking assistance, blind-spot identification, and collision avoidance. Additionally, wide-band ISAC is essential in high-resolution security systems, such as biometric sensing for fingerprint or facial identification, where detailed sensing must be combined with secure, reliable communication.

6. Near-Field ISAC Literature Survey

As mentioned earlier, the shift towards higher frequency bands, combined with the incorporation of extensive antenna elements, in the future of wireless communication (6G), leads to a broadened NF region in contrast to previous generations (1G–5G). This transition to a higher band and the utilization of massive antenna arrays yield high-accuracy and high-resolution sensing capabilities, thereby facilitating the implementation of ISAC. Hence, various previous studies have addressed diverse facets such as beamforming and waveform design, system optimization, and performance analysis within the NF integrated with the ISAC system.

For instance, in the work by Luo et al. [57], researchers investigate the NF user localization by leveraging the beam squint effect. Here, the BS utilizes orthogonal frequency-division multiplexing (OFDM) signals to detect users in the frequency domain, thereby diminishing the overhead associated with beam sweeping. Similarly, the research presented in [58] illustrates that employing OFDM waveforms alongside large arrays in full-duplex (FD) massive MIMO BS facilitates precise localization even within a narrow bandwidth. Examining the self-interference bottleneck within the FD system, ref. [59] delves into the transceiver design of the ISAC system within the THz frequency band, featuring FD-MIMO metasurface antenna panels at the BS. The objective is to devise dynamic metasurface antenna beamforming matrices alongside digital transmit beamforming matrices and digital self-interference cancellation matrices. In a similar vein, ref. [60] tries to minimize the weighted summation of radar and communication beamforming error in a multi-user single-target system. The Cramer–Rao bound (CRB) for target parameter estimate is derived to confirm the effectiveness of NF beamforming. The CRB is formulated and minimized in the study by Wang et al. for NF joint distance and angle sensing [12]. The investigation explores two scenarios: (1) employing fully digital antennas, where each antenna at the BS is linked to a dedicated RF chain, and (2) utilizing hybrid digital and analog antennas, incorporating large-dimensional analog phase shifters alongside low-dimensional digital components. The findings indicate that NF favors ISAC more than FF. In NF multi-target detection, ref. [61] pioneered a beamforming design for an NF-ISAC system with multi-user, multi-target, and FD BS. It presented an iterative beamforming approach to reduce BS transmit power by designing optimal transmit (communication and sensing) and reception beamforming at the BS.

WPT is a significant application of NF communication, facilitating the charging of electronic devices like smartphones, smartwatches, and electric toothbrushes, as well as enabling wireless charging pads and mats. Notably, the distinction between NF WPT and FF WPT lies in the efficiency drop of power transfer as the system transitions from NF to FF conditions [62]. In light of this, ref. [63] delves into the study of transmit beam pattern matching and max-min beam pattern gain in radar sensing and WPT systems. The findings underscore NF's superior transmit beam pattern gain compared to the FF scenario.

To address the multi-functionality demands of future wireless communication systems, ref. [64] investigates a NOMA-assisted ISAC multi-tier computing (MTC) system. In this proposed framework, a multi-functional BS conducts target sensing while concurrently offering edge computing services to nearby users, with the ability to offload computation tasks to a robust cloud server. Results demonstrate the superiority of the MTC system over single-tier computing systems, with NOMA playing a crucial role in ensuring high-quality computing services.

Furthermore, RIS technology is closely intertwined with 6G applications. While RIS comprises numerous passive reconfigurable elements, its integration with cellular networks can dynamically alter the wireless propagation environment [65,66]. As a result, ISAC can harness the additional spatial degrees of freedom facilitated by integration with RIS. In study [65], the focus lies on RIS-assisted NF localization. The polar-domain gradient descent algorithm and multiple signal classification (MUSIC) demonstrate superior angle accuracy compared to existing algorithms. In [66], the RIS enhances the transmission link by establishing LoS paths between the transmitter and receiver in NF scenarios. The continuous RIS design aims to transform planar waves into cylindrical or spherical waves, thus focusing energy on the ULA or the single antenna receiver. Furthermore, the maximum likelihood method and focal scanning are utilized for receiver localization, with corresponding position error analysis conducted.

7. Case Study and Discussion

The NF-ISAC design philosophies were outlined in Section 4.1. Based on them, this section develops several case studies.

In addition to the design philosophies, we consider NOMA to facilitate multiple access for communication users. NOMA has been extensively studied as an efficient multiple-access technique, primarily in FF scenarios. However, when considering NF-NOMA, as opposed to its well-established use in FF, the transition is not merely about repurposing existing FF concepts. Instead, NF presents unique opportunities and challenges that can lead to innovations and adaptations for NOMA. In particular, NOMA could develop in the NF in the following aspects:

- *Exploiting spatial resolution:* Unlike conventional FF, NF spherical wavefronts enable finer spatial resolution. Thus, NOMA can use this to effectively separate users in the spatial domain, resulting in enhanced user differentiation even when users are closely located. Additionally, the NF environment enables more precise user clustering based on spatial features such as distance, angle, and position. NOMA can be designed to exploit these features, optimizing power allocation and interference management more effectively than in the FF.
- *Advanced beamforming:* NF beamforming is much more precise due to the ability to focus beams on a specific point in space rather than just a direction. NOMA can leverage this by enabling multiple users to share the same spatial resources with carefully designed power levels, leading to higher spectral efficiency.
- *Enhanced interference management:* NF beam-focusing improves interference management. NOMA can exploit this to develop new interference cancellation strategies that utilize the precise spatial information offered by NF, minimizing inter-user interference more effectively than FF. Furthermore, NF NOMA can employ power control techniques that include the user's distance, spatial location, and angle, resulting in more efficient interference mitigation and resource utilization.

While some concepts from FF NOMA can be repurposed for NF, NF presents unique opportunities that require NOMA to evolve and adapt. The NF's ability to utilize fine spatial resolution, beam focusing, and wavefront modification offers new avenues for NOMA in user distinction, beamforming, and interference management. Thus, NOMA in NF is more than just repeating FF notions; it is about utilizing NF's distinctive characteristics to establish more intelligent, efficient, and secure communication systems.

7.1. Preliminaries

7.1.1. System and Channel Models

Figure 5 investigates a downlink (DL) NOMA ISAC network. It comprises an M ULA antenna BS, K single-antenna NOMA users, and a sensing target. Without loss of generality, the ULA antennas are at a spacing of d, and the coordinate system's origin is at the center of the ULA.

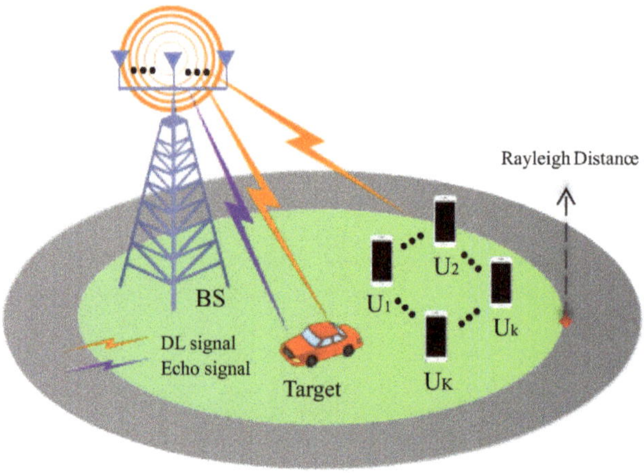

Figure 5. System model of DL NOMA empowered ISAC.

NF-channel model: The NF spherical wave channel model for a ULA in Section 3.1 is adopted here. The coordinates of the m-th element of ULA, the k-th user, and the target are $\mathbf{s}_m = [md, 0]^{\mathrm{T}}$ for $m \in \{-\hat{M}, \ldots, \hat{M}\}$, $\mathbf{u}_k = [r_k \cos(\theta_k), r_k \sin(\theta_k)]^{\mathrm{T}}$ for $k \in \mathcal{K}$, and $\mathbf{u}_t = [r_t \cos(\theta_t), r_t \sin(\theta_t)]^{\mathrm{T}}$, respectively. In addition, θ_k and θ_t are the azimuth angle of the k-th user and the target, respectively, and $r_k = \|\mathbf{u}_k - \mathbf{s}_0\|$ denotes the distance between k-th user and center of ULA BS antenna. Using (6) and (5), the NF-LoS channel between ULA BS and k-th user can be written as [2]:

$$\mathbf{h}_{k,\mathrm{NF}}^{\mathrm{LoS}} = \beta e^{-j\frac{2\pi}{\lambda}r_k}\left[e^{-j\frac{2\pi}{\lambda}\left(\hat{M}d\cos(\theta_k)+\frac{\hat{M}^2d^2\sin(\theta_k)^2}{2r_k}\right)}, \ldots, e^{-j\frac{2\pi}{\lambda}\left(-\hat{M}d\cos(\theta_k)+\frac{\hat{M}^2d^2\sin(\theta_k)^2}{2r_k}\right)}\right], \quad (12)$$

where β is the channel gain, which is assumed to have an equal value for all antenna elements [2]. Assuming all users and the target are in the ULA's NF region, only one line-of-sight (LoS) channel between the users/target and the BS is considered. Nonetheless, our framework is adaptable to multi-path and non-LoS scenarios.

While this study primarily focuses on NF channels due to the extended range of NF–FF boundaries at higher frequencies, the FF channel model is also employed as a comparative benchmark.

FF-channel model: The distance between the BS and the users/target for the FF becomes $r \gg D$. Therefore, the NF channel model in (12) reduces into following model:

$$\mathbf{h}_{k,\mathrm{FF}}^{\mathrm{LoS}} = \beta e^{-j\frac{2\pi}{\lambda}r_k}\left[e^{-j\frac{2\pi}{\lambda}\hat{M}d\cos(\theta_k)}, \ldots, e^{j\frac{2\pi}{\lambda}\hat{M}d\cos(\theta_k)}\right]. \quad (13)$$

In the sequel, the channel between the k-th user and the BS is $\mathbf{h}_k \in \mathbb{C}^{M\times 1}$, and the channel between the target and the BS is $\mathbf{g}_t \in \mathbb{C}^{M\times 1}$.

7.1.2. Transmit Signal Model

The BS transmits the information-bearing signal $s_k \sim \mathcal{CN}(0,1)$ for the k-th user ($k \in \mathcal{K}$), along with the dedicated sensing signal $\mathbf{s}_t \sim \mathcal{CN}(0, \mathbf{R}_t)$. Here, $\mathbf{R}_t = \mathbb{E}[\mathbf{s}_t\mathbf{s}_t^{\mathrm{H}}] \succeq 0$ is the covariance matrix of the sensing signal, which is designed to extend the degrees-of-freedom of the BS transmit signal to achieve enhanced sensing performance [67]. Signals

$\{\{s_k\}_{k\in\mathcal{K}}, \mathbf{s}_t\}$ are statistically independent from each other [67]. Thus, the BS broadcast signal may be expressed as:

$$\mathbf{x} = \sum_{k\in\mathcal{K}} \mathbf{w}_k s_k + \mathbf{s}_t, \tag{14}$$

where $\mathbf{w}_k$ denotes the transmit beamforming vector assigned for signal s_k.

Each user receives the BS signal $\mathbf{x}$ and must extract the information sent for itself. In this process, the sensing and other user signals act as interference. Consequently, the received signal at the k-th user can be written as:

$$y_k = \mathbf{h}_k^{\mathrm{H}}\mathbf{w}_k s_k + \underbrace{\sum_{i\in\mathcal{K}_k} \mathbf{h}_k^{\mathrm{H}}\mathbf{w}_i s_i}_{\text{inter-user interference}} + \underbrace{\mathbf{h}_k^{\mathrm{H}}\mathbf{s}_t}_{\text{sensing interference}} + n_k, \tag{15}$$

where $n_k \sim \mathcal{CN}(0,\sigma_k^2)$ is the additive white Gaussian noise (AWGN) at the k-th user with zero mean and σ_k^2 variance.

Conversely, the BS uses the received echo signal from the target for sensing. The received echo signal at the BS from the target is thus expressed as:

$$\mathbf{y}_t = \mathbf{G}_t\mathbf{x} + \mathbf{n}_t, \tag{16}$$

where $\mathbf{G}_t \triangleq \mathbf{g}_t\mathbf{g}_t^{\mathrm{H}} \in \mathbb{C}^{M\times M}$ due to the reciprocity in the time division duplex mode [68,69], and $\mathbf{n}_t \sim \mathcal{CN}(0,\sigma_t^2\mathbf{I}_M)$ is the AWGN at the BS. The BS then applies the receiver beamforming $\mathbf{u} \in \mathbb{C}^{M\times 1}$ to the received signal (16) to capture the sensing information from the target. The post-processed signal for the target's sensing information can be given as:

$$y_t = \mathbf{u}^{\mathrm{H}}\mathbf{y}_t = \mathbf{u}^{\mathrm{H}}\mathbf{G}_t\mathbf{x} + \mathbf{u}^{\mathrm{H}}\mathbf{n}_t. \tag{17}$$

7.2. Communication and Sensing Performance

The sensing and communication rates/signal-to-noise-to-interference ratios (SINRs) significantly affect the performance of both systems. Here, the users' and the target's sensing rates are used to analyze and optimize the NF-ISAC system.

7.2.1. Communication Performance

From Equation (15), the received SINR at the k-th user is given as:

$$\gamma_k = \frac{|\mathbf{h}_k^{\mathrm{H}}\mathbf{w}_k|^2}{\sum_{i\in\mathcal{K}_k} |\mathbf{h}_k^{\mathrm{H}}\mathbf{w}_i|^2 + \mathbf{h}_k^{\mathrm{H}}\mathbf{R}_t\mathbf{h}_k + \sigma_k^2}. \tag{18}$$

Thereby, the rate of the k-th user can be approximated as:

$$\mathcal{R}_k = \log_2(1+\gamma_k). \tag{19}$$

7.2.2. Sensing Performance

The BS learns and obtains environmental information using the post-processed reflected signal from the target in Equation (17). Thus, the sensing SINR of the target at the BS is given by:

$$\gamma_t = \frac{\mathbb{E}\{|\mathbf{u}^{\mathrm{H}}\mathbf{G}_t\mathbf{x}|^2\}}{\mathbb{E}\{|\mathbf{u}^{\mathrm{H}}\mathbf{n}_t|^2\}} = \frac{\mathbf{u}^{\mathrm{H}}\mathbf{G}_t\mathbf{R}_x\mathbf{G}_t^{\mathrm{H}}\mathbf{u}}{\mathbf{u}^{\mathrm{H}}\sigma_t^2\mathbf{I}_M\mathbf{u}}, \tag{20}$$

where $\mathbf{R}_x \triangleq \mathbb{E}\{\mathbf{x}\mathbf{x}^{\mathrm{H}}\} = \sum_{k\in\mathcal{K}} \mathbf{w}_k\mathbf{w}_k^{\mathrm{H}} + \mathbf{R}_t$ is the covariance matrix of the BS transmitted signal [67].

Remark 1. *The associated SINRs primarily determine the communication and sensing performance of ISAC systems. In particular, the probability of detecting a communication symbol increases monotonically with SINR [70,71]. Maximizing SINR eventually minimizes the symbol error probability. Hence, the communication SINR (or rate) is a standard performance metric. Similarly, in sensing, the detection probability of a target is proportional to its sensing SINR [67,72]. The sensing SINR enables target detection using both transmit and receiver beamforming. It also aids in reducing interference between targets. However, the standard mean squared error of the transmit beam pattern does not account for the receiver beam pattern or target interference. Given the benefits of sensing SINR, it is a popular metric for sensing performance.*

As mentioned in Section 4.1, the NF-ISAC systems can also be classified into three categories based on the design target. This section examines those and defines an optimization problem tailored to the criteria of each class.

7.3. Case 1: Sensing Performance Maximization

In Case 1, the priority is maximizing sensing performance over communication. Subsequently, the optimization problem can be expressed as:

$$\begin{aligned}
(\text{P1}): \max_{\{\mathbf{w}_k\},\mathbf{R}_t,\mathbf{u}} \quad & \gamma_t(\{\mathbf{w}_k\},\mathbf{R}_t,\mathbf{u}), && \text{(21a)}\\
\text{s.t.} \quad & \gamma_k(\{\mathbf{w}_k\},\mathbf{R}_t) \geq \gamma_{th}, \ \forall k \in \mathcal{K}, && \text{(21b)}\\
& \sum_{k\in\mathcal{K}} \|\mathbf{w}_k\|^2 + \text{Tr}(\mathbf{R}_t) \leq P_b, && \text{(21c)}\\
& \mathbf{R}_t \succeq 0, && \text{(21d)}
\end{aligned}$$

where γ_{th} is the minimum SINR threshold required by the k-th user and P_b is the maximum allowable transmit power of the BS. In (P1), (21b) is to satisfy the minimum SINR required for each user, (21c) is the total transmit power constraint at the BS, and (21d) is the inherent constraint of the sensing signal. Problem (P1) is non-convex because of its objective and constraints, including the optimization variables' products. To address this, alternating optimization (AO) is utilized [73].

Proposed Solution

To solve problem (P1), it is decomposed into two distinct subproblems involving two-variable blocks using AO techniques: $\mathbf{w}_k, \mathbf{R}_t$ and $\mathbf{u}$. Specifically, the subsequent subsection focuses on maximizing sensing SINR by jointly optimizing the design of $\mathbf{w}_k$ and $\mathbf{R}_t$ while keeping $\mathbf{u}$ constant. Next, $\mathbf{u}$ is optimized while maintaining the fixed values of $\mathbf{w}_k, \mathbf{R}_t$. This procedure repeats until convergence or the stopping criterion is satisfied. This strategy is useful when direct or simultaneous optimization of all variables is difficult or computationally costly [73]. The details are as follows:

Optimizing $\{\mathbf{w}_k\}$ *and* $\mathbf{R}_t$: To solve this subproblem, the semidefinite relaxation (SDR) technique can be applied. Denote matrix $\mathbf{W}_k = \mathbf{w}_k\mathbf{w}_k^H$, where $\mathbf{W}_k$ is semidefinite and satisfies rank-one constraint. Consequently, problem (P1) can be written as:

$$\begin{aligned}
(\text{P1.1}): \max_{\{\mathbf{W}_k\},\mathbf{R}_t} \quad & \frac{\mathbf{u}^{\text{H}}\mathbf{G}_t(\sum_{k\in\mathcal{K}}\mathbf{W}_k+\mathbf{R}_t)\mathbf{G}_t^{\text{H}}\mathbf{u}}{\mathbf{u}^{\text{H}}\sigma_t^2\mathbf{I}_M\mathbf{u}}, && \text{(22a)}\\
\text{s.t.} \quad & \frac{\text{Tr}(\mathbf{h}_k\mathbf{h}_k^{\text{H}}\mathbf{W}_k)}{\sum_{i\in\mathcal{K}_k}\text{Tr}(\mathbf{h}_k\mathbf{h}_k^{\text{H}}\mathbf{W}_i)+\mathbf{h}_k^{\text{H}}\mathbf{R}_t\mathbf{h}_k+\sigma_k^2} \geq \gamma_{th}, \ \forall k\in\mathcal{K}, && \text{(22b)}\\
& \sum_{k\in\mathcal{K}}\text{Tr}(\mathbf{W}_k)+\text{Tr}(\mathbf{R}_t) \leq P_b, && \text{(22c)}\\
& \mathbf{R}_t \succeq 0, \{\mathbf{W}_k\} \succeq 0, \ \forall k\in\mathcal{K}, && \text{(22d)}\\
& \text{rank}(\mathbf{W}_k) = 1, \ \forall k\in\mathcal{K}. && \text{(22e)}
\end{aligned}$$

Note that P1.1 is still non-convex due to the constraint of Equation (22e). Thus, relaxing this rank-one constraint, (P1.1) is transformed into a convex optimization problem, i.e., a conventional semi-definite programming (SDP) problem, and can be solved via Matlab CVX [73]. To optimize $\{\mathbf{w}_k\}$, the Gaussian randomization (GR) can be utilized to enforce the desired implicit rank-one constraint on the matrix solution. In particular, let (P1.1)'s optimal solution be $\mathbf{W}_k^*$, which is not necessarily rank-one. If it is, the sole eigenvalue is the trace of $\mathbf{W}_k^*$, and the related eigenvector defines the optimal beamforming vector. Otherwise, a rank-one solution is extracted from $\mathbf{W}_k^*$. To do this, a random vector $\mathbf{r}$ can be selected from $\mathcal{CN}(0, \mathbf{W}_k^*)$ and projected to the feasible region of P1.1. This random sampling is repeated numerous times, and only the best approximate solution is selected. This method is referred to as GR [74].

Optimizing $\mathbf{u}$*:* It can be noted that only γ_t is dependent on $\mathbf{u}$. Therefore, for given $\{\{\mathbf{w}_k\}, \mathbf{R}_t\}$, P1 can be expressed as:

$$(\text{P1.2}) : \max_{\mathbf{u}} \quad \frac{\mathbf{u}^{\mathrm{H}}\mathbf{G}_t\mathbf{R}_x\mathbf{G}_t^{\mathrm{H}}\mathbf{u}}{\mathbf{u}^{\mathrm{H}}\sigma_t^2\mathbf{I}_M\mathbf{u}}. \tag{23}$$

This is a standard Rayleigh quotient problem [75]. Therefore, the optimal $\mathbf{u}^*$ is the eigenvector corresponding to the maximum eigenvalue of matrix $\mathbf{B}^{-1}\mathbf{A}$, where $\mathbf{B} = \sigma_t^2\mathbf{I}_M$ and $\mathbf{A} = \mathbf{G}_t\mathbf{R}_x\mathbf{G}_t^{\mathrm{H}}$. The implementation steps are detailed in Algorithm 1.

Algorithm 1 Algorithm for solving Case 1.

Input: Set the iteration counter $x = 0$, the convergence tolerance $\epsilon > 0$, initial feasible solution $\{\mathbf{W}_k^{(0)}\}, \mathbf{R}_t^{(0)}$. Initialize the objective function value $\gamma_t^{(0)}(\mathbf{Z}) = 0$.
while $\frac{\gamma_t^{(x+1)}(\mathbf{Z}) - \gamma_t^{(x)}(\mathbf{Z})}{\gamma_t^{(x+1)}(\mathbf{Z})} \geq \epsilon$ **do**
  Solve (P1.1) and obtain the optimal solution for $\{\mathbf{w}_k^{(x+1)}\}$ by recovering a rank-one solution via GR in Algorithm 2.
  Solve Equation (23) and obtain the optimal solution as $\mathbf{u}^{(x+1)}$.
  Calculate the objective function value $\gamma_t^{(x+1)}(\mathbf{Z})$.
  Set $x \leftarrow x + 1$;
end while
Output: $\mathbf{Z}^* = \{\{\mathbf{w}_k^*\}, \mathbf{R}_t^*, \mathbf{u}^*\}$.

Algorithm 2 Gaussian Randomization.

Input: matrix $\mathbf{F}$, number of randomization sample N, and initial objective value $\mathcal{O}^0 = -\mathrm{Inf}$.
Do Apply eigenvalue decomposition to obtain $\mathbf{F}_{op} = \mathbf{U}\mathbf{\Lambda}\mathbf{U}^{\mathrm{H}}$.
for $n \leq N$ **do**
  Generate N independent unit-variance zero-mean circularly symmetric complex Gaussian vector $\mathbf{z}$.
  Set $\mathbf{f}_{op} = \mathbf{U}\mathbf{\Lambda}^{0.5}\mathbf{z}$ and calculate $\mathcal{O}^{n+1}(\mathbf{f}_{op})$.
  if $\mathcal{O}^{n+1} \geq \mathcal{O}^n$ **then**
    $\mathcal{O}^n \leftarrow \mathcal{O}^{n+1}, \mathbf{f}_{op} = \mathbf{U}\mathbf{\Lambda}^{0.5}\mathbf{z}$
  end if
end for
Output: $\mathbf{f}_{op}$.

7.4. Case 2: Communication Performance Maximization

If communication functionality is of higher priority, the design strategies may delve into maximizing communication performance. The communication-centric optimization problem can be expressed as follows:

$$
\begin{aligned}
(\text{P2}):\ \max_{\{\mathbf{w}_k\},\mathbf{R}_t,\mathbf{u}}\quad & \sum_{k\in\mathcal{K}} \mathcal{R}_k(\{\mathbf{w}_k\},\mathbf{R}_t), && \text{(24a)}\\
\text{s.t.}\quad & \gamma_t(\{\mathbf{w}_k\},\mathbf{R}_t,\mathbf{u}) \geq \eta, && \text{(24b)}\\
& \sum_{k\in\mathcal{K}} \|\mathbf{w}_k\|^2 + \mathrm{Tr}(\mathbf{R}_t) \leq P_b, && \text{(24c)}\\
& \mathbf{R}_t \succeq 0, && \text{(24d)}
\end{aligned}
$$

where η is the minimum required sensing SINR. Adopting a method like the one described in Section 7.3 for optimizing the combiner vector $\mathbf{u}$, the optimal combiner vector for this problem can be obtained by solving (23). Consequently, to maintain conciseness, specific details are excluded from this discussion.

Therefore, the optimization problem over beamformers and sensing covariance matrix can be written as:

$$
\begin{aligned}
(\text{P2.1}):\ \max_{\{\mathbf{w}_k\},\mathbf{R}_t}\quad & \sum_{k\in\mathcal{K}} \mathcal{R}_k(\{\mathbf{w}_k\},\mathbf{R}_t), && \text{(25a)}\\
\text{s.t.}\quad & \text{(24b)}, \text{(24c)}, \text{(24d)}. && \text{(25b)}
\end{aligned}
$$

The objective of (P2.1) is non-convex due to the fractional form terms and quadratic form of $\mathbf{w}_k$. To address this issue and transform the problem into a convex formulation, use the SDR with matrix definition $\mathbf{W}_k = \mathbf{w}_k\mathbf{w}_k^H$ satisfying rank-one constraint. Then, fractional programming (FP) [76] is employed. To invoke FP techniques, one utilizes an auxiliary variable and the Lagrangian dual transform $\boldsymbol{\gamma} = [\gamma_1, \ldots, \gamma_K]^T$. Therefore, the objective in (P2.1) can be rewritten as:

$$
\begin{aligned}
\bar{R}_{\text{sum}}(\boldsymbol{\gamma}, \{\mathbf{W}_k\}, \mathbf{R}_t) = \sum_{k\in\mathcal{K}} \log_2\Bigg[1 + 2\gamma_k\sqrt{\mathbf{h}_k^H\mathbf{W}_k\mathbf{h}_k} - \\
\gamma_k^2\left(\sum_{i\in\mathcal{K}_k} \mathbf{h}_k^H\mathbf{W}_i\mathbf{h}_k + \mathbf{h}_k^{\mathrm{H}}\mathbf{R}_t\mathbf{h}_k + \sigma_k^2\right)\Bigg].
\end{aligned} \tag{26}
$$

Subsequently, problem (P2.1) can be rewritten as:

$$
\begin{aligned}
(\text{P2.2}):\ \max_{\{\mathbf{W}_k\},\mathbf{R}_t,\boldsymbol{\gamma}}\quad & \bar{R}_{\text{sum}}, && \text{(27a)}\\
\text{s.t.}\quad & \frac{\mathbf{u}^{\mathrm{H}}\mathbf{G}_t(\mathbf{W}_k + \mathbf{R}_t)\mathbf{G}_t^{\mathrm{H}}\mathbf{u}}{\mathbf{u}^{\mathrm{H}}\sigma_t^2\mathbf{I}_M\mathbf{u}} \geq \gamma_{th},\ \forall k \in \mathcal{K}, && \text{(27b)}\\
& \sum_{k\in\mathcal{K}} \mathrm{Tr}(\mathbf{W}_k) + \mathrm{Tr}(\mathbf{R}_t) \leq P_b, && \text{(27c)}\\
& \mathbf{W}_k \succeq 0,\ \forall k,\quad \mathbf{R}_t \succeq 0, && \text{(27d)}\\
& \mathrm{rank}(\mathbf{W}_k) = 1,\ \forall k. && \text{(27e)}
\end{aligned}
$$

Problem (P2.2) remains non-convex due to the rank-one constraint. To resolve this, the constraint is dropped, resulting in a convex optimization problem for $\{\mathbf{W}_k\}$, and $\mathbf{R}_t$, which can be solved using Matlab CVX [73]. However, the AO algorithm must be deployed to update the auxiliary variable γ and optimize $\{\mathbf{W}_k\}$, and $\mathbf{R}_t$ iteratively. The optimal γ in a x iteration of AO algorithm can be determined by solving $\frac{\partial \bar{R}_{\text{sum}}}{\partial \gamma_k} = 0$ and which yields

$$\gamma_k = \frac{\sqrt{\mathbf{h}_k^H \mathbf{W}_k^{(x)} \mathbf{h}_k}}{\sum_{i \in \mathcal{K}_k} \mathbf{h}_k^H \mathbf{W}_i^{(x)} \mathbf{h}_k + \mathbf{h}_k^H \mathbf{R}_t \mathbf{h}_k + \sigma_k^2}. \tag{28}$$

Afterward, the Gaussian randomization process [77] can be deployed to ensure the implicit satisfaction of the rank-one constraint. Algorithm 3 presents the steps for solving (P2).

Algorithm 3 Algorithm for solving Case 2.

Input: Set the iteration counter $x = 0$, the convergence tolerance $\epsilon > 0$, initial feasible solution $\{\mathbf{W}_k^{(0)}\}, \mathbf{R}_t^{(0)}, \gamma^{(0)}$. Initialize the objective function value $\bar{R}_{\text{sum}}^{(0)}(\mathbf{Z}) = 0$.
while $\frac{\bar{R}_{\text{sum}}^{(x+1)}(\mathbf{Z}) - \bar{R}_{\text{sum}}^{(x)}(\mathbf{Z})}{\bar{R}_{\text{sum}}^{(x+1)}(\mathbf{Z})} \geq \epsilon$ **do**
  Do: Solve (P2.2) and obtain solution as $\{\mathbf{W}_k^{(x+1)}\}, \mathbf{R}_t^{(x+1)}$.
  Update: $\gamma^{(x+1)}$ using Equation (28).
  Until: There is no increment in the objective.
  Solve Equation (23) and obtain the optimal solution as $\mathbf{u}^{(x+1)}$.
  Calculate the objective function value $\bar{R}_{\text{sum}}^{(x+1)}(\mathbf{Z})$.
  Obtain $\{\mathbf{w}_k^{(x+1)}\}$ by recovering a rank-one solution via GR and update $\{\mathbf{W}_k^{(x+1)}\}$.
  Set $x \leftarrow x + 1$;
end while
Output: $\mathbf{Z}^* = \{\{\mathbf{w}_k^*\}, \mathbf{R}_t^*, \mathbf{u}^*\}$.

7.5. Case 3: Joint Sensing and Communication Performance Maximization

In this class, sensing and communication priorities can be adjusted. Therefore, the design can be tailored to favor performance trade-offs. Consequently, the optimization issue can be stated as a weighted sum user rate and target as follows:

$$\begin{aligned}
\text{(P3)}: \max_{\{\mathbf{w}_k\}, \mathbf{R}_t, \mathbf{u}} \quad & \rho \sum_{k \in \mathcal{K}} \mathcal{R}_k(\{\mathbf{w}_k\}, \mathbf{R}_t) + (1-\rho) \log_2(1+\gamma_t), && \text{(29a)} \\
\text{s.t.} \quad & \sum_{k \in \mathcal{K}} \|\mathbf{w}_k\|^2 + \text{Tr}(\mathbf{R}_t) \leq P_b, && \text{(29b)} \\
& \mathbf{R}_t \succeq 0. && \text{(29c)}
\end{aligned}$$

Here, $0 \leq \rho \leq 1$ represents the constant weight that determines the priority of the sum user rate, while $1 - \rho$ indicates the priority of the target sensing rate. By following the analogous procedure to previous problems, it becomes evident that since γ_t solely relies on the combiner vector $\mathbf{u}$ and logarithm in a monotonically increasing function of its argument, the optimal $\mathbf{u}^*$ that maximizes the rate can be derived by solving Equation (23). Consequently, this variable is excluded from (P3). Subsequently, similar to case 2, to reformulate this problem in a tractable form, the sum user rate transformation outlined in Equation (26) is employed. This transformation is then substituted into the sum user rate expression in Equation (29a) and SDR is used. The final result is as follows:

$$\begin{aligned}
\text{(P3.1)}: \max_{\{\mathbf{W}_k\}, \mathbf{R}_t} \quad & \rho \bar{R}_{\text{sum}} + (1-\rho) \bar{R}_t, && \text{(30a)} \\
\text{s.t.} \quad & \sum_{k \in \mathcal{K}} \text{Tr}(\mathbf{W}_k) + \text{Tr}(\mathbf{R}_t) \leq P_b, && \text{(30b)} \\
& \mathbf{W}_k \succeq 0, \ \forall k, \quad \mathbf{R}_t \succeq 0, && \text{(30c)} \\
& \text{rank}(\mathbf{W}_k) = 1, \ \forall k, && \text{(30d)}
\end{aligned}$$

where

$$\bar{R}_t = \log_2\left(1 + \frac{\mathbf{u}^{\mathrm{H}}\mathbf{G}_t(\mathbf{W}_k + \mathbf{R}_t)\mathbf{G}_t^{\mathrm{H}}\mathbf{u}}{\mathbf{u}^{\mathrm{H}}\sigma_t^2\mathbf{I}_M\mathbf{u}}\right). \tag{31}$$

However, (P3.1) exhibits non-convexity, attributed to the presence of a non-convex constraint in Equation (30d). Consequently, dropping this constraint makes (P3.1) a convex problem that can be solved via CVX. To ensure a rank-one solution, Gaussian randomization is deployed. The steps for solving (P3) are outlined in Algorithm 4.

Algorithm 4 Algorithm for solving Case 3.

Input: Set the iteration counter $x = 0$, the convergence tolerance $\epsilon > 0$, initial feasible solution $\{\mathbf{W}_k^{(0)}\}, \mathbf{R}_t^{(0)}, \gamma^{(0)}, \boldsymbol{\rho}^{(0)}$. Initialize C_0, C_1, λ and the objective function value $\chi^{(0)}(\mathbf{Z}) = 0$.
while $\frac{\chi_{\text{sum}}^{(x+1)}(\mathbf{Z}) - \chi_{\text{sum}}^{(x)}(\mathbf{Z})}{\chi_{\text{sum}}^{(x+1)}(\mathbf{Z})} \geq \epsilon$ **do**
  Do: Solve (P3.2) and obtain solution as $\{\mathbf{W}_k^{(x+1)}\}, \mathbf{R}_t^{(x+1)}$.
  Update: $\gamma^{(x+1)}$ using Equation (28).
  Until: There is no increment in the objective.
  Solve Equation (23) and obtain the optimal solution as $\mathbf{u}^{(x+1)}$.
  Calculate the objective function value $\chi_{\text{sum}}^{(x+1)}(\mathbf{Z})$.
  Obtain $\{\mathbf{w}_k^{(x+1)}\}$ by recovering a rank-one solution via GR and update $\{\mathbf{W}_k^{(x+1)}\}$.
  Set $x \leftarrow x + 1$;
end while
Output: $\mathbf{Z}^* = \{\{\mathbf{w}_k^*\}, \mathbf{R}_t^*, \mathbf{u}^*\}$.

7.6. Simulation Results and Discussion

This section confirms the effectiveness of the proposed designs by analyzing the simulation results for three specific case studies. Unless otherwise specified, the simulation setup is as follows: $M = 129$ BS antennas [12], uniformly distributed in lines at half-wavelength spacing; the carrier frequency is set to 28 GHz [12]; and there are $K = 4$ [12] users assumed to be randomly located within the Rayleigh distance from the BS. Additionally, a single target is positioned within the Rayleigh distance from the BS and at an angle of $-60°$. The sensing threshold is $\eta = -10$ dB, the user's QoS threshold is $\gamma_{th} = 10$ dB, and the noise power is $\sigma^2 = -80$ dBm.

7.6.1. Sensing Performance Maximization (Case 1)

Figure 6 plots the achieved sensing SNR as a function of the transmit power at the BS. To comprehensively compare the performance of the proposed system design, the following three baseline schemes are considered. First, the FF performance is provided along with the NF performance for better insight. In the "communication only" scheme, the dedicated signal for sensing is eliminated, allowing the communication signal to perform target sensing. In the "sensing only" scheme, the BS performs sensing exclusively by sending a sensing signal and listening to its echo without serving DL users.

This figure indicates that the NF-proposed scheme performs better than the FF counterpart by approximately 41 %. Indeed, this is owing to the NF beam-focusing effect with ELAA at BS and a high operating frequency, which results in a narrow and concentrated beam. However, there is no noticeable difference between the proposed scheme and "communication only" for both NF and FF, as most of the power is assigned to a communication system that can also support the target sensing. Conversely, in "sensing only", as BS only performs target sensing tasks, BS adjusts its single beam toward the target. Therefore, in most regimes, specifically higher transmit power, better performance is achieved than the proposed scheme, in which BS has dual functionalities, i.e., communication and sensing.

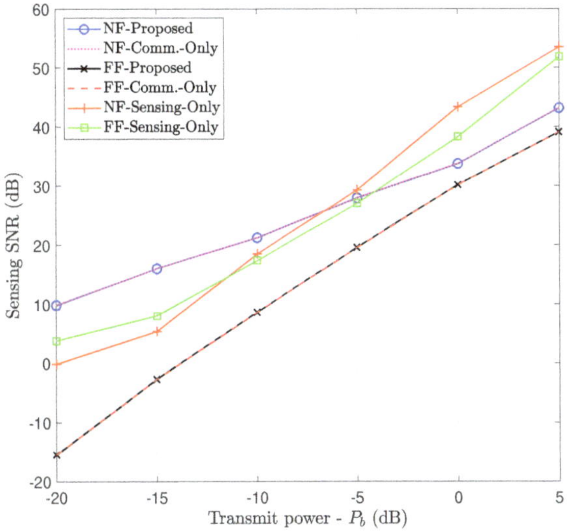

Figure 6. Case 1: Achieved sensing SNR versus BS transmit power.

Figure 7 illustrates the achieved beam pattern gain for case study 1, utilizing both NF and FF channel models. In both scenarios, the beams are effectively aligned towards the target angle, aligning to maximize the target's SNR. However, compared to the FF counterpart, the NF configuration demonstrates a higher beam pattern gain with a narrow beam, i.e., the NF configuration yields a pencil beam, resulting in a narrower and more focused beam due to the beam-focusing effect in NF propagation. Consequently, this enhanced beam gain in the NF scenario can effectively meet the QoS requirements for the user's SINR.

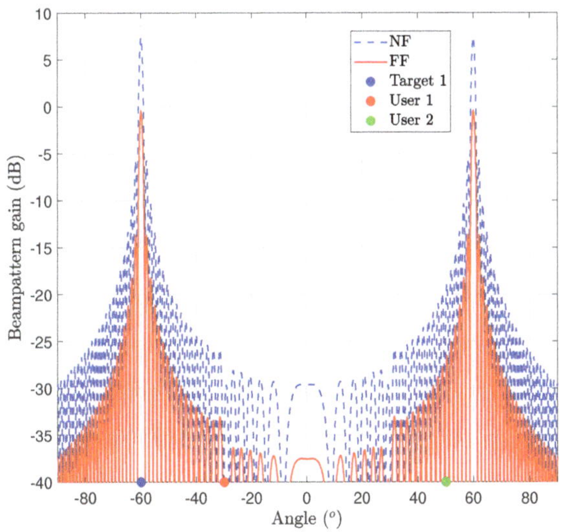

Figure 7. Case 1: Beam pattern gain.

7.6.2. Communication Performance Maximization (Case 2)

Figure 8 illustrates the achieved sum rate for users by solving Algorithm 3. For comparison, this figure also displays the performance of the FF and communication-only schemes, where no dedicated signal is used for target sensing in the latter. As demonstrated, the NF proposed scheme outperforms the FF scheme by 21 % regarding the users' achieved sum rate. This highlights the advantage of focused and narrow beams in NF, effectively reducing user interference. Conversely, the NF communication-only scheme performs slightly better or the same as proposed. However, as shown in the figure, the FF proposed scheme performs marginally better than its communication-only counterpart.

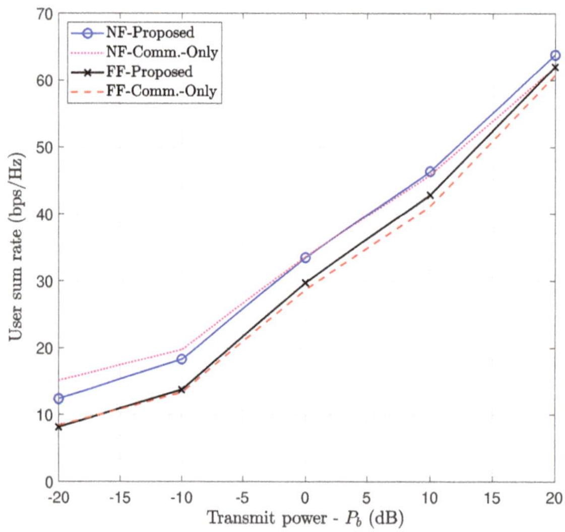

Figure 8. Case 2: Achieved user sum rate versus BS transmit power.

Figure 9 illustrates the beam pattern gain for Case 2 in a two-user scenario. The beam pattern aligns with the users' angles to maximize the total user rate. However, the NF gain outperforms the FF gain. Consistent with the findings in Figure 8, the communication-only scheme exhibits lower beam gain because removing the sensing signal causes power to spread toward the users' beamformers, potentially increasing inter-user interference.

7.6.3. Joint Sensing and Performance Optimization (Case 3)

Figure 10 shows the achieved sum user and sensing rate when both sensing and communication have the same priority, i.e., $\rho = 0.5$. As validated in the previous cases, NF performs much better than the FF, by a magnitude of 52% on average. In the high transmit power regime, the communication-only scheme works better, to an approximate degree of 13 % in the NF scheme and 12 % in FF.

Figure 11 illustrates the achieved beam pattern gain for case study 3. The beam is clearly aligned towards both the target and user angles as the objective in this case prioritizes both sensing and communication. Comparatively, the communication-only scheme exhibits slightly better beam gain for both NF and FF. Nonetheless, as observed in the previous case, NF outperforms FF regarding beam gain and alignment toward the users and target.

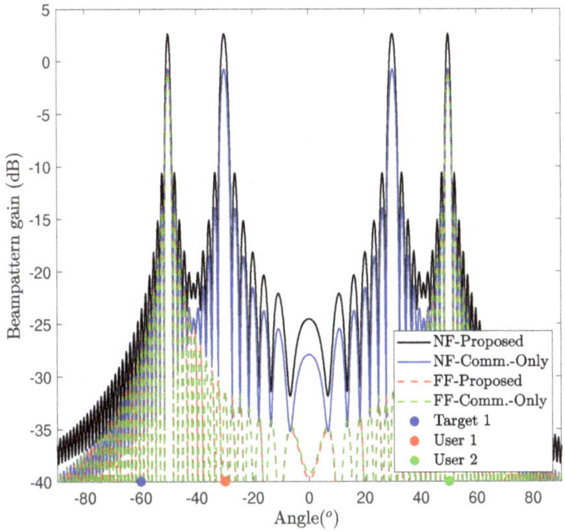

Figure 9. Case 2: Beam pattern gain.

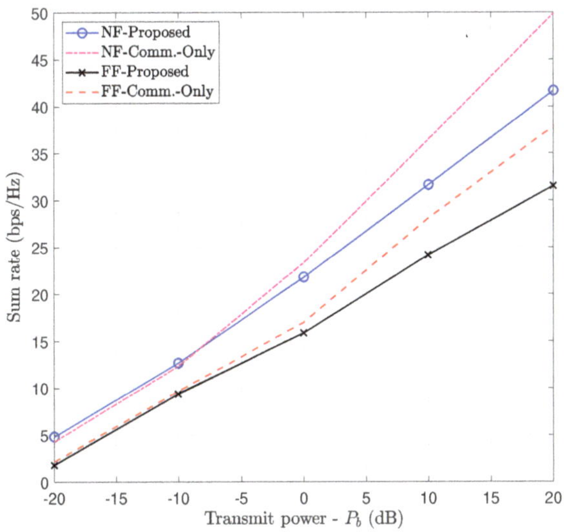

Figure 10. Case 3: Achieved user sum and sensing rate versus BS transmit power.

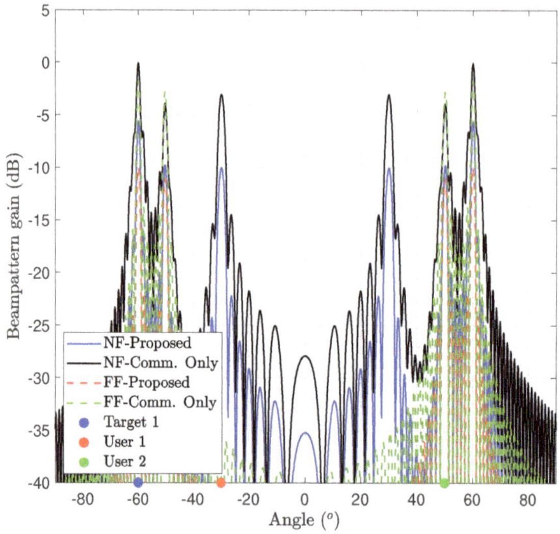

Figure 11. Case 3: Beam pattern gain.

8. Future Research Directions

This section addresses open issues, trends, and opportunities for widespread implementation of future NF-ISAC networks.

8.1. NF-FF Distance Improvement

The boundary between the NF and FF is not strictly confined. The current phase error-based (i.e., Rayleigh distance) and channel gain error-based approaches fail to encompass all performance metrics comprehensively [78]. On the other hand, when the transmission distance is less than the threshold distance, i.e., the NF–FF boundary, the NF distance is a valuable reference and can be a deciding factor in using NF intelligent spectrum access and control approaches. Consequently, establishing an adequate NF–FF border that considers channel characteristics, signal propagation properties, and various performance criteria is essential to enhancing the overall effectiveness of NF-ISAC.

8.2. Accurate NF Channel Models

The NF channel models must be more precise in representing the actual NF EM phenomena [79]. Some existing models frequently treat the NF as equal to the FF counterpart, only accounting for spherical propagation curvature, ignoring the range-dependent amplitude of the received signal, and focusing entirely on phase considerations. Furthermore, they overlook critical NF signal source characteristics such as transmit antenna type, size, and orientation. Consequently, these oversights might considerably influence the signal received by the array [79]. It is thus critical to determine if existing mathematical models effectively describe authentic NF features and to investigate innovative techniques that better capture the intricacies of real-world NF EM behaviors. This could lead to more accurate and reliable results in practical applications.

8.3. Channel Estimation

Accurate channel estimation benefits communication and sensing by enabling advanced signal processing, improving spectrum efficiency, increasing localization accuracy, and enhancing reliability [80,81]. In contrast to FF, NF channel estimation offers improved angular resolution and higher channel diversity, allowing more precise and robust estimates. Also, the impacts of multi-path propagation are less prominent than those of FF. This reduced multi-path interference simplifies channel estimation, resulting in more accurate and reliable estimations. However, the structural changes in EM waves caused by ELAAs render conventional channel estimation methods inadequate for NF, prompting the development of revised and unique channel estimation strategies [82,83].

8.4. Signal Processing and Low-Complexity Beam-Focusing Designs

Direction-of-arrival (DoA) estimation and beam-focusing (beamforming) are two of the most critical fundamental ISAC signal-processing tasks. DoA estimation determines the angles at which signals arrive at the antenna array. In contrast, beamforming focuses the transmitted and received signals to a specified location while suppressing interference from other directions. Additionally, precise DoA estimation benefits from accurate beamforming and vice versa. With many antenna elements, the scale of beamforming and NF-ISAC optimization challenges becomes significantly large. Standard FF DoA estimation and beamforming designs cannot be directly adapted to NF because they do not account for complicated spatial variations, such as varying signal strengths and phase shifts across the antenna array. Furthermore, FF beams become divergent and wider at close ranges, resulting in higher user interference and angle estimation errors.

8.5. Multiple Access

An ELAA introduces an additional spatial resolution based on distance and angle. This and the NF beam-focusing effect can improve the spatial resolution of traditional space-division multiple access (SDMA). Thus, the spatial selectivity of SDMA becomes even more prominent in the NF, effectively exploiting spatial resources for simultaneous data transmission to multiple users [84]. However, it also poses challenges in beam steering and interference management that must be addressed by specifically designed multiple access techniques, such as location division multiple access (LDMA) [84].

8.6. ML for NF-ISAC

Machine learning (ML) is a branch of artificial intelligence (AI) that focuses on developing algorithms and statistical models that enable computers to learn and make predictions or decisions based on data. Instead of being explicitly programmed to perform a task, ML algorithms improve their performance through experience and by identifying patterns in data. ML techniques have recently been used in channel estimation and other communication tasks [80,85,86]. ML-based techniques are capable of handling a variety of challenges with NF-ISAC systems. For example, conventional hybrid/digital beamforming techniques, such as zero-forcing with ELAAs, result in excessively high signal processing complexity due to the channel inversion operation of large channel matrices [80]. One solution is to develop tailored deep neural networks that learn effective beamforming using unsupervised deep learning and an appropriate loss function. On the other hand, for NF sensing, ML algorithms can be used to develop NF predictive beam tracking using only a few echo signals [13]. This can significantly increase beam tracking efficiency and reliability while lowering design complexity. Furthermore, ML approaches can be used for NF gesture recognition and motion prediction applications, which rely on NF-received signal features for estimation/recognition.

8.7. Integration of NF-ISAC with Other Technologies

Integrating NF-ISAC with other technologies, such as cell-free (CF) architecture, backscatter communication (BackCom), RISs, and others, is essential for unlocking its full potential.

8.7.1. CF Architecture

CF networks utilize many distributed access points (APs) over a coverage area, significantly mitigating inter-cell interference and reducing transmission distances. For example, cell-free massive MIMO systems can deliver approximately a five-fold boost in 95 %-likely per-user throughput compared to small-cell systems under uncorrelated shadow fading conditions, and a ten-fold improvement when shadow fading is correlated [87,88]. This design approach thus enhances the coverage probability and provides macro-diversity gains against large-scale fading, improving overall network performance by up to 30% [87–93]. These characteristics of CF systems can effectively mitigate severe path loss effects in NF-ISAC systems.

Moreover, CF-ISAC can overcome the drawbacks of single AP or BS-based conventional ISAC systems, such as obstructed observation angles in complex urban settings, which can reduce target detection accuracy [94–97]. In addition, CF-ISAC systems improve communication and sensing performance by coordinating distributed APs to execute both tasks. Thus, the diversity advantages of uncorrelated sensing observations at distributed APs/receivers will increase sensing performance. For instance, this coordination leads to an estimated 15–20 % improvement in communication reliability and a 30 % enhancement in sensing accuracy [94–97]. Despite its potential, only a few studies have characterized the performance of CF architecture in traditional ISAC systems thus far [94–97]. On the other hand, integrating CF and NF-ISAC will be an intriguing research opportunity for improving network performance.

CF network performance can be significantly degraded by impairments such as hardware imperfections in transceivers and SIC imperfections. One critical issue is the non-linearity of non-ideal RF front-ends [98]. These impairments impact reliability and security. Incorporating these impairments into NF-ISAC system models is crucial for analytical results to provide valuable insights into real-world conditions.

8.7.2. BackCom

Due to the simple RF design, backscatter tags are cost-effective and have ultra-low energy requirements (a few nW-µW) [99–103]. They are typically equipped with sensors and perform passive reflection or backscattering rather than active transmission to convey data. Thus, they will facilitate future ambient-powered or battery-free (i.e., EH enabled) IoT networks and applications such as smart homes, cities, autonomous vehicles, industrial IoT, healthcare, and so forth [99–102]. These applications necessitate low-power communication and exceptional sensing capability. Thus, incorporating sensing into BackCom networks, also known as integrated sensing and BackCom (ISABC), enables IoT networks to extract crucial sensing/state information such as range, velocity, and angle for tracking, as well as environment learning and mapping [104]. EH is critical for BackCom, and non-linear models [105] could be used for design purposes.

On the other hand, as the backscatter devices are low-power EH devices with limited processing capabilities, their operational range and data rates are limited, i.e., the range is typically less than 6 m and the data rate is less than 1 bps/Hz [99–102]. Hence, many of these systems operate in the NF region of the transmitter/RF source [106–109]. However, except for a few works [106–109], the majority of the current literature focuses on FF BackCom systems. Nonetheless, almost all of these studies are measurement-based field trials designed to evaluate the NF impact of BackCom [106–109]. Only ref. [106] briefly explores RF sensing in the NF BC system, focusing on the interaction between NF signals and a shape-changing backscatter object, especially one with a low permittivity contrast to the background media. With minimal research on NF BC and no research on NF ISABC, there are plenty of opportunities to explore the feasibility of the NF BC and NF ISABC systems. Theoretical frameworks, in particular, must be established, since backscatter devices with sensing capabilities will enable identification, monitoring, and tracking in a wide range of future applications.

8.7.3. RIS

Due to RIS's energy and hardware efficiency in manipulating the wireless environment, RIS-integrated ISAC can unlock numerous new applications. Deploying RIS near the transmitter or receiver enhances passive beamforming gains for either interference management in ISAC or enhancing the performance of both communication and sensing [110]. For example, RIS can achieve over a 40 % improvement in an ISAC system as compared to a no-RIS system [111]. Additionally, RISs operating at high frequencies can expand the NF region. Consequently, NF RIS-integrated ISAC is expected to become increasingly common [110].

Furthermore, the conventional RIS can only reflect signals that necessitate the users and the target to be at one side of BS and RIS [110]. To address this limitation, simultaneously transmitting and reflecting reconfigurable intelligent surfaces (STAR-RIS) have been developed, offering 360-degree coverage [68]. Consequently, STAR-RIS enhances the adaptability of network deployment with its full spatial coverage while optimizing signal propagation for both sensing targets and communication users. However, beamforming design and channel estimation are challenging in such systems, specifically as the number of RIS/STAR-RIS elements or BS antenna increases. Therefore, further in-depth research is needed to address these challenges.

9. Conclusions

In conclusion, the shift towards higher bandwidths and frequencies presents exciting opportunities for technologies like ISAC, facilitating applications that demand high accuracy and resolution. However, high-frequency systems combined with massive antenna arrays enlarge the NF region. Therefore, this work provides a comprehensive review and case study covering various aspects of NF and ISAC and their integration. Three different NF-ISAC case studies bolster this effort. Additionally, the existing challenges and open issues in this area are outlined. Finally, a broad perspective on future research directions is provided.

Author Contributions: Conceptualization, A.H. and D.G.; methodology, A.H. and D.G.; software, A.H.; validation, A.H. and D.G.; formal analysis, A.H. and D.G.; investigation, A.H. and D.G.; resources, A.H. and D.G.; data curation, A.H. and D.G.; writing—original draft preparation, A.H. and D.G.; writing—review and editing, C.T.; visualization, A.H. and D.G.; supervision, C.T.; project administration, D.G.; funding acquisition, C.T. All authors have read and agreed to the published version of the manuscript.

Funding: This research received no external funding.

Institutional Review Board Statement: Not applicable.

Data Availability Statement: No new data were created or analyzed in this study. Data sharing is not applicable to this article.

Conflicts of Interest: The authors declare no conflicts of interest.

Abbreviations

The following abbreviations are used in this manuscript:

4G	Fourth Generation
5G	Fifth Generation
6G	Sixth Generation
AI	Artificial Intelligence
AoA	Angle of Arival
AWGN	Additive White Gaussian Noise
BS	Base Station
DoF	Degree of Freedom
EH	Energy Harvesting
ELAA	Extremely Large-Scale Antenna Array
EM	Electromagnetic
eMBB	Enhanced Mobile Broadband
FF	Far-Field

GHz	Gigaherts
IoT	Internet of Things
ISAC	Integrated Sensing and Communication
LoS	Line-of-Sight
MIMO	Multiple-Input Multiple-Output
mMTC	Massive Machine Type Communication
mmWave	Millimeter Wave
NF	Near-Field
NOMA	Non-orthogonal Multiple Access
NUSW	Non-uniform Spherical Wave
RF	Radio Frequency
SNR	Signal-to-Noise-Ratio
SPD	Spatially discrete
THz	Terahertz
UAV	Unmanned Aerial Vehicle
ULA	Uniform Linear Array
UPA	Uniform Planar Array
UPW	Uniform Planar Wave
URLLC	Ultra Low-Latency Communication
USW	Uniform Spherical Wave
V2X	Vehicle-to-Everything
WPT	Wireless Power Transfer
XR	Extended Reality

References

1. Wang, C.X.; You, X.; Gao, X.; Zhu, X.; Li, Z.; Zhang, C.; Wang, H.; Huang, Y.; Chen, Y.; Haas, H.; et al. On the Road to 6G: Visions, Requirements, Key Technologies, and Testbeds. *IEEE Commun. Surv. Tuts.* **2023**, *25*, 905–974. [CrossRef]
2. Liu, Y.; Wang, Z.; Xu, J.; Ouyang, C.; Mu, X.; Schober, R. Near-Field Communications: A Tutorial Review. *IEEE Open J. Commun. Soc.* **2023**, *4*, 1999–2049. [CrossRef]
3. Weissberger, A. Framework and Overall Objectives of the Future Development of IMT for 2030 and Beyond. 2023. Available online: https://techblog.comsoc.org/tag/imt-framework-for-2030-and-beyond/ (accessed on 4 August 2024).
4. Serghiou, D.; Khalily, M.; Brown, T.W.C.; Tafazolli, R. Terahertz Channel Propagation Phenomena, Measurement Techniques and Modeling for 6G Wireless Communication Applications: A Survey, Open Challenges and Future Research Directions. *IEEE Commun. Surv. Tuts.* **2022**, *24*, 1957–1996. [CrossRef]
5. Liu, A.; Huang, Z.; Li, M.; Wan, Y.; Li, W.; Han, T.X.; Liu, C.; Du, R.; Tan, D.K.P.; Lu, J.; et al. A Survey on Fundamental Limits of Integrated Sensing and Communication. *IEEE Commun. Surv. Tutor.* **2022**, *24*, 994–1034. [CrossRef]
6. Wang, J.; Varshney, N.; Gentile, C.; Blandino, S.; Chuang, J.; Golmie, N. Integrated Sensing and Communication: Enabling Techniques, Applications, Tools and Data Sets, Standardization, and Future Directions. *IEEE Internet Things J.* **2022**, *9*, 23416–23440. [CrossRef]
7. Zhang, J.A.; Rahman, M.L.; Wu, K.; Huang, X.; Guo, Y.J.; Chen, S.; Yuan, J. Enabling Joint Communication and Radar Sensing in Mobile Networks—A Survey. *IEEE Commun. Surv. Tuts.* **2022**, *24*, 306–345. [CrossRef]
8. Björnson, E.; Sanguinetti, L.; Wymeersch, H.; Hoydis, J.; Marzetta, T.L. Massive MIMO is a Reality—What is Next?: Five Promising Research Directions for Antenna Arrays. *Digit. Signal Process.* **2019**, *94*, 3–20. [CrossRef]
9. Lu, H.; Zeng, Y.; You, C.; Han, Y.; Zhang, J.; Wang, Z.; Dong, Z.; Jin, S.; Wang, C.X.; Jiang, T.; et al. A Tutorial On Near-Field XL-MIMO Communications towards 6G. *IEEE Commun. Surv. Tutor.* 2024, *early access*. [CrossRef]
10. Liu, F.; Cui, Y.; Masouros, C.; Xu, J.; Han, T.X.; Eldar, Y.C.; Buzzi, S. Integrated Sensing and Communications: Toward Dual-Functional Wireless Networks for 6G and Beyond. *IEEE J. Sel. Areas Commun.* **2022**, *40*, 1728–1767. [CrossRef]
11. Akyildiz, I.F.; Jornet, J.M.; Han, C. TeraNets: Ultra-Broadband Communication Networks in the Terahertz Band. *IEEE Wirel. Commun.* **2014**, *21*, 130–135. [CrossRef]
12. Wang, Z.; Mu, X.; Liu, Y. Near-Field Integrated Sensing and Communications. *IEEE Commun. Lett.* **2023**, *27*, 2048–2052. [CrossRef]
13. Cong, J.; You, C.; Li, J.; Chen, L.; Zheng, B.; Liu, Y.; Wu, W.; Gong, Y.; Jin, S.; Zhang, R. Near-Field Integrated Sensing and Communication: Opportunities and Challenges. *arXiv* **2023**, arXiv:2310.01342. [CrossRef]
14. Qu, K.; Guo, S.; Ye, J.; Saeed, N. Near-Field Integrated Sensing and Communications: Unlocking Potentials and Shaping the Future. *arXiv* **2023**, arXiv:2308.00252. [CrossRef]
15. Liu, F.; Masouros, C.; Eldar, Y. (Eds.) *Integrated Sensing and Communications*; Springer: Singapore, 2023. [CrossRef]
16. Nemati, M.; Kim, Y.H.; Choi, J. Toward Joint Radar, Communication, Computation, Localization, and Sensing in IoT. *IEEE Access* **2022**, *10*, 11772–11788. [CrossRef]
17. Wang, Z.; Mu, X.; Liu, Y. Rethinking Integrated Sensing and Communication: When Near Field Meets Wideband. *arXiv* **2024**, arXiv:2311.11416. [CrossRef]

18. Selvan, K.T.; Janaswamy, R. Fraunhofer and Fresnel Distances: Unified Derivation for Aperture Antennas. *IEEE Antennas Propag. Mag.* **2017**, *59*, 12–15. [CrossRef]
19. Heath, R.W., Jr.; Lozano, A. *Foundations of MIMO Communication*; Cambridge University Press: Cambridge, UK, 2018. [CrossRef]
20. de Lima, C.H.M.; Belot, D.; Berkvens, R.; Bourdoux, A.; Dardari, D.; Guillaud, M.; Isomursu, M.; Lohan, E.S.; Miao, Y.; Barreto, A.N.; et al. Convergent Communication, Sensing and Localization in 6G Systems: An Overview of Technologies, Opportunities and Challenges. *IEEE Access* **2021**, *9*, 26902–26925. [CrossRef]
21. 5G Channel Model for Bands Up to100 GHz; Technical Report. 2016. Available online: http://www.5gworkshops.com/2016/5G_Channel_Model_for_bands_up_to100_GHz(2015-12-6).pdf (accessed on 10 July 2024).
22. Xiong, Y.; Liu, F.; Cui, Y.; Yuan, W.; Han, T.X. Flowing the Information from Shannon to Fisher: Towards the Fundamental Tradeoff in ISAC. In Proceedings of the IGLOBECOM 2022—2022 IEEE Global Communications Conference, Rio de Janeiro, Brazil, 4–8 December 2022; pp. 5601–5606. [CrossRef]
23. Wei, Z.; Xu, R.; Feng, Z.; Wu, H.; Zhang, N.; Jiang, W.; Yang, X. Symbol-Level Integrated Sensing and Communication Enabled Multiple Base Stations Cooperative Sensing. *IEEE Trans. Veh. Technol.* **2024**, *73*, 724–738. [CrossRef]
24. Ali, A.; Gonzalez-Prelcic, N.; Heath, R.W.; Ghosh, A. Leveraging Sensing at the Infrastructure for mmWave Communication. *IEEE Commun. Mag.* **2020**, *58*, 84–89. [CrossRef]
25. Zhou, W.; Zhang, R.; Chen, G.; Wu, W. Integrated Sensing and Communication Waveform Design: A Survey. *IEEE Open J. Commun. Soc.* **2022**, *3*, 1930–1949. [CrossRef]
26. Zheng, K.; Zheng, Q.; Yang, H.; Zhao, L.; Hou, L.; Chatzimisios, P. Reliable and Efficient Autonomous Driving: The Need for Heterogeneous Vehicular Networks. *IEEE Commun. Mag.* **2015**, *53*, 72–79. [CrossRef]
27. Chang, B.; Tang, W.; Yan, X.; Tong, X.; Chen, Z. Integrated Scheduling of Sensing, Communication, and Control for mmWave/THz Communications in Cellular Connected UAV Networks. *IEEE J. Sel. Areas Commun.* **2022**, *40*, 2103–2113. [CrossRef]
28. Stoica, P.; Li, J.; Xie, Y. On Probing Signal Design For MIMO Radar. *IEEE Trans. Signal Process.* **2007**, *55*, 4151–4161. [CrossRef]
29. Xiao, Z.; Zeng, Y. Waveform Design and Performance Analysis for Full-Duplex Integrated Sensing and Communication. *IEEE J. Sel. Areas Commun.* **2022**, *40*, 1823–1837. [CrossRef]
30. Tang, B.; Li, J. Spectrally Constrained MIMO Radar Waveform Design Based on Mutual Information. *IEEE Trans. Signal Process.* **2019**, *67*, 821–834. [CrossRef]
31. Zhang, R.; Shim, B.; Yuan, W.; Renzo, M.D.; Dang, X.; Wu, W. Integrated Sensing and Communication Waveform Design with Sparse Vector Coding: Low Sidelobes and Ultra Reliability. *IEEE Trans. Veh. Technol.* **2022**, *71*, 4489–4494. [CrossRef]
32. Ahmadipour, M.; Wigger, M.; Kobayashi, M. Coding for Sensing: An Improved Scheme for Integrated Sensing and Communication over MACs. In Proceedings of the 2022 IEEE International Symposium on Information Theory (ISIT), Espoo, Finland, 26 June–1 July 2022; pp. 3025–3030. [CrossRef]
33. Zeng, Y.; Xiang, L.; Yang, K. A Polarisation Coding Scheme Based on An Integrated Sensing and Communication System. *J. Inf. Intell.* **2024**, *2*, 289–301. [CrossRef]
34. Xu, T.; Liu, F.; Masouros, C.; Darwazeh, I. An Experimental Proof of Concept for Integrated Sensing and Communications Waveform Design. *IEEE Open J. Commun. Soc.* **2022**, *3*, 1643–1655. [CrossRef]
35. Temiz, M.; Horne, C.; Peters, N.J.; Ritchie, M.A.; Masouros, C. An Experimental Study of Radar-Centric Transmission for Integrated Sensing and Communications. *IEEE Trans. Microw. Theory Tech.* **2023**, *71*, 3203–3216. [CrossRef]
36. Ji, K.; Zhang, Q.; Wei, Z.; Feng, Z.; Zhang, P. Networking Based ISAC Hardware Testbed and Performance Evaluation. *IEEE Commun. Mag.* **2023**, *61*, 76–82. [CrossRef]
37. Zhang, J.A.; Liu, F.; Masouros, C.; Heath, R.W.; Feng, Z.; Zheng, L.; Petropulu, A. An Overview of Signal Processing Techniques for Joint Communication and Radar Sensing. *IEEE J. Sel. Top. Signal Process.* **2021**, *15*, 1295–1315. [CrossRef]
38. Ma, D.; Shlezinger, N.; Huang, T.; Liu, Y.; Eldar, Y.C. Joint Radar-Communication Strategies for Autonomous Vehicles: Combining Two Key Automotive Technologies. *IEEE Signal Process. Mag.* **2020**, *37*, 85–97. [CrossRef]
39. Ma, D.; Shlezinger, N.; Huang, T.; Liu, Y.; Eldar, Y.C. FRaC: FMCW-Based Joint Radar-Communications System via Index Modulation. *IEEE J. Sel. Top. Signal Process.* **2021**, *15*, 1348–1364. [CrossRef]
40. Chen, L.; Wang, Z.; Du, Y.; Chen, Y.; Yu, F.R. Generalized Transceiver Beamforming for DFRC with MIMO Radar and MU-MIMO Communication. *IEEE J. Sel. Areas Commun.* **2022**, *40*, 1795–1808. [CrossRef]
41. Liu, X.; Huang, T.; Shlezinger, N.; Liu, Y.; Zhou, J.; Eldar, Y.C. Joint Transmit Beamforming for Multiuser MIMO Communications and MIMO Radar. *IEEE Trans. Signal Process.* **2020**, *68*, 3929–3944. [CrossRef]
42. Ma, Y.; Zhou, G.; Wang, S. WiFi Sensing with Channel State Information: A Survey. *ACM Comput. Surv.* **2019**, *52*, 1–36. [CrossRef]
43. del Peral-Rosado, J.A.; Raulefs, R.; López-Salcedo, J.A.; Seco-Granados, G. Survey of Cellular Mobile Radio Localization Methods: From 1G to 5G. *IEEE Commun. Surv. Tutor.* **2018**, *20*, 1124–1148. [CrossRef]
44. Zeng, T.; Semiari, O.; Saad, W.; Bennis, M. Joint Communication and Control for Wireless Autonomous Vehicular Platoon Systems. *IEEE Trans. Commun.* **2019**, *67*, 7907–7922. [CrossRef]
45. Milanes, V.; Villagra, J.; Godoy, J.; Simo, J.; Perez, J.; Onieva, E. An Intelligent V2I-Based Traffic Management System. *IEEE Trans. Intell. Transp. Syst.* **2012**, *13*, 49–58. [CrossRef]
46. Popovski, P.; Stefanovic, C.; Nielsen, J.J.; de Carvalho, E.; Angjelichinoski, M.; Trillingsgaard, K.F.; Bana, A.S. Wireless Access in Ultra-Reliable Low-Latency Communication (URLLC). *IEEE Trans. Commun.* **2019**, *67*, 5783–5801. [CrossRef]
47. Messer, H.; Zinevich, A.; Alpert, P. Environmental monitoring by wireless communication networks. *Science* **2006**, *312*, 713.

48. Wikstrom, G.; Persson, P.; Parkvall, S.; Mildh, G.; Dahlman, E.; Balakrishnan, B.; Ohlen, P.; Trojer, E. *White Paper: 6G—Connecting a Cyber-Physical World*; Technical Report; Ericsson: Stockholm, Sweden, 2022.
49. Cordeiro, C. *Next-Generation Wi-Fi: Wi-Fi 7 and Beyond*; Technical Report; Intel Corporation: Santa Clara, CA, USA, 2020.
50. Pin Tan, D.K.; He, J.; Li, Y.; Bayesteh, A.; Chen, Y.; Zhu, P.; Tong, W. Integrated Sensing and Communication in 6G: Motivations, Use Cases, Requirements, Challenges and Future Directions. In Proceedings of the 2021 1st IEEE International Online Symposium on Joint Communications & Sensing (JC&S), Dresden, Germany, 23–24 February 2021; pp. 1–6. [CrossRef]
51. Waring, J. Huawei Readies 5.5G as Bosses Push Evolution. 2020. Available online: https://www.mobileworldlive.com/home-banner/huawei-readies-5-5g-as-bosses-push-evolution/ (accessed on 4 August 2024).
52. Alloulah, M.; Huang, H. Future Millimeter-Wave Indoor Systems: A Blueprint for Joint Communication and Sensing. *Computer* **2019**, *52*, 16–24. [CrossRef]
53. Meneghello, F.; Chen, C.; Cordeiro, C.; Restuccia, F. Toward Integrated Sensing and Communications in IEEE 802.11bf Wi-Fi Networks. *IEEE Commun. Mag.* **2023**, *61*, 128–133. [CrossRef]
54. Chepuri, S.P.; Shlezinger, N.; Liu, F.; Alexandropoulos, G.C.; Buzzi, S.; Eldar, Y.C. Integrated Sensing and Communications with Reconfigurable Intelligent Surfaces: From Signal Modeling to Processing. *IEEE Signal Process. Mag.* **2023**, *40*, 41–62. [CrossRef]
55. Cui, M.; Wu, Z.; Lu, Y.; Wei, X.; Dai, L. Near-Field MIMO Communications for 6G: Fundamentals, Challenges, Potentials, and Future Directions. *IEEE Commun. Mag.* **2023**, *61*, 40–46. [CrossRef]
56. Cui, M.; Dai, L.; Wang, Z.; Zhou, S.; Ge, N. Near-Field Rainbow: Wideband Beam Training for XL-MIMO. *IEEE Trans. Wirel. Commun.* **2023**, *22*, 3899–3912. [CrossRef]
57. Luo, H.; Gao, F.; Yuan, W.; Zhang, S. Beam Squint Assisted User Localization in Near-Field Integrated Sensing and Communications Systems. *IEEE Trans. Wirel. Commun.* **2023**, *23*, 4504–4517. [CrossRef]
58. Sakhnini, A.; De Bast, S.; Guenach, M.; Bourdoux, A.; Sahli, H.; Pollin, S. Near-Field Coherent Radar Sensing Using a Massive MIMO Communication Testbed. *IEEE Trans. Wirel. Commun.* **2022**, *21*, 6256–6270. [CrossRef]
59. Gavras, I.; Islam, M.A.; Smida, B.; Alexandropoulos, G.C. Full Duplex Holographic MIMO for Near-Field Integrated Sensing and Communications. In Proceedings of the 2023 31st European Signal Processing Conference (EUSIPCO), Helsinki, Finland, 4–8 September 2023; pp. 700–704. [CrossRef]
60. Qu, K.; Guo, S.; Saeed, N. Near-Field Integrated Sensing and Communication: Performance Analysis and Beamforming Design. *arXiv* **2023**, arXiv:2308.06455. [CrossRef]
61. Galappaththige, D.; Zargari, S.; Tellambura, C.; Li, G.Y. Near-Field ISAC: Beamforming for Multi-Target Detection. *IEEE Wirel. Commun. Lett.* **2024**, *13*, 1938–1942. [CrossRef]
62. Jawad, A.M.; Nordin, R.; Gharghan, S.K.; Jawad, H.M.; Ismail, M. Opportunities and Challenges for Near-Field Wireless Power Transfer: A Review. *Energies* **2017**, *10*, 1022. [CrossRef]
63. Sun, P.; Dai, H.; Wang, B. Optimal Transmit Beamforming for Near-Field Integrated Sensing and Wireless Power Transfer systems. *Intell. Converg. Netw.* **2022**, *3*, 378–386. [CrossRef]
64. Wang, Z.; Mu, X.; Liu, Y.; Xu, X.; Zhang, P. NOMA-Aided Joint Communication, Sensing, and Multi-Tier Computing Systems. *IEEE J. Sel. Areas Commun.* **2023**, *41*, 574–588. [CrossRef]
65. Li, Z.; Wan, Z.; Ying, K.; Mei, Y.; Ke, M.; Gao, Z. Reconfigurable Intelligent Surface Assisted Localization Over Near-Field Beam Squint Effect. In Proceedings of the 2022 International Symposium on Wireless Communication Systems (ISWCS), Hangzhou, China, 19–22 October 2022; pp. 1–6. [CrossRef]
66. Jiang, Y.; Gao, F.; Jian, M.; Zhang, S.; Zhang, W. Reconfigurable Intelligent Surface for Near Field Communications: Beamforming and Sensing. *IEEE Trans. Wirel. Commun.* **2023**, *22*, 3447–3459. [CrossRef]
67. He, Z.; Xu, W.; Shen, H.; Ng, D.W.K.; Eldar, Y.C.; You, X. Full-Duplex Communication for ISAC: Joint Beamforming and Power Optimization. *IEEE J. Sel. Areas Commun.* **2023**, *41*, 2920–2936. [CrossRef]
68. Liu, Z.; Li, X.; Ji, H.; Zhang, H.; Leung, V.C.M. Toward STAR-RIS-Empowered Integrated Sensing and Communications: Joint Active and Passive Beamforming Design. *IEEE Trans. Veh. Technol.* **2023**, *72*, 15991–16005. [CrossRef]
69. Hua, H.; Xu, J.; Han, T.X. Optimal Transmit Beamforming for Integrated Sensing and Communication. *IEEE Trans. Veh. Technol.* **2023**, *72*, 10588–10603. [CrossRef]
70. Goldsmith, A. *Wireless Communications*; Cambridge University Press: Cambridge, UK, 2005.
71. Tse, D.; Viswanath, P. *Fundamentals of Wireless Communication*; Cambridge University Press: Cambridge, UK, 2005.
72. Cui, G.; Li, H.; Rangaswamy, M. MIMO Radar Waveform Design with Constant Modulus and Similarity Constraints. *IEEE Trans. Signal Process.* **2014**, *62*, 343–353. [CrossRef]
73. Boyd, S.; Vandenberghe, L. *Convex Optimization*; Cambridge University Press: Cambridge, UK, 2004.
74. Wu, Q.; Zhang, R. Intelligent Reflecting Surface Enhanced Wireless Network via Joint Active and Passive Beamforming. *IEEE Trans. Wirel. Commun.* **2019**, *18*, 5394–5409. [CrossRef]
75. Hakimi, A.; Zargari, S.; Tellambura, C.; Herath, S. Sum Rate Maximization of MIMO Monostatic Backscatter Networks by Suppressing Residual Self-Interference. *IEEE Trans. Commun.* **2023**, *71*, 512–526. [CrossRef]
76. Shen, K.; Yu, W. Fractional Programming for Communication Systems—Part II: Uplink Scheduling via Matching. *IEEE Trans. Signal Process.* **2018**, *66*, 2631–2644. [CrossRef]
77. Sidiropoulos, N.D.; Davidson, T.N.; Luo, Z. Transmit Beamforming for Physical-Layer Multicasting. *IEEE Trans. Signal Process.* **2006**, *54*, 2239–2251. [CrossRef]

78. Guerra, A.; Guidi, F.; Dardari, D.; Djurić, P.M. Near-Field Tracking with Large Antenna Arrays: Fundamental Limits and Practical Algorithms. *IEEE Trans. Signal Process.* **2021**, *69*, 5723–5738. [CrossRef]
79. Friedlander, B. Localization of Signals in the Near-Field of an Antenna Array. *IEEE Trans. Signal Process.* **2019**, *67*, 3885–3893. [CrossRef]
80. Liu, Y.; Al-Nahhal, I.; Dobre, O.A.; Wang, F. Deep-Learning Channel Estimation for IRS-Assisted Integrated Sensing and Communication System. *IEEE Trans. Veh. Technol.* **2023**, *72*, 6181–6193. [CrossRef]
81. Zhou, L.; Dai, J.; Xu, W.; Chang, C. Joint Target Detection and Channel Estimation for Distributed Massive MIMO ISAC Systems. *IEEE Trans. Cogn. Commun. Netw.* 2024, *early access*. [CrossRef]
82. Han, Y.; Jin, S.; Wen, C.K.; Ma, X. Channel Estimation for Extremely Large-Scale Massive MIMO Systems. *IEEE Wirel. Commun. Lett.* **2020**, *9*, 633–637. [CrossRef]
83. Cui, M.; Dai, L. Channel Estimation for Extremely Large-Scale MIMO: Far-Field or Near-Field? *IEEE Trans. Commun.* **2022**, *70*, 2663–2677. [CrossRef]
84. Wu, Z.; Dai, L. Multiple Access for Near-Field Communications: SDMA or LDMA? *IEEE J. Sel. Areas Commun.* **2023**, *41*, 1918–1935. [CrossRef]
85. Zargari, S.; Hakimi, A.; Tellambura, C.; Maaref, A. Enhancing AmBC Systems with Deep Learning for Joint Channel Estimation and Signal Detection. *IEEE Trans. Commun.* **2024**, *72*, 1716–1731. [CrossRef]
86. Zargari, S.; Galappaththige, D.; Tellambura, C. Refined-Deep Reinforcement Learning for MIMO Bistatic Backscatter Resource Allocation. *arXiv* **2024**, arXiv:2405.14046. [CrossRef]
87. Ngo, H.Q.; Ashikhmin, A.; Yang, H.; Larsson, E.G.; Marzetta, T.L. Cell-Free Massive MIMO Versus Small Cells. *IEEE Trans. Wirel. Commun.* **2017**, *16*, 1834–1850. [CrossRef]
88. Demir, Ö.T.; Björnson, E.; Sanguinetti, L. Foundations of User-Centric Cell-Free Massive MIMO. *Found. Trends Signal Process.* **2021**, *14*, 162–472. [CrossRef]
89. Galappaththige, D.; Baduge, G.A.A. Exploiting Underlay Spectrum Sharing in Cell-Free Massive MIMO Systems. *IEEE Trans. Commun.* **2021**, *69*, 7470–7488. [CrossRef]
90. Nayebi, E.; Ashikhmin, A.; Marzetta, T.L.; Yang, H.; Rao, B.D. Precoding and Power Optimization in Cell-Free Massive MIMO Systems. *IEEE Trans. Wirel. Commun.* **2017**, *16*, 4445–4459. [CrossRef]
91. Ngo, H.Q.; Tran, L.N.; Duong, T.Q.; Matthaiou, M.; Larsson, E.G. On the Total Energy Efficiency of Cell-Free Massive MIMO. *IEEE Trans. Green Commun. Netw.* **2018**, *2*, 25–39. [CrossRef]
92. Galappaththige, D.; Amarasuriya, G. NOMA-Aided Cell-Free Massive MIMO with Underlay Spectrum-Sharing. In Proceedings of the ICC 2020—2020 IEEE International Conference on Communications (ICC), Dublin, Ireland, 7–11 June 2020; pp. 1–6. [CrossRef]
93. Galappaththige, D.; Amarasuriya, G. Cell-Free Massive MIMO with Underlay Spectrum-Sharing. In Proceedings of the ICC 2019—2019 IEEE International Conference on Communications (ICC), Shanghai, China, 20–24 May 2019; pp. 1–7. [CrossRef]
94. Mao, W.; Lu, Y.; Liu, J.; Ai, B.; Zhong, Z.; Ding, Z. Beamforming Design in Cell-Free Massive MIMO Integrated Sensing and Communication Systems. In Proceedings of the GLOBECOM 2023—2023 IEEE Global Communications Conference, Kuala Lumpur, Malaysia, 4–8 December 2023; pp. 546–551. [CrossRef]
95. Demirhan, U.; Alkhateeb, A. Cell-Free Joint Sensing and Communication MIMO: A Max-Min Fair Beamforming Approach. In Proceedings of the 57th Asilomar Conference on Signals, Systems, and Computers, Pacific Grove, CA, USA, 29 October 2023–1 November 2023; pp. 381–386. [CrossRef]
96. Demirhan, U.; Alkhateeb, A. Cell-Free ISAC MIMO Systems: Joint Sensing and Communication Beamforming. *arXiv* **2024**, arXiv:2301.11328. [CrossRef]
97. Liu, S.; Liu, R.; Li, M.; Liu, Q. Cooperative Cell-Free ISAC Networks: Joint BS Mode Selection and Beamforming Design. *arXiv* **2024**, arXiv:2305.10800. [CrossRef]
98. Mohammadian, A.; Tellambura, C. RF Impairments in Wireless Transceivers: Phase Noise, CFO, and IQ Imbalance—A Survey. *IEEE Access* **2021**, *9*, 111718–111791. [CrossRef]
99. Shafique, K.; Khawaja, B.A.; Sabir, F.; Qazi, S.; Mustaqim, M. Internet of Things (IoT) for Next-Generation Smart Systems: A Review of Current Challenges, Future Trends and Prospects for Emerging 5G-IoT Scenarios. *IEEE Access* **2020**, *8*, 23022–23040. [CrossRef]
100. Galappaththige, D.; Rezaei, F.; Tellambura, C.; Herath, S. Link Budget Analysis for Backscatter-Based Passive IoT. *IEEE Access* **2022**, *10*, 128890–128922. [CrossRef]
101. Rezaei, F.; Galappaththige, D.; Tellambura, C.; Herath, S. Coding Techniques for Backscatter Communications—A Contemporary Survey. *IEEE Commun. Surv. Tutor.* **2023**, *25*, 1020–1058. [CrossRef]
102. Galappaththige, D.; Rezaei, F.; Tellambura, C.; Maaref, A. Cell-Free Bistatic Backscatter Communication: Channel Estimation, Optimization, and Performance Analysis. *IEEE Trans. Commun.* 2024, *early access*. [CrossRef]
103. Rezaei, F.; Tellambura, C.; Herath, S. Large-Scale Wireless-Powered Networks with Backscatter Communications—A Comprehensive Survey. *IEEE Open J. Commun. Soc.* **2020**, *1*, 1100–1130. [CrossRef]
104. Galappaththige, D.; Tellambura, C.; Maaref, A. Integrated Sensing and Backscatter Communication. *IEEE Wirel. Commun. Lett.* **2023**, *12*, 2043–2047. [CrossRef]

105. Wang, D.; Rezaei, F.; Tellambura, C. Performance Analysis and Resource Allocations for a WPCN with a New Nonlinear Energy Harvester Model. *IEEE Open J. Commun. Soc.* **2020**, *1*, 1403–1424. [CrossRef]
106. Zhou, J.; Abdelrahman, M.I.; Conroy, T.B.; Monticone, F.; Kan, E.C. A Backscatter Field Model for Near-Field RF Sensing. *IEEE Trans. Microw. Theory Techn.* **2024**, *72*, 1254–1265. [CrossRef]
107. Nikitin, P.V.; Rao, K.; Lazar, S. An Overview of Near Field UHF RFID. In Proceedings of the IEEE International Conference on RFID, Grapevine, TX, USA, 26–28 March 2007; pp. 167–174. [CrossRef]
108. Kuo, N.C.; Zhao, B.; Niknejad, A.M. Near-field Power Transfer and Backscattering Communication to Miniature RFID Tag in 65 nm CMOS Technology. In Proceedings of the IEEE MTT-S International Microwave Symposium (IMS), San Francisco, CA, USA, 22–27 May 2016; pp. 1–4. [CrossRef]
109. Grosinger, J.; Pachler, W.; Bosch, W. Tag Size Matters: Miniaturized RFID Tags to Connect Smart Objects to the Internet. *IEEE Microw. Mag.* **2018**, *19*, 101–111. [CrossRef]
110. Liu, R.; Li, M.; Luo, H.; Liu, Q.; Swindlehurst, A.L. Integrated Sensing and Communication with Reconfigurable Intelligent Surfaces: Opportunities, Applications, and Future Directions. *IEEE Wirel. Commun.* **2023**, *30*, 50–57. [CrossRef]
111. Luo, H.; Liu, R.; Li, M.; Liu, Q. RIS-Aided Integrated Sensing and Communication: Joint Beamforming and Reflection Design. *IEEE Trans. Veh. Technol.* **2023**, *72*, 9626–9630. [CrossRef]

entropy

Article

Coverage Analysis for High-Speed Railway Communications with Narrow-Strip-Shaped Cells over Suzuki Fading Channels

Shenghong Lin [1,*], Hongyan Wang [1], Weiyong Li [1] and Jinyuan Wang [2,3]

[1] Jiangsu Province Service Customization Network Application Engineering Research Center, Nanjing Vocational College of Information Technology, Nanjing 210023, China; wanghy@njcit.cn (H.W.); liwy@njcit.cn (W.L.)
[2] School of Communications and Information Engineering, Nanjing University of Posts and Telecommunications, Nanjing 210003, China; jywang@njupt.edu.cn
[3] Chuan and Zang Smart Tourism Engineering Research Center of Colleges and Universities of Sichuan Province, Sichuan Tourism University, Chengdu 610100, China
* Correspondence: linsh@njcit.cn

Abstract: Unlike circular cell coverage in public land mobile communications, narrow-strip-shaped cell coverage should be considered in high-speed railway (HSR) communications. Moreover, for the coverage analysis in HSR communications, most works ignore the effect of small-scale fading, which results in an inaccurate coverage performance evaluation. In this paper, we focus on the coverage analysis for HSR communications with narrow-strip-shaped cells over the Suzuki fading channel, where the composite channel fading includes path loss, lognormal shadowing, and Rayleigh-distributed small-scale fading. Based on the channel model, we first analyze the statistical characteristic of the received signal-to-noise ratio. Then, we derive analytical expressions of the edge coverage probability (ECP) and the percentage of cell coverage area (CCA). To link the edge coverage performance and the average coverage performance of a cell, we express the percentage of CCA as a summation of the ECP and a positive increment. As special cases, we also obtain the coverage performance expressions for the systems without small-scale fading. Through Monte Carlo simulations, the accuracy of the derived expressions is verified. Numerical results also show that the small-scale fading has a strong effect on coverage performance and cannot be ignored. In addition, the effects of key parameters are also discussed.

Keywords: high-speed railway communications; coverage performance; narrow-strip-shaped cell; Suzuki fading; edge coverage probability; percentage of cell coverage area

1. Introduction

As a comfortable and fast transportation tool, the high-speed railway (HSR) has been making great progress globally in recent decades. Thanks to its excellent performance, the HSR has improved people's travel efficiency profoundly, expanding people's living scope and promoting economic development. At present, China has the world's largest HSR network, which is still expanding [1]. By the end of the medium–long-term railway network plan (2016–2030), China's HSR network is expected to expand to around 200,000 km. To provide various mobile services for passengers on the high-speed train, advanced wireless communication techniques should also be explored.

For wireless communications, a good coverage scheme can significantly improve the system performance. To evaluate a certain coverage scheme, some coverage performance metrics, such as the percentage of cell coverage area (CCA), the edge coverage probability (ECP), and the edge outage probability, have been investigated. For traditional public land cellular communication systems with circular cells, the theoretical expression of the percentage of CCA was derived [2,3]. Moreover, the relationship between the percentage of CCA and the ECP was obtained [4], which links the edge coverage performance and the average coverage performance of a cell. Note that the overlap region between adjacent cells is not considered in [2–4]. To facilitate the handoff between different base stations (BSs), the percentage of CCA was further analyzed by considering the overlap region [5]. By considering the handoff, the relationship between the percentage of CCA and the ECP was also further analyzed [6]. However, due to the linear user distribution along the railway track, the circular cells used in [2–6] will lead to low coverage efficiency and radio resource waste in the coverage of HSR communication systems. Moreover, the coverage performance of HSR communication systems depends on the propagation environments (such as urban, suburban, rural, viaduct, cutting, station mountain, and river), which is different from that of conventional public land cellular communication (PLCC) systems.

Recently, some works have evaluated the coverage performance for HSR communication systems. To improve the coverage efficiency of circular cells, a coverage approach by decreasing the cell size for the HSR systems was proposed [7]. However, the coverage performance may be further improved by replacing circular cells with narrow-strip-shaped cells (also named linear cells, or elliptical cells). For a single-frequency-based HSR communication system in China, the narrow-strip-shaped coverage along the railway track was studied [8]. Moreover, the railway coverage in Spanish peninsular municipalities in 2018 and planned for 2024 was evaluated [9]. However, the specific performance indicators are not analyzed in [8,9]. By employing the modified Hata model [10] and the cooperative transmission scheme [11], the coverage probability, outage probability, transmission capacity, and number of BSs for HSR communication systems were evaluated. For HSR systems with narrow-strip-shaped cells, the percentage of CCA and the cell edge outage probability were analyzed [12]. For HSR systems with elliptical cells, a beamforming design problem was proposed to maximize the percentage of rail coverage [13]. However, only isolated cells were considered [12,13]. In practical systems, the handoff over the overlap region should be considered. To reduce handoff frequency, free-space optical communication frameworks were proposed for HSR communications [14,15]. By considering the handoff scheme, the percentage of CCA and the ECP were analyzed for HSR systems [16]. Moreover, the coverage analysis for HSR systems was also extended to the carrier aggregation scenario [17] and the cell-free massive multi-input multi-output scenario [18]. It should be emphasized that almost all the above works focus on large-scale fading and ignore small-scale fading. However, the small-scale fading has also a notable effect on coverage performance.

In this paper, we investigate the coverage performance of an HSR communication system. Unlike the circular cells used in PLCC systems [2–6], the narrow-strip-shaped cells are deployed along the railway track in this paper. Unlike previous studies that only consider the large-scale fading for HSR coverage [7–18], this paper considers the Suzuki fading channel, which includes both large-scale fading and small-scale fading. The main contributions of this paper are summarized as follows:

- We analyze the coverage performance at the cell edge. First, we analyze the statistical characteristics of the received signal-to-noise ratio (SNR). Then, according to the definition of ECP and the Gaussian–Hermite integral, we derive an analytical expression of the ECP. The ECP expression is shown to be a function of transmit power, cell radius, noise variance, standard deviation of shadow fading, HSR propagation environment, and SNR threshold.
- We analyze the average coverage performance of a cell, which can be characterized by the percentage of CCA. According to its mathematical definition, we derive its analytical expression. The percentage of CCA is also expressed as a function of system key parameters.
- We obtain a theoretical expression to link the ECP and the percentage of CCA. The percentage of CCA is expressed as the summation of the ECP and a positive increment. Thus, the relationship between the ECP and the percentage of CCA is established. As special cases, we also derive the theoretical expressions for the system without considering the small-scale fading.

- Some numerical results are provided. It is shown that the theoretical results match well with the simulation results, which verifies the accuracy of the derived theoretical expressions. Moreover, the small-scale fading has a strong effect on coverage performance, and thus it cannot be ignored. Furthermore, the effects of cell radius, transmit power, SNR threshold, propagation environment, and the shadow fading standard derivation on coverage performance are also provided.

The remainder of this paper is organized as follows. In Section 2, the system model of the HSR communication system is provided. Section 3 analyzes theoretical expressions of the ECP and percentage of CCA. Then, the relationship between the ECP and percentage of CCA is obtained in Section 4. Numerical results are presented in Section 5. Finally, conclusions are drawn in Section 6.

Notation: In this paper, italicized symbols denote scalar values; $\sqrt{\alpha}$ denotes the square root of a number α; $\log_{10}(\cdot)$ denotes the base-10 logarithm; $\ln(\cdot)$ denotes the natural logarithm; $\exp(\cdot)$ denotes the standard exponential function; $|\cdot|$ denotes the modulus of a complex number; $f_X(x)$ denotes the probability density function (PDF) of the variable X; $\operatorname{erfc}(x) = \frac{2}{\sqrt{\pi}} \int_x^{\infty} e^{-y^2} \mathrm{d}y$ denotes the complementary error function; $\gamma(a, b) = \int_0^b t^{a-1} e^{-t} \mathrm{d}t$ denotes the lower incomplete Gamma function; and $Q(x) = \int_x^{\infty} \frac{1}{\sqrt{2\pi}} \exp\left(\frac{t^2}{2}\right) \mathrm{d}t$ denotes the Gaussian Q-function.

2. System Model

As shown in Figure 1, we consider a classic HSR communication system in this paper. The system framework is applied to a high-speed train running on a straight railway track. The BSs are uniformly deployed along the railway track. To enlarge the coverage area, two directional antennas are installed on each BS, which point to each side of the BS. Thus, the whole system is covered by narrow-strip-shaped cells. Similar to [16–18], a train access point (TAP) is installed on the front roof of the train and taken as a mobile relay to deal with the rapid channel variation between the BS and the users in the train. As a result, the communication link between the BS and the users is divided into a backhaul link (i.e., the BS-TAP link) and an access link (i.e., the TAP-user link). The access link is similar to typical cellular communications, which has been widely studied in the literature. Therefore, the main challenge lies in the backhaul link. Without loss of generality, we concentrate on the backhaul link only in this paper.

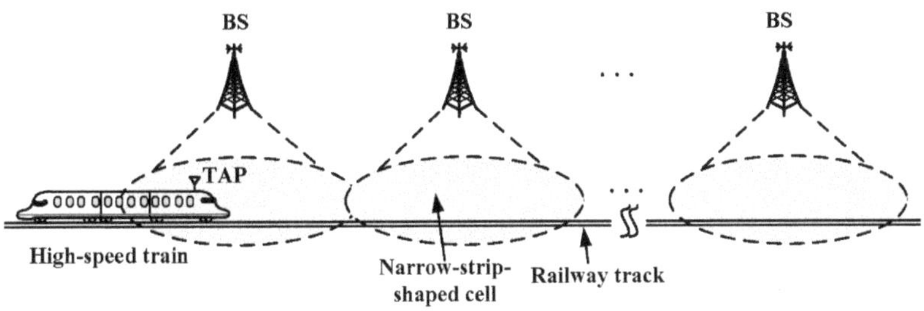

Figure 1. An HST communication system with narrow-strip-shaped cells.

Due to the deployment of BSs near the track, the vertical distance between each BS and the track can be ignored. Therefore, the horizontal distance between each BS and the TAP on the train can be considered the distance of these two nodes. We consider the downlink transmission from the BS to the TAP, and the cell radius is set to be D. When the horizontal distance between the BS and the TAP is l, the received signal y at the TAP can be written as [2]

$$y = \sqrt{P_{\mathrm{T}}} h(l) x + z, \; 0 \le l \le D \tag{1}$$

where P_T is the transmit power of the BS; x is the transmit electrical signal with unit power; z is the additive white Gaussian noise with zero mean and variance P_N; and $h(l)$ denotes the channel gain between the TAP and the BS.

In this paper, we consider a composite channel for HSR communications, which includes path loss, shadow fading, and small-scale fading. We assume that the encountered channel exhibits non-frequency-selective fading (i.e., flat-fading only). For a frequency-selective channel, orthogonal frequency division multiplexing schemes can be used to convert it into multiple flat-fading channels. Moreover, the effect of Doppler shift is ignored in the channel model because we can effectively eliminate it by employing some advanced compensation methods. Therefore, $h(l)$ in (1) is given by

$$h(l) = g\sqrt{PL(l) \cdot S} \tag{2}$$

where $PL(l)$ is the path loss, S is the shadow fading, and g is the small-scale fading.

In (2), the path loss $PL(l)$ for HSR communications can be modeled as [19]

$$PL(l) = 10^{-\frac{A+B\log_{10}(l)}{10}} \tag{3}$$

where A and B are defined as

$$\begin{cases} A = \Delta_1 + 74.52 + 26.16\log_{10}(f) - 13.82\log_{10}(h_b) - 3.2(\log_{10}(11.75h_T))^2 \\ B = 44.9 - 6.55\log_{10}(h_b) + \Delta_2 \end{cases} \tag{4}$$

where f (in MHz) is the adopted frequency and h_b (in m) and h_T (in m) are effective antenna heights of the BS and the TAP. Moreover, Δ_1 and Δ_2 are the correction factors, which varies with the HSR propagation environment. Referring to [19], the values of Δ_1 and Δ_2 for various HSR propagation environments are provided in Table 1. Moreover, Figure 2 shows the path loss values (in dB) versus the transmit distance l (in km) under different HSR propagation environments when $f = 800$ MHz, $h_b = 20$ m, and $h_T = 4$ m. As can be seen, propagation environments have strong effects on path loss performance. The larger the transmit distance is, the larger the path loss becomes. For each fixed transmit distance, the mountain environment always achieves the largest path loss, while the station environment always achieves the smallest path loss.

Table 1. Values of correction factors for different HSR propagation environments.

Scenarios	Δ_1	Δ_2
Urban	-20.47	-1.82
Suburban	$5.74\log_{10}(h_b) - 30.42$	-6.72
Rural	$6.43\log_{10}(h_b) - 30.44$	-6.71
Cutting	-18.78	$51.34\log_{10}(h_b) - 78.99$
Station	$34.29\log_{10}(h_b) - 70.75$	-8.86
Mountain	$-66.26\log_{10}(h_b) + 73.77$	$-84.14\log_{10}(h_b) + 106.79$
River	$8.79\log_{10}(h_b) - 33.99$	-2.93

In (2), the shadow fading S in HSR communications follows a lognormal distribution. Thus, the PDF of S is given by

$$f_S(s) = \frac{\xi}{\sqrt{2\pi}\sigma s}\exp\left(-\frac{(10\log_{10}s)^2}{2\sigma^2}\right), s > 0 \tag{5}$$

where $\xi = 10/\ln 10$ and σ (in dB) is the standard deviation of $10\log_{10}S$.

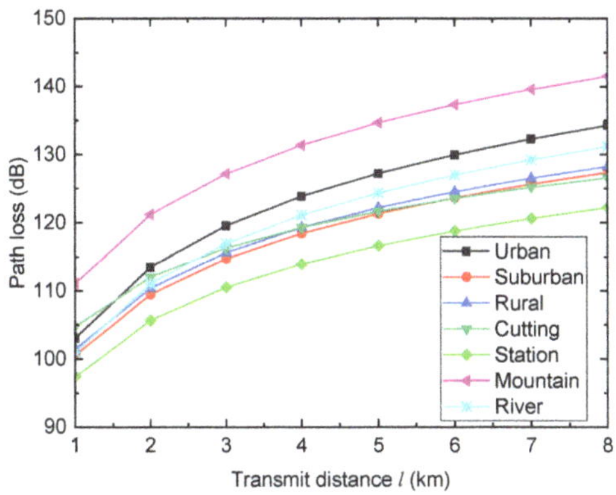

Figure 2. Path loss versus transmit distance under different HSR propagation environments when $f = 800$ MHz, $h_{\mathrm{b}} = 20$ m, and $h_{\mathrm{T}} = 4$ m.

In HSR communications, the small-scale fading follows the Rice distribution for viaduct scenarios over a plain area [20,21]. However, in mountain, hilly, or urban areas, or some other environment that is rich in scattering, Rayleigh fading is a reasonable model to capture the channel characteristics [22]. In this paper, the HSR scenario with rich scattering components is considered, and the small-scale fading g in (2) is modeled as the Rayleigh distribution [22]. As is known, the term $|g|^2$ follows an exponential distribution. Assuming that the mean value of $|g|^2$ is one, the PDF of $|g|^2$ is given by

$$f_{|g|^2}(t) = \exp(-t), t \geq 0 \tag{6}$$

According to the above analysis, the considered composite channel follows a Suzuki distribution (i.e., Rayleigh–lognormal distribution). The measurement results in [23] showed that the Suzuki distribution best describes the composite multipath/shadowing channels in HSR communications.

Moreover, the received SNR at the location l is expressed as

$$\gamma(l) = \frac{P_{\mathrm{T}}|h(l)|^2}{P_{\mathrm{N}}} = \Omega(l)|g|^2 \tag{7}$$

where $\Omega(l) \triangleq P_{\mathrm{T}} \cdot PL(l) \cdot S / P_{\mathrm{N}}$.

3. Coverage Performance Analysis

In this section, the statistical characteristics of the SNR are first analyzed. Then, the coverage performance (i.e., the ECP and the percentage of CCA) of the HSR communication system is analyzed.

3.1. Statistical Characteristics of SNR

Since S follows a lognormal distribution, $\Omega(l)$ also follows a lognormal distribution. Therefore, we can obtain the PDF of $\Omega(l)$ as

$$\begin{aligned} f_{\Omega(l)}(w) &= \frac{P_{\mathrm{N}}}{P_{\mathrm{T}} \cdot PL(l)} f_S\left(\frac{P_{\mathrm{N}}}{P_{\mathrm{T}} \cdot PL(l)} w\right) \\ &= \frac{\xi}{\sqrt{2\pi}\sigma w} \exp\left(-\frac{\left(10\log_{10}\left(\frac{P_{\mathrm{N}} w}{P_{\mathrm{T}} \cdot PL(l)}\right)\right)^2}{2\sigma^2}\right), w > 0 \end{aligned} \tag{8}$$

According to (6) and (7), we can obtain the following conditional PDF as

$$f_{\gamma(l)|\Omega(l)}(r|w) = \frac{1}{w}f_{|g|^2}\left(\frac{r}{w}\right)$$
$$= \frac{1}{w}\exp\left(-\frac{r}{w}\right),\ r \geq 0 \quad (9)$$

From (8) and (9), we can obtain the PDF of the SNR as

$$f_{\gamma(l)}(r) = \int_0^\infty f_{\Omega(l)}(w)f_{\gamma(l)|\Omega(l)}(r|w)\mathrm{d}w$$
$$= \int_0^\infty \frac{\xi}{\sqrt{2\pi}\sigma w}\exp\left(-\frac{\left(10\log_{10}\frac{P_\mathrm{N}w}{P_\mathrm{T}\cdot PL(l)}\right)^2}{2\sigma^2}\right)\frac{1}{w}\exp\left(-\frac{r}{w}\right)\mathrm{d}w \quad (10)$$

Note that the SNR $\gamma(l)$ follows a Gamma–lognormal distribution. It is challenging to obtain a closed-from expression of $f_{\gamma(l)}(r)$. To solve the integral in (10), we let $t = 10\log_{10}\left(\frac{P_\mathrm{N}w}{P_\mathrm{T}\cdot PL(l)}\right)/(\sqrt{2}\sigma)$ and rewrite (10) as

$$f_{\gamma(l)}(r) = \int_{-\infty}^\infty \frac{\xi}{\sqrt{2\pi}\sigma}\exp(-t^2)\frac{\exp\left(-\frac{r}{\frac{P_\mathrm{T}\cdot PL(l)}{P_\mathrm{N}}10^{\frac{\sqrt{2}\sigma t}{10}}}\right)}{\left(\frac{P_\mathrm{T}\cdot PL(l)}{P_\mathrm{N}}10^{\frac{\sqrt{2}\sigma t}{10}}\right)^2}\frac{P_\mathrm{T}\cdot PL(l)}{P_\mathrm{N}}10^{\frac{\sqrt{2}\sigma t}{10}}\frac{\ln 10}{10}\sqrt{2}\sigma\mathrm{d}t$$
$$= \frac{P_\mathrm{N}}{\sqrt{\pi}P_\mathrm{T}\cdot PL(l)}\int_{-\infty}^\infty \exp\left(-t^2\right)10^{-\frac{\sqrt{2}\sigma t}{10}}\exp\left(-\frac{P_\mathrm{N}r}{P_\mathrm{T}\cdot PL(l)}10^{-\frac{\sqrt{2}\sigma t}{10}}\right)\mathrm{d}t \quad (11)$$

According to [24], the Gauss–Hermite integral is given by

$$\int_{-\infty}^\infty e^{-x^2}f(x)dx \cong \sum_{i=1}^{N_\mathrm{p}} H_i f(x_i) \quad (12)$$

where t_i and H_i are the base point and weight factor of the N_p-order Gauss–Hermite approximation.

Solving (11) by using (12), we have

$$f_{\gamma(l)}(r) \cong \frac{P_\mathrm{N}}{\sqrt{\pi}P_\mathrm{T}\cdot PL(l)}\sum_{i=1}^{N_\mathrm{p}} H_i 10^{-\frac{\sqrt{2}\sigma t_i}{10}}\exp\left(-\frac{P_\mathrm{N}r}{P_\mathrm{T}\cdot PL(l)}10^{-\frac{\sqrt{2}\sigma t_i}{10}}\right) \quad (13)$$

Remark 1. *If the small-scale fading is not considered, the Suzuki channel (i.e., path loss + shadow fading + Rayleigh fading) becomes the lognormal shadowing channel (i.e., path loss + shadow fading). In this case, the SNR $\gamma(l)$ in (7) reduces to $\gamma'(l) = \Omega(l)$. Therefore, we can conclude that, if small-scale fading is ignored, we will overestimate or underestimate the quality of the received signal, resulting in inaccurate coverage performance analysis. Moreover, in this case, the PDF of SNR (13) reduces to $f_{\gamma'(l)}(r) = f_{\Omega(l)}(r)$.*

3.2. ECP Analysis

In this subsection, the ECP is analyzed. As is known, the ECP P_e is defined as the probability that the SNR at the cell edge (i.e., $l = D$) exceeds a given threshold γ_th. Therefore, we have

$$P_\mathrm{e} = \Pr\{\gamma(D) \geq \gamma_\mathrm{th}\}$$
$$= \int_{\gamma_\mathrm{th}}^\infty f_{\gamma(D)}(r)\mathrm{d}r \quad (14)$$

By solving (14), we obtain the following theorem.

Theorem 1. *For the considered HSR communication system over the Suzuki fading channels, the ECP is given by*

$$P_e = \frac{1}{\sqrt{\pi}} \sum_{i=1}^{N_\mathrm{p}} H_i \exp\left(-\frac{P_\mathrm{N}\gamma_\mathrm{th}}{P_\mathrm{T} \cdot PL(D)} 10^{-\frac{\sqrt{2}\sigma t_i}{10}} \right) \tag{15}$$

Proof. See Appendix A. □

Corollary 1. *If the small-scale fading is not considered, the Suzuki channel becomes the lognormal shadowing channel. In this case, the ECP in Theorem 1 reduces to*

$$\begin{aligned} P_\mathrm{e}' &= \int_{\gamma_\mathrm{th}}^{\infty} \frac{\xi}{\sqrt{2\pi}\sigma w} \exp\left(-\frac{\left(10\log_{10}\frac{P_\mathrm{N}w}{P_\mathrm{T}\cdot PL(D)}\right)^2}{2\sigma^2} \right) \mathrm{d}w \\ &= \frac{1}{2}\mathrm{erfc}\left(\frac{10\log_{10}\left(\frac{P_\mathrm{N}\gamma_\mathrm{th}}{P_\mathrm{T}\cdot PL(D)}\right)}{\sqrt{2}\sigma} \right) \end{aligned} \tag{16}$$

3.3. Percentage of CCA Analysis

The percentage of CCA is another important indicator for coverage analysis. For the HSR communication with narrow-strip-shaped cells, the percentage of CCA is defined as

$$\begin{aligned} P_\mathrm{a} &= \frac{1}{D}\int_0^D \Pr\{\gamma(l) \geq \gamma_\mathrm{th}\}\mathrm{d}l \\ &= \frac{1}{D}\int_0^D \int_{\gamma_\mathrm{th}}^{\infty} f_{\gamma(l)}(r)dr\mathrm{d}l \end{aligned} \tag{17}$$

By solving (17), we obtain the following theorem.

Theorem 2. *For the considered HSR communication system over the Suzuki fading channels, the percentage of CCA is given by*

$$P_\mathrm{a} = \frac{1}{\sqrt{\pi}D}\frac{10}{B}\sum_{i=1}^{N_\mathrm{p}} H_i \left(\frac{P_\mathrm{T}}{P_\mathrm{N}\gamma_\mathrm{th}} 10^{\frac{\sqrt{2}\sigma t_i - A}{10}} \right)^{\frac{10}{B}} \gamma\left(\frac{10}{B}, \frac{P_\mathrm{N}\gamma_\mathrm{th}}{P_\mathrm{T}} 10^{\frac{A-\sqrt{2}\sigma t_i}{10}} D^{\frac{B}{10}} \right) \tag{18}$$

where $\gamma(\cdot,\cdot)$ is the lower incomplete Gamma function.

Proof. See Appendix B. □

Corollary 2. *If the small-scale fading is not considered, the Suzuki channel becomes the lognormal shadowing channel. In this case, the percentage of CCA in Theorem 2 reduces to*

$$\begin{aligned} P_a' = {} & \frac{1}{2D} 10^{\frac{10\log_{10}\left(\frac{P_\mathrm{T}}{P_\mathrm{N}\gamma_\mathrm{th}}\right)-A}{B}} \left[e^{\frac{\ln 10}{B}\cdot 10\log_{10}\left(\frac{P_\mathrm{N}\gamma_\mathrm{th}}{P_\mathrm{T}\cdot 10^{\frac{A+B\log_{10}(D)}{10}}}\right)} \mathrm{erfc}\left(\frac{10\log_{10}\left(\frac{P_\mathrm{N}\gamma_\mathrm{th}}{P_\mathrm{T}\cdot 10^{\frac{A+B\log_{10}(D)}{10}}}\right)}{\sqrt{2}\sigma} \right) \right. \\ & \left. + \; e^{\frac{\sigma^2(\ln 10)^2}{2B^2}} \mathrm{erfc}\left(\frac{\sigma\ln 10}{\sqrt{2}B} - \frac{10\log_{10}\left(\frac{P_\mathrm{N}\gamma_\mathrm{th}}{P_\mathrm{T}\cdot 10^{\frac{A+B\log_{10}(D)}{10}}}\right)}{\sqrt{2}\sigma} \right) \right] \end{aligned} \tag{19}$$

Proof. See Appendix C. □

4. Relationship Between ECP and Percentage of CCA

In Section 3, the derived ECP and percentage of CCA are isolated expressions, but the relationship between them is not provided. In this section, we try to link the ECP and the percentage of CCA to show their relationship.

According to the definition of the percentage of CCA, we can further obtain

$$\begin{aligned} P_\mathrm{a} &= \frac{1}{D}\int_0^D \left[\int_{\gamma_\mathrm{th}}^{\infty} f_{\gamma(l)}(r)\mathrm{d}r\right]\mathrm{d}l \\ &= \frac{1}{D}\int_0^D \left[1-\int_0^{\gamma_\mathrm{th}} f_{\gamma(l)}(r)\mathrm{d}r\right]\mathrm{d}l \\ &= 1-\frac{1}{D}\left[l\int_0^{\gamma_\mathrm{th}} f_{\gamma(l)}(r)\mathrm{d}r\Big|_0^D - \int_0^D l\int_0^{\gamma_\mathrm{th}} \frac{\partial}{\partial l} f_{\gamma(l)}(r)\mathrm{d}r\mathrm{d}l\right] \\ &= P_\mathrm{e} + \frac{1}{D}\int_0^D l\int_0^{\gamma_\mathrm{th}} \frac{\partial}{\partial l} f_{\gamma(l)}(r)\mathrm{d}r\mathrm{d}l \end{aligned} \tag{20}$$

By solving (20), we obtain the following theorem.

Theorem 3. *For the considered HSR communication system under the Suzuki fading channels, the relationship between the percentage of CCA P_a and the ECP P_e is given by*

$$P_\mathrm{a} = P_\mathrm{e} + \frac{1}{\sqrt{\pi}D}\sum_{i=1}^{N_\mathrm{p}} H_i\left(10^{\frac{\sqrt{2}\sigma t_i - A}{10}}\frac{P_\mathrm{T}}{P_\mathrm{N}\gamma_\mathrm{th}}\right)^{\frac{10}{B}}\gamma\left(\frac{10}{B}+1, \frac{P_\mathrm{N}\gamma_\mathrm{th}}{P_\mathrm{T}}10^{\frac{A-\sqrt{2}\sigma t_i}{10}}D^{\frac{B}{10}}\right) \tag{21}$$

Proof. See Appendix D. □

Remark 2. *Theorem 3 links the ECP and the percentage of CCA. As can be seen, the percentage of CCA is equal to the ECP plus a positive increment. As is known, the worst coverage performance is achieved at the cell edge, and thus the average coverage performance of the cell is better than the coverage performance at the cell edge.*

Corollary 3. *If the small-scale fading is not considered, the Suzuki channel becomes the lognormal shadowing channel. In this case, the relationship between the percentage of CCA P'_a and the ECP P'_e becomes*

$$\begin{aligned} P'_\mathrm{a} &= P'_\mathrm{e} + \frac{1}{D}\left(\frac{P_\mathrm{N}\gamma_\mathrm{th}}{P_\mathrm{T}}\right)^{-\frac{10}{B}} 10^{-\frac{A}{B}} \exp\left(\frac{\sigma^2(\ln 10)^2}{2B^2}\right) \\ &\times \left[1 - Q\left(\frac{10\log_{10}\left(\frac{P_\mathrm{N}\gamma_\mathrm{th}}{P_\mathrm{T}\cdot PL(D)}\right)}{\sigma} - \frac{\sigma\ln 10}{B}\right)\right] \end{aligned} \tag{22}$$

Proof. See Appendix E. □

5. Numerical Results

In this section, some classic numerical results are shown to validate the cell coverage analysis for HSR communications. Monte Carlo simulation results are also given to verify the correctness of the derived expressions. In the simulation figures of this section, the simulation (or theoretical) results with g correspond to the Suzuki channel, while the simulation (or theoretical) results without g correspond to the lognormal shadowing channel. The main simulation parameters are listed in Table 2.

Table 2. Main simulation parameters.

Parameters	Symbols	Values
Carrier frequency	f	800 MHz
BS antenna height	h_b	20 m
TAP antenna height	h_T	4 m
Noise variance	P_N	10^{-14}
Order of Gauss–Hermite approximation	N_p	40

5.1. Results of ECP

In this subsection, we provide some results of ECP in Figures 3–6. To facilitate the comparison, the coverage performance for the system without considering small-scale fading g is also provided.

Figure 3 shows the ECP P_e versus the cell radius D for the urban environment with different transmit power values P_T when $\gamma_{th} = 1$ and $\sigma = 6$. It can be observed that the ECP value reduces with D, which indicates that the larger coverage region will result in a lower coverage probability at the cell edge. For a large transmit power P_T, a large SNR can be obtained, and thus a better coverage performance can be achieved. Moreover, when the cell radius D is small, the ECP of the system with the small-scale fading is smaller than that of the system without the small-scale fading. However, when D is large, the ECP of the system with the small-scale fading is larger than that of the system without the small-scale fading. This indicates that the ECP performance is overestimated at small D and underestimated at large D when the small-scale fading is not considered, resulting in an inaccurate performance evaluation. Therefore, the small-scale fading has a strong effect on ECP performance and cannot be ignored in practical coverage analysis.

Figure 4 plots the ECP P_e versus the cell radius D for the urban environment with different shadow fading standard derivations σ when $\gamma_{th} = 1$ and $P_T = 15$ dBm. For a small cell radius D, the ECP reduces with the increase in standard derivation σ. However, for a large cell radius D, the ECP increases with σ. Similar to Figure 3, the ECP performance of the system with g is worse than that of the system without g when D is small, and the performance of the system with g is better than that of the system without g when D is large. Therefore, the small-scale fading cannot be ignored.

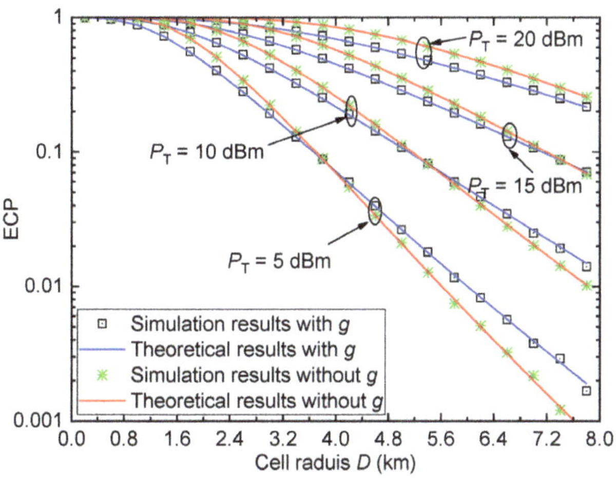

Figure 3. ECP versus cell radius for the urban environment with different P_T when $\gamma_{th} = 1$ and $\sigma = 6$.

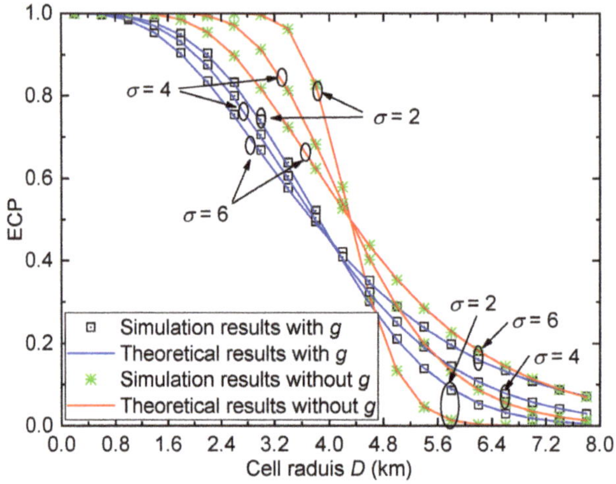

Figure 4. ECP versus cell radius for the urban environment with different σ when $\gamma_{th} = 1$, and $P_T = 15$ dBm.

Figure 5 provides the relationship between the ECP P_e and the transmit power P_T for the urban environment with different SNR thresholds γ_{th} when $\sigma = 6$ and $D = 5$ km. As can be seen, the ECP values increase with the increase in P_T, which is the same as that in Figure 3. With the increase in γ_{th}, the ECP values reduce. This is obvious because a larger SNR threshold will result in a worse coverage performance. Moreover, the ECP performance of VLC with small-scale fading is better than that without small-scale fading for small values of P_T; however, the observation is opposite for large values of P_T.

Figure 6 shows the ECP P_e versus the transmit power P_T for different HSR propagation environments when $\gamma_{th} = 1, \sigma = 6$, and $D = 5$ km. As can be observed, the coverage performance varies with the HSR propagation environment. The station environment achieves the best ECP performance, the suburban environment achieves the second best ECP performance, the urban environment achieves the second worst ECP performance, and the mountain environment achieves the worst ECP performance. These conclusions are obvious by observing the path loss values of different HSR propagation environments in Figure 2.

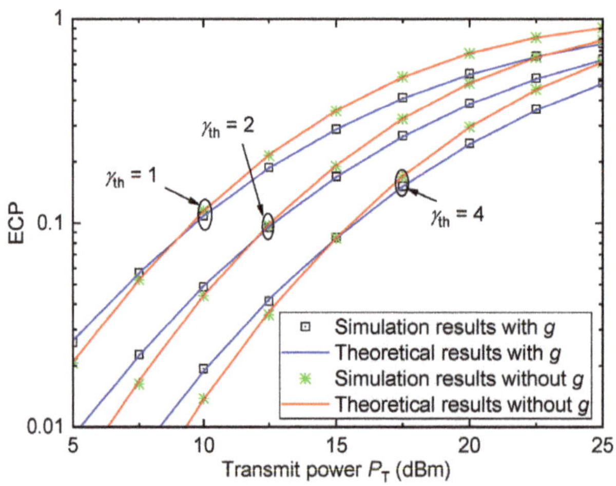

Figure 5. ECP versus transmit power for the urban environment with different γ_{th} when $\sigma = 6$ and $D = 5$ km.

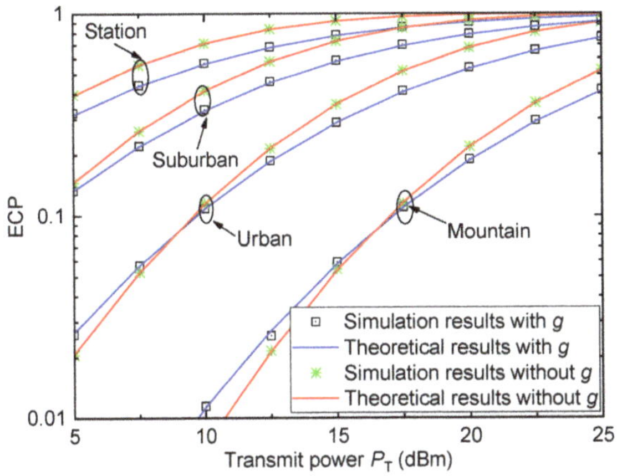

Figure 6. ECP versus transmit power for different HSR propagation environments when $\gamma_{th} = 1$, $\sigma = 6$, and $D = 5$ km.

For Figures 3–6, all theoretical results match well with the simulation results, which verify the accuracy of the derived ECP expression.

5.2. Results of Percentage of CCA

In this subsection, we provide some results of the percentage of CCA in Figures 7–10. To facilitate the comparison, the coverage performance for the system without considering small-scale fading is also provided.

Figure 7 shows the percentage of CCA P_a versus the cell radius D for the urban environment with different transmit power values P_T when $\gamma_{th} = 1$ and $\sigma = 6$. Obviously, the values of the percentage of CCA reduce with the increase in D, which is similar to the conclusion in Figure 3. Moreover, compared with Figure 3, we can observe that for fixed P_T and D, the ECP values are always smaller than the percentage of CCA values, which indicates that the average coverage performance of the cell outperforms the cell edge coverage performance. Furthermore, when D changes from 0 to 8 km, the performance of the system without considering small-scale fading is always better than that of the system with the small-scale fading. Therefore, the average coverage performance will be overestimated if the small-scale fading is ignored.

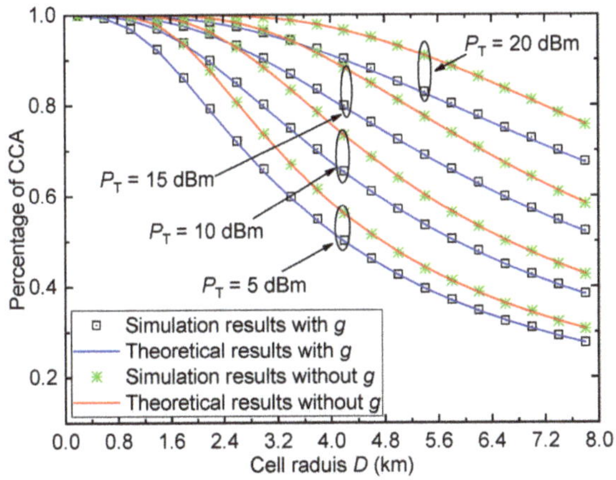

Figure 7. Percentage of CCA versus cell radius for the urban environment with different P_T when $\gamma_{th} = 1$, and $\sigma = 6$.

Figure 8 shows the percentage of CCA P_a versus the transmit power P_T for the urban environment with different SNR thresholds γ_{th} when $\sigma = 6$ and $D = 5$ km. Similar to Figure 5, the percentage of CCA also increases with the increase in transmit power P_T or the decrease in SNR threshold γ_{th}. Moreover, the percentage of CCA in this figure is larger than the ECP in Figure 5 for fixed P_T and γ_{th}. As can be seen, the performance of the system with small-scale fading is worse than that without considering the small-scale fading. These conclusions are similar to that in Figure 7.

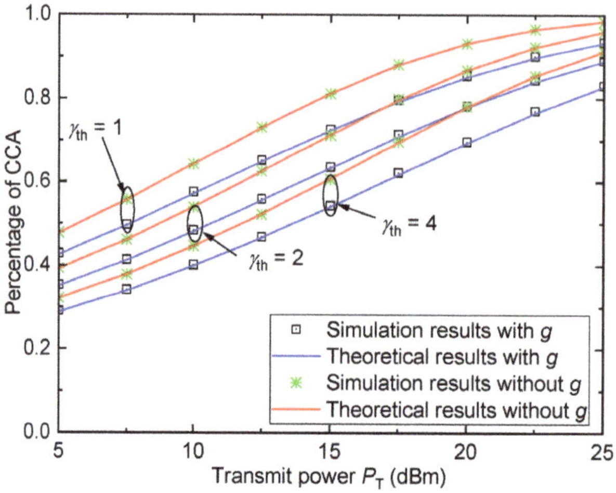

Figure 8. Percentage of CCA versus transmit power for the urban environment with different γ_{th} when $\sigma = 6$ and $D = 5$ km.

Figure 9 shows the percentage of CCA P_e versus the transmit power P_T for different HSR propagation environments when $\gamma_{th} = 1$, $\sigma = 6$, and $D = 5$ km. As can be seen, the HSR propagation environments have strong effects on coverage performance. For different HSR environments, the average coverage performance from best to worst is ranked as station > suburban > urban > mountain, which is the same as that in Figure 6. We can also observe that the average coverage performance is overestimated if the small-scale fading is not considered.

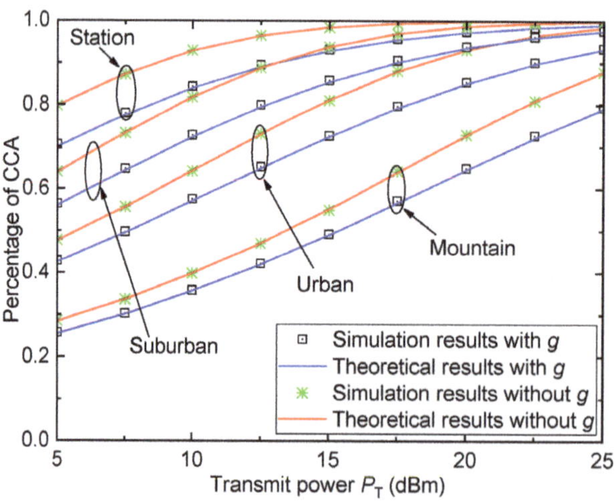

Figure 9. Percentage of CCA versus transmit power for different HSR propagation environments when $\gamma_{th} = 1$, $\sigma = 6$, and $D = 5$ km.

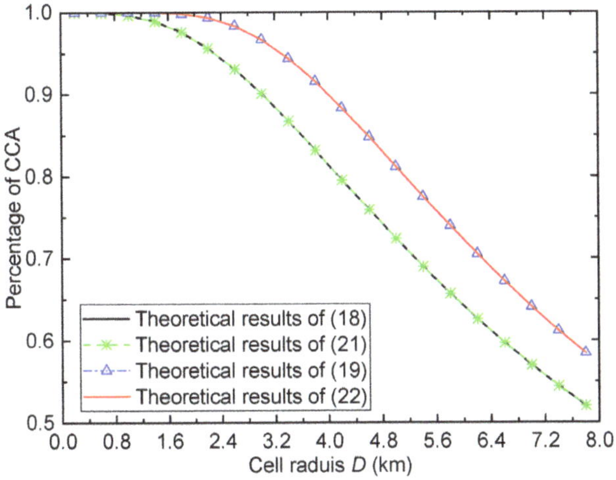

Figure 10. Different results of the percentage of CCA versus cell radius for the urban environment when $\gamma_{th} = 1$, $\sigma = 6$, and $P_T = 15$ dBm.

For Figures 7–9, the gaps between all theoretical results and simulation results are sufficiently small and can be ignored, which verify the accuracy of the derived percentage of CCA expression.

Moreover, as stated in the previous section, Equations (21) and (22) provide the relationships between the ECP and the percentage of CCA. To verify the accuracy of the derived expressions ((21) and (22)), Figure 10 shows different results of the percentage of CCA versus the cell radius for the urban environment when $\gamma_{th} = 1, \sigma = 6$, and $P_T = 15$ dBm. To facilitate the comparison, Equations (18) and (19) are also provided, which can be used to evaluate the percentage of CCA directly. However, Equations (21) and (22) are the functions of P_e and P_e', respectively. The value of P_e in (21) is determined by (15), while the value of P_e' in (22) is determined by (16). From the figure, it can be observed that the theoretical results of (21) match well with those of (18), and the theoretical results of (22) match well with those of (19). These results verify the accuracy of the derived expressions (21) and (22).

6. Conclusions

This paper investigates the coverage performance for HSR communications with narrow-strip-shaped cells. Unlike previous works, the small-scale fading is considered in the channel model, and thus a composite channel named the Suzuki fading channel is considered. The main conclusions are summarized as follows:

- For the HSR system with small-scale fading, analytical expressions of the ECP and the percentage of CCA are derived, respectively. To facilitate the comparison, the coverage performance indicator expressions for the system without considering the small-scale fading are also derived. Numerical results verify the accuracy of the derived expressions.
- To link the edge coverage performance and the average coverage performance of the whole cell, we derive the relationship between the ECP and the percentage of CCA. Specifically, the percentage of CCA is expressed as the summation of the ECP and a positive increment. Therefore, the average coverage performance of a cell is always better than the edge coverage performance.
- The HSR propagation environments have strong effects on coverage performance. For example, the station scenario has a smaller path loss and thus has a better coverage performance, while the mountain scenario has a larger path loss and thus has a worse coverage performance.

- It is shown that the ECP or the percentage of CCA will be overestimated or underestimated if the small-scale fading is not considered. This indicates that, for the coverage analysis in HSR communications, the small-scale fading cannot be ignored.
- The cell radius, transmit power, SNR threshold, and shadow fading standard derivation also have strong effects on coverage performance. Specifically, the coverage performance improves with the decrease in cell radius, the increase in transmit power, the decrease in SNR threshold, or the decrease in shadow fading standard derivation.

Author Contributions: Conceptualization, S.L.; methodology, J.W.; software, S.L.; validation, H.W., W.L. and J.W.; investigation, H.W.; resources, W.L.; writing—original draft preparation, S.L. and J.W.; writing—review and editing, H.W. and W.L. All authors have read and agreed to the published version of the manuscript.

Funding: This research was funded by the open research fund of the Chuan and Zang Smart Tourism Engineering Research Center of Colleges and Universities of Sichuan Province under Grant ZLGC2022A01, the Scientific Research Foundation for High-level Talents in NJCIT under Grant YB20210501, the Natural Science Foundation of the Jiangsu Higher Education Institutions of China under Grant 23KJD510008, and the Project entrusted by NJUPT Communication Network Industry Research Institute Co., Ltd.

Institutional Review Board Statement: Not applicable.

Data Availability Statement: The data presented in this study are available on request from the corresponding author.

Acknowledgments: The authors thank the anonymous reviewers and the Editor for their helpful comments and review of this article.

Conflicts of Interest: The authors declare no conflicts of interest.

Appendix A. Proof of Theorem 1

Substituting (13) into (14), we have

$$P_{\mathrm{e}} = \int_{\gamma_{\mathrm{th}}}^{\infty} \frac{P_{\mathrm{N}}}{\sqrt{\pi}P_{\mathrm{T}} \cdot PL(D)} \sum_{i=1}^{N_{\mathrm{p}}} H_i 10^{-\frac{\sqrt{2}\sigma t_i}{10}} \exp\left(-\frac{P_{\mathrm{N}} r}{P_{\mathrm{T}} \cdot PL(D)} 10^{-\frac{\sqrt{2}\sigma t_i}{10}}\right) \mathrm{d}r \tag{A1}$$

Change the order of integral and summation in (A1), we can further express P_{e} as

$$P_{\mathrm{e}} = \frac{P_{\mathrm{N}}}{\sqrt{\pi}P_{\mathrm{T}} \cdot PL(D)} \sum_{i=1}^{N_{\mathrm{p}}} H_i 10^{-\frac{\sqrt{2}\sigma t_i}{10}} \int_{\gamma_{\mathrm{th}}}^{\infty} \exp\left(-\frac{P_{\mathrm{N}} r}{P_{\mathrm{T}} \cdot PL(D)} 10^{-\frac{\sqrt{2}\sigma t_i}{10}}\right) \mathrm{d}r \tag{A2}$$

Then, let $m = \frac{P_{\mathrm{N}} r}{P_{\mathrm{T}} \cdot PL(D)} 10^{-\frac{\sqrt{2}\sigma t_i}{10}}$, Equation (A2) can be written as

$$\begin{aligned} P_{\mathrm{e}} &= \frac{P_{\mathrm{N}}}{\sqrt{\pi}P_{\mathrm{T}} \cdot PL(D)} \sum_{i=1}^{N_{\mathrm{p}}} H_i 10^{-\frac{\sqrt{2}\sigma t_i}{10}} \int_{\frac{P_{\mathrm{N}}\gamma_{\mathrm{th}}}{P_{\mathrm{T}} \cdot PL(D)} 10^{-\frac{\sqrt{2}\sigma t_i}{10}}}^{\infty} \exp(-m) \frac{P_{\mathrm{T}} \cdot PL(D)}{P_{\mathrm{N}}} 10^{\frac{\sqrt{2}\sigma t_i}{10}} \mathrm{d}m \\ &= \frac{1}{\sqrt{\pi}} \sum_{i=1}^{N_{\mathrm{p}}} H_i \exp\left(-\frac{P_{\mathrm{N}}\gamma_{\mathrm{th}}}{P_{\mathrm{T}} \cdot PL(D)} 10^{-\frac{\sqrt{2}\sigma t_i}{10}}\right) \end{aligned} \tag{A3}$$

Therefore, Equation (15) holds.

Appendix B. Proof of Theorem 2

According to (14) and (15), we can obtain

$$\int_{\gamma_{\mathrm{th}}}^{\infty} f_{\gamma(l)}(r) dr = \frac{1}{\sqrt{\pi}} \sum_{i=1}^{N_{\mathrm{p}}} H_i \exp\left(-\frac{P_{\mathrm{N}}\gamma_{\mathrm{th}}}{P_{\mathrm{T}} \cdot PL(l)} 10^{-\frac{\sqrt{2}\sigma t_i}{10}}\right) \tag{A4}$$

Then, substituting (A4) into (17), we have

$$P_{\mathrm{a}} = \frac{1}{D}\int_0^D \frac{1}{\sqrt{\pi}} \sum_{i=1}^{N_{\mathrm{p}}} H_i \exp\left(-\frac{P_{\mathrm{N}}\gamma_{\mathrm{th}}}{P_{\mathrm{T}} \cdot PL(l)} 10^{-\frac{\sqrt{2}\sigma t_i}{10}}\right) \mathrm{d}l \tag{A5}$$

Changing the order of integral and summation, we can further obtain

$$P_{\mathrm{a}} = \frac{1}{\sqrt{\pi}D} \sum_{i=1}^{N_{\mathrm{p}}} H_i \int_0^D \exp\left(-\frac{P_{\mathrm{N}}\gamma_{\mathrm{th}} 10^{\frac{A-\sqrt{2}\sigma t_i}{10}}}{P_{\mathrm{T}}} l^{\frac{B}{10}}\right) \mathrm{d}l \tag{A6}$$

Let $m = \frac{P_{\mathrm{N}}\gamma_{\mathrm{th}} 10^{\frac{A-\sqrt{2}\sigma t_i}{10}}}{P_{\mathrm{T}}} l^{\frac{B}{10}}$, Equation (A6) can be further written as

$$\begin{aligned} P_{\mathrm{a}} &= \frac{1}{\sqrt{\pi}D} \sum_{i=1}^{N_{\mathrm{p}}} H_i \int_0^{\frac{P_{\mathrm{N}}\gamma_{\mathrm{th}} 10^{\frac{A-\sqrt{2}\sigma t_i}{10}}}{P_{\mathrm{T}}} D^{\frac{B}{10}}} e^{-m} \frac{10}{B} \left(\frac{P_{\mathrm{T}} \cdot 10^{\frac{\sqrt{2}\sigma t_i - A}{10}}}{P_{\mathrm{N}}\gamma_{\mathrm{th}}} m\right)^{\frac{10}{B}-1} \frac{P_{\mathrm{T}}}{P_{\mathrm{N}}\gamma_{\mathrm{th}}} 10^{\frac{\sqrt{2}\sigma t_i - A}{10}} \mathrm{d}m \\ &= \frac{1}{\sqrt{\pi}D} \frac{10}{B} \sum_{i=1}^{N_{\mathrm{p}}} H_i \left(\frac{P_{\mathrm{T}}}{P_{\mathrm{N}}\gamma_{\mathrm{th}}} 10^{\frac{\sqrt{2}\sigma t_i - A}{10}}\right)^{\frac{10}{B}} \int_0^{\frac{P_{\mathrm{N}}\gamma_{\mathrm{th}} 10^{\frac{A+B\log_{10}(D)-\sqrt{2}\sigma t_i}{10}}}{P_{\mathrm{T}}}} e^{-m} \cdot m^{\frac{10}{B}-1} \mathrm{d}m \end{aligned} \tag{A7}$$

Moreover, the lower incomplete Gamma function is defined as [24], Equation (6.5.2).

$$\gamma(s, x) = \int_0^x t^{s-1} e^{-t} \mathrm{d}t \tag{A8}$$

By applying (A8) for (A7), we can obtain (18).

Appendix C. Proof of Corollary 2

When the small-scale fading is not considered, substituting $f_{\gamma'(l)}(r) = f_{\Omega(l)}(r)$ into (17) and using Corollary 1, we obtain the percentage of CCA as

$$\begin{aligned} P'_{\mathrm{a}} &= \frac{1}{D}\int_0^D \int_{\gamma_{\mathrm{th}}}^{\infty} \frac{\xi}{\sqrt{2\pi}\sigma r} \exp\left(-\frac{\left(10\log_{10}\left(\frac{P_{\mathrm{N}} r}{P_{\mathrm{T}} \cdot PL(l)}\right)\right)^2}{2\sigma^2}\right) \mathrm{d}r\mathrm{d}l \\ &= \frac{1}{D}\int_0^D \frac{1}{2}\mathrm{erfc}\left(\frac{10\log_{10}\left(\frac{P_{\mathrm{N}}\gamma_{\mathrm{th}}}{P_{\mathrm{T}} \cdot PL(l)}\right)}{\sqrt{2}\sigma}\right) \mathrm{d}l \end{aligned} \tag{A9}$$

Then, substituting (3) into (A9) and letting $t = 10\log_{10}\left(\frac{P_{\mathrm{N}}\gamma_{\mathrm{th}}}{P_{\mathrm{T}} \cdot 10^{-\frac{A+B\log_{10}(l)}{10}}}\right)/(\sqrt{2}\sigma)$, we have

$$\begin{aligned} P'_{\mathrm{a}} &= \frac{\sigma \ln 10}{\sqrt{2}DB} 10^{\frac{10\log_{10}\left(\frac{P_{\mathrm{T}}}{P_{\mathrm{N}}\gamma_{\mathrm{th}}}\right) - A}{B}} \int_{-\infty}^{\frac{10\log_{10}\left(\frac{P_{\mathrm{N}}\gamma_{\mathrm{th}}}{P_{\mathrm{T}} \cdot 10^{\frac{A+B\log_{10}(D)}{10}}}\right)}{\sqrt{2}\sigma}} \mathrm{erfc}(t) 10^{\frac{\sqrt{2}\sigma t}{B}} \mathrm{d}t \\ &= \frac{\sigma \ln 10}{\sqrt{2}DB} 10^{\frac{10\log_{10}\left(\frac{P_{\mathrm{T}}}{P_{\mathrm{N}}\gamma_{\mathrm{th}}}\right) - A}{B}} \int_{-\infty}^{\frac{10\log_{10}\left(\frac{P_{\mathrm{N}}\gamma_{\mathrm{th}}}{P_{\mathrm{T}} \cdot 10^{\frac{A+B\log_{10}(D)}{10}}}\right)}{\sqrt{2}\sigma}} \mathrm{erfc}(t) e^{\frac{\sqrt{2}\sigma \ln 10}{B} t} \mathrm{d}t \end{aligned} \tag{A10}$$

According to [25], we have the following identity as

$$\int e^{bt}\mathrm{erfc}(at)\mathrm{d}t = \frac{1}{b}\left[e^{bt}\mathrm{erfc}(at) - e^{\frac{b^2}{4a^2}}\mathrm{erf}\left(\frac{b}{2a} - at\right)\right] \tag{A11}$$

By using (A11), we have

$$\begin{aligned} P_a' = \frac{1}{2D}10^{\frac{10\log_{10}\left(\frac{P_\mathrm{T}}{P_\mathrm{N}\gamma_\mathrm{th}}\right)-A}{B}}\Bigg[& e^{\frac{\ln 10}{B}\cdot 10\log_{10}\left(\frac{P_\mathrm{N}\gamma_\mathrm{th}}{P_\mathrm{T}\cdot 10^{\frac{A+B\log_{10}(D)}{10}}}\right)}\mathrm{erfc}\left(\frac{10\log_{10}\left(\frac{P_\mathrm{N}\gamma_\mathrm{th}}{P_\mathrm{T}\cdot 10^{\frac{A+B\log_{10}(D)}{10}}}\right)}{\sqrt{2}\sigma}\right) \\ & - e^{\frac{\sigma^2(\ln 10)^2}{2B^2}}\mathrm{erf}\left(\frac{\sigma\ln 10}{\sqrt{2}B} - \frac{10\log_{10}\left(\frac{P_\mathrm{N}\gamma_\mathrm{th}}{P_\mathrm{T}\cdot 10^{\frac{A+B\log_{10}(D)}{10}}}\right)}{\sqrt{2}\sigma}\right) + e^{\frac{\sigma^2(\ln 10)^2}{2B^2}}\Bigg] \end{aligned} \tag{A12}$$

Then, using $\mathrm{erfc}(x) = 1 - \mathrm{erf}(x)$, we can obtain (19).

Appendix D. Proof of Theorem 3

According to (13) and (3), $\partial f_{\gamma(l)}(r)/\partial l$ can be written as

$$\begin{aligned} \frac{\partial f_{\gamma(l)}(r)}{\partial l} &= \frac{\partial}{\partial l}\left[\frac{P_\mathrm{N}}{\sqrt{\pi}P_\mathrm{T}}\sum_{i=1}^{N_\mathrm{p}} H_i 10^{\frac{A+B\log_{10}(l)-\sqrt{2}\sigma t_i}{10}}\exp\left(-\frac{P_\mathrm{N}r}{P_\mathrm{T}}10^{\frac{A+B\log_{10}(l)-\sqrt{2}\sigma t_i}{10}}\right)\right] \\ &= \frac{P_\mathrm{N}}{\sqrt{\pi}P_\mathrm{T}}\sum_{i=1}^{N_\mathrm{p}} H_i 10^{\frac{A-\sqrt{2}\sigma t_i}{10}}\cdot\frac{\partial}{\partial l}\left[l^{\frac{B}{10}}\exp\left(-\frac{P_\mathrm{N}r}{P_\mathrm{T}}10^{\frac{A-\sqrt{2}\sigma t_i}{10}}l^{\frac{B}{10}}\right)\right] \\ &= \frac{P_\mathrm{N}B}{10\sqrt{\pi}P_\mathrm{T}}\sum_{i=1}^{N_\mathrm{p}} H_i 10^{\frac{A-\sqrt{2}\sigma t_i}{10}}\left(l^{\frac{B}{10}-1} - \frac{P_\mathrm{N}r}{P_\mathrm{T}}10^{\frac{A-\sqrt{2}\sigma t_i}{10}}l^{\frac{B}{5}-1}\right)e^{-\frac{P_\mathrm{N}r}{P_\mathrm{T}}10^{\frac{A-\sqrt{2}\sigma t_i}{10}}l^{\frac{B}{10}}} \end{aligned} \tag{A13}$$

Then, we have

$$\begin{aligned} \int_0^D l\int_0^{\gamma_\mathrm{th}}\frac{\partial}{\partial l}f_{\gamma(l)}(r)drdl &= \frac{P_\mathrm{N}B}{10\sqrt{\pi}P_\mathrm{T}}\sum_{i=1}^{N_\mathrm{p}} H_i 10^{\frac{A-\sqrt{2}\sigma t_i}{10}} \\ &\times \int_0^D l\underbrace{\int_0^{\gamma_\mathrm{th}}\left(l^{\frac{B}{10}-1} - \frac{P_\mathrm{N}}{P_\mathrm{T}}10^{\frac{A-\sqrt{2}\sigma t_i}{10}}l^{\frac{B}{5}-1}r\right)\exp\left(-\frac{P_\mathrm{N}}{P_\mathrm{T}}10^{\frac{A-\sqrt{2}\sigma t_i}{10}}l^{\frac{B}{10}}r\right)\mathrm{d}r\,\mathrm{d}l}_{\triangleq I_1} \end{aligned} \tag{A14}$$

For the integral term I_1 in (A14), we let $m = \frac{P_\mathrm{N}}{P_\mathrm{T}}10^{\frac{A-\sqrt{2}\sigma t_i}{10}}l^{\frac{B}{10}}r$ and obtain

$$\begin{aligned} I_1 &= \frac{P_\mathrm{T}}{P_\mathrm{N}}10^{\frac{\sqrt{2}\sigma t_i - A}{10}}l^{-1}\int_0^{\frac{P_\mathrm{N}}{P_\mathrm{T}}10^{\frac{A-\sqrt{2}\sigma t_i}{10}}l^{\frac{B}{10}}\gamma_\mathrm{th}}(1-m)\exp(-m)dm \\ &= l^{\frac{B}{10}-1}\gamma_\mathrm{th}\exp\left(-\frac{P_\mathrm{N}}{P_\mathrm{T}}10^{\frac{A-\sqrt{2}\sigma t_i}{10}}l^{\frac{B}{10}}\gamma_\mathrm{th}\right) \end{aligned} \tag{A15}$$

Substituting (A15) into (A14), we have

$$\int_0^D l \int_{\gamma_{\text{th}}}^{\infty} \frac{\partial}{\partial l} f_{\gamma(l)}(r) dr dl = \frac{BP_{\text{N}}\gamma_{\text{th}}}{10\sqrt{\pi}P_{\text{T}}} \sum_{i=1}^{N_{\text{p}}} H_i 10^{\frac{A-\sqrt{2}\sigma t_i}{10}} \times \underbrace{\int_0^D l^{\frac{B}{10}} \exp\left(-\frac{P_{\text{N}}}{P_{\text{T}}} 10^{\frac{A-\sqrt{2}\sigma t_i}{10}} l^{\frac{B}{10}} \gamma_{\text{th}}\right) \text{d}l}_{\triangleq I_2} \tag{A16}$$

For the integral term I_2 in (A16), we let $m = \frac{P_{\text{N}}}{P_{\text{T}}} 10^{\frac{A-\sqrt{2}\sigma t_i}{10}} l^{\frac{B}{10}} \gamma_{\text{th}}$ and obtain

$$\begin{aligned} I_2 &= \int_0^{\frac{P_{\text{N}}}{P_{\text{T}}} 10^{\frac{A-\sqrt{2}\sigma t_i}{10}} D^{\frac{B}{10}} \gamma_{\text{th}}} \left(10^{\frac{\sqrt{2}\sigma t_i - A}{10}} \frac{P_{\text{T}} m}{P_{\text{N}}\gamma_{\text{th}}}\right) \exp(-m) \left(10^{\frac{\sqrt{2}\sigma t_i - A}{10}} \frac{P_{\text{T}}}{P_{\text{N}}\gamma_{\text{th}}}\right)^{\frac{10}{B}} \frac{10}{B} m^{\frac{10}{B}-1} \text{d}m \\ &= \frac{10}{B}\left(10^{\frac{\sqrt{2}\sigma t_i - A}{10}} \frac{P_{\text{T}}}{P_{\text{N}}\gamma_{\text{th}}}\right)^{\frac{10}{B}+1} \int_0^{\frac{P_{\text{N}}}{P_{\text{T}}} 10^{\frac{A-\sqrt{2}\sigma t_i}{10}} D^{\frac{B}{10}} \gamma_{\text{th}}} m^{\frac{10}{B}} \exp(-m) \text{d}m \\ &= \frac{10}{B}\left(10^{\frac{\sqrt{2}\sigma t_i - A}{10}} \frac{P_{\text{T}}}{P_{\text{N}}\gamma_{\text{th}}}\right)^{\frac{10}{B}+1} \gamma\left(\frac{10}{B}+1, \frac{P_{\text{N}}}{P_{\text{T}}} 10^{\frac{A-\sqrt{2}\sigma t_i}{10}} D^{\frac{B}{10}} \gamma_{\text{th}}\right) \end{aligned} \tag{A17}$$

Substituting (A16) and (A17) into (20), we can obtain (21).

Appendix E. Proof of Corollary 3

Similar to (20), we can obtain the relationship between the ECP and the percentage of CCA as

$$P_{\text{a}}' = P_{\text{e}}' + \frac{1}{D} \int_0^D l \int_0^{\gamma_{\text{th}}} \frac{\partial}{\partial l} f_{\gamma'(l)}(r) dr \text{d}l \tag{A18}$$

In this case, $\partial f_{\gamma'(l)}(r)/\partial l$ can be written as

$$\begin{aligned} \frac{\partial f_{\gamma'(l)}(r)}{\partial l} &= \frac{\partial f_{\Omega(l)}(r)}{\partial l} \\ &= \frac{\xi}{\sqrt{2\pi}\sigma r} \frac{\partial}{\partial l} \exp\left(-\frac{\left(10\log_{10}\left(\frac{P_{\text{N}} r}{P_{\text{T}} \cdot PL(l)}\right)\right)^2}{2\sigma^2}\right) \\ &= -\frac{B\xi^2 \log_{10}\left(\frac{P_{\text{N}} r}{P_{\text{T}} \cdot PL(l)}\right)}{\sigma^3 l \sqrt{2\pi} r} \cdot \exp\left(-\frac{\left(10\log_{10}\left(\frac{P_{\text{N}} r}{P_{\text{T}} \cdot PL(l)}\right)\right)^2}{2\sigma^2}\right) \end{aligned} \tag{A19}$$

Then, we have

$$\begin{aligned} \int_0^{\gamma_{\text{th}}} \frac{\partial}{\partial l} f_{\gamma'(l)}(r) \text{d}r &= -\frac{B\xi^2}{\sigma^3 l \sqrt{2\pi}} \int_0^{\gamma_{\text{th}}} \frac{\log_{10}\left(\frac{P_{\text{N}} r}{P_{\text{T}} \cdot PL(l)}\right)}{r} \cdot \exp\left(-\frac{\left(10\log_{10}\left(\frac{P_{\text{N}} r}{P_{\text{T}} \cdot PL(l)}\right)\right)^2}{2\sigma^2}\right) \text{d}r \\ &= \frac{B}{\sigma l \sqrt{2\pi} \ln 10} \exp\left(-\frac{\left(10\log_{10}\left(\frac{P_{\text{N}}\gamma_{\text{th}}}{P_{\text{T}} \cdot PL(l)}\right)\right)^2}{2\sigma^2}\right) \end{aligned} \tag{A20}$$

Furthermore, let $m = 10\log_{10}\left(\frac{P_{\text{N}}\gamma_{\text{th}}}{P_{\text{T}}} 10^{\frac{A+B\log_{10}(l)}{10}}\right)/\sigma$, we have

$$\int_0^D l\int_0^{\gamma_{\rm th}}\frac{\partial}{\partial l}f_{\gamma'(l)}(r){\rm d}r{\rm d}l = \frac{B}{\sigma\sqrt{2\pi}\ln 10}\int_0^D \exp\left(-\frac{\left(10\log_{10}\left(\frac{P_{\rm N}\gamma_{\rm th}}{P_{\rm T}}10^{\frac{A+B\log_{10}(l)}{10}}\right)\right)^2}{2\sigma^2}\right){\rm d}l$$

$$= \frac{1}{\sqrt{2\pi}}\left(\frac{P_{\rm N}\gamma_{\rm th}}{P_{\rm T}}\right)^{-\frac{10}{B}}10^{-\frac{A}{B}}\int_{-\infty}^{\frac{10\log_{10}\left(\frac{P_{\rm N}\gamma_{\rm th}}{P_{\rm T}}10^{\frac{A+B\log_{10}(D)}{10}}\right)}{\sigma}} e^{-\frac{m^2}{2}}10^{\frac{\sigma m}{B}}{\rm d}m \tag{A21}$$

Then, we use $a^x = e^{\ln a^x} = e^{x\ln a}$, and obtain

$$\int_0^D l\int_0^{\gamma_{\rm th}}\frac{\partial}{\partial l}f_{\gamma'(l)}(r){\rm d}r{\rm d}l = \frac{1}{\sqrt{2\pi}}\left(\frac{P_{\rm N}\gamma_{\rm th}}{P_{\rm T}}\right)^{-\frac{10}{B}}10^{-\frac{A}{B}}\int_{-\infty}^{\frac{10\log_{10}\left(\frac{P_{\rm N}\gamma_{\rm th}}{P_{\rm T}}10^{\frac{A+B\log_{10}(D)}{10}}\right)}{\sigma}} e^{-\frac{m^2}{2}+\frac{\sigma\ln 10}{B}m}{\rm d}m$$

$$= \left(\frac{P_{\rm N}\gamma_{\rm th}}{P_{\rm T}}\right)^{-\frac{10}{B}}10^{-\frac{A}{B}}\exp\left(\frac{\sigma^2(\ln 10)^2}{2B^2}\right)\left[1-Q\left(\frac{10\log_{10}\left(\frac{P_{\rm N}\gamma_{\rm th}}{P_{\rm T}\cdot PL(D)}\right)}{\sigma}-\frac{\sigma\ln 10}{B}\right)\right] \tag{A22}$$

Substituting (A22) into (A18), we can obtain (22).

References

1. Li, P.; Zhu, Y.; Liu, Y.; Jin, C.Research on system architecture, basic platform, and development path of autonomous intelligent high-speed railway systems. *IEEE Intell. Transport. Syst. Mag.* **2024**, *16*, 62–73. [CrossRef]
2. Goldsmith, A. *Wireless Communications*; Cambridge University Press: Cambridge, UK, 2005.
3. Borralho, R.; Quddus, A.U.; Mohamed, A.; Vieira, P.; Tafazolli, R. Coverage and data rate analysis for a novel cell-sweeping-based RAN deployment. *IEEE Trans. Wireles Commun.* **2024**, *23*, 217–230. [CrossRef]
4. Reudink, D.O. Properties of mobile radio propagation above 400 MHz. *IEEE Trans. Veh. Technol.* **1974**, *VT-23*, 143–159. [CrossRef]
5. Homnan, B. On the cell coverage area of overlap area-based cellular model in wireless cellular systems: Base station locations. In Proceedings of the International Conference on Future Networks, Bangkok, Thailand, 7–9 March 2009; pp. 113–116.
6. Yang, H. Simple formulas for area coverage probability of cellular wireless networks. In Proceedings of the 2012 IEEE 75th Vehicular Technology Conference, Yokohama, Japan, 6–9 May 2012; pp. 1–5.
7. Zhang, J.-Y.; Tan, Z.-H.; Yu, Z.-Z. Coverage efficiency of radio-over-fiber network for high-speed railways. In Proceedings of the 6th International Conference on Wireless Communications, Networking and Mobile Computing, Chengdu, China, 23–25 September 2010; pp. 1–4.
8. Gui, L.; Ma, W.; Liu, B.; Shen, P. Single frequency network system coverage and trial testing of high speed railway television system. *IEEE Trans. Broadcast.* **2010**, *56*, 160–170.
9. Gómez, J.M.N.; Castanho, R.A.; Fernández, J.C.; Loures, L.C. Assessment of high-speed rail service coverage in municipalities of peninsular Spain. *Infrastructures* **2020**, *5*, 11. [CrossRef]
10. Munjal, M.; Singh, N.P. Low cost communication for high speed railway. *Wireless Pers. Commun.* **2020**, *111*, 163–178. [CrossRef]
11. Munjal, M.; Singh, N.P. Group mobility by cooperative communication for high speed railway. *Wireless Netw.* **2019**, *25*, 3857–3866. [CrossRef]
12. Ai, B.; He, R.; Li, G.; Guan, K.; He, D.; Zhing, Z. Determination of cell coverage area and its applications in high-speed railway environments. *IEEE Trans. Veh. Technol.* **2017**, *66*, 3515–3525. [CrossRef]
13. Liu, X.; Qiao, D. Design and coverage analysis of elliptical cell based beamforming for HSR communication systems. In Proceedings of the IEEE International Conference on Communications, Shanghai, China, 20–24 May 2019; pp. 1–6.
14. Taheri, M.; Ansari, N.; Feng, J.; Rojas-Cessa, R.; Zhou, M. Provisioning internet access using FSO in high-speed rail networks. *IEEE Netw.* **2017**, *31*, 96–101. [CrossRef]
15. Fan, Q.; Taheri, M.; Ansari, N.; Feng, J.; Rojas-Cessa, R.; Zhou, M.; Zhang, T. Reducing the impact of handovers in ground-to-train free space optical communications. *IEEE Trans. Veh. Technol.* **2018**, *67*, 1292–1301. [CrossRef]
16. Lin, S.-H.; Zu, Y.; Wang, J.-Y. Coverage analysis and optimization for high-speed railway communication systems with narrow-strip-shaped cells. *IEEE Trans. Veh. Technol.* **2020**, *69*, 11544–11556. [CrossRef]
17. Lin, S.-H.; Xu, Y.; Wang, L.; Wang, J.-Y. Coverage analysis and chance-constrained optimization for HSR communications with carrier aggregation. *IEEE Trans. Intell. Transp. Syst.* **2022**, *23*, 15107–15120. [CrossRef]
18. Lin, S.-H.; Xu, Y.; Wang, J.-Y. Coverage analysis for cell-free massive MIMO high-speed railway communication systems. *IEEE Trans. Veh. Technol.* **2022**, *71*, 10499–10511. [CrossRef]

19. He, R.; Zhing, Z.; Ai, B.; Guan, K. Reducing the cost of high-speed railway communications: From the propagation channel view. *IEEE Trans. Intell. Transp. Syst.* **2015**, *16*, 2050–2060. [CrossRef]
20. Liu, L.; Tao, C.; Qiu, J.; Chen, H.; Yu, L.; Dong, W.; Yuan, Y. Position-based modeling for wireless channel on high-speed railway under a viaduct at 2.35 GHz. *IEEE J. Sel. Areas Commun.* **2012**, *30*, 834–845. [CrossRef]
21. Li, X.; Shen, C.; Bo, A.; Zhu, G. Finite-state Markov modeling of fading channels: A field measurement in high-speed railways. In Proceedings of the 2013 IEEE/CIC International Conference on Communications in China, Xi'an, China, 12–14 August 2013; pp. 577–582.
22. Li, T.; Wang, X.; Fan, P.; Riihonen, T. Position-aided large-scale MIMO channel estimation for high-speed railway communication systems. *IEEE Trans. Veh. Technol.* **2017**, *66*, 8964–8978. [CrossRef]
23. He, R.; Zhong, Z.; Ai, B.; Wang, G.; Ding, J.; Molisch, A.F. Measurements and analysis of propagation channels in high-speed railway viaducts. *IEEE Trans. Wireless Commun.* **2013**, *12*, 794–805. [CrossRef]
24. Abramowitz, M.; Stegun, I.A. *Handbook of Mathematical Functions with Formulas, Graphs, and Mathematical Tables*, 9th ed.; Dover Publications: New York, NY, USA, 1970.
25. Farid, A.A.; Hranilovic, S. Outage capacity optimization for free-space optical links with pointing errors. *J. Lightwave Technol.* **2007**, *25*, 1702–1710. [CrossRef]

entropy

MDPI

Article

A Three-Dimensional Time-Varying Channel Model for THz UAV-Based Dual-Mobility Channels

Kai Zhang [1,†], Fenglei Zhang [1], Yongjun Li [1,*], Xiang Wang [1,*], Zhaohui Yang [2], Yuanhao Liu [1], Changming Zhang [3] and Xin Li [1]

[1] School of Information and Navigation, Air Force Engineering University, Xi'an 710077, China; 13262733921@163.com (K.Z.); 15029889024@163.com (F.Z.); liu_yuanhao@outlook.com (Y.L.); 18829443976@163.com (X.L.)

[2] College of Information Science and Electronic Engineering, Zhejiang University, Hangzhou 310027, China; yang_zhaohui@zju.edu.cn

[3] Research Center of Novel Computational Sensing and Intelligent Processing, Zhejiang Lab, Hangzhou 311121, China; zhangcm@zhejianglab.com

[*] Correspondence: tz_228@163.com (Y.L.); lleafwx626@126.com (X.W.)

[†] Current address: 1 Fenghao East Road, Lianhu District, Xi'an 710082, China.

Abstract: Unmanned aerial vehicle (UAV) as an aerial base station or relay device is a promising technology to rapidly provide wireless connectivity to ground device. Given UAV's agility and mobility, ground user's mobility, a key question is how to analyze and value the performance of UAV-based wireless channel in the terahertz (THz) band. In this paper, a three-dimensional (3D) time-varying channel model is proposed for UAV-based dual-mobility wireless channels based on geometric channel model theory in THz band. In this proposed channel model, the small-scale fading (e.g., scattering fading and reflection fading) on rough surfaces of communication environment and the atmospheric molecule absorption attenuations are considered in THz band. Moreover, the statistical properties of the proposed channel model, including path loss, time autocorrelation function (T-ACF) and Doppler power spectrum density (DPSD), have been derived and the impact of several important UAV-related and vehicle-related parameters have been investigated and compared to millimeter wave (mm-wave) band. Furthermore, the correctness of the proposed channel model has been verified via simulation, and some useful observations are provided for the system design of THz UAV-based dual-mobility wireless communication systems.

Keywords: Time-varying channel; terahertz (THz); unmanned aerial vehicle (UAV); channel model; path loss; time auto-correlation function (T-ACF); Doppler power spectrum density (DPSD)

1. Introduction

The space-air-ground integrate networks (SAGINs) and THz communication technology as the key technologies of the sixth-generation (6G) mobility communication technology, they have gained significant attention in recent years. SAGIN can provide seamless connectivity for future communication systems, THz communication technology can provide ultra-high data rate and ultra-wideband. With the advantages of high flexibility, rapid 3-D deployment, low cost and high mobility, unmanned aerial vehicles (UAVs) can be deployed in the air to increase the probability of Line-of-Sight (LoS) transmissions when working as a relay to establish wireless communication links between the mobile devices (MDs) on the ground and BSs. In SAGINs, UAV is widely used in many scenarios [1] serving as air platforms (e.g., base station and relay station) for the wireless communication systems, and UAV is a promising technology to rapidly provide wireless connectivity to ground mobile/stationary devices or air device. Therefore, it is foreseen that the future UAV-based communication systems can support ultra-high data rate [2], which requires ultra-large transmission bandwidth. For the THz band, it has rich spectrum resources, which can meet the needs of ultra-high date rate. The channel model is essential for wireless communication technology in the THz bands. It accurately characterizes UAV-based wireless channels and is helpful to system design and performance evaluation of UAV-based THz wireless communications. However, so far, there are limited studies on channel modeling for UAV-based communications in the THz band.

The UAV has several important characteristics that make it the ideal choice for some application scenarios. These include rapid three-dimensional (3D) deployment, low cost, and high mobility. It can serve as both an air base station and a relay station, and can be used in various fields including commercial and military domains [3–8]. In general, UAVs have two main applications in wireless communications: air-to-ground (A2G) [9–13] and air-to-air (A2A) [11,14]. The existing literature clearly shows that the time-varying channel properties of the wireless channel of UAV-based communication, including the number of paths, average fading duration (AFD), space-time correlation (STC), path gain, Doppler power spectrum density (DPSD), and dynamic angle parameters, have been extensively investigated in different scenarios and carrier frequencies (e.g., microwave [9,10] and millimeter wave [11–13] bands).

Meanwhile, the data rate requirement for wireless communication via UAVs is increasing. The terahertz (THz) band, ranging from 0.1 THz to 10 THz, offers significant benefits due to its huge bandwidth. This facilitates ultra-high data transmission rates in various wireless applications. The peak data rates of 160 Gb/s and 1.176 Tbps have been achieved for THz wireless communications in [14,15]. The existing literature clearly shows that studies of THz and sub-THz wireless channels have been carried out both indoors and outdoors. These studies have involved measurements [16–18], channel modeling [16,17,19,20], and performance analysis [18–20]. Then, the statistical properties of the wireless channel have been investigated, including the path loss, scattering attenuation [18], and space-time-frequency correlation (STFC) [19]. The hybrid dynamic channel model and non-stationary space-time-frequency channel model in the THz band were proposed in [20]. Additionally, the authors from [20] take into consideration atmospheric absorption attenuations. The authors in [21] proposed a THz channel model for indoor scenarios based on ray tracing. They studied the characteristics of propagation paths with LoS paths, reflection paths, scattering paths, and diffraction paths.

In addition, several studies have been conducted on the channel modeling of UAV-based wireless communication in sub-THz bands [22–24]. The existing literature has investigated the three-dimensional (3D) time-varying channel model of UAV-based wireless communication for the A2G and A2A scenarios. However, there have been limited studies on the UAV-based wireless channels in the THz band. Furthermore, these studies only consider the impact of reflection and scattering fading on rough surfaces, but not the impact of atmospheric absorption attenuation.

In this paper, we present a novel channel model for THz UAV-based dual-mobility channels. This proposed model accurately characterizes the time-varying dual-mobility scenario in the THz UAV-based channels by combining the geometry-based stochastic model (GBSM), reflection and scattering fading, and atmospheric absorption attenuation. In addition, we have investigated the impact of some important time-varying UAV-related parameters on the channel characteristics, including the path loss, time autocorrelation function (T-ACF), and DPSD.

2. Modeling of Uav-Based Dual-Mobility Wireless Channels in the Thz Band

2.1. Description of the Wireless Communication System

This paper assumes that the antennas at the transmitter (Tx) and receiver (Rx) for the considered UAV-based wireless communication system are isotropic [18]. The antennas at Tx and Rx are located on the UAV and vehicle, respectively. The UAV-based dual-mobility communication system exhibit several characteristics, including 3D random deployment and mobility, flexible environment adaptability, and high space-time variation. The real UAV-based dual-mobility wireless communications scenario in the THz band is shown in Figure 1, where the propagation paths between UAV and vehicle are time-varying. In a time-varying wireless channel, the multipaths are informed in wireless environments with free space, obstructions, and rough surfaces. In general, the propagation path can be categorized into line-of-sight (LoS), and non-line-of-sight (NLoS) cases, as shown in Figure 2.

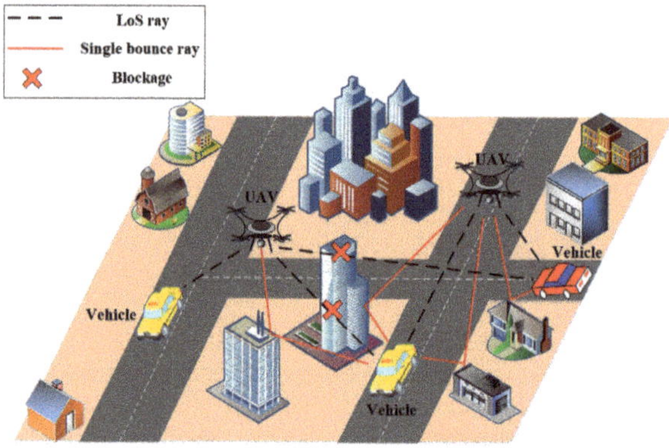

Figure 1. Real UAV-based dual-mobility wireless communications scenario in the THz band.

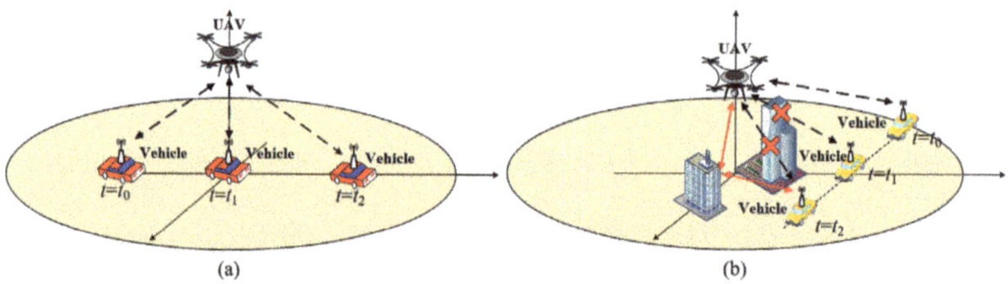

Figure 2. Different propagation paths between UAV and vehicle in time-varying UAV-based wireless communication system in the THz band: (**a**) LoS propagation path survival; (**b**) LoS propagation path death and NLoS propagation path birth.

The height of Tx is always higher than that of Rx, so the scattering and reflection propagations around the vehicles are more than those around the UAV. Furthermore, the wavelength of the THz band is smaller than that of the mm-wave and microwave bands. This means that the probability of scattering and reflection on the rough surface of different materials in the THz band is greater than that in the mm-wave and microwave bands. Besides, the atmospheric molecule absorption attenuation of time-varying wireless channel $A_{\mathrm{ma}}(f_c, t)$ increases rapidly with the increasing of the radio-wave frequency. In particular, $A_{\mathrm{ma}}(f_c, t)$ in the THz band is higher than that in the mm-wave and micro-wave bands obviously. In this paper, f_c denotes radio-wave frequency.

In signal processing and wireless communications [13], the output signal $y(t)$ of a wireless communciation system is a function of the channel impulse response (CIR) $h(t)$ and an input signal $x(t)$, plus some random noise $n(t)$, which can be expressed as $y(t) = x(t) * h(t) + n(t)$.

The CIR in the time-varying channels can be further expressed as a function of delay $\tau(t)$, and Doppler frequency shift $f_D(t)$, which can be written as:

$$h(t) = \sum_{i=1}^{I(t)} \alpha_i(t) e^{j[2\pi t f_{D,i}(t) - 2\pi f_c \tau_i(t)]} \quad (1)$$

where $I(t)$ is the number of the time-varying multipath components (MPCs), $\alpha_i(t)$ is the channel gain value of the time-varying path, $\tau_i(t)$ is the delay of the time-varying path. In our modeling process, we consider the LoS, reflection, and scattering paths in the channel model to be separate.

2.2. Time-Varying Small-Scale Fading Coefficient of the NLoS Path on the Rough Surfaces

This paper precisely describes the time-varying small-scale fading coefficients of the NLoS path on rough surfaces for UAV-based dual-mobility wireless communication in the THz band. It considers both reflection and scattering paths on rough surfaces.

According to the Kirchhoff scattering theory [19] and the modified Beckmann-Kirchhoff theory [19,21], the reflection coefficient $R(f_c,t)$ and scattering coefficient $S(f_c,t)$ for a rough surface can be calculated as:

$$R(f_c,t) = -e^{\frac{-2\cos\theta_i(t)}{\sqrt{n_{t,\text{ref}}^2-1}}} \cdot e^{-\frac{8\pi^2 f_c^2 \sigma_{h,\text{ref}}^2 \cos^2[\theta_i(t)]}{c^2}} \tag{2}$$

$$\begin{aligned} S(f_c,t) = &-e^{\frac{-2\cos\theta_1(t)}{\sqrt{n_{t,\text{sca}}^2-1}}} \cdot \left[1+g(f_c,t)+\frac{g^2(f_c,t)}{2}+\frac{g^3(f_c,t)}{6}\right]^{-\frac{1}{2}} \\ &\times \left\{\rho_0^2(f_c,t)+\frac{\pi\cos\theta_1(t)}{100}\cdot\left[g(f_c,t)e^{-v_{xy}(f_c,t)}+\frac{g^2(f_c,t)}{4}\cdot e^{-\frac{v_{xy}(f_c,t)}{2}}\right]\right\}^{\frac{1}{2}} \end{aligned} \tag{3}$$

where $n_{t,\text{ref}}$ and $n_{t,\text{sca}}$ are the refractive indexes in reflection and scattering propagations respectively, $\sigma_{h,\text{ref}}$ is the standard deviation in the height of rough surface in reflection propagation, which is commonly considered to be Gaussian distributed, c denotes the light speed, $\theta_i(t)$ and $\theta_1(t)$ are the angles of incident wave in reflection and scattering propagations respectively [19,21]. Detailed computation for the parameters including $g(f_c,t)$, $\rho_0(f_c,t)$ and $v_{xy}(f_c,t)$ can be referred to [19].

2.3. Atmospheric Molecule Absorption Attenuation of Radio-Wave for the Wireless Channel

It is clear that water vapor and oxygen molecules have a much greater impact on the absorption attenuation of radio waves than other molecules [25]. This paper considers only the absorption attenuation caused by oxygen and water vapor molecules in the atmosphere. Then, according to the attenuation experience model of oxygen and water vapor molecules in ITU-R P.676-11 [25], the atmospheric molecule absorption attenuation can be described as:

2.3.1. Attenuation Coefficient of Oxygen Molecule to Radio-Wave $\gamma_o(f_c)$

The frequency of radio-wave f_c is from 100 GHz to 350 GHz, the atmospheric pressure P is 1013 hpa, and the temperature of environment T is 15 °C, then $\gamma_o(f_c)$[dB/km] can be calculated by [25]:

$$\begin{aligned} \gamma_o(f_c) = &\left[3.79\times10^{-7}f_c+\frac{0.265}{(f_c-63)^2+1.59}+\frac{0.028}{(f_c-118)^2+1.47}\right] \\ &\times(f_c+198)^2\times10^{-3} \end{aligned} \tag{4}$$

2.3.2. Attenuation Coefficient of Water Vapor Molecule to Radio-Wave $\gamma_w(f_c)$

The frequency of radio-wave f_c is from 100 GHz to 350 GHz, the atmospheric pressure P is 1013 hpa, and the temperature of environment T is 15 °C, then $\gamma_w(f_c)$[dB/km] can be calculated by [25]:

$$\begin{aligned} \gamma_w(f_c) = &\left[0.067+\frac{3}{(f_c-22.3)^2+7.3}+\frac{9}{(f_c-183.3)^2+6}+\frac{4.3}{(f_c-323.8)^2+10}\right] \\ &\times f_c^2\rho_w\times10^{-4} \end{aligned} \tag{5}$$

When the environment temperature T is from -20 °C to +40 °C, the attenuation coefficient of oxygen and water vapor molecules can be expressed as [25]:

$$\gamma_o^{'}(f_c) = \gamma_o(f_c)[1 - 0.01 \times (T - 15)] \tag{6}$$

$$\gamma_w^{'}(f_c) = \gamma_w(f_c)[1 - 0.006 \times (T - 15)] \tag{7}$$

In UAV-based dual-mobility wireless communications, the heights of the transmitter and receiver are always different. To calculate atmospheric absorption attenuation, multiply the equivalent height of oxygen by the attenuation coefficient of the oxygen molecule and the equivalent height of water vapor by the attenuation coefficient of the water vapor. This is described in the ITU-R P.676-11 model theory [25]. Besides, the expression of atmospheric absorption attenuation varies with the elevations angle of the propagation path.

When the frequency range of radio-wave is from 100 GHz to 350 GHz, the equivalent heights of oxygen molecule and water vapor can be calculated as [25]:

(i) Oxygen molecule O_2:

$$h_o(f_c) = 5.542 - 1.76414 \times 10^{-3} f_c + 3.05354 \times 10^{-6} f_c^2 + \frac{6.815}{(f_c - 118.75)^2 + 0.321} \tag{8}$$

(ii) Water vapor H_2O:

$$h_w(f_c) = 1.65 \times \left[1 + \frac{1.61}{(f_c - 22.23)^2 + 2.91} + \frac{3.33}{(f_c - 183.3)^2 + 4.58} + \frac{1.9}{(f_c - 3235.1)^2 + 3.34}\right] \tag{9}$$

where the unit of f_c is GHz and the unit of height is km in Equations (4)–(9).

When the elevation angle range of the propagation path is from 0° to 5°, the atmospheric absorption attenuation can be calculated as [25]:

$$\begin{aligned} A_{\text{ma}}^{\text{LoS}}(f_c, t) = {} & \gamma_o^{'}(f_c)\sqrt{h_o(f_c)}\left\{\frac{\sqrt{r_e + h_1} \cdot F[x_1(t)]e^{-\frac{h_1}{h_o(f_c)}}}{\cos[\theta(t)]} - \frac{\sqrt{r_e + h_2} \cdot F[x_2(t)]e^{-\frac{h_2}{h_o(f_c)}}}{\cos[\theta'(t)]}\right\} \\ & + \gamma_w^{'}(f_c)\sqrt{h_w(f_c)}\left\{\frac{\sqrt{r_e + h_1} \cdot F[x_1(t)]e^{-\frac{h_1}{h_w(f_c)}}}{\cos[\theta(t)]} - \frac{\sqrt{r_e + h_2} \cdot F[x_2(t)]e^{-\frac{h_2}{h_w(f_c)}}}{\cos[\theta'(t)]}\right\} \end{aligned} \tag{10}$$

where r_e denotes the efficient radius of the earth in km, h_1 denotes the smaller height between Tx and Rx, h_2 denotes the higher height between Tx and Rx, and θ denotes the elevation from h_1 to h_2. The functions of $F(\cdot)$, $\theta(t)$, $\theta^{'}(t)$, $x_i(t)$ and $x_i^{'}(t)$ are referred to [25].

When the elevation angle range of the propagation path is from 5° to 90°, the atmospheric absorption attenuation can be calculated as [25]:

$$A_{\text{ma}}^{\text{LoS}}(f_c, t) = \frac{\gamma_o^{'}(f_c)h_o^{'}(f_c) + \gamma_w^{'}(f_c)h_w^{'}(f_c)}{\sin[\theta(t)]} \tag{11}$$

where $h_o^{'}(f_c)$ and $h_w^{'}(f_c)$ denote the modified equivalent heights of the absorption attenuation of oxygen molecule and water vapor respectively, which are calculated as [25]:

$$h_o^{'}(f_c) = h_o(f_c)[e^{-\frac{h_1}{h_o(f_c)}} - e^{-\frac{h_2}{h_o(f_c)}}] \tag{12}$$

$$h_w^{'}(f_c) = h_w(f_c)[e^{-\frac{h_1}{h_w(f_c)}} - e^{-\frac{h_2}{h_w(f_c)}}] \tag{13}$$

2.4. Propagation Scenarios for the Time-Varying Channel

This paper categorizes propagation paths into two main types: LoS and NLoS (i.e., reflection and scattering propagation paths) scenarios, as shown in Figure 2a,b.

2.4.1. LoS Propagation Path

In a time-varying LoS propagation channel of this paper, there are no obstructions between the propagation paths of Tx and Rx, and as shown in Figures 1 and 2a. Then, the path loss PL_{LoS} and delay τ_{LoS} of the LoS propagation path are calculated as:

$$PL_{\text{LoS}}(t) = 20\log_{10}\left[\frac{4\pi d_{\text{LoS}}(t)}{\lambda_c}\right] \tag{14}$$

$$\tau_{\text{LoS}}(t) = \frac{d_{\text{LoS}}(t)}{c} \tag{15}$$

where $d_{\text{LoS}}(t)$ is the 3D distance between Tx and Rx for the LoS propagation path, $\lambda_c = f_c/c$ is the wavelength of radio-wave, and c is the speed of light.

2.4.2. NLoS Propagation Path

In a time-varying channel of this paper, besides the direct LoS propagation path, the NLoS propagation paths usually undergo reflection and scattering on the rough surface and obstruction before arriving at the Rx antenna array as shown in Figures 1 and 2b. In this paper, we focus on the reflection and scattering on the rough surface for the NLoS propagation paths. Besides, every NLoS propagation path only experience single reflection or scattering before arriving at the Rx antenna, i.e., single bounce propagation. The properties of time-varying in NLoS channels can be characterized using the Cartesian coordinates of the Tx and Rx antenna, points of scattering and reflection. In addition, the moving directions of the Tx (i.e., moving angles of elevation and azimuth β_{Tx} and α_{Tx}) and Rx (i.e., moving angle of azimuth α_{Rx}), the locations of point of reflection and scattering are assumed to be arbitrary in this paper.

Accordingly, the angles of reflection and scattering will also be time-varying in this paper. Given the Cartesian coordinates of Tx, Rx, points of reflection and scattering, the time-varying path length and delay of each reflected and scattered ray can be calculated respectively, in both elevation and azimuth planes. Then, the time-varying propagation path lengths of reflection at the n-th ($n = 1, ..., N(t)$, $N(t)$ denotes the time-varying number of efficient reflection points) reflection point and scattering path at the m-th ($m = 1, ..., M(t)$, $M(t)$ denotes the time-varying number of efficient scattering points) scattering point are calculated as $d^{\text{ref}}_{\text{NLoS},n}(t) = |d^{\text{ref}}_{\text{Tx},n}(t)| + |d^{\text{ref}}_{\text{Rx},n}(t)|$ and $d^{\text{sca}}_{\text{NLoS},m}(t) = |d^{\text{sca}}_{\text{Tx},m}(t)| + |d^{\text{sca}}_{\text{Rx},m}(t)|$ respectively, where $d^{\text{ref}}_{\text{Tx},n}(t)$ and $d^{\text{ref}}_{\text{Rx},n}(t)$ are the time-varying path lengths seen by the Tx and Rx for the reflection propagation path at the n-th reflection point, respectively; $d^{\text{sca}}_{\text{Tx},m}(t)$ and $d^{\text{sca}}_{\text{Rx},m}(t)$ are the time-varying path lengths seen by the Tx and Rx for the scattering propagation path at the m-th scattering point, respectively. Hence, the delays of arrival of the n-th reflected and m-th scattered rays can be calculated as $\tau^{\text{ref}}_{\text{NLoS},n}(t) = d^{\text{ref}}_{\text{NLoS},n}(t)/c$ and $\tau^{\text{sca}}_{\text{NLoS},m}(t) = d^{\text{sca}}_{\text{NLoS},m}(t)/c$, and the path loss of n-th reflection propagation path and m-th scattering propagation path can also be calculated as:

$$PL^{\text{ref}}_{\text{NLoS},n}(t) = 20\log_{10}\left[\frac{16\pi^2 d^{\text{ref}}_{\text{Tx},n}(t)\cdot d^{\text{ref}}_{\text{Rx},n}(t)}{\lambda_c^2\cdot R_n(f_c,t)}\right] \tag{16}$$

$$PL^{\text{sca}}_{\text{NLoS},m}(t) = 20\log_{10}\left[\frac{16\pi^2 d^{\text{sca}}_{\text{Tx},m}(t)\cdot d^{\text{sca}}_{\text{Rx},m}(t)}{\lambda_c^2\cdot S_m(f_c,t)}\right] \tag{17}$$

where $R_n(f_c,t)$ and $S_m(f_c,t)$ are the time-varying reflection coefficient at the n-th reflection propagation path and the time-varying scattering coefficient at the m-th scattering propagation path, respectively.

2.4.3. Time Auto-Correlation Function

According to the Equation (2) and description of propagation path above this paper, we can assume that the LoS and NLoS paths are independent of each other, and the reflection and scattering paths in NLoS path are independent of each other, then the time-varying CIR from Tx to Rx can be written as:

$$h(t) = h_{\rm LoS}(t) + h^{\rm ref}_{\rm NLoS}(t) + h^{\rm sca}_{\rm NLoS}(t) \tag{18}$$

where

$$h_{\rm LoS}(t) = \sqrt{\frac{\Omega(t)K(t)}{K(t)+1}} \cdot e^{j\{2\pi t[f^{\rm Tx}_{\rm D,LoS}(t)+f^{\rm Rx}_{\rm D,LoS}(t)]-2\pi f_c \tau_{\rm LoS}(t)\}} \tag{19}$$

$$\begin{aligned} h^{\rm ref}_{\rm NLoS}(t) = & \sqrt{\frac{\Omega(t)\eta_{\rm ref}(t)}{N(t)[K(t)+1]}} \sum_{n=1}^{N(t)} \alpha^{\rm ref}_{\rm NLoS,n}(t) \cdot e^{j\theta_n} \\ & \times e^{j\{2\pi t[f^{\rm Tx,ref}_{\rm D,NLoS,n}(t)+f^{\rm Rx,ref}_{\rm D,NLoS,n}(t)]-2\pi f_c \tau^{\rm ref}_{\rm NLoS,n}(t)\}} \end{aligned} \tag{20}$$

$$\begin{aligned} h^{\rm sca}_{\rm NLoS}(t) = & \sqrt{\frac{\Omega(t)\eta_{\rm sca}(t)}{M(t)[K(t)+1]}} \sum_{m=1}^{M(t)} \alpha^{\rm sca}_{\rm NLoS,m}(t) \cdot e^{j\theta_m} \\ & \times e^{j\{2\pi t[f^{\rm Tx,sca}_{\rm D,NLoS,m}(t)+f^{\rm Rx,sca}_{\rm D,NLoS,m}(t)]-2\pi f_c \tau^{\rm sca}_{\rm NLoS,m}(t)\}} \end{aligned} \tag{21}$$

with the time-varying Doppler frequency shifts terms $f^{\rm Tx}_{\rm D,LoS}(t)$, $f^{\rm Tx,ref}_{\rm D,NLoS,n}(t)$, $f^{\rm Rx,ref}_{\rm D,NLoS,n}(t)$, $f^{\rm Tx,sca}_{\rm D,NLoS,m}(t)$, and $f^{\rm Rx,sca}_{\rm D,NLoS,m}(t)$.

In Equations (19)–(21), $K(t)$ denotes the Rician factor, and we assumed that $K(t) = K$ in this paper. $\Omega(t) = {\rm E}[|h(t)|^2]$ (where ${\rm E}[\cdot]$ is the statistical expectation operation) is the propagation path power from Tx to Rx. $\eta_{\rm ref}(t)$ and $\eta_{\rm sca}(t)$ are the power allocation factors of the refletion propagation path and scattering propagation path and satisfying $\eta_{\rm ref}(t) + \eta_{\rm sca}(t) = 1$. The random phases (i.e., θ_n and θ_m) introduced by reflection and scattering paths are assumed to be independently and uniform randomly distributed in the interval of $[-\pi, \pi)$.

According to Equations (2), (3), (16) and (17), the time-varying channel coefficients of the reflection component and the scattering component can be calculated as:

$$\alpha^{\rm ref}_{\rm NLoS,n}(t) = \frac{\frac{\lambda_c^2 \cdot R_n(f_c,t)}{16\pi^2 d^{\rm ref}_{\rm Tx,n}(t)\cdot d^{\rm ref}_{\rm Rx,n}(t)}}{\sqrt{\frac{1}{N(t)}\sum_n^{N(t)}\left[\frac{\lambda_c^2 \cdot R_n(f_c,t)}{16\pi^2 d^{\rm ref}_{\rm Tx,n}(t)\cdot d^{\rm ref}_{\rm Rx,n}(t)}\right]^2}} \tag{22}$$

$$\alpha^{\rm sca}_{\rm NLoS,m}(t) = \frac{\frac{\lambda_c^2 \cdot S_m(f_c,t)}{16\pi^2 d^{\rm sca}_{\rm Tx,m}(t)\cdot d^{\rm sca}_{\rm Rx,m}(t)}}{\sqrt{\frac{1}{M(t)}\sum_m^{M(t)}\left[\frac{\lambda_c^2 \cdot S_m(f_c,t)}{16\pi^2 d^{\rm sca}_{\rm Tx,m}(t)\cdot d^{\rm sca}_{\rm Rx,m}(t)}\right]^2}} \tag{23}$$

For the stochastic time-varying channel model of this paper, the large-scale parameters in two adjacent time intervals are highly correlated, thus this paper use an time auto-correlation function (T-ACF) $\rho(t, \Delta t)$ to express the relationship between the large-scale parameters in two adjacent time intervals, which can be written as:

$$\begin{aligned} \rho(t,\Delta t) &= \frac{{\rm E}[h(t)h(t+\Delta t)^*]}{\sqrt{{\rm E}[|h(t)|^2]{\rm E}[|h(t+\Delta t)|^2]}} \\ &= \frac{{\rm E}[h(t)h(t+\Delta t)^*]}{\sqrt{\Omega(t)\cdot\Omega(t+\Delta t)}} \end{aligned} \tag{24}$$

where $[\cdot]^*$ denotes the complex conjugate.

2.5. Doppler Power Spectrum Density

In this subsection, the DPSD is derived by the Fourier transform of T-ACF, which is given as:

$$S_{\text{Doppler}}(t,f) = \int_{-\infty}^{\infty} \rho(t,\Delta t) e^{-j2\pi f \Delta t} d\Delta t \tag{25}$$

3. Numerical Results and Analysis

In this section, the performance of the proposed channel model is analyzed in terms of the path loss, T-ACF, and DPSD with different channel model parameters, including mobile properties of Tx and Rx (e.g., moving speed and moving time), horizontal distance from Tx to Rx, and carrier frequencies (e.g., mm-wave and THz bands). By comparison the simulation and theoretical results, the correctness of proposed channel model can be validated, in which the simulation of the proposed UAV-based dual-mobility wireless channel model are carried out by ray-tracing method. The simulation and theoretical parameters are set as follows: $\alpha_{\text{Tx}} = \pi, \alpha_{\text{Rx}} = \beta_{\text{Tx}} = 0$ (where α_{Tx}, α_{Rx}, and β_{Tx} are the azimuth and elevation of moving angle for Tx and Rx respectively), $\eta_{\text{ref}} = 0.7$, $\eta_{\text{sca}} = 0.3$, $n_{t,\text{ref}} = n_{t,\text{sca}} = 2.2$ (i.e., the material of the rough surface is concrete), $\sigma_{h,\text{ref}} = 0.05$ mm, $\sigma_{h,\text{sca}} = 0.15$ mm (where $\sigma_{h,\text{sca}}$ is the standard deviation in the height of rough surface in reflection propagation) [22,23,25]. For the analysis of the performance difference between mm-wave and THz, 60 GHz, 200 GHz, and 300 GHz are chosen as the carrier frequencies of mm-wave and THz bands respectively.

Figure 3 shows the simulation and theoretical results of the T-ACF with different moving speeds of Tx and Rx for the NLoS propagation path, where $d_{0,V}$ and $d_{0,H}$ are the initial vertical and horizontal distances between UAV and MS (i.e., mobile station, MS denotes the moving vehicle on the ground), respectively. The results of Figure 3 clearly demonstrate that the faster speeds of the Tx and Rx, the quicker drop of T-ACF value, this conclusion is consistent with [11]. The reason is clear: the changing of CIR's phase increases with the increase of the moving speed for the Tx and Rx. This means that the drop rate of T-ACF increases with the increase of Δt. Furthermore, the theoretical and simulation results for T-ACF are in close alignment. This is evident when comparing the theoretical results for T-ACF with the simulation results for T-ACF under the same configurations.

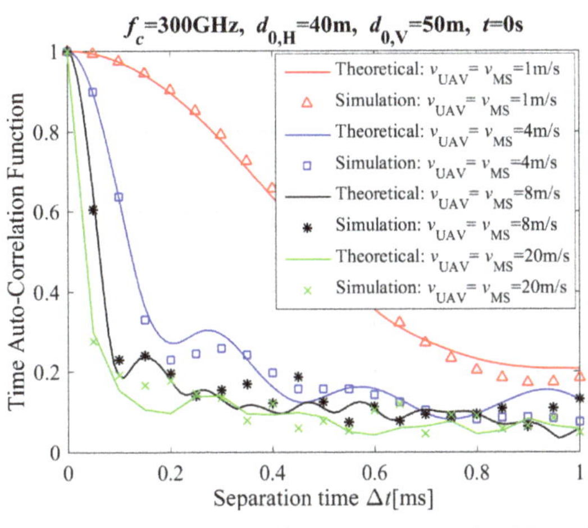

Figure 3. The T-ACF with different moving speeds of Tx and Rx for the NLoS path (including reflection and scattering paths).

Figure 4 shows the theoretical results of the T-ACF with different vertical distances between UAV and MS for the NLoS propagation path. The results of Figure 4 show that the T-ACF increases with an increase slowly of vertical distances between UAV and MS. The reason is clear: the slow changing of CIR's phase increases with an increase slowly of vertical distances between UAV and MS.

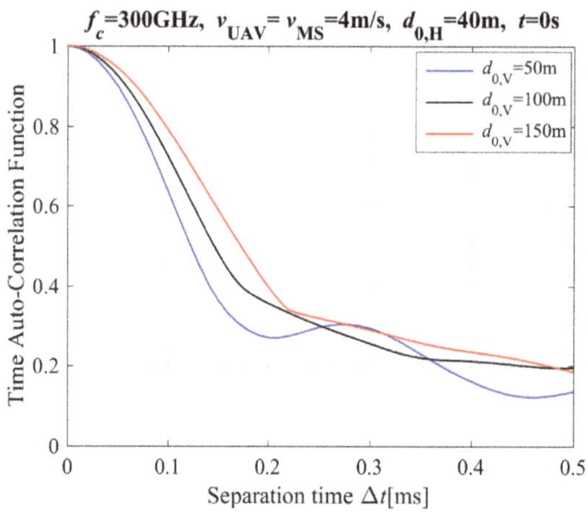

Figure 4. The T-ACF with different vertical distance of Tx and Rx for the NLoS path (including reflection and scattering paths).

Figure 5 depicts the theoretical results of the T-ACF with different power ratio between UAV and MS for the NLoS propagation path. According to Figure 4, it is found that the value of T-ACF with $\eta_{\text{ref}} \geq 0.5$ is more than that with $\eta_{\text{ref}} < 0.5$. Besides, the influence of η_{ref} and η_{sca} on T-ACF is not obvious.

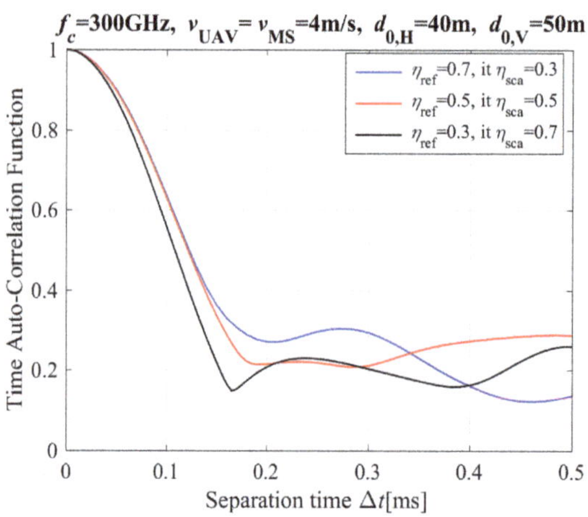

Figure 5. The T-ACF with different power ratio of reflection and scattering propagations for the NLoS path (including reflection and scattering paths).

Figure 6 presents the T-ACF with different Ricican-K factors, where $d_{0,\text{V}}$ and $d_{0,\text{H}}$ are the initial vertical and horziontal distances between UAV and MS, respectively. Figure 6 clearly shows that the T-ACF increases with the Rician K-factor. These conclusions are in line with those in [11]. The Rician K-factor is defined as the ratio between the powers LoS path and NLoS path (i.e., reflection and scattering paths). Though not shown in this paper, it is found that the T-ACF's value of LoS path is much higher than that of NLoS path. In addition, when the Rician *K*-factor decreases gradually, the power of NLoS path increases gradually and becomes more dominant. Thus, the value of T-ACF decreases.

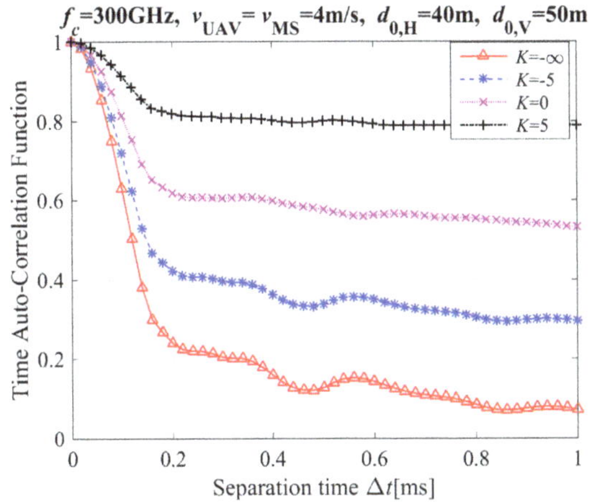

Figure 6. The T-ACF with different Rican *K*-factor.

Figure 7 depicts the path loss of the MPCs (including LoS, reflection, and scattering paths) with different carrier frequencies, where $d_{0,V}$ and $d_{0,H}$ are the initial vertical and horizontal distances between UAV and MS, respectively. Figure 7 clearly shows that path loss increases with an increase in carrier frequency. The reason is that: (i) the atmospheric absorption attenuation is more noticeable in the higher frequency band than in the lower frequency band; (ii) the reflection and scattering fading in the higher frequency band is more serious than that in the lower frequency band. Therefore, the path loss in the higher frequency band is greater than that in the lower frequency band, as shown in Figure 7.

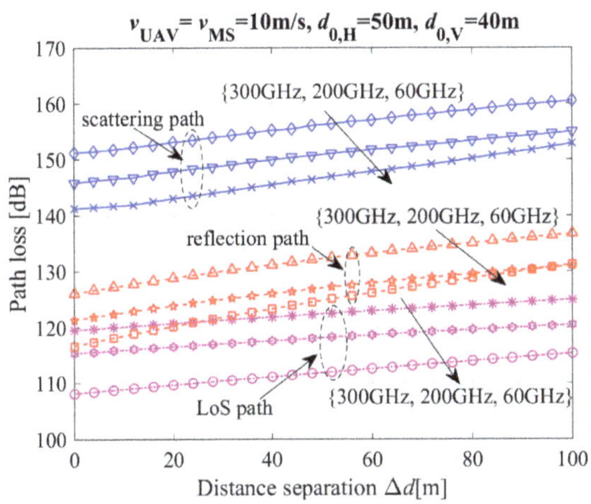

Figure 7. Path loss of the MPCs (including LoS, reflection, and scattering paths) with different carrier frequencies.

Besides, Figure 7 clearly shows that the path losses of NLoS paths (including scattering and reflection paths) are significantly higher than those of LoS paths in variant carrier frequencies. This is because that: (i) the propagation lengths for scattering and reflection paths are longer than those for the LoS path; (ii) the path loss of a line-of-sight (LoS) path is solely due to the spread attenuation in free space, the path loss of a non-line-of-sight (NLoS) path is the result of both the spread attenuation in free space and reflection and scattering fading on rough surfaces.

Figure 8 depicts the DPSD with different moving times and propagation paths (inluding NLoS and LoS+NLoS paths), where $d_{0,\mathrm{V}}$ and $d_{0,\mathrm{H}}$ are the initial vertical and horziontal distances between UAV and MS, respectively. From the Figure 8, we can see that the values of DPSDs for the LoS+NLoS path with variant moving times are greater than that the NLoS path in 300GHz band. In addition, the DPSDs' values of both the LoS+NLoS and NLoS paths in t = 4 s are more than that in t = 0 s.

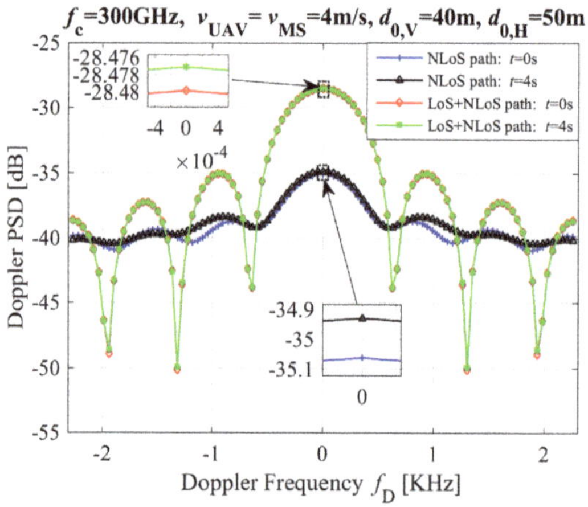

Figure 8. The DPSD with different moving times and different paths.

4. Conclusions

In this paper, we proposed a 3D time-varying channel model for UAV-based A2G dual-mobility wireless channels in the THz band, where the reflection and scattering fading on rough surfaces are considered, as well as atmospheric absorption attenuation of wireless propagation channels. In order to value assess the impact of UAV's movement in 3D free space and the MS's movement on the ground on performance of dual-mobility wireless channels, the path loss, T-ACF, and the DPSD are derived and subjected to comprehensive analysis. The correctness of the proposed model in this paper has been validated via simulation (i.e., ray tracing method). Numerical results have shown that the channel model parameters (e.g., moving speed, moving time, carrier frequency, and propagation path) at THz bands have significant impacts on path loss, T-ACF, and DPSD. The numerical results demonstrate that the path loss for different propagation paths (i.e., LoS, reflection, and scattering paths) exhibits an increase with the increase of carrier frequencies. Besides, the T-ACF decreases with an increase of the moving speed of UAV and MS. The influence of η_{ref} and η_{sca} on T-ACF is not obvious based on the results of T-ACF for the NLoS propagation path. In addition, the DPSD of NLoS+LoS path exhibits greater variability than that of LoS path for different Doppler frequencies.

Author Contributions: Conceptualization, K.Z.; Methodology, K.Z.; Writing—original draft, K.Z.; Writing—review & editing, K.Z., F.Z., Y.L. (Yongjun Li), X.W., Z.Y., Y.L. (Yuanhao Liu), C.Z. and X.L. All authors have read and agreed to the published version of the manuscript.

Funding: This work was supported by the National Key Research and Development Program of China (2021YFB2800800), National Funded Postdoctoral Researcher Program (GZC20233578), National Natural Science Foundation of China (NSFC) (62394292, 62394290).

Data Availability Statement: The original contributions presented in this study are included in the article. Further inquiries can be directed to the corresponding authors.

Conflicts of Interest: The authors declare no conflicts of any interest or personal circumstances.

Abbreviations

Abbreviations used in this Paper

Abbreviation	Description
UAV	Unmanned aerial vehicle
THz	Terahertz
3D	Three-dimensional
T-ACF	Time autocorrelation function
DPSD	Doppler power spectrum density
mm-wave	Millimeter wave
SAGNs	Space–air–ground integrate networks
6G	Sixth generation
A2G	Air to ground
A2A	Air to air
AFD	Average fading duration
STC	Space–time correlation
STFC	Space–time–frequency correlation
GBSM	Geometry-based stochastic model
Tx	Transmitter
Rx	Receiver
LoS	Line-of-sight
NLoS	Non-line-of-sight
CIR	Channel impulse response
MPCs	Multipath components

References

1. Wang, C.X.; Huang, J.; Wang, H.; Gao, X.; You, X.; Hao, Y. 6G wireless channel measurements and models trends and challenges. *IEEE Veh. Technol. Mag.* **2020**, *15*, 22–32. [CrossRef]
2. Yuan, Y.; He, R.; Ai, B.; Ma, Z.; Miao, Y.; Niu, Y.; Zhang, J.; Chen, R.; Zhong, Z. A 3D geometry-based THz channel model for 6G ultra massive MIMO systems. *IEEE Trans. Veh. Technol.* **2022**, *71*, 2251–2266. [CrossRef]
3. Valavanis, K.P.; Vachtsevanos, G.J. *Handbook of Unmanned Aerial Vehicle*; Springer: Berlin, Gemany, 2015; p. 2077.
4. Namuduri, K.; Chaumette, S.; Kim, J.H.; Sterbenz, J.P. *UAV Networks and Communications*; Cambridge University Press: Cambridge, UK, 2017.
5. Mozaffari, M.; Saad, W.; Bennis, M.; Nam, Y.H.; Debbah, M. A tutorial on UAVs for wireless networks: Applications, challenges, and open problems. *IEEE Commun. Surv. Tut.* **2019**, *21*, 2334–2360. [CrossRef]
6. Klaine, P.V.; Nadas, J.P.; Souza, R.D.; Imran, M.A. Distributed drone base station positioning for emergency cellular networks using reinforcement learning. *Cogn. Comput.* **2018**, *10*, 790–804. [CrossRef] [PubMed]
7. Jayakody, D.N.K.; Perera, T.D.P.; Ghrayeb, A.; Hasna, M.O. Selfenergized UAV-assisted scheme for cooperative wireless relay networks. *IEEE Trans Veh. Technol.* **2020**, *69*, 578–592. [CrossRef]
8. Song, Y.; Lim, S.H.; Jeon, S.W.; Baek, S. On cooperative achievable rates of UAV assisted cellular networks. *IEEE Trans. Veh. Technol.* **2020**, *69*, 9882–9895. [CrossRef]
9. Bian, J.; Wang, C.X.; Liu, Y.; Tian, J.; Qiao, J.; Zheng, X. 3D Non-Stationary Wideband UAV-to-Ground MIMO Channel Models Based on Aeronautic Random Mobility Model. *IEEE Trans. Veh. Technol.* **2021**, *70*, 11154–11168. [CrossRef]
10. Lian, Z.; Su, Y.; Wang, Y.; Jiang, L.; Zhang, Z.; Xie, Z.; Li, S. A Non-Stationary 3-D Wideband Channel Model for Low Altitude UAV-MIMO Communication Systems. *IEEE Internet Things J.* **2022**, *9*, 5290–5303. [CrossRef]
11. Bai, L.; Huang, Z.; Zhang, X.; Cheng, X. A Non-Stationary 3D Model for 6G Massive MIMO mmWave UAV Channels. *IEEE Trans. Wireless Commun.* **2021**, 1–14. [CrossRef]
12. Ma, Z.; Ai, B.; He, R.; Zhong, Z.; Yang, M. A Non-Stationary Geometry-Based MIMO Channel Model for Millimeter-Wave UAV Networks. *IEEE J. Sel. Areas Commun.* **2021**, *39*, 2960–2974. [CrossRef]
13. Bai, L.; Han, R.; Liu, J.; Yu, Q.; Choi, J.; Zhang, W. Air-to-Ground Wireless Links for High-Speed UAVs. *IEEE J. Sel. Areas Commun.* **2020**, *38*, 2918–2930. [CrossRef]
14. Yu, X.; Jia, S.; Hu, H.; Galili, M.; Morioka, T.; Jepsen, P.U.; Oxenløwe, L.K. 160 Gbit/s photonics wireless transmission in the 300–500 GHz band. *APL Photon.* **2016**, *1*, 081301. [CrossRef]
15. Zhang, H.; Zhang, L.; Wang, S.; Lu, Z.; Yang, Z.; Liu, S.; Qiao, M.; He, Y.; Pang, X.; Zhang, X.; et al. Tbit/s multi-dimensional multiplexing THz-over-fiber for 6G wireless communication. *J. Lightw. Technol.* **2021**, *39*, 5783–5790. [CrossRef]

16. Wang, C.X.; Bian, J.; Sun, J.; Zhang, W.; Zhang, M. A survey of 5G channel measurements and models. *IEEE Commun. Surveys Tuts.* **2018**, *20*, 3142–3168. [CrossRef]
17. Rappaport, T.S.; Xing, Y.; Kanhere, O.; Ju, S.; Madanayake, A.; Mandal, S.; Alkhateeb, A.; Trichopoulos, G.C. Wireless communications and applications above 100 GHz: Opportunities and challenges for 6G and beyond. *IEEE Access* **2019**, *7*, 78729–78757. [CrossRef]
18. Wang, J.; Wang, C.X.; Huang, J.; Wang, H.; Gao, X. A General 3D Space-Time-Frequency Non-Stationary THz Channel Model for 6G Ultra-Massive MIMO Wireless Communication Systems. *IEEE J. Sel. areas Commun.* **2021**, *39*, 1576–1589. [CrossRef]
19. Han, C.; Bicen, A.O.; Akyildiz, I.F. Multiray channel modeling and wideband characterization for wireless communications in the terahertz band. *IEEE Trans. Wireless Commun.* **2015**, *14*, 2402–2412. [CrossRef]
20. Zhang, Y.; Zhao, L.; He, Z. A 3-D hybrid dynamic channel model for indoor THz communications. *China Commun.* **2021**, *18*, 50–65. [CrossRef]
21. Wang, J.; Wang, C.X.; Huang, J.; Wang, H.; Gao, X.; You, X.; Hao, Y. A novel 3D non-stationary GBSM for 6G THz ultra-massive MIMO wireless systems. *IEEE Trans. Veh. Technol.* **2021**, *70*, 12312–12324. [CrossRef]
22. Zhang, K.; Wang, H.; Zhang, C.; Yu, X.; Dong, Y. Three-dimensional non-stationary geometry-based modeling of sub-THz MIMO channels for UAV air-to-ground communications. In Proceedings of the IEEE International Conference on Communications, Rome, Italy, 28 May–1 June 2023.
23. Zhang, K.; Wang, H.; Zhang, C.; Yu, X.; Dong, Y. Three dimensional geometry based stochastic model for sub-terahertz air to air UAV-MIMO channels. In Proceedings of the IEEE/CIC lnternational Conference on Communications, Dalian, China, 10–12 August 2023.
24. Zhang, K.; Zhang, C.; Wang, H.; Yang, Z.; Yu, X.; Li, Y. Nonstationary Channel Modeling for Wireless Communications Underlaying UAV-Based RelayAssisted IIoT Networks in the Subterahertz Band. *IEEE Internet Things J.* **2024**, *11*, 26888–26900. [CrossRef]
25. International Telecommunication Union, Radio Communication Sector (ITU-R), Recommendation P.676-11, Attenuation by Atmospheric GASES. 2016. Available online: https://www.itu.int/rec/R-REC-P.676-11-201609-I (accessed on 20 July 2024).

entropy

Article

Robustness of Entanglement for Dicke-W and Greenberger-Horne-Zeilinger Mixed States

Ling-Hui Zhu [1], Zhen Zhu [1,2], Guo-Lin Lv [1,3], Chong-Qiang Ye [1] and Xiao-Yu Chen [1,*]

[1] School of Information and Electrical Engineering, Hangzhou City University, Hangzhou 310015, China; 2230201010@stu.hzcu.edu.cn (L.-H.Z.); 221122030296@zjut.edu.cn (Z.Z.); 22231168@zju.edu.cn (G.-L.L.); chongqiangye@aliyun.com (C.-Q.Y.)

[2] College of Information Engineering, Zhejiang University of Technology, Hangzhou 310014, China

[3] College of Information Science & Electronic Engineering, Zhejiang University, Hangzhou 310027, China

* Correspondence: chenxiaoyu@hzcu.edu.cn

Abstract: Quantum entanglement is a fundamental characteristic of quantum mechanics, and understanding the robustness of entanglement across different mixed states is crucial for comprehending the entanglement properties of general quantum states. In this paper, the robustness of entanglement of Dicke–W and Greenberger–Horne–Zeilinger (GHZ) mixed states under different mixing ratios is calculated using the entanglement witness method. The robustnesses of entanglement of Dicke–W and GHZ mixed states are different when the probability ratio of Dicke to W is greater than $\frac{3}{2}$ and less than $\frac{3}{2}$. For the probability of Dicke and W states greater than or equal to $\frac{3}{2}$, we study the robustness of entanglement of Dicke and GHZ mixed states and analyze and calculate their upper and lower bounds. For the probability of Dicke and W states less than $\frac{3}{2}$, we take the equal probability ratio of Dicke and W states as an example and calculate and analyze the upper and lower bounds of their robustness of entanglement in detail.

Keywords: quantum entanglement; robustness of entanglement; Dicke state; W state; GHZ state

1. Introduction

In quantum information theory, quantum entanglement [1] plays a fundamental role as a resource in quantum communication between two parties separated by macroscopic distances [2]. A state is called separable if it can be written as the probability mixture of product states [3]; otherwise, it is entangled. However, detecting entanglement remains an NP-hard problem.

To address this challenge, considerable work has been conducted on developing criteria for judging entanglement or separability. The most widely used is the positive partial transpose (PPT) criterion [4,5] since it provides a complete characterization of entanglement for two-qubit systems, and studies have shown that violating PPT conditions can be used to quantify entanglement [6,7]. The computable cross norm or realignment criterion (CCNR criterion) [8,9] is also a strong criterion that can be considered as a supplement to the PPT criterion. Entanglement witnesses are Hermitian operators for detecting entanglement [10–13] and are a necessary entanglement criterion in terms of directly measurable observables. It is a very useful tool for the analysis of entanglement in experiments, but it still cannot solve all entanglement problems because the construction of witnesses is still difficult. The range criterion [14] is used for determining the separability of mixed states based on the range of the density matrix. This criterion is generally used when the PPT criterion fails. Compared with the PPT criterion, the range criterion has wider applicability and can detect more quantum entanglement states. However, the range criterion also has limitations and does not apply to situations where quantum states are disturbed by noise.

The theory of entanglement has been gradually improved and has become a core theme of quantum information science. At the same time, the development of quantum entanglement has also driven the development of many fields such as quantum dense coding [15,16], quantum teleportation [17,18], entanglement purification [19], quantum error correction [2,20], and quantum memory [21].

Entanglement in multipartite systems is a key resource for quantum information and communication protocols [1,22]. In experiments, different multipartite entangled states have been prepared [23–25]. Determining whether the state produced in the experiment is a multipartite (partial) entangled state has become a highly relevant topic in quantum information theory [26]. The Greenberger–Horne–Zeilinger (GHZ) state is also called the maximally entangled state. Due to its maximum entanglement characteristics and measurement accuracy approaching the Heisenberg limit [27], it has considerable application prospects in the fields of quantum information, quantum communication, and precision measurement. At the same time, due to the special properties of the GHZ state, it can be used in multipartite secret sharing protocols. In addition, when the entangled GHZ state is used as a resource, the photonic architecture of measurement-based quantum computing becomes more efficient [28]. Chen et al. [29] proposed a road map for finding the separability criteria of multipartite entangled states and derived a set of tripartite separability criteria for the four-qubit GHZ diagonal states. The Dicke state [30], proposed by R.H. Dicke in 1954, is of great significance for studying the properties of multi-particle quantum entanglement and building multi-user quantum networks because its entanglement is robust to particle loss and is attractive in practical applications such as multi-party quantum networks [31] and quantum metrology. The W state corresponds to the Dicke state $|Dicke_{3,1}\rangle$, and therefore, examples of a Dicke state [6] and W states have already been prepared in many experiments [32–34]. Chen et al. [35] demonstrated an entanglement criterion for any four-qubit state which is necessary and sufficient for the generalized noisy four-qubit Dicke states. Carvalho [36] showed that there is a clear scaling of the entanglement decay rates for the GHZ and W states for various environments. Zhang [37] found that the scar state encompasses both GHZ and W states, which provides the possibility of thermal-free quantum information processing in finite-sized quantum spin clusters. Chen et al. [38] showed that for a mixture of a four-qubit GHZ state with a Dicke state and white noise, there exists a new Wootters formula. Therefore, the measurement and detection of the robustness of entanglement of Dicke–W and GHZ mixed states is of great significance for quantum information applications and the preparation of related quantum states.

In this paper, we give the numerical analysis and theoretical interpretation of the robustness of entanglement of the full separability of Dicke–W and GHZ mixed states under two different mixing ratios. By plotting the figure of the robustness of entanglement, we find two different types of figures and then analyze the mixing ratio of the boundary between the two types.

The paper is organized as follows: In Section 2, we introduce the concept of robustness of entanglement and some concepts involved in the paper. Section 3 elaborates on the numerical and theoretical analysis methods for the robustness of entanglement in the fully separable case of Dicke–GHZ mixed states. Section 4 shows the results of the robustness of entanglement for Dicke–W and GHZ when the mixing ratio of Dicke–W is 45° in the case of full separability and also gives the results for the upper and lower bounds. Section 5 gives the boundary of the mixing ratio. Section 6 is the conclusions.

2. Preliminary

A multipartite state ρ is separable when it can be written as

$$\rho = \sum_i p_i \rho_i^{A_1} \otimes \rho_i^{A_2} \cdots \otimes \rho_i^{A_N}, \tag{1}$$

where $\rho_i^{A_j}$ is the state of A_j, and p_i is the probability distribution.

In our work, we explore the robustness of entanglement of Dicke–W and GHZ mixed states in different ratios. The density matrix for a mixed state consisting of Dicke–W and GHZ is given by

$$\rho = p_1|GHZ_4\rangle\langle\cdot| + (1-p_1)[p_2|Dicke_{4,2}\rangle\langle\cdot| + (1-p_2)|W_4\rangle\langle\cdot|], \tag{2}$$

where $\{p_1, (1-p_1)p_2, (1-p_1)(1-p_2)\}$ denote the probability distributions, and we abbreviate $|\chi\rangle\langle\chi| = |\chi\rangle\langle\cdot|$. $|GHZ_4\rangle = \frac{1}{\sqrt{2}}(|0000\rangle + |1111\rangle)$, $|Dicke_{4,2}\rangle = \frac{1}{\sqrt{6}}\sum_{|k|=2}|\mathbf{k}\rangle$, $|W_4\rangle = \frac{1}{2}\sum_{|k|=1}|\mathbf{k}\rangle$, with $\mathbf{k} \in \{0,1\}^{\otimes 4}$ being a binary vector and $|k|$ being the Hamming weight of $\mathbf{k}$. We denote $\tan\Theta = \frac{p_1}{1-p_1}$ and $\tan\Phi = \frac{p_2}{1-p_2}$, and we call Φ the mixing angle of Dicke and W states. We will omit the subscripts of GHZ, Dicke, and W for simplicity.

The definition of robustness is given by Lami et al. [39] for any state ρ and is expressed as follows:

$$R_{\mathcal{F}}(\omega) := \inf\left\{1+\lambda \,\middle|\, \frac{\omega+\lambda\tau}{1+\lambda} \in \mathcal{F}, \tau \in \mathfrak{D}(\mathcal{H})\right\}, \tag{3}$$

where $\tau \in \mathfrak{D}(\mathcal{H})$ are noise states defined on Hilbert space $\mathcal{H}$, $\mathfrak{D}(\mathcal{H})$ is the density matrix in it, and the separable state set $\mathcal{F}$ is generally assumed to be convex and closed. Equation (3) leads to the upper bound of the robustness of entanglement, and for any states $\rho, \omega \in \mathfrak{D}(\mathcal{H})$ it holds that

$$\underline{R}_{\mathcal{F}}(\rho) \geq \frac{\langle\rho,\omega\rangle}{\sup_{\sigma\in\mathcal{F}}\langle\sigma,\omega\rangle}, \tag{4}$$

where σ is a separable state and $\underline{R}$ represents the lower bound.

The minimum quantity of noise state $1+\lambda$ is the upper bound of robustness, and the free state reaches the optimum, denoted as σ. We let $\tilde{\rho} = \frac{\rho}{1+\lambda}$, which can be expressed as

$$\tilde{\rho} = \frac{p_1}{1+\lambda}|GHZ\rangle\langle\cdot| + \frac{1-p_1}{1+\lambda}[p_2|Dicke\rangle\langle\cdot| + (1-p_2)|W\rangle\langle\cdot|]. \tag{5}$$

We denote $g = \frac{p_1}{1+\lambda}$, $d = \frac{1-p_1}{1+\lambda}p_2$ and $w = \frac{1-p_1}{1+\lambda}(1-p_2)$. Then g, d, and w represent the coefficients of $|GHZ\rangle$ state, $|Dicke\rangle$ state, and $|W\rangle$ state in σ, respectively. For ease of presentation, we use $d_w = \sqrt{d^2+w^2}$ as the figure's horizontal axis. It can be seen that $g + \frac{d_w}{\sqrt{p_2^2+(1-p_2)^2}} = \frac{1}{1+\lambda} = \frac{1}{R}$ and $\frac{g}{d_w} = \frac{p_1}{1-p_1} = \tan\Theta$. Thus, the relationship between d_w, g, and R will be displayed in the figure shown later by Θ.

3. Robustness of Fully Separable Four-Qubit Dicke–GHZ Mixed States ($\Phi = 90°$)

In this section, $\Phi = 90°$, so the density matrix ρ_1 in Equation (2) will be given by

$$\rho_1 = p|GHZ\rangle\langle GHZ| + (1-p)|Dicke\rangle\langle Dicke|, \tag{6}$$

where p and $1-p$ denote the probability distributions. A density matrix ρ can be expressed using the 2×2 identity matrix σ_0 and Pauli matrices σ_1, σ_2, and σ_3 as follows:

$$\rho = \frac{1}{2^4}\sum_{i,j,j,l=0}^{3} R_{ijkl}\sigma_i\otimes\sigma_j\otimes\sigma_k\otimes\sigma_l, \tag{7}$$

where $R_{ijkl} = Tr(\rho_s)\sigma_i\otimes\sigma_j\otimes\sigma_k\otimes\sigma_l$. The mixed states ρ_1 in Equation (6) can be expressed as

$$\begin{aligned}\rho_1 = \frac{1}{16}[&IIII + R_1IIIZ + R_2IIZI + R_3IZII + R_4ZIII\\ &+R_5IZZZ_p + R_6(IIXX_p + IIYY_p) + R_7(XXZZ_p\\ &+YYZZ_p) + R_8IXXZ_p + M_9IYYZ_p + R_{10}ZZZZ\\ &+R_{11}IIZZ_p + R_{12}XXYY_p + R_{13}XXXX + M_{14}YYYY],\end{aligned} \tag{8}$$

where the subscript p denotes the summation over all permutations of Pauli matrices, and I, X, Y, and Z denote the Pauli matrices σ_0, σ_1, σ_2, and σ_3, respectively. The characteristic vector is expressed in the form of

$$\mathbf{R} = (R_1, R_2, R_3, R_4, 4R_5, 12R_6, 12R_7, 12R_8, 12R_9, R_{10}, 6R_{11}, 6R_{12}, R_{13}, R_{14})^T. \tag{9}$$

For mixed states of Dicke and GHZ, the witness ω of Equation (4) can be expressed in terms of the parameters M_i and the tensor product of Pauli matrices as follows:

$$\omega = M_0 IIII + \hat{M}', \tag{10}$$

and the trace-free matrix $\hat{M}'$ can be expressed as

$$\begin{aligned} \hat{M}' =& M_1 IIIZ + M_2 IIZI + M_3 IZII + M_4 ZIII + M_5 IZZZ_p \\ &+ M_6(IIXX_p + IIYY_p) + M_7(XXZZ_p + YYZZ_p) \\ &+ M_8 IXXZ_p + M_9 IYYZ_p + M_{10} ZZZZ + M_{11} IIZZ_p \\ &+ M_{12} XXYY_p + M_{13} XXXX + M_{14} YYYY, \end{aligned} \tag{11}$$

where the subscript p denotes the summation over all permutations of Pauli matrices.

To find the maximum value of the denominator in Equation (4), we transform it into finding the maximum of $\langle\psi_s|\omega|\psi_s\rangle$ and $|\psi_s\rangle = |\psi_1\rangle|\psi_2\rangle|\psi_3\rangle|\psi_4\rangle$. The product state $|\psi_k\rangle$ can be expressed in the Bloch representation as $|\psi_k\rangle = \cos\frac{\theta_k}{2}|0\rangle + \sin\frac{\theta_k}{2}e^{i\varphi_k}|1\rangle$, $(k = 1,2,3,4)$, where θ is the polar angle, and φ is the azimuthal angle. The density matrix of $\rho_k = |\psi_k\rangle\langle\psi_k|$ can be represented by Pauli matrices as

$$\rho_k = \frac{1}{2}(I + x_k\sigma_1 + y_k\sigma_2 + z_k\sigma_3), \tag{12}$$

where $x_k = \sin\theta_k\cos\varphi_k$, $y_k = \sin\theta_k\sin\varphi_k$, $z_k = \cos\theta_k$, $(k = 1,2,3,4)$. $Tr(\hat{M}|\psi_s\rangle\langle\psi_s|) = M_0 + \Lambda$, and $\Lambda = \Lambda_1 + \Lambda_2 x_4 + \Lambda_3 y_4 + \Lambda_4 z_4$, then

$$\Lambda = \max_{\theta_1\theta_2\theta_3\varphi_1\varphi_2\varphi_3} \Lambda_1 + \sqrt{\Lambda_2^2 + \Lambda_3^2 + \Lambda_4^2}, \tag{13}$$

where $\Lambda_i (i = 1,2,3,4)$ is a function with variables M_j, x_k, y_k, z_k. More details are given in the Appendix A. The numerator of Equation (4) can be expressed as

$$\langle\rho,\omega\rangle = M_0 + \mathbf{M}\cdot\frac{\mathbf{R}}{\cos\Theta + \sin\Theta}. \tag{14}$$

The denominator of Equation (4) can be expressed as

$$\langle\sigma,\omega\rangle = M_0 + \Lambda, \tag{15}$$

where the vector $\mathbf{M} = (M_1, \ldots, M_{14})$. Given a Dicke–GHZ state ρ, we adjust the parameter vector $\mathbf{M}$ to achieve the minimum value of $\mathcal{L}$,

$$\mathcal{L} = \frac{M_0 + \Lambda}{M_0(\cos\Theta + \sin\Theta) + \mathbf{M}\cdot\mathbf{R}}. \tag{16}$$

The robustness of entanglement of four-qubit Dicke–GHZ mixed states is described by the curve in Figure 1.

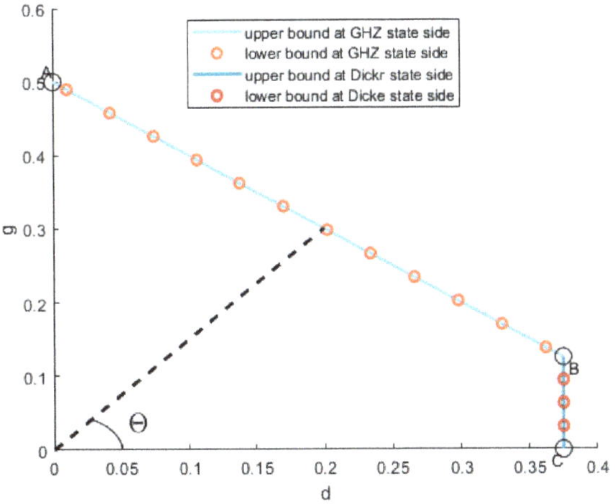

Figure 1. The robustness of entanglement for four-qubit Dicke–GHZ mixed states. The curve can be divided into two parts, and each part consists of a lower bound and an upper bound. The angle between the horizontal axis of the coordinate and the dotted line is called Θ, and it satisfies $\tan\Theta = \frac{p}{1-p}$. Points A, B, and C divide the curve into two parts. Line $\overline{AB}$ corresponds to $\Theta \in [18.4349^\circ, 90^\circ]$, while line $\overline{BC}$ corresponds to $\Theta \in [0^\circ, 18.4349^\circ]$. The coordinates of points A, B, and C are $\left(0, \frac{1}{2}\right)$, $\left(\frac{3}{8}, \frac{1}{8}\right)$, and $\left(\frac{3}{8}, 0\right)$, respectively.

3.1. Segment AB

For Dicke–GHZ mixed states at the GHZ state side, numerical calculation suggests

$$
\begin{aligned}
M_1 = M_2 = M_3 = M_4 = M_5 = M_8 = M_9 = 0 \\
M_{13} = M_{14}.
\end{aligned}
\tag{17}
$$

By analyzing and calculating, we get the relationship among M_i:

$$
\begin{aligned}
M_{10} &= 6(M_6 - M_{11}) + M_{13} \\
M_{12} &= M_7 + M_{11} - M_6.
\end{aligned}
\tag{18}
$$

and we find that when $\theta = 90^\circ$ and $\varphi = \frac{k\pi}{2}$ where $k = 0, 1, 2, 3$, the Λ of Equation (13) will achieve the same maximum value:

$$
\Lambda = 6M_6 + M_{13}. \tag{19}
$$

The M_0 in Equation (15) can be expressed as

$$
M_0 = 6(-M_7 - M_{11}) + M_{13}, \tag{20}
$$

and the rest of Equation (15) can be expressed as

$$
\begin{aligned}
\mathbf{M} \cdot \mathbf{R} = & M_6 8\cos\Theta + M_7(-8\cos\Theta) + M_{10}(\cos\Theta + \sin\Theta) \\
& + M_{11}(-2\cos\Theta + 6\sin\Theta) + M_{12}(2\cos\Theta - 6\sin\Theta) \\
& + 2M_{13}(\cos\Theta + \sin\Theta).
\end{aligned}
\tag{21}
$$

Combining the analysis among parameters and Equation (16), we get $\mathcal{L} = \frac{1}{2(\cos\Theta+\sin\Theta)}$. Comparing Equation (4) and Equation (16), we can easily get $\mathcal{L} = \frac{1}{R(\cos\Theta+\sin\Theta)}$, and by converting it to the Descartes coordinate system, we get

$$g + d = \frac{1}{2}, \tag{22}$$

where $g = \frac{\sin\Theta}{R(\sin\Theta+\cos\Theta)}$ and $d = \frac{\cos\Theta}{R(\sin\Theta+\cos\Theta)}$. The criterion is shown in Figure 1, with the blue line indicating Line AB and accounting well for the numerical necessary criterion when $\Theta \in [\Theta_1, 90^\circ]$ with the angle $\Theta_1 = \tan^{-1}\frac{1}{3} \approx 18.4349^\circ$.

Next, we will present the upper bound of this segment. A product state $|\psi_s\rangle = (c|0\rangle + se^{i\varphi}|1\rangle)^{\otimes 4}$ can be represented as

$$\begin{aligned} |\psi_s\rangle =& c^4|0000\rangle + s^4e^{i4\varphi}|1111\rangle \\ &+ c^3se^{i\varphi}(|0001\rangle + |0010\rangle + |0100\rangle + |1000\rangle) \\ &+ c^2s^2e^{i2\varphi}(|0011\rangle + |0101\rangle + |0110\rangle + |1001\rangle + |1010\rangle + |1100\rangle) \\ &+ cs^3e^{i3\varphi}(|0111\rangle + |1011\rangle + |1101\rangle + |1110\rangle), \end{aligned} \tag{23}$$

where $c = \cos\frac{\theta}{2}$, $s = \sin\frac{\theta}{2}$. Let $\varphi = \frac{k\pi}{2}$; we define the separable state $\sigma = \frac{1}{4}\sum_{k=0}^{3}|\psi_s\rangle\langle\psi_s|$. The separable state σ can be expressed as

$$\begin{aligned} \sigma =& c^8|0000\rangle\langle 0000| + s^8|1111\rangle\langle 1111| \\ &+ c^4s^4(|0000\rangle\langle 1111| + |1111\rangle\langle 0000|) \\ &+ 4c^6s^2|W\rangle\langle W| \\ &+ 6c^4s^4|Dicke\rangle\langle Dicke| \\ &+ 4c^2s^6|\bar{W}\rangle\langle\bar{W}|, \end{aligned} \tag{24}$$

where $|\bar{W}\rangle = \frac{1}{2}(|0111\rangle + |1011\rangle + |1101\rangle + |1110\rangle)$.

When $\Phi = 90^\circ$, ρ is a Dicke–GHZ mixed state, $\theta = \frac{\pi}{2}$, $\varphi = \frac{k\pi}{2}$, $k = 0, 1, 2, 3$; thus, the separable state σ_B for point B in Figure 1 can be expressed as

$$\sigma_B = \frac{1}{8}|GHZ\rangle\langle GHZ| + \frac{3}{8}|Dicke\rangle\langle Dicke| + \frac{1}{4}|W\rangle\langle W| + \frac{1}{4}|\bar{W}\rangle\langle\bar{W}|. \tag{25}$$

The point B is located at $(\frac{3}{8}, \frac{1}{8})$ with $\Theta = 18.4349^\circ$. When $\Theta = 90^\circ$, the separable state σ_A of point A is

$$\sigma_A = \frac{1}{2}(|GHZ\rangle\langle GHZ| + |GHZ_-\rangle\langle GHZ_-|), \tag{26}$$

where $|GHZ_-\rangle = \frac{1}{\sqrt{2}}(|0000\rangle - |1111\rangle)$. Any separable state σ_{AB} on segment AB can be expressed as

$$\sigma_{AB} = p\sigma_A + (1-p)\sigma_B, \tag{27}$$

where p and $1-p$ denote the probability distributions. We have $g = \frac{p}{2} + \frac{1-p}{8}$, $d = \frac{3}{8}(1-p)$; thus, the expression of segment AB can be represented as

$$g + d = \frac{1}{2}. \tag{28}$$

3.2. Segment BC

By analyzing and calculating, we get the relationship among M_i:

$$\begin{aligned} -M_{10} - 6M_{11} &= M_{10} - 6(M_6 + M_7) \\ M_{12} &= \frac{1}{3}M_{13}. \end{aligned} \tag{29}$$

and we find that when $\theta = 90^\circ$ and $\varphi = \frac{k\pi}{2}$ where $k = 0, 1, 2, 3$, the Λ of Equation (13) will achieve the same maximum value:

$$\Lambda = 6M_6 + M_{13}. \tag{30}$$

The M_0 in Equation (15) can be expressed as

$$M_0 = M_{10} - 6(M_6 + M_7), \tag{31}$$

and the rest of Equation (15) is equal to that of Equation (21).

Combining the above analysis and conclusions, we get $\mathcal{L} = \frac{3}{8}\frac{1}{\cos\Theta}$. According to the relationship between $\mathcal{L}$ and R, in the Descartes coordinate system, we can easily get

$$d = \frac{3}{8}, \tag{32}$$

where $d = \frac{\cos\Theta}{R(\sin\Theta + \cos\Theta)}$.

The criterion is shown in Figure 1, with the blue line representing Line $\overline{BC}$ and accounting for the numerical necessary criterion well when $\Theta \in [0^\circ, \Theta_1]$ with the angle $\Theta_1 = \tan^{-1}\frac{1}{3} \approx 18.4349^\circ$.

Next, we will present the upper bound of this segment. As Θ gradually approaches 0°, the $|Dicke\rangle$ state in σ_B is gradually classified as a noise state. The separable state for point C can be represented as

$$\sigma_C = \frac{3}{8}|Dicke\rangle\langle Dicke| + \frac{5}{8}\tau \tag{33}$$

where τ is the noise state. Hence, the coordinate of point C can easily be obtained, which is $(\frac{3}{8}, 0)$. Segment BC can be expressed as

$$d = \frac{3}{8}. \tag{34}$$

4. Robustness of Fully Separable Four-Qubit Dicke–W Mixed with GHZ ($\Phi = 56.31^\circ$)

When ρ is a Dicke–W and GHZ mixed state, the component of the $|W\rangle$ state in Equation (24) should be taken into account. However, for $\Phi \in [56.31^\circ, 90^\circ]$, the only difference from the result at $\Phi = 90^\circ$ is that the abscissa is transformed from d to d_w. A schematic representation is depicted in Figure 2.

The reason for this is that when $\theta = \frac{\pi}{2}$, the ratio of the $|Dicke\rangle$ to $|W\rangle$ components is $\frac{6c^4s^4}{4c^6s^2} = \frac{3}{2}$, and $\arctan\frac{3}{2} = 56.31^\circ$. The green point in Figure 2 located at $(w, g, d) = (\frac{1}{4}, \frac{3}{8}, \frac{1}{8})$ represents the unnormalized state $\tilde{\rho} = \frac{1}{8}|GHZ\rangle\langle GHZ| + \frac{3}{8}|Dicke\rangle\langle Dicke| + \frac{1}{4}|W\rangle\langle W|$. Clearly, it is a part of the separable state σ_B in Equation (25). Thus, the robustness of a state with $\Phi = 56.31^\circ$ can easily be obtained. When $\Phi < 56.31^\circ$, due to the insufficient components of the $|W\rangle$ state, θ begins to vary. We will show this case in the next section with an example.

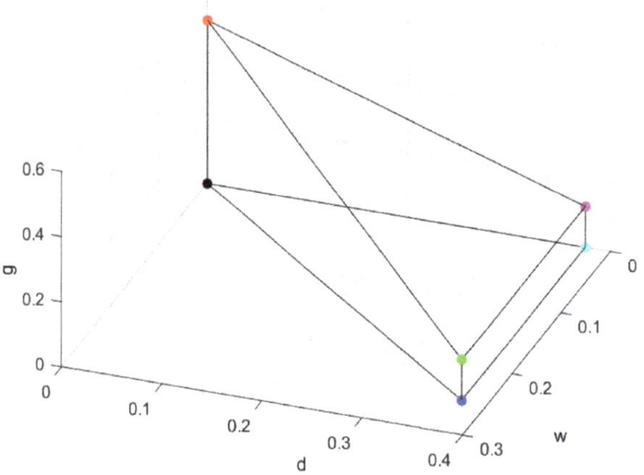

Figure 2. The robustness of fully separable four-qubit Dicke–W and GHZ mixed states. The coordinates of the green point in the figure are $(\frac{1}{4}, \frac{3}{8}, \frac{1}{8})$.

5. Robustness of Fully Separable Four-Qubit Dicke–W Mixed with GHZ ($\Phi = 45°$)

In this section, $\Phi = 45°$, so the density matrix ρ can be expressed as Equation (2). In this case, the entanglement witness for four-qubit Dicke–W and GHZ mixed states is represented using Pauli matrices in the following form:

$$\begin{aligned} \omega =& M_1 IIII + M_2 IIIZ_p + M_3 IIXX_p + M_4 IIYY_p \\ &+ M_5 IXXZ_p + M_6 IYYZ_p + M_7 IZZZ_p + M_8 XXZZ_p \\ &+ M_9 YYZZ_p + M_{10} ZZZZ + M_{11} IIZZ_p + M_{12} XXXX \\ &+ M_{13} XXYY_p + M_{14} YYYY, \end{aligned} \tag{35}$$

where the subscript p denotes the summation over all permutations of Pauli matrices.

After analysis and calculation, we divide the robustness of the mixed state of four-qubit Dicke–W mixed with GHZ at a 45° ratio into four segments, the specific contents of which are shown in Table 1.

Table 1. Segmentation of the four-qubit Dicke–W and GHZ mixed states with a Dicke–W mixing ratio of $\Phi = 45°$.

Segment Number	Θ	ω	θ	Line Type
Segment I	$[20.9576°, 90°]$	Type 1	$0°, 78.4630°, \pi$	Straight line
Segment II	$[13.2627°, 20.9576°]$	Type 2	$78.4630°, \pi$	Curve
Segment III	$[5.9803°, 13.2627°]$	Type 3	$78.4630°, \pi$	Straight line
Segment IV	$[0°, 5.9803°]$	Type 4	$78.4630°$	Straight line

The robustness of entanglement for four-qubit of Dicke–W and GHZ mixed states with a 45° Dicke–W mixing ratio is shown in Figure 3.

In this section, we employ $\Phi = 45°$. The upper bound of the robustness is depicted in Figure 3. Based on the characteristics of the figure, several points are labeled as A, B, C, D, and E. To avoid misunderstanding, the points A, B, C, D, and E mentioned in this section refer to the points in Figure 3 rather than the points in Figure 1 of Section 3. The analytical solutions for each segment with the deductions are presented in the following text.

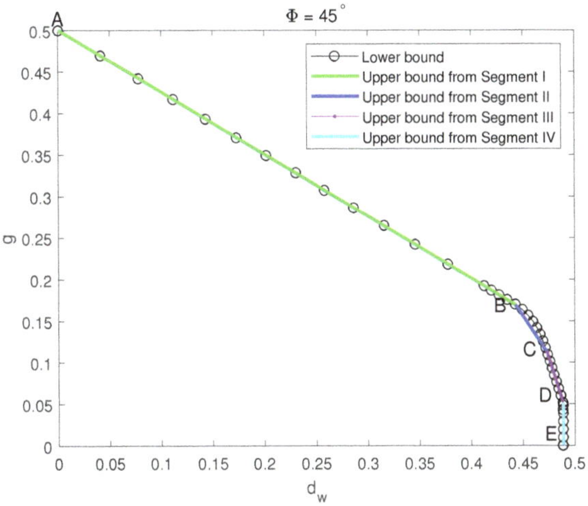

Figure 3. Robustness of entanglement for four-qubit of Dicke–W and GHZ mixed states with a 45° Dicke–W ratio.

5.1. Segment I

In this segment, the entanglement witness matrix ω in Equation (4) can be divided into three parts—ω_1, ω_2, and ω_3—which are as follows:

$$\omega = \omega_1 \oplus \omega_2 \oplus \omega_3, \tag{36}$$

where ω_1 represents the witness matrix in the corresponding subspace of the W state, ω_2 represents the witness matrix in the corresponding subspace of the Dicke state, and ω_3 represents the witness matrix in the corresponding subspace of the GHZ state. In the subspace formed by $|0001\rangle$, $|0010\rangle$, $|0100\rangle$, and $|1000\rangle$, ω_1 can be represented as

$$\omega_1 = bI_4 + aD_4, \tag{37}$$

where a and b are positive parameters, I_4 is a 4×4 identity matrix, and D_4 is a 4×4 matrix with all elements equal to one. In the subspace formed by $|0011\rangle$, $|0101\rangle$, $|0110\rangle$, $|1001\rangle$, $|1010\rangle$, and $|1100\rangle$, ω_2 can be represented as

$$\omega_2 = dI_6 + eD_6 + fG_6, \tag{38}$$

where d, e, and f are positive parameters, I_6 is a 6×6 identity matrix, D_6 is a 6×6 matrix with all elements equal to one, and G_6 is a 6×6 matrix and can be expressed as follows:

$$\begin{aligned} G_6 = &|0011\rangle\langle 1100| + |0101\rangle\langle 1010| + |0110\rangle\langle 1001| \\ &+ |1001\rangle\langle 0110| + |1010\rangle\langle 0101| + |1100\rangle\langle 0011|. \end{aligned} \tag{39}$$

In the subspace formed by $|0000\rangle$ and $|1111\rangle$, ω_3 can be represented as

$$\omega_3 = hI_2, \tag{40}$$

where h is a positive parameter, and I_2 is a 2×2 identity matrix.

After completing the construction of the density matrix, we can calculate and analyze the lower bound of the robustness of entanglement through Equation (4). The denominator in Equation (4) can be represented as

$$\langle \psi_4 | \omega | \psi_4 \rangle = hc^8 + hs^8 + 6(d + 6e + f)c^4 s^4 + 4(4b + a)c^6 s^2, \tag{41}$$

where $|\psi_4\rangle = |\psi\rangle^{\otimes 4}$, $c = \cos\frac{\theta}{2}$, and $s = \sin\frac{\theta}{2}$. According to the analytical calculation shown in Table 1, we have determined that in this part, there are three θ that simultaneously cause Equation (41) to reach the maximum value, one of which is $\theta = \pi$, and the maximum is h. Substituting the maximum value h into Equation (41), we can easily obtain

$$c^8 + s^8 + \frac{6(d+6e+f)c^4s^4 + 4(4b+a)c^6s^2}{h} - 1 = 0. \tag{42}$$

Apart from $\theta = 0$ and $\theta = \pi$, we substitute $\theta = 78.4630^\circ$ into the formula and can easily get the relationship between other parameters and h. The numerator in Equation (4) can be expressed as

$$\langle \rho, \omega \rangle = \frac{1-p_1}{2}[p_2(d+6e+f) + (1-p_2)(4a+b)] + 2p_1 h. \tag{43}$$

Combining Equation (4), Equation (43), and the relationship between h and the other parameters derived from Equation (42), we can get the lower bound result of robustness of entanglement in this segment:

$$2g + \frac{19}{9\sqrt{2}}d_w = 1, \tag{44}$$

where $g = \frac{p_1}{R}$ and $d_w = (1-p_1)\frac{\sqrt{(p_2^2+(1-p_2)^2)}}{R}$. Equation (44) is the expression of Segment I of the lower bound of robustness for the Dicke–W and GHZ mixed states with a Dicke–W mixing ratio of 45°.

Next, we will present the upper bound of this segment. When $\Phi = 45^\circ$, the coefficients of $|W\rangle$ and $|Dicke\rangle$ are equal; thus, we have $4c^6s^2 = 6c^4s^4$, $c^2 = \frac{3}{5}$, and $s^2 = \frac{2}{5}$. Substituting c and s back into Equation (24), σ can be expressed as

$$\begin{aligned}\sigma =& \frac{81}{625}|0000\rangle\langle 0000| + \frac{16}{625}|1111\rangle\langle 1111| \\ &+ \frac{36}{625}(|0000\rangle\langle 1111| + |1111\rangle\langle 0000|) \\ &+ \frac{216}{625}(|W\rangle\langle W| + |Dicke\rangle\langle Dicke|) \\ &+ \frac{96}{625}|\bar{W}\rangle\langle \bar{W}|.\end{aligned} \tag{45}$$

When $\theta = 0^\circ$ or 180°, $\sigma_1 = |0000\rangle\langle 0000|$ and $\sigma_2 = |1111\rangle\langle 1111|$ based on Equation (24). Therefore, the component of $|GHZ\rangle$ can be increased by adding σ_1 and σ_2 to σ in Equation (45). Actually, it suffices to add σ_2 in σ. The separable state σ_B for point B can be expressed as

$$\begin{aligned}\sigma_B =& \frac{625}{690}(\sigma + \frac{65}{625}\sigma_2) \\ =& \frac{39}{230}|GHZ\rangle\langle GHZ| + \frac{72}{230}(|W\rangle\langle W| + |Dicke\rangle\langle Dicke|) \\ &+ \frac{32}{230}|\bar{W}\rangle\langle \bar{W}| + \frac{15}{230}|GHZ_-\rangle\langle GHZ_-|.\end{aligned} \tag{46}$$

Point B is $(\frac{72}{230}\sqrt{2}, \frac{39}{230})$ with $\Theta = 20.96^\circ$. For $\Theta = 90^\circ$, for example, ρ is a $|GHZ\rangle$ pure state, and point A is known to be $(0, \frac{1}{2})$. Thus, segment AB can be expressed as

$$g = -\frac{19\sqrt{2}}{36}d_w + \frac{1}{2}. \tag{47}$$

5.2. Segment II

The entanglement witness matrix of this segment is similar to that of Segment I. The difference is ω_3, and it is represented as follows:

$$\omega_3 = u\begin{bmatrix} \frac{1}{h} & 1 \\ 1 & h \end{bmatrix}, \tag{48}$$

where h and u are positive parameters. Similar to the process of Segment I, the denominator in Equation (4) can be represented as

$$\langle \psi_4|\omega|\psi_4\rangle = \frac{u}{h}c^8 + 2uc^4s^4 + hus^8 + 6(d+6e+f)c^4s^4 + 4(4b+a)c^6s^2. \tag{49}$$

According to the analytical calculation shown in Table 1, we have obtained that in this part, there are two θ that simultaneously cause Equation (49) to reach the maximum value, one of which is $\theta = \pi$, and the maximum is hu. Substituting the maximum value hu into Equation (49), we can easily obtain

$$\frac{1}{h^2}c^8 + 2\frac{1}{h}c^4s^4 + s^8 + \frac{6(d+6e+f)c^4s^4 + 4(4b+a)c^6s^2}{hu} - 1 = 0. \tag{50}$$

Apart from $\theta = \pi$, we substitute $\theta = 78.4630°$ into the formula and can easily get the relationship between other parameters and h. The numerator in Equation (4) can be expressed as

$$\langle \rho, \omega\rangle = \frac{1-p}{2}[(d+6e+f) + (4a+b)] + \frac{p}{2}(\frac{u}{h} + 2u + hu). \tag{51}$$

According to the conditions for obtaining the maximum value, we get the relationship between h and Θ as

$$h = \frac{\frac{\tan\Theta}{\sqrt{2}} - \frac{1}{6}}{-\frac{\tan\Theta}{\sqrt{2}} + \frac{3}{8}}, \tag{52}$$

where $\tan\Theta = \frac{p}{\frac{1}{\sqrt{2}}(1-p)}$. Substituting Equation (49), Equation (51), and the relationship between parameters into Equation (4), we get the lower bound of the robustness of entanglement in this segment:

$$g = \frac{\sin\Theta}{R(\sin\Theta + \sqrt{2}\cos\Theta)} = \frac{2x}{\frac{(x-\frac{1}{6})^2}{\frac{3}{8}-x} + x + \frac{203}{72}}, \tag{53}$$

$$d_w = \frac{\sqrt{2}\cos\Theta}{R(\sin\Theta + \sqrt{2}\cos\Theta)} = \frac{1}{\frac{(x-\frac{1}{6})^2}{\frac{3}{8}-x} + x + \frac{203}{72}}, \tag{54}$$

where $x = \frac{1}{\sqrt{2}}\tan\Theta$.

The upper bound of this segment is shown in Figure 3 with a blue line.

5.3. Segment III

The entanglement witness matrix of this segment is similar to that of Segment II. The difference is ω_3, and it is represented as follows:

$$\omega_3 = u\begin{bmatrix} 0 & 0 \\ 0 & h \end{bmatrix}, \tag{55}$$

where h is a positive parameter. Similar to the process of Segment I, the denominator in Equation (4) can be represented as

$$\langle\psi_4|\omega|\psi_4\rangle = hs^8 + 6(d+6e+f)c^4s^4 + 4(4b+a)c^6s^2. \tag{56}$$

According to the analytical calculation shown in Table 1, we have determined that in this part, there are two θ that simultaneously cause Equation (56) to reach the maximum value, one of which is $\theta = \pi$, and the maximum is h. Substituting the maximum value h into Equation (56), we can easily obtain

$$s^8 + \frac{6(d+6e+f)c^4s^4 + 4(4b+a)c^6s^2}{h} - 1 = 0. \tag{57}$$

Apart from $\theta = \pi$, we substitute $\theta = 78.4630^\circ$ into the formula and can easily get the relationship between other parameters and h, and it is expressed as follows:

$$\frac{16}{625} + \frac{1}{h}\frac{216}{625}(A+B) - 1 = 0, \tag{58}$$

where $A = d + 4e + f$, and $B = 4a + b$. The numerator in Equation (4) can be expressed as

$$\langle\rho,\omega\rangle = \frac{1-p_1}{2}[p_2A + (1-p_2)B] + \frac{p_1}{2}h. \tag{59}$$

Substituting Equation (56), Equation (59), and the relationship between parameters analyzed above into Equation (4), we get the lower bound result of the robustness of entanglement in this segment:

$$\frac{203}{72\sqrt{2}}d_w + \frac{1}{2}g = 1. \tag{60}$$

Next, we will present the upper bound of this segment. σ_D corresponding to point D can be expressed as

$$\sigma_D = \frac{216}{625}(|W\rangle\langle W| + |Dicke\rangle\langle Dicke|) + \frac{32}{625}|GHZ\rangle\langle GHZ| + \frac{377}{625}\tau, \tag{61}$$

where τ is the noise state.

The expression of the separable state σ_C contains a higher proportion of $|GHZ\rangle$ than σ_D. Based on Equation (45) and σ_D, we still have $\frac{20}{625}$ of $|0000\rangle\langle 1111| + |1111\rangle\langle 0000|$ that can be transformed into GHZ states. By adding $\frac{20}{625}|1111\rangle\langle 1111|$ and normalizing, which is give by $\sigma_C = \frac{1}{1+p}(\sigma_D + p|1111\rangle\langle 1111|)$, the expression of σ_C is obtained as follows:

$$\begin{aligned}\sigma_C =& \frac{24}{215}|GHZ\rangle\langle GHZ| + \frac{72}{215}(|W\rangle\langle W| + |Dicke\rangle\langle Dicke|)\\ &+ \frac{32}{215}|\bar{W}\rangle\langle\bar{W}| + \frac{15}{215}|0000\rangle\langle 0000|.\end{aligned} \tag{62}$$

Therefore, the point C is $(\frac{72}{215}\sqrt{2}, \frac{24}{215})$with $\Theta = 13.26^\circ$. Similarly, $\sigma_{CD} = (1-p)\sigma_C + p\sigma_D$, the expression of segment CD can be represented as

$$g = -\frac{203\sqrt{2}}{72}d_w + 2. \tag{63}$$

5.4. Segment IV

The entanglement witness matrix of this segment only has two parts: ω_1 and ω_2. The denominator in Equation (4) can be represented as

$$\langle\psi_4|\omega|\psi_4\rangle = 6(d+6e+f)c^4s^4 + 4(4b+a)c^6s^2. \tag{64}$$

According to the analytical calculation shown in Table 1, we determine that in this part, there is only one θ that causes Equation (64) to reach the maximum value, and we can easily obtain

$$\langle\sigma,\omega\rangle = \frac{216}{625}[(d+6e+f)+(4b+a)]. \tag{65}$$

The numerator in Equation (4) can be expressed as

$$\langle\rho,\omega\rangle = \frac{1-p_1}{2}[p_2(d+6e+f)+(1-p_2)(4b+a)]. \tag{66}$$

Combining the above analysis structures, we get the lower bound result of robustness of entanglement in this segment:

$$d_w = \frac{216}{625}\sqrt{2}. \tag{67}$$

Therefore, σ_E corresponding to point E can be expressed as

$$\sigma_E = \frac{216}{625}(|W\rangle\langle W| + |Dicke\rangle\langle Dicke|) + \frac{409}{625}\tau, \tag{68}$$

where τ is the noise state. It is evident that $w = d = \frac{216}{625}$ for segment DE, and g increases from 0 to $\frac{32}{625}$ as Θ increases. Therefore, D is $(\frac{216}{625}\sqrt{2}, \frac{32}{625})$ with $\Theta = 5.98^\circ$, and E is $(\frac{216}{625}\sqrt{2}, 0)$ with $\Theta = 0^\circ$. The separable state σ_{DE} for segment DE can be expressed as $\sigma_{DE} = (1-p)\sigma_D + p\sigma_E$, which can represent an arbitrary separable state on the segment DE. By respectively transforming the variables p and $(1-p)$ into g and d_w, the expression for the upper bound is derived as follows:

$$d_w = \frac{216}{625}\sqrt{2}. \tag{69}$$

6. Conclusions

In this paper, we propose a method to quantify the multipartite entangled states of quantum resources. We quantify full separability for Dicke–W and GHZ mixed states by analyzing the upper bound and the lower bound of the robustness of entanglement. When Φ is 90°, that is, in the case of Dicke and GHZ mixed states, we obtain the analytical expression of the lower bound and upper bound, and the results show that the lower bound coincides with the upper bound. When Φ is 45°, that is, in the case of equal proportions of Dicke and W states mixed with GHZ state, we obtain analytical solutions for the straight segments and a tractable solution for the curved segment. Finally, we determine that the bound of the Dicke–W mixing ratio is $\Phi = 56.31^\circ$, which divides the curve of the robustness of entanglement into two types.

Author Contributions: Formal analysis, L.-H.Z., Z.Z. and G.-L.L.; Writing—original draft, L.-H.Z.; Supervision, C.-Q.Y. and X.-Y.C. All authors have read and agreed to the published version of the manuscript.

Funding: This work was supported by the National Natural Science Foundation of China (Grant No: 61871347).

Data Availability Statement: No new data were created or analyzed in this study. Data sharing is not applicable to this article.

Acknowledgments: This work was also supported by the advanced computing resources provided by the Supercomputing Center of Hangzhou City University.

Conflicts of Interest: The authors declare no conflicts of interest.

Appendix A

According to Equations (11) and (12), we get

$$\begin{aligned}
&Tr(\hat{M}'|\psi_s\rangle\langle\psi_s|) \\
&= Tr\sum_{ijkl=0}^{3} M_{ijkl}\sigma_i\otimes\sigma_j\otimes\sigma_k\otimes\sigma_l\cdot\rho_1\otimes\rho_2\otimes\rho_3\otimes\rho_4 \\
&= \sum_{ijkl=0}^{3} M_{ijkl}[Tr(\rho_1\sigma_i)Tr(\rho_2\sigma_j)Tr(\rho_3\sigma_k)Tr(\rho_4\sigma_l)] \\
&= \Lambda_1+\Lambda_2x_4+\Lambda_3y_4+\Lambda_4z_4 \\
&\leq \Lambda_1+\sqrt{\Lambda_2^2+\Lambda_3^2+\Lambda_4^2},
\end{aligned} \tag{A1}$$

$$\begin{aligned}
\Lambda_1 =& M_2z_3+M_3z_2+M_4z_1+M_5z_1z_2z_3+M_6(x_1x_2+x_1x_3+x_2x_3 \\
&+y_1y_2+y_1y_3+y_2y_3)+M_8(x_1x_2z_3+x_Az_2x_3+z_1x_2x_3) \\
&+M_9(y_1y_2z_3+y_1z_2y_3+z_1y_2y_3)+M_{11}(z_1z_2+z_1z_3+z_2z_3), \\
\Lambda_2 =& M_6(x_1+x_2+x_3)+M_7(z_1z_2x_3+z_1x_2z_3+x_1z_2z_3) \\
&+M_8(x_1z_2+x_1z_3+x_2z_3+z_1x_2+z_1x_3+z_2x_3) \\
&+M_{12}(x_1y_2y_3+y_1x_2y_3+y_1y_2x_3)+M_{13}x_1x_2x_3 \\
\Lambda_3 =& M_6(y_1+y_2+y_3)+M_7(z_1z_2y_3+z_1y_2z_3+y_1z_2z_3) \\
&+M_9(y_1z_2+y_1z_3+y_2z_3+z_1y_2+z_1y_3+z_2y_3) \\
&+M_{12}(x_1x_2y_3+x_1y_2x_3+y_1x_2x_3)+M_{14}y_1y_2y_3, \\
\Lambda_4 =& M_1+M_5(z_1z_2+z_1z_3+z_2z_3)+M_7(z_1x_2x_3+x_1z_2x_3 \\
&+x_1x_2z_3+z_1y_2y_3+y_1z_2y_3+y_1y_2z_3)+M_8(x_1x_2+x_1x_3 \\
&+x_2x_3)+M_9(y_1y_2+y_1y_3+y_2y_3)+M_{10}z_1z_2z_3 \\
&+M_{11}(z_1+z_2+z_3).
\end{aligned} \tag{A2}$$

Thus, we get Equation (13).

References

1. Horodecki, R.; Horodecki, P.; Horodecki, M.; Horodecki, K. Quantum entanglement. *Rev. Mod. Phys.* **2009**, *81*, 865–942. [CrossRef]
2. Bennett, C.H.; DiVincenzo, D.P.; Smolin, J.A.; Wootters, W.K. Mixed-state entanglement and quantum error correction. *Phys. Rev. A* **1996**, *54*, 3824. [CrossRef] [PubMed]
3. Werner, R.F. Quantum states with Einstein-Podolsky-Rosen correlations admitting a hidden-variable model. *Phys. Rev. A* **1989**, *40*, 4277. [CrossRef] [PubMed]
4. Horodecki, M.; Horodecki, P.; Horodecki, R. On the necessary and sufficient conditions for separability of mixed quantum states. *Phys. Lett. A* **1996**, *223*. [CrossRef]
5. Horodecki, P. Separability criterion and inseparable mixed states with positive partial transposition. *Phys. Lett. A* **1997**, *232*, 333–339. [CrossRef]
6. Zyczkowski, K.; Horodecki, P.; Sanpera, A.; Lewenstein, M. On the volume of the set of mixed entangled states. *Phys. Rev. A* **1998**, *58*, 883. [CrossRef]
7. Vidal, G.; Werner, R.F. Computable measure of entanglement. *Phys. Rev. A* **2002**, *65*, 032314. [CrossRef]
8. Chen, K.; Wu, L.A. A matrix realignment method for recognizing entanglement. *Quantum Inf. Comput.* **2003**, *3*, 193–202. [CrossRef]
9. Rudolph, O. Further results on the cross norm criterion for separability. *Quantum Inf. Process.* **2005**, *4*, 219–239. [CrossRef]
10. Terhal, B.M. Bell inequalities and the separability criterion. *Phys. Lett. A* **2000**, *271*, 319–326. [CrossRef]
11. Bourennane, M.; Eibl, M.; Kurtsiefer, C.; Gaertner, S.; Weinfurter, H.; Gühne, O.; Hyllus, P.; Bruß, D.; Lewenstein, M.; Sanpera, A. Experimental detection of multipartite entanglement using witness operators. *Phys. Rev. Lett.* **2004**, *92*, 087902. [CrossRef] [PubMed]
12. Lewenstein, M.; Kraus, B.; Cirac, J.I.; Horodecki, P. Optimization of entanglement witnesses. *Phys. Rev. A* **2000**, *62*, 052310. [CrossRef]
13. Chen, X.Y.; Jiang, L.Z.; Xu, Z.A. Matched witness for multipartite entanglement. *Quantum Inf. Process.* **2017**, *16*, 1–22. [CrossRef]

14. Ren, Z.; Li, W.; Smerzi, A.; Gessner, M. Metrological detection of multipartite entanglement from young diagrams. *Phys. Rev. Lett.* **2021**, *126*, 080502. [CrossRef] [PubMed]
15. Bruß, D.; D'Ariano, G.M.; Lewenstein, M.; Macchiavello, C.; Sen, A.; Sen, U. Distributed quantum dense coding. *Phys. Rev. Lett.* **2004**, *93*, 210501. [CrossRef]
16. Braunstein, S.L.; Kimble, H.J. Dense coding for continuous variables. *Phys. Rev. A* **2000**, *61*, 042302. [CrossRef]
17. Bennett, C.H.; Brassard, G.; Crépeau, C.; Jozsa, R.; Peres, A.; Wootters, W.K. Teleporting an unknown quantum state via dual classical and Einstein-Podolsky-Rosen channels. *Phys. Rev. Lett.* **1993**, *70*, 1895. [CrossRef]
18. Bouwmeester, D.; Pan, J.W.; Mattle, K.; Eibl, M.; Weinfurter, H.; Zeilinger, A. Experimental quantum teleportation. *Nature* **1997**, *390*, 575–579. [CrossRef]
19. Bennett, C.H.; Brassard, G.; Popescu, S.; Schumacher, B.; Smolin, J.A.; Wootters, W.K. Purification of noisy entanglement and faithful teleportation via noisy channels. *Phys. Rev. Lett.* **1996**, *76*, 722. [CrossRef]
20. Ekert, A.; Macchiavello, C. Quantum error correction for communication. *Phys. Rev. Lett.* **1996**, *77*, 2585. [CrossRef]
21. Biham, E.; Huttner, B.; Mor, T. Quantum cryptographic network based on quantum memories. *Phys. Rev. A* **1996**, *54*, 2651. [CrossRef] [PubMed]
22. Gühne, O.; Tóth, G. Entanglement detection. *Phys. Rep.* **2009**, *474*, 1–75. [CrossRef]
23. Yao, X.C.; Wang, T.X.; Xu, P.; Lu, H.; Pan, G.S.; Bao, X.H.; Peng, C.Z.; Lu, C.Y.; Chen, Y.A.; Pan, J.W. Observation of eight-photon entanglement. *Nat. Photonics* **2012**, *6*, 225–228. [CrossRef]
24. Sackett, C.A.; Kielpinski, D.; King, B.E.; Langer, C.; Meyer, V.; Myatt, C.J.; Rowe, M.; Turchette, Q.; Itano, W.M.; Wineland, D.J.; et al. Experimental entanglement of four particles. *Nature* **2000**, *404*, 256–259. [CrossRef]
25. Zhao, Z.; Chen, Y.A.; Zhang, A.N.; Yang, T.; Briegel, H.J.; Pan, J.W. Experimental demonstration of five-photon entanglement and open-destination teleportation. *Nature* **2004**, *430*, 54–58. [CrossRef]
26. Chen, X.Y.; Jiang, L.Z.; Xu, Z.A. Necessary and sufficient criterion for k-separability of N-qubit noisy GHZ states. *Int. J. Quantum Inf.* **2018**, *16*, 1850037. [CrossRef]
27. Bollinger, J.J.; Itano, W.M.; Wineland, D.J.; Heinzen, D.J. Optimal frequency measurements with maximally correlated states. *Phys. Rev. A* **1996**, *54*, R4649. [CrossRef]
28. Gimeno-Segovia, M.; Shadbolt, P.; Browne, D.E.; Rudolph, T. From three-photon Greenberger-Horne-Zeilinger states to ballistic universal quantum computation. *Phys. Rev. Lett.* **2015**, *115*, 020502. [CrossRef]
29. Chen, X.Y.; Jiang, L.Z.; Xu, Z.A. Precise detection of multipartite entanglement in four-qubit Greenberger–Horne–Zeilinger diagonal states. *Front. Phys.* **2018**, *13*, 130317. [CrossRef]
30. Dicke, R.H. Coherence in spontaneous radiation processes. *Phys. Rev.* **1954**, *93*, 99. [CrossRef]
31. Kiesel, N.; Schmid, C.; Tóth, G.; Solano, E.; Weinfurter, H. Experimental observation of four-photon entangled Dicke state with high fidelity. *Phys. Rev. Lett.* **2007**, *98*, 063604. [CrossRef] [PubMed]
32. Eibl, M.; Kiesel, N.; Bourennane, M.; Kurtsiefer, C.; Weinfurter, H. Experimental realization of a three-qubit entangled W state. *Phys. Rev. Lett.* **2004**, *92*, 077901. [CrossRef]
33. Zou, X.; Pahlke, K.; Mathis, W. Generation of an entangled four-photon W state. *Phys. Rev. A* **2002**, *66*, 044302. [CrossRef]
34. Tashima, T.; Özdemir, Ş.K.; Yamamoto, T.; Koashi, M.; Imoto, N. Elementary optical gate for expanding an entanglement web. *Phys. Rev. A Atomic, Mol. Opt. Phys.* **2008**, *77*, 030302. [CrossRef]
35. Chen, X.Y.; Jiang, L.Z. Noise tolerance of Dicke states. *Phys. Rev. A* **2020**, *101*, 012308. [CrossRef]
36. Carvalho, A.R.; Mintert, F.; Buchleitner, A. Decoherence and multipartite entanglement. *Phys. Rev. Lett.* **2004**, *93*, 230501. [CrossRef] [PubMed]
37. Zhang, G.; Song, Z. Quantum scars in spin-1/2 isotropic Heisenberg clusters. *New J. Phys.* **2023**, *25*, 053025. [CrossRef]
38. Chen, X.Y.; Jiang, L.Z. What criterion can we get from precise entanglement witnesses? *IEEE J. Sel. Areas Commun.* **2020**, *38*, 557–567. [CrossRef]
39. Lami, L.; Regula, B.; Takagi, R.; Ferrari, G. Framework for resource quantification in infinite-dimensional general probabilistic theories. *Phys. Rev. A* **2021**, *103*, 032424. [CrossRef]

entropy

MDPI

Article

Knowledge-Assisted Actor Critic Proximal Policy Optimization-Based Service Function Chain Reconfiguration Algorithm for 6G IoT Scenario

Bei Liu [1,*], Shuting Long [1] and Xin Su [2]

1 School of Communication and Information Engineering, Chongqing University of Posts and Telecommunications, Chongqing 400065, China
2 Department of Electronic Engineering, Tsinghua University, Beijing 100084, China
* Correspondence: liubei@tsinghua.edu.cn or l.bei@foxmail.com

Abstract: Future 6G networks will inherit and develop Network Function Virtualization (NFV) architecture. With the NFV-enabled network architecture, it becomes possible to establish different virtual networks within the same infrastructure, create different Virtual Network Functions (VNFs) in different virtual networks, and form Service Function Chains (SFCs) that meet different service requirements through the orderly combination of VNFs. These SFCs can be deployed to physical entities as needed to provide network functions that support different services. To meet the highly dynamic service requirements in the future 6G Internet of Things (IoT) scenario, the highly flexible and efficient SFC reconfiguration algorithm is the key research direction. Deep-learning-based algorithms have shown their advantages in solving this type of dynamic optimization problem. Considering that the efficiency of the traditional Actor Critic (AC) algorithm is limited, the policy does not directly participate in the value function update. In this paper, we use the Proximal Policy Optimization (PPO) clip function to restrict the difference between the new policy and the old policy, to ensure the stability of the updating process. We combine PPO with AC, and further bring the historical decision information as the network knowledge to offer better initial policies, to accelerate the training speed. We also propose the Knowledge = Assisted Actor Critic Proximal Policy Optimization (KA-ACPPO)-based SFC reconfiguration algorithm to ensure the Quality of Service (QoS) of end-to-end services. Simulation results show that the proposed KA-ACPPO algorithm can effectively reduce computing cost and power consumption.

Keywords: 6G IoT; SFC reconfiguration; knowledge assisted; ACPPO algorithm

1. Introduction

Looking towards 2030 and beyond, 6G will provide lower latency and more reliable end-to-end connectivity to power Industry 4.0 to Industry 5.0 [1]. The network environment and service requirements are highly dynamic in future 6G IoT scenarios. To cope with highly dynamic service requirements, 6G networks will inherit and develop NFV architecture. With the NFV-enabled network, it becomes possible to establish different virtual networks within the same infrastructure and create different VNFs in different virtual networks. When the service requirements arrive, VNFs are combined in an orderly manner to form SFCs to meet different service requirements. In this way, SFC deployment based on mobile edge computing and network function virtualization can be a viable solution for providing flexible and controlled network services [2]. When the service requirement changes, the corresponding SFC needs to be reconfigured to guarantee the QoS requirements. To meet the low latency requirements of 6G services, highly flexible and efficient SFC orchestration and reconfiguration algorithms have become a crucial problem.

The SFC reconfiguration problem involves the study of how to modify, re-organize, and monitor the deployed SFC to cope with network dynamic changes and meet new service requirements. SFC reconfiguration is always implemented by VNF migration. Reference [3] modeled the SFC reconfiguration problem as an Integer Linear Programming (ILP) formulation and proposed a greedy-based heuristic algorithm to solve the problem Reference [4] proposed a threshold-dependent scalable cluster VNF migration algorithm to minimize the cost of embedding VNF under the condition of satisfying delay constraints. Reference [5] described the VNF migration problem as a new graph theory problem and proposed an efficient heuristic algorithm based on dynamic programming to effectively alleviate dynamic traffic and reduce total traffic costs. Reference [6] studied the reconfiguration problem of a set of SFCs with different priorities. A polynomial time heuristic algorithm was proposed to quickly deploy emergency SFCs while meeting the requirements of emergency SFC input traffic and resource constraints. Another precise algorithm was also proposed to achieve maximum profit for service providers. Reference [7] aimed to minimize the total migration cost and used a temporal conventional network to predict network traffic, proposing a fast and efficient heuristic VNF migration algorithm.

Nowadays, with the rapid development of Artificial Intelligence (AI), deep learning has been widely investigated and has shown advantages in coping with this type of dynamic problem. Reference [8] proposed a DRL-based algorithm to provide fast VNF migration decisions in highly dynamic environments, with the goal of minimizing the weighted total latency and cost of VNF migration. Reference [9] used a mixed-density neural network to accurately model complex user migration patterns in reality, in order to support the prediction of user edge cloud access probabilities and minimize the sum of operational costs and potential losses caused by downtime. Reference [10] proposed a deep Dyna-Q approach to solve the SFC reconfiguration problem under the premise of guaranteeing QoS and resource constraints. Reference [11] proposed a SFC management scheme based on the prediction of service. Reference [12] proposed the Dueling Double Deep Q Network (D3QN)-based SFC orchestration scheme to ensure QoE of users in the highly dynamic and resource-constrained Unmanned Aerial Vehicle (UAV) scenario.

The aforementioned research demonstrates the tremendous potential of deep learning algorithms. The AC algorithm, as a policy-based algorithm, has been used in many dynamic optimization problems. However, the efficiency of classical AC is limited in that the policy does not directly participate in the value function update. So, we use the principle of PPO, that is, proximal ratio pruning, to limit the magnitude of policy updates, and form the ACPPO algorithm. Based on this, we further utilize the historical decision information as the network knowledge to offer better initial policies and accelerate the training speed.

The main contributions of this paper are synthesized as follows:

- We formulate the SFC reconfiguration as a VNF migration problem, aiming to minimize the migration cost, to cope with the highly dynamic service requirements in 6G IoT scenarios.
- We use proximal ratio pruning to limit the magnitude of policy updates, reducing the impact of coupling relationship in the AC algorithm, and combine the advantages of the AC and PPO algorithm to solve the VNF migration problem. And we further introduce the 6G knowledge base, which is dynamic updating by saving and updating the excellent historical policies, to expand the action space and formulate the KA-ACPPO algorithm.
- Simulation results show that our proposed KA-ACPPO can effectively reduce computing cost and power consumption.

This paper is organized as follows. In Section 1, the system model is presented and the problem is formulated. Section 2 proposes the KA-ACPPO algorithm. Section 3 gives the simulation results. Finally, Section 4 concludes the paper and gives future research directions.

2. System Model and Problem Formulation

2.1. Network Model

Consider the 6G IoT scenario. Service requirements are satisfied by SFC orchestration. SFC orchestration is complemented by global control with the help of the 6G knowledge base. When the device produces a service request, it is assigned to the closed-edge node and communicates with the top-level SDN controller and NFV manager. The VNF manager complements the VNF orchestration, forming an ordered set of VNFs, that is, the SFC deployed policy. Finally, the policy is issued to the infrastructure layer to deploy the network functions to meet the service requirements. And when the service requirements change, SFC can be reconfigured through VNF migration, for example, VNF is migrated from one edge node to another edge node.

It is shown in Figure 1 that the edge nodes and the physical links in the infrastructure layer are expressed as the undirected graph $G = (X, L)$, in which X and L, respectively, describe the set of the edge nodes and the physical links. The available computing resource of the edge node $x \in X$ can be expressed as C_x^{CPU}. $l_{x,y}$ describes the physical link between the edge node x and the edge node y, in which $x, y \in X$ and $x \neq y$. SFC reconfiguration is implemented through VNF migration, so the physical link needs to reserve bandwidth B for VNF migration. In fact, every SFC can be expressed as an ordered set of VNF, denoted as $V = \{v_1, v_2, \cdots, v_{|V|}\}$. We divide IoT services into two categories: delay-sensitive services and delay-tolerant services. The corresponding SFC requests can be denoted as delay-sensitive request $j \in J$ and delay-tolerant request $k \in K$, and the corresponding computing resource provided by the edge nodes can be, respectively, denoted as b_j and b_k. This paper ensures the QoS of service requests by controlling the reconstruction costs of VNF migration in three aspects: computing cost, power consumption cost, and delay cost.

For clear presentation, we summarize the notations used in the following formulation in Table 1.

Table 1. Notations used in this paper.

Notation	Description
X	the set of edge nodes
L	the set of physical links
l_{xy}	the link between edge node x and edge node y
$j \in J$	the index of delay-sensitive requests
$k \in K$	the index of delay-tolerant requests
b_j	computing resource allocated to request j
b_k	computing resource allocated to request k
z_x	the computing resource consumption of edge node x
U_x^V	the variable of SFC deployment
W_x^V	the variable of SFC reconfiguration
P_x	the power consumption cost
λ_j, λ_k	the arriving rates of delay-sensitive and delay-tolerant requests
λ'	total arriving rate of requests
μ'	the serving rate of VNF
ρ'	the traffic intensity
D_q	the queuing delay of SFC requests
D_j	the delay of the delay-sensitive SFC requests
β_j	the data size of jth delay-sensitive SFC requests
D_m	the total delay of VNF migration caused by the position change of VNFs
D_o	the additional delay caused by prolonged downtime
Q_j	QoS degradation rate of jth delay-sensitive SFC requests

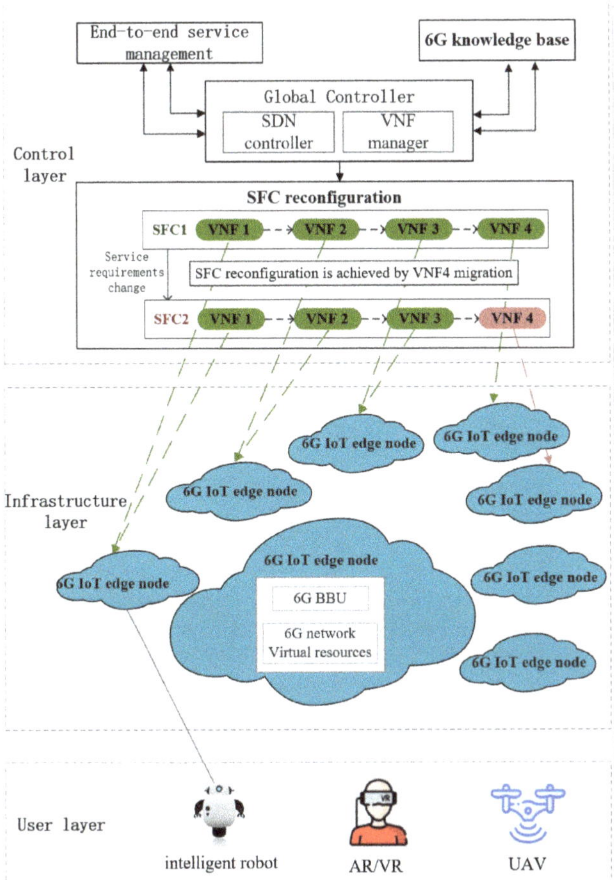

Figure 1. 6G IoT scenario and system model.

2.2. The Computing Cost of VNF Migration

SFCs are deployed in the edge node and are allocated computing resources from the edge node to meet the service requirements. To represent the computing resource consumption of node x, we denote the decision-making variables of SFC reconfiguration as follows:

$$\theta_{x,j} = \begin{cases} 1 & \text{if resource } b_j \text{ is allocates to request } j \\ 0 & \text{otherwise} \end{cases}, \tag{1}$$

$$\theta_{x,k} = \begin{cases} 1 & \text{if resource } b_k \text{ is allocates to request } k \\ 0 & \text{otherwise} \end{cases}. \tag{2}$$

The computing resource consumption of the edge node x can be denoted as follows:

$$z_k = \sum_{j=1}^{J} \theta_{x,j} b_j + \sum_{k=1}^{K} \theta_{x,k} b_k. \tag{3}$$

During the VNF migration process, only the CPU state between the initial node and the target node is considered. Therefore, the computing cost of VNF migration can be denoted as follows

$$Cost_1 = \sum_{x=1}^{X} \sum_{v=1}^{V} W_x^V (1 - U_x^V) z_k, \tag{4}$$

where U_x^V is the variable of SFC deployment, and W_x^V is the variable of SFC reconfiguration. For example, if the first VNF of the current SFC is not deployed at edge node 1, it can be denoted as $U_1^1 = 0$. If the VNF is migrated to the edge node 1, it can be denoted as $W_1^1 = 1$, and $W_1^{1,1}(1 - U_1^{1,1}) = 1$, which means the VNF is migrated.

2.3. The Power Consumption Cost of VNF Migration

VNF migration can cause changes in the working status of certain edge nodes, resulting in additional power consumption. Power consumption is related to the usage of computing resources at edge nodes. Based on the current SFC deployment policy, the power consumption of the edge node x can be denoted as [13]:

$$P_x = P(0\%) + (P(100\%) - P(0\%))(2z_x - (z_x)^{1.4}), \tag{5}$$

where $P(0\%)$ denotes the power consumption of the edge nodes in idle modes, and $P(100\%)$ denotes the power consumption of the edge nodes in full-load modes.

Therefore, the power consumption cost of VNF migration can be denoted as:

$$Cost_2 = \sum_{x=1}^{X} \sum_{v=1}^{V} W_x^V (1 - U_x^V) P_x. \tag{6}$$

2.4. The Delay Cost of VNF Migration

In this paper, the delay-sensitive services and the delay-tolerate services are both modeled as $M/M/1$ queuing, where the arriving rates are, respectively, depicted as λ_j and λ_k. And the total arriving rate at edge node x can be denoted as follows:

$$\lambda' = \sum_{j=1}^{J} \lambda_j + \sum_{k=1}^{K} \lambda_k. \tag{7}$$

The CPU working frequency of the edge node x is denoted as F, and the serving rate of VNF can be denoted as:

$$\mu' = z_x F. \tag{8}$$

And the traffic intensity can be denoted as:

$$\rho' = \frac{\lambda'}{\mu'}. \tag{9}$$

Therefore, the queuing delay of the SFC requests can be denoted as:

$$D_q = \frac{\rho'^2}{\lambda(1 - \rho')}. \tag{10}$$

Simply, only the queuing delay of the delay-sensitive SFC requests is considered, which is denoted as:

$$D_j = \frac{\beta_j}{b_j}, \tag{11}$$

where β_j denotes the data size of the jth delay-sensitive SFC request.

Considering the network overhead of the infrastructure layer and the stability of network services, the impact of VNF migration latency and downtime should be minimized to the greatest extent possible. The VNF migration latency is related to the position change of VNF, and the number of migrated VNF at edge node x can be denoted as:

$$\mu_x = \sum_{v=1}^{V} |W_x^V - U_x^V|. \tag{12}$$

And the number of VNF migration between edge node x and edge node y can be denoted as:

$$\mu_{x,y} = \sum_{v=1}^{V} W_x^V (1 - U_y^V). \tag{13}$$

The corresponding delay of VNF migration between edge node x and edge node y can be denoted as:

$$D_{x,y}^v = \frac{d_j}{B} W_x^v U_y^v \max_{(x,y)\in X} (\mu_x, \mu_{x,y}, \mu_y). \tag{14}$$

The total delay of VNF migration caused by the position change of VNFs can be denoted as:

$$D_m = \max_{(v,x,y)} D_{x,y}^v. \tag{15}$$

Downtime refers to the time period during which the migrated VNF is unresponsive, as the VNF state is either migrating from the initial node to the target node or the network has not yet aggregated. Therefore, it is necessary to minimize the possibility of network service interruption caused by prolonged downtime. Using the current resource utilization rate of network edge nodes as a penalty factor to calculate the additional delay caused by excessive downtime, the CPU utilization rate of current network edge node x can be expressed as:

$$\rho_x = \frac{z_x}{C_x^{CPU}}. \tag{16}$$

The additional delay caused by prolonged downtime is denoted as

$$D_o = \sum_{x=1}^{X} \sum_{v=1}^{V} (D_j + D_q) W_x^{v,s} (1 - U_x^{v,s}) \rho_x. \tag{17}$$

When the downtime of VNF migration is too long, it can lead to delay-sensitive SFC requests not being processed in a timely manner during migration, resulting in QoS degradation of delay-sensitive SFC requests. The QoS degradation rate of delay-sensitive SFC requests caused by long VNF downtime can be expressed as:

$$Q_j = \frac{D_o}{D_q + D_j + D_m + D_o}. \tag{18}$$

Taking into account the total latency of VNF migration and the additional latency caused by excessive downtime, the latency cost of VNF migration can be expressed as:

$$Cost_3 = D_m + D_o. \tag{19}$$

2.5. Problem Formulation

The total cost of SFC reconfiguration in 6G IoT scenario can be denoted as:

$$Cost = \frac{Cost_1}{\sigma_1} + \frac{Cost_2}{\sigma_2} + \frac{Cost_3}{\sigma_3}, \tag{20}$$

where σ_1, σ_2, and σ_3 are normalized constants. In this paper, we minimize the total cost of SFC reconfiguration, and formulate the problem as follows

$$\min_{U_x^V, W_x^V} (Cost = \frac{Cost_1}{\sigma_1} + \frac{Cost_2}{\sigma_2} + \frac{Cost_3}{\sigma_3}), \tag{21}$$

Satisfy the constraint:

$$\begin{cases} C_1 : z_x \leq C_x^{CPU}, \forall j \in X \\ C_2 : \sum_{x=1}^{X} W_x^V = v, \forall v \in V \\ C_3 : D_q + D_j + D_m + D_o \leq D, \forall v \in V, \forall a, y \in X \\ C_4 : z_x F - (\sum_{j=1}^{J} \lambda_j + \sum_{k=1}^{K} \lambda_k) \geq 0, \forall x \in X, \forall j \in J, \forall k \in K \end{cases}, \tag{22}$$

where C_1 ensures that the edge node x would not be overloaded. C_2 ensures that reasonable quantity of VNFs are deployed in every SFC. C_3 is the constraint of delay constraint of the delay-sensitive services. C_4 ensures the stability of the SFC request queueing.

3. Knowledge-Assisted ACPPO Algorithm

In this section, the Markov Decision Process (MDP) is used to model the above problem. The MDP can be simply defined as $< S, A, R >$, where state S, action A, and reward R are, respectively, defined as follows.

3.1. State

Define the state S as system state at time t, which can be written as follows:

$$s(t) = \{c_j(t), b_k(t), z_x(t)\}, \tag{23}$$

where $c_j(t)$ and $b_k(t)$, respectively, denote the resource requirements of delay-sensitive and delay-tolerate SFC requests, which are detailed as follows:

$$c_j(t) = \{c_1(t), c_2(t), \cdots, c_J(t)\}, \tag{24}$$

$$b_k(t) = \{b_1(t), b_2(t), \cdots, b_K(t)\}, \tag{25}$$

$$z_x(t) = \{z_1(t), z_2(t), \cdots, z_X(t)\}. \tag{26}$$

3.2. Action

Action A denotes the sets of VNF migration, and the VNF migration at time t can be denoted as

$$a(t) = \{a_1(t), a_2(t), \cdots, a_i(t), \cdots, a_V(t)\}, \forall i \in V, \tag{27}$$

where $a_i(t)$ describes that if the ith VNF is migrated. $a_i(t) = 1$ means that ith VNF is migrated, and otherwise, $a_i(t) = 0$. In order to better adapt to the network changes, the knowledge-molded VNF transfer method is introduced into the 6G network knowledge base as prior knowledge to expand the action space, allowing for more choices in VNF transfer decision-making and optimization processes, thereby improving the flexibility and adaptability of the algorithm. At the same time, it is possible to explore more fully, thus having the opportunity to find better solutions and improve algorithm performance and efficiency.

3.3. Reward

Since the section aims to minimize the configuration cost of SFC, the reward function can be denoted as

$$r(t) = -\alpha Cost, \tag{28}$$

where $\alpha > 0$ is the reward factor, and the larger $r(t)$, the smaller SFC reconfiguration cost.

Based on the above MDP model, the KA-ACPPO algorithm is proposed to solve the SFC reconfiguration problem. Combining the advantages of the AC algorithm and the PPO algorithm, the ACPPO algorithm is efficient, stable, and adaptable. However, the selection of hyperparameters and the efficiency of sample processing are still issues that need to be considered when using the ACPPO algorithm. Therefore, under the 6G autonomous control framework, the KA-ACPPO algorithm introduces network knowledge related to the algorithm from the 6G network knowledge base, initializes the Actor network parameters and the Critical network parameters, and accelerates the learning process and reduces training time by providing a good initial strategy in the initial stage, avoiding a lot of exploration. Meanwhile, another VNF transfer method is introduced from the 6G network knowledge base as prior knowledge to expand the action space, helping the algorithm converge and learn faster, improving its performance in complex tasks, and enabling it to cope with a wider range of 6G scenarios.

The framework of the KA-ACPPO algorithm is shown in Figure 2. Input the current network topology G and the VNF deployment state, then the Actor network factor and the Critic network factor are initialized according to the module of the 6G knowledge base. The historic policies in the module knowledge base are updated based on the SFC reconfiguration policy. The Actor network is responsible for selecting VNF migration actions, while the Critic network evaluates the value of each action. The gradient descent method is used to update the parameters of the Critic network and the Actor network, and the optimal SFC reconstruction strategy can be denoted as:

$$\pi^* = \arg\max_a V^{\pi}(s,a), \forall s,a. \tag{29}$$

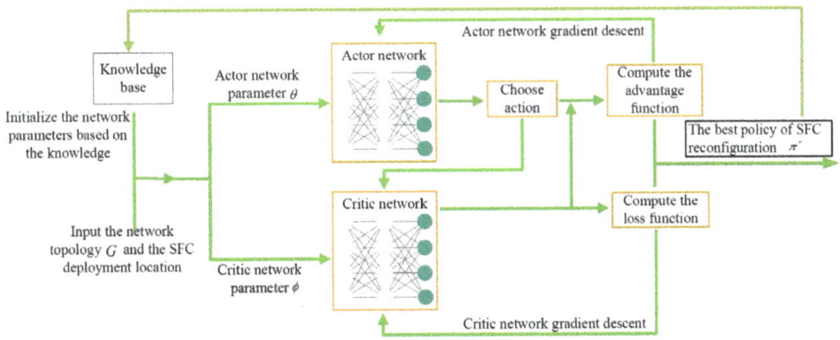

Figure 2. The framework of the proposed KA-ACPPO algorithm.

The Critical network evaluates the Actor network by using the advantage function to select actions based on the current output strategy, thereby assisting the Actor in policy updates. In the current state $s(t)$, the advantage function of selecting action $a(t)$ based on the strategy of the Actor network can be denoted as:

$$\hat{A}_t = Q_{\pi old}(s_t, a_t) - V_{\pi old}(s_t), \tag{30}$$

where $Q_{\pi old}(s_t, a_t)$ denotes the value $V_{\pi old}(s_t)$ is the value at state s_t.

According to generalized advantage estimation [14], the advantage function can be denoted as:

$$\hat{A}_t = \theta_t + (\gamma\lambda)\theta_{(t+1)} + \cdots + (\gamma\lambda)^{T-t+1}\theta_{(T-1)}, \tag{31}$$

where $\theta_t = r(t) + \gamma V_{(s_{t+})} - V_{(s_t)}$ is the error of Temporal Difference (TD), and γ is the discount factor, Let $\lambda = 1$, the advantage function can be denoted as:

$$\hat{A}_t = -V(s_t) + r_t + \gamma r_{t+1} + \cdots + \gamma^{T-t+1} r_{t-1} + \gamma^{T-t} V(s_T). \tag{32}$$

The larger the value of the advantage function, the better the selected action. The advantage function is designed to increase the stability of policy and avoid unstable optimization processes caused by excessive policy updates. At the same time, by limiting the amplitude of action policy updates, the Actor network's action update amplitude will not be too large. The update amplitude of the Actor network action strategy represents the ratio of the probability of the current strategy taking action $a(t)$ in state $s(t)$ to the probability of the old strategy taking action $a(t)$ in state $s(t)$, which is denoted as:

$$r_t(\theta) = \frac{\pi_\theta(a_t|s_t)}{\pi_{\theta_{old}}(a_t|s_t)}. \tag{33}$$

The clip function limits the size of the policy update step, preventing excessive policy updates and thus avoiding instability issues. The clip function can be denoted as:

$$clip(r_t(\theta), 1-\varepsilon, 1+\varepsilon) = \begin{cases} 1+\varepsilon & r_t(\theta) > 1+\varepsilon \\ 1-\varepsilon & r_t(\theta) > 1+\varepsilon. \\ r_t(\theta) & otherwise \end{cases} \tag{34}$$

The clip function restrict the difference between the new policy and the old policy, to ensure the stability of the updating process. The proximal clip loss function in this section can be denoted as:

$$L^{clip}(\theta) = E[\min r_t(\theta)\hat{A}_t, clip(r_t(\theta), 1-\varepsilon, 1+\varepsilon)\hat{A}_t]. \tag{35}$$

Critic network updates as follows:

$$\phi = \phi + \beta \nabla_\phi L(\phi). \tag{36}$$

where β is the learning rate of the Critic network, and $L(\phi)$ is the loss function of the Critic network to minimize TD error, which can be denoted as:

$$L(\phi) = r_t + \gamma r_{t+1} + \cdots + \gamma^{T-t+1} r_{T-1} + \gamma^{T-1} V(s(T+1) - V(s(T)))^2. \tag{37}$$

The Actor network updates as follows:

$$\theta = \theta + \beta' \nabla_\theta L^{clip}(\theta), \tag{38}$$

where β' is the learning rate of the Actor network. Algorithm 1 detailed the training process of KA-ACPPO.

Algorithm 1 The Training Process of KA-ACPPO.

```
Input: the current network topology G and current VNF deployment location
Output: the best SFC reconfiguration policy π*
Initialize the Actor network factor θ and the Critic network factor φ
for thread − number = 1, ··· , N do
    for interation − number = 1, ··· , K do
        Obtain the current network state s(t), and bring into the knowledge-based VNF migration to expand the action space
        Select the action a(t) according to the policy of Actor network
        if the constraints C1 C4 are satisfied then
            Execute the action a(t), get the instant reward r(t), and turn to the next state s(t + 1)
        else
            Reselect the action a(t)
        end if
    end for
    Calculate Â_t and L^clip(θ)
    Update the Critic network based on Equation (35)
    Update the Actor network based on Equation (37)
end for
```

4. Simulation and Result Analysis

4.1. Parameter Setting

In this section, we set up the simulation environment to evaluate the performance of the proposed KA-ACPPO algorithm. The key parameters are listed in Table 2 [15]. And it is worth noting that the parameter setting of the service requests are generated by random selection from the sum of Gaussian functions with different parameters [15].

Table 2. Simulation parameters.

Simulation Parameters	Simulation Values
number of edge nodes	10
computation resource of edge nodes	[100,200] (units)
power consumption of the edge node at idle state	87 (W)
power consumption of the edge node at full-load state	145 (W)
reserve bandwidth for VNF	[5,15] (Mbps)
required computing resource for delay-sensitive SFC	[10,30] (units)
required computing resource for delay-tolerate SFC	[5,15] (units)
length of SFC	[4,5,6,7]
end-to-end delay constraint	60 (ms)
the parameter of clip function	0.2
learning rate of Actor network	0.001
learning rate of Critic network	0.01
discount factor	0.99

4.2. Results Analysis

Figure 3 shows KA-ACPPO algorithm rewards for different number of neurons. It can be seen that, as the number of neurons increases, the reward of the KA-ACPPO algorithm also increases. The reason is that increasing the number of neurons can improve the representation ability of neural networks, enabling them to better learn and fit complex functional relationships, and enable the network to better express the complex relationships between states and actions, thereby improving rewards. However, increasing the number of hidden layer neurons can increase the complexity of the model and may lead to a slower training process. Therefore, when selecting the number of hidden layer neurons, it is necessary to balance training efficiency and increasing rewards. Considering training efficiency and rewards comprehensively, the number of neurons for the KA-ACPPO algorithm is determined to be 128 in the following simulation in this paper.

Figure 4 shows the normalized reconfiguration costs for different algorithms, which is our optimization object in Equation (21). From the figure, we can see that the AC algorithm can quickly converge into a higher convergence value, but it fluctuates greatly throughout the entire training cycle, resulting in a lower convergence value. This is because the strategies and value functions in the AC algorithm are interdependent, and their updates affect each other. This coupling relationship leads to fluctuations in the training process of the AC algorithm. ACPPO algorithm uses proximal ratio pruning to limit the magnitude of policy updates, reducing the impact of this coupling relationship to ensure a stable training process. Moreover, the proposed KA-ACPPO algorithm introduces 6G knowledge on the basis of the ACPPO algorithm, which helps the algorithm learn and converge quickly, ultimately achieving smaller fluctuations and higher convergence.

Figure 5 shows the computing cost for VNF migration using different algorithms, which is related to Equation (4). The length of SFC affects the computational resource cost of VNF migration from two aspects: the current usage of computing resources of edge nodes, and the number of VNF migrations. As the length of SFC increases, the number of VNFs per SFC increases, and the current usage of edge node computing resources becomes more complex. With the goal of minimizing reconstruction costs, the number of VNF migrations should be controlled while migrating VNFs that use fewer computing resources. From the figure, it can be also observed that as the length of SFC increases, the computational resource cost of VNF migration for all three algorithms increases. The computational resource cost of VNF migration using the proposed KA-ACPPO algorithm is always lower than the other two algorithms, ultimately controlling the computational resource cost of VNF migration at 34%.

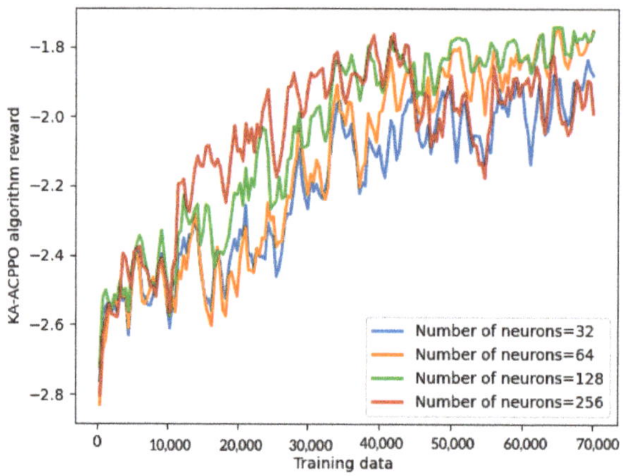

Figure 3. Rewards of KA-ACPPO algorithm for different number of neurons.

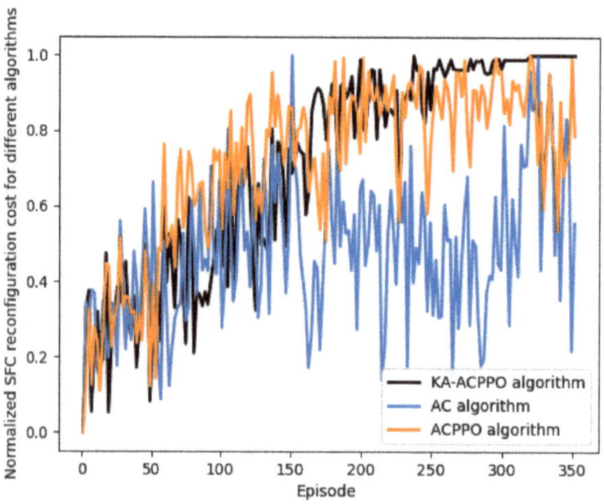

Figure 4. The normalized reconfiguration cots for different algorithm.

Figure 6 shows the power costs of VNF migration for different algorithms, which is related to Equation (6), and the comparison algorithm is Reinforcement Learning (RL)-based VNF deployment scheme proposed in [16]. The power cost of VNF migration is related to the usage of network node computing resources. Therefore, the length of SFC also affects the power cost of VNF migration in two aspects: the current usage of edge node computing resources and the number of VNF migrations. As the length of SFC increases, the number of VNFs per SFC increases, and the computing resources of the network edge nodes currently used increase, The power consumption increases accordingly. Considering that VNF migration may change the working mode of edge nodes, with the goal of minimizing reconstruction costs, the number of VNF migrations should be controlled to minimize the additional power consumption caused by such changes. From Figure 6, it can be observed that as the length of SFC increases, the power cost of VNF migration for all three algorithms increases. The power cost of VNF migration for the KA-ACPPO algorithm is always lower than the other two algorithms.

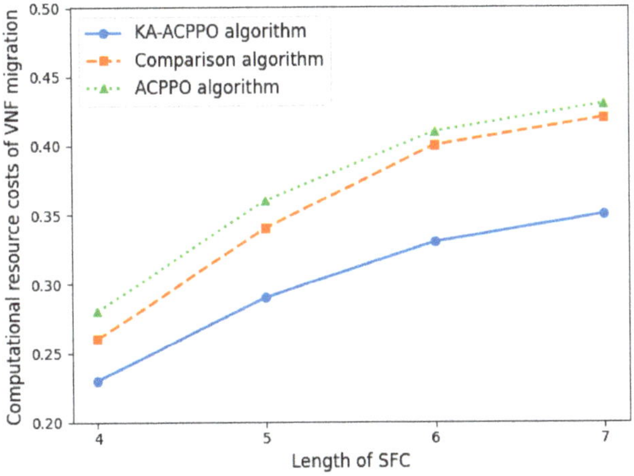

Figure 5. The computing cost of VNF migration for different algorithms.

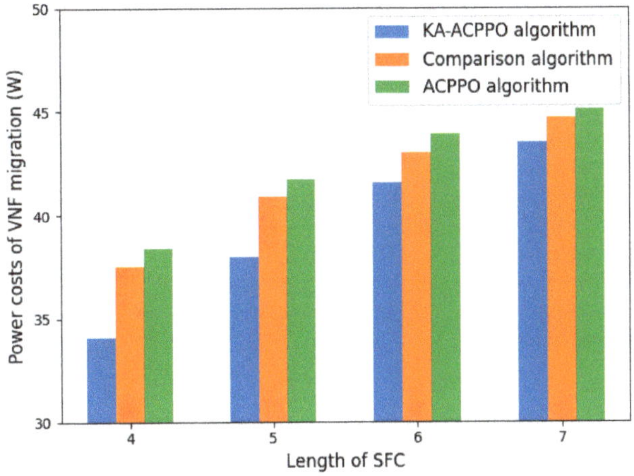

Figure 6. The power consumption cost of VNF migration for different algorithms.

Figure 7 shows the QoS degradation rates, which is related to Equation (18), of delay-sensitive SFC requests using different algorithms. According to Formula (18), the QoS degradation rate of delay-sensitive SFC is defined as the proportion of additional delay caused by excessive downtime to the total end-to-end delay. The length of SFC affects the QoS degradation rate of delay-sensitive SFCs from two aspects. Firstly, as the length of the SFC increases, the total end-to-end latency also increases. Secondly, as the length of the SFC increases, the number of VNF migrations per SFC may also increase. The negative impact of migration latency and downtime on the QoS of delay-sensitive SFC will also increase. From Figure 7, it can be observed that as the length of the SFC increases, the QoS degradation rate of the delay-sensitive SFC of all three algorithms increases. The QoS degradation rate of the delay-sensitive SFC of the KA-ACPPO algorithm is always lower than the other two algorithms, and the QoS degradation rate of the delay-sensitive SFC is ultimately controlled at 39%.

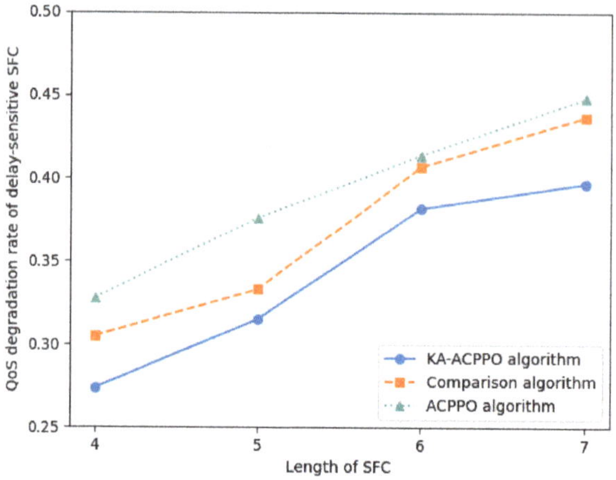

Figure 7. QoS degradation rate of delay-sensitive SFC for different algorithms.

5. Conclusions and Discussion

6G will inherit and develop NFV architecture. With this SFC-enabled flexible network architecture, SFC orchestrates different VNFs to improve maintaining the QoS of different services. Considering that the coupling relationship leads to fluctuations in the training process of the traditional AC algorithm, this paper brings proximal ratio pruning to limit the magnitude of policy updates and combine the advantages of the AC algorithm and the PPO algorithm. Moreover, we propose to introduce the 6G knowledge base to provide a better initial policy to accelerate training speed and form the KA-ACPPO-based VNF migration algorithm. Simulation results show that the proposed algorithm can effectively reduce computing cost and power consumption.

However, in this paper, we consider using historic policies as network knowledge to improve performance. In future work, more dimensions of knowledge, such as model-based algorithms and expert experiences, could be researched to further improve the performance of deep learning algorithms.

Author Contributions: Conceptualization, X.S. and B.L.; methodology, B.L.; software, S.L.; validation, B.L. and S.L.; formal analysis, B.L.; writing—original draft preparation, B.L.; writing—review and editing, S.L. and X.S.; supervision, X.S.; funding acquisition, X.S. All authors have read and agreed to the published version of the manuscript.

Funding: This work was supported by National Key R&D Program of China (No. 2020YFB1806702).

Institutional Review Board Statement: Not applicable.

Informed Consent Statement: Not applicable.

Data Availability Statement: No new data were created.

Conflicts of Interest: The authors declare no conflicts of interest.

Abbreviations

The following abbreviations are used in this manuscript:

NFV	Network Function Virtualization
VNF	Virtual Network Function
SFC	Service Function Chains
IoT	Internet of Things
AC	Actor Critic
PPO	Proximal Policy Optimization
QoS	Quality of Service

AI	Artificial Intelligence
DQN	Deep Q Network
D3QN	Dueling Double Deep Q Network
UAV	Unmanned Aerial Vehicle
RL	Reinforcement Learning

References

1. Giordani, M.; Polese, M.; Mezzavilla, M.; Rangan, S.; Zorzi, M. Towards 6G Networks: Use Cases and Technologies. *IEEE Commun. Mag.* **2020**, *35*, 55–61. [CrossRef]
2. Liao, J.; Fu, X.; Wang, J.; Sun, H. 6G-ADM: Knowledge based 6G network management and control architecture. *J. Commun.* **2022**, *43*, 6.
3. Wen, T.; Yu, H.; Sun, G.; Liu, L. Network function consolidation in service function chaining orchestration. In Proceedings of the 2016 IEEE International Conference on Communications (ICC), Kuala Lumpur, Malaysia, 22–27 May 2016; pp. 1–6. [CrossRef]
4. Afrasiabi, S.N.; Ebrahimzadeh, A.; Promwongsa, N.; Mouradian, C.; Li, W.; Recse, Á.; Szabó, R.; Glitho, R.H. Cost-Efficient Cluster Migration of VNFs for Service Function Chain Embedding. *IEEE Trans. Netw. Serv. Manag.* **2024**, *21*, 979–993. [CrossRef]
5. Tran, V.; Sun, J.; Tang, B.; Pan, D. Traffic-Optimal Virtual Network Function Placement and Migration in Dynamic Cloud Data Centers. In Proceedings of the 2022 IEEE International Parallel and Distributed Processing Symposium (IPDPS), Lyon, France, 30 May–3 June 2022; pp. 919–929.
6. Farkiani, B.; Bakhshi, B.; MirHassani, S.A.; Wauters, T.; Volckaert, B.; De Turck, F. Prioritized Deployment of Dynamic Service Function Chains. *IEEE/ACM Trans. Netw.* **2021**, *29*, 979–993. [CrossRef]
7. Zhang, F.; Lu, H.; Guo, F.; Gu, Z. Traffic Prediction Based VNF Migration with Temporal Convolutional Network. In Proceedings of the 2021 IEEE Global Communications Conference (GLOBECOM), Madrid, Spain, 7–11 December 2021.
8. Afrasiabi, S.N.; Ebrahimzadeh, A.; Mouradian, C.; Malektaji, S.; Glitho, R.H. Reinforcement Learning-Based Optimization Framework for Application Component Migration in NFV Cloud-Fog Environments. *IEEE Trans. Netw. Serv. Manag.* **2023**, *20*, 1866–1883. [CrossRef]
9. Ibrahimpasic, A.L.; Han, B.; Schotten, H.D. AI-Empowered VNF Migration as a Cost-Loss-Effective Solution for Network Resilience. In Proceedings of the 2021 IEEE Wireless Communications and Networking Conference Workshops (WCNCW), Nanjing, China, 29 March 2021; pp. 1–6.
10. Liu, Y.; Zhang, C.; Yang, H.; Zhang, S.; Wang, X.; Li, F. A Deep Reinforcement Learning-Based Approach for Adaptive SFC Deployment in Multi-Domain Networks. In Proceedings of the 2023 15th International Conference on Communication Software and Networks (ICCSN), Shenyang, China, 21–23 July 2023; pp. 199–203. [CrossRef]
11. Hirayama, T.; Jibiki, M. SFC Path Selection based on Combination of Topological Analysis and Demand Prediction. In Proceedings of the 2022 23rd Asia-Pacific Network Operations and Management Symposium (APNOMS), Takamatsu, Japan, 28–30 September 2022; pp. 1–4.
12. Wu, Y.; Jia, Z.; Wu, Q.; Lu, Z. Adaptive QoE-Aware SFC Orchestration in UAV Networks: A Deep Reinforcement Learning Approach. *IEEE Trans. Netw. Sci. Eng.* **2024**. [CrossRef]
13. Liu, N.; Li, Z.; Xu, J.; Xu, Z.; Lin, S.; Qiu, Q.; Tang, J.; Wang, Y. A hierarchical framework of cloud resource allocation and power management using deep reinforcement learning. In Proceedings of the IEEE 37th International Conference on Distributed Computing Systems, Atlanta, GA, USA, 5–8 June 2017; pp. 372–382.
14. Zhang, Z.; Luo, X.; Liu, T.; Xie, S.; Wang, J.; Wang, W.; Li, Y.; Peng, Y. Proximal Policy Optimization with Mixed Distributed Training. In Proceedings of the 2019 IEEE 31st International Conference on Tools with Artificial Intelligence (ICTAI), Portland, OR, USA, 4–6 November 2019; pp. 1452–1456.
15. Fu, X.; Yu, F.R.; Wang, J.; Qi, Q.; Liao, J. Dynamic Service Function Chain Embedding for NFV-Enabled IoT: A Deep Reinforcement Learning Approach. *IEEE Trans. Wirel. Commun.* **2020**, *19*, 507–519. [CrossRef]
16. Yao, J.; Chen, M. A Flexible Deployment Scheme for Virtual Network Function Based on Reinforcement Learning. In Proceedings of the 2020 IEEE 6th International Conference on Computer and Communications (ICCC), Chengdu, China, 11–14 December 2020; pp. 1505–1510.

entropy

MDPI

Article

Ultra-Reliable and Low-Latency Wireless Hierarchical Federated Learning: Performance Analysis †

Haonan Zhang [1,2,3], Peng Xu [2,4] and Bin Dai [1,2,3,*]

[1] School of Information Science and Technology, Southwest JiaoTong University, Chengdu 611756, China; zhanghaonan@my.swjtu.edu.cn
[2] Chongqing Key Laboratory of Mobile Communications Technology, Chongqing 400065, China; xupeng@cqupt.edu.cn
[3] Provincial Key Lab of Information Coding and Transmission, Southwest Jiaotong University, Chengdu 611756, China
[4] School of Communications and Information Engineering, Chongqing University of Posts and Telecommunications, Chongqing 400065, China
* Correspondence: daibin@home.swjtu.edu.cn; Tel.: +86-13548053724

† An earlier version of this paper was presented in part at the IEEE Conference on Computer Communications Workshops (INFOCOM WKSHPS), New York, NY, USA, 17–20 May 2023 and in part at the IEEE International Conference on Communications Workshops (ICC Workshops), Roma, Italy, 28 May–1 June 2023

Abstract: Wireless hierarchical federated learning (WHFL) is an implementation of wireless federated Learning (WFL) on a cloud–edge–client hierarchical architecture that accelerates model training and achieves more favorable trade-offs between communication and computation. However, due to the broadcast nature of wireless communication, the WHFL is susceptible to eavesdropping during the training process. Apart from this, recently ultra-reliable and low-latency communication (URLLC) has received much attention since it serves as a critical communication service in current 5G and upcoming 6G, and this motivates us to study the URLLC-WHFL in the presence of physical layer security (PLS) issue. In this paper, we propose a secure finite block-length (FBL) approach for the multi-antenna URLLC-WHFL, and characterize the relationship between privacy, utility, and PLS of the proposed scheme. Simulation results show that when the eavesdropper's CSI is perfectly known by the edge server, our proposed FBL approach not only almost achieves perfect secrecy but also does not affect learning performance, and further shows the robustness of our schemes against imperfect CSI of the eavesdropper's channel. This paper provides a new method for the URLLC-WHFL in the presence of PLS.

Keywords: finite block-length coding; physical layer security; privacy-utility relationship; wireless federated learning

1. Introduction

Wireless federated learning (WFL), which allows the training of machine learning (ML) models on a large corpus of decentralized data stored on mobile devices [1], has attracted significant research interest. Currently, WFL primarily aims to enhance communication efficiency [2–6], improve privacy and security [7–9], investigate a balance between privacy and utility [10–14], manage power control for wireless devices [15,16], and design effective beamforming strategies [17]. To optimize the processing power of edge and cloud servers, a hierarchical federated learning (HFL) system involving clients, edge servers, and cloud servers has been proposed in [18]. Compared to FL systems relying on a single server, HFL reduces the computational load [19–21], lowers user-to-cloud server communication costs [18–23], decreases FL processing time [18], and improves privacy and security in FL [22,23]. Specifically, since the convergence performance of the HFL system is theoretically proved in [18], joint user scheduling and wireless resource allocation are established to improve both communication and energy efficiency [19–21]. To enhance privacy in wireless hierarchical federated learning (WHFL), ref. [22] introduced a method based on local differential privacy (LDP) that involves artificial noise into the shared model parameters at two stages. Additionally, ref. [24] considered the influence of device mobility on the learning performance of WHFL systems.

The broadcast nature of wireless communication renders WFL susceptible to eavesdropping. As a result, tackling the challenge of WFL in the presence of physical layer security (PLS) is a significant issue. Different from the privacy requirement of the FL that the information leakage between users and servers does not need to be arbitrarily small due to the accuracy of data analysis, the information leakage to the eavesdropper should vanish, which is also known as the PLS requirement [25]. The current research in WFL in the presence of PLS [26–29] primarily focuses on enhancing the security of data through resource allocation and artificial jamming techniques. Specifically, reference[26] focuses on optimizing the power control of drones to enhance the security rate of the WFL system, considering constraints such as WFL training time and battery capacity of the drone. In [27], a method for achieving secrecy in WFL was proposed via using cooperative jamming, which involves the cooperative provision of jamming signals by users to counteract eavesdropping attempts and enhance security. Ref. [28] proposed the method of using conventional wireless devices to form a non-orthogonal multiple access (NOMA) transmission group with an edge device for secrecy-enhanced mobile edge computing, and the devices provide cooperative jamming to an eavesdropper while transmitting data to a cellular base station. In [29], a power allocation algorithm is proposed for WFL, where the transmitting power is divided proportionally between the transmitted signal and artificial noise to maximize the secrecy rate while satisfying the model performance requirement. Apart from this, ref. [30] proposed a PLS measure while considering the privacy-utility constraints in WFL.

Very recently, ultra-reliable and low-latency communication (URLLC) has attracted significant attention, as it serves as a critical communication service in fifth-generation (5G) and sixth-generation (6G) cellular networks. One essential technology for URLLC is short-sized packet communication [31], which indicates that the coding block length should be finite, and finite block-length (FBL) coding [32] provides an effective way for this scenario. Currently, the study of WFL combined with URLLC includes the design of a multi-level architecture to satisfy URLLC requirements [33], and the application of WFL in vehicular networks while considering URLLC constraints [34]. To the best of the authors' knowledge, the practical FBL scheme for the WFL remains unknown. Then it is natural to ask: *is there any practical FBL scheme for the WHFL in the presence of PLS, if yes, what is the relationship between PLS, privacy, and utility in WHFL systems while considering URLLC requirements?*

One possible solution to the aforementioned question is a channel feedback coding scheme. The study of channel feedback scheme started from [35], where an elegant feedback coding scheme called the Schalkwijk–Kailath (SK) scheme was proposed for additive white Gaussian noise (AWGN) channel with noiseless feedback. In this scheme, the transmitter sends the original message only in the initial transmission. In the subsequent round, the receiver sends an estimate of the original message to the transmitter via a noiseless feedback channel. The transmitter sends an amplified version of the estimation error back to the receiver, and the receiver obtains an estimate of the estimation error by using the minimum mean square error (MMSE). After a predetermined number of rounds, the receiver uses the minimum distance rule to decode the message. It was shown that the SK scheme [35] is not only capacity-achieving but also its decoding error probability *doubly exponentially decays* to zero as the coding block length increases, which indicates that the SK scheme requires an extremely short coding block length to achieve a desired decoding error probability. Furthermore, ref. [36] showed that the SK scheme achieves perfect weak secrecy by itself, i.e., the SK scheme satisfies PLS requirement by itself. Recently, ref. [37] showed that the

SK scheme [35] is almost the optimal FBL scheme for the AWGN channel with feedback, which indicates that it may be a good choice for URLLC.

However, note that the application of the SK scheme to the wireless fading channel still has a long way to go since it is based on the assumption that the feedback channel is a noiseless channel. Apart from this, in wireless communication, the channel feedback is often utilized to transmit channel state information (CSI) back to the device for each uplink transmission [17], and this allows the device to adjust its transmission parameters based on the received feedback. Then it is natural to ask: Can we utilize the channel feedback not only for CSI transmission but also for designing an FBL approach for the multi-antenna URLLC-WHFL in the presence of PLS, i.e., is it possible to extend the classical SK scheme to the multi-antenna URLLC-WHFL in the presence of PLS?

In this paper, we answer the aforementioned questions by studying the WHFL in the presence of PLS. Figure 1 illustrates the collaborative training of a learning model by users, edge servers, and cloud servers. To preserve privacy, a local differential privacy (LDP) mechanism [38] is utilized by adding Gaussian noise to each user's gradient before aggregating all gradients to the edge servers. Furthermore, communication between each edge server and the cloud server over a quasi-static fading duplex channel, which, due to the inherent broadcast characteristics of wireless communication, is eavesdropping by an external eavesdropper. Our primary objective is to ensure that the polluted gradient data retains a certain amount of utility while minimizing privacy leakage to the cloud server and protecting the gradient data transmitted from edge servers to the cloud server from eavesdropping. A straightforward way to achieve the above goal is for the edge servers to securely encode the polluted data gradients as codewords and transmit them into wireless duplex fading channels. The cloud server can successfully decode the polluted data gradients, while the eavesdropper obtains no information about them. In this way, the PLS and the privacy of the data can simultaneously be guaranteed since the real data gradients are protected by the LDP mechanism.

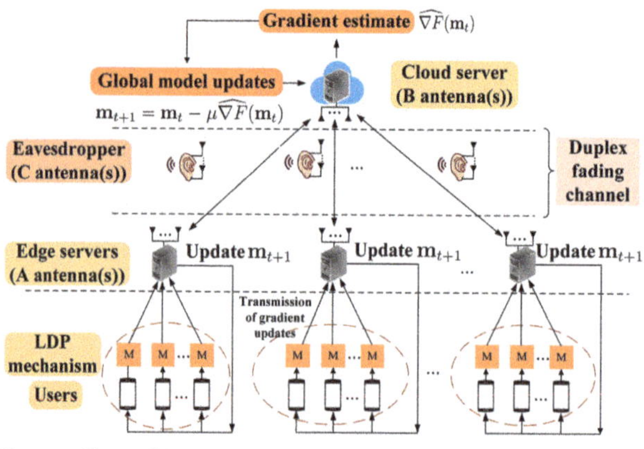

Figure 1. The multi-antenna WHFL in the presence of PLS.

Our key contributions to this paper are summarized as follows:

- We propose an FBL approach for multi-antenna WHFL in the presence of PLS. In this approach, the feedback link is not only utilized for CSI transmission but also used to send the cloud server's MMSE about the transmitted polluted data gradient back to the edge server. The key idea of the proposed scheme is to apply the modulo-lattice operation (MLO) [39] to eliminate the impact of feedback channel noise on the performance of the SK scheme [35], and further extend the SK-type scheme to a two-dimensional situation, which performs well in the SISO fading channel. Then further applying pre-coding, beamforming, and singular value decomposition (SVD) techniques to the extended scheme for the SISO case, the FBL coding scheme for the multi-antenna WHFL is obtained.

- We derive the achievable secrecy rate of our proposed scheme and characterize the relationship between PLS, privacy, and utility of our scheme. Moreover, given fixed decoding error probability and coding block length, we establish lower and upper bounds on the LDP noise variance that ensure certain privacy, utility, and secrecy levels of PLS.

To obtain a better understanding of the contribution of this paper and the related works studied in the literature, the following Table 1 summarizes the study of WFL in the presence of privacy, utility, PLS, and URLLC in the literature.

Table 1. Summarizing all results in WFL in the presence of privacy, utility, PLS and URLLC.

Related Work	Privacy	Utility	PLS	Relationship between PLS, Privacy, and Utility	URLLC
[7–9,16,23]	✓	–	–	–	–
[10–14,22]	✓	✓	–	Relationship between Privacy and Utility	–
[26–28]	–	–	✓	–	–
[29]	–	✓	✓	Relationship between PLS and Utility	–
[30]	✓	✓	✓	Relationship between PLS-Privacy-Utility	–
[33,34]	–	–	–	–	✓
This Work	✓	✓	✓	Relationship between PLS-Privacy-Utility	✓

The remainder of this paper is organized as follows. In Section 2, the definitions, system model, and main results are given. The FBL approach for the MIMO case is shown in Section 3. FBL approaches for the SIMO/MISO cases are proposed in Section 4. Simulation results are shown in Section 5. Section 6 summarizes all results in this paper and discusses future work.

2. Definitions, System Model and Main Results

2.1. WHFL System

Figure 1 illustrates a system composed of K_{tot} users, L edge servers indexed by ℓ and a cloud server. The disjoint user sets are denoted as $\{\mathcal{C}_\ell\}_{\ell=1}^{L}$, and $K = |\mathcal{C}_\ell|$ representing the number of users in edge server ℓ. The distributed datasets are represented by $\{\mathcal{S}_{\ell,k}\}_{k=1}^{|\mathcal{C}_\ell|}$, where $S_{\ell,k} = |\mathcal{S}_{\ell,k}|$ is the size of $\mathcal{S}_{\ell,k}$. Each dataset $\mathcal{S}_{\ell,k}$ is defined as $\{(\mathbf{u}_{k,j}, v_{k,j})\}_{j=1}^{|\mathcal{S}_{\ell,k}|}$, where $\mathbf{u}_{k,j}$ represents the j-th input sample and $v_{k,j}$ is the corresponding label. $\mathcal{S}_\ell$ is the aggregated dataset of edge server ℓ, and the gradients from each user are aggregated by their corresponding edge server. The global loss function $F(\mathbf{m})$ is defined as follows:

$$F(\mathbf{m}) = \frac{1}{S}\sum_{\ell=1}^{L}\sum_{k=1}^{|\mathcal{C}_\ell|} S_{\ell,k}F_{\ell,k}(\mathbf{m}), \tag{1}$$

where model vector $\mathbf{m} \in \mathbb{R}^q$ and $S = \sum_\ell \sum_k S_{\ell,k}$. The local loss function is given by the following:

$$F_{\ell,k}(\mathbf{m}) = \frac{1}{S_{\ell,k}} \sum_{(\mathbf{u}_{k,j},v_{k,j})\in\mathcal{S}_{\ell,k}} f(\mathbf{m};\mathbf{u}_{k,j},v_{k,j}), \tag{2}$$

where $f(\mathbf{m};\mathbf{u}_{k,j},v_{k,j})$ represents the sample-wise loss function. The goal of model training is to minimize the global loss function, as follows:

$$\mathbf{m}^\star = \arg\min_{\mathbf{m}} F(\mathbf{m}). \tag{3}$$

To achieve this, we employ a distributed gradient descent iterative algorithm. Specifically, in the t-th ($t \in \{1,2,...,T\}$) communication round, the cloud server broadcasts the current global model vector $\mathbf{m}_t$ to all users, and every user has perfect knowledge of $\mathbf{m}_t$. Each user k then computes its local gradient $\nabla F_{\ell,k}(\mathbf{m}_t)$ using its dataset $\mathcal{S}_{\ell,k}$ and the current model $\mathbf{m}_t$. Once the edge server ℓ receives all the noisy local gradients from its users, which have been perturbed by Gaussian noise for LDP, it computes an estimation of the partial gradient as follows:

$$\nabla F_\ell(\mathbf{m}_t) = \frac{1}{S_\ell} \sum_{k \in \mathcal{C}^\ell} S_{\ell,k} \nabla F_{\ell,k}(\mathbf{m}_t), \tag{4}$$

where $S_\ell = |\mathcal{S}_\ell|$ denotes the size of $\mathcal{S}_\ell$. Then, the cloud server aggregates the partial gradient estimates from all edge servers to compute the estimation $\widehat{\nabla F}(\mathbf{m}_t)$ of the global gradient, as follows:

$$\nabla F(\mathbf{m}_t) = \frac{1}{S} \sum_{\ell=1}^{L} S_\ell \nabla F_\ell(\mathbf{m}_t), \tag{5}$$

and updates the global model $\mathbf{m}_{t+1}$ by the following:

$$\mathbf{m}_{t+1} = \mathbf{m}_t - \mu \widehat{\nabla F}(\mathbf{m}_t), \tag{6}$$

where μ denotes the learning rate.

2.2. Model Formulation

An information-theoretic model of WHFL system is shown in Figure 2. Without loss of generality, we adopt the following assumptions:

Assumption 1. *The communication of any individual edge server to the cloud server is not affected by other edge servers, and the downlink transmission from the cloud server to the edge servers is reliable [16]. Furthermore, we consider that an external eavesdropper targets the information transmitted during the uplink communication from the edge servers to the cloud server. Consequently, this paper primarily focuses on the PLS of the T rounds of uplink communication from one edge server to the cloud server.*

Assumption 2. *The channels are quasi-static fading.*

Assumption 3. *Following similar arguments in [3,9,16,17], we assume that the perfect CSI of the feedforward and feedback channels is known by both the cloud server and the edge server. Here note that this assumption is well-justified from a practical standpoint. For the feedforward channel, the channel training for estimating CSI at the cloud server can be achieved by transmitting pilot sequences from the edge servers, and the channel estimation is perfect when the length of the pilot sequences is sufficiently large [40]. On the other hand, when the cloud server transmits the perfectly estimated CSI to the edge server through the feedback channel, only a few feedback bits are required. By using a code with a low coding rate and high error-correcting capability, the probability of feedback error can be negligible [41] and, hence, the CSI of the feedforward channel is perfectly known by the transceiver. For the feedback channel, the perfect CSI sharing between transceivers can be realized in a similar way.*

2.2.1. Privacy-Utility

In Figure 2, let $\mathbf{W}_{t,k} = \sum_{j=1}^{S_{\ell,k}} \nabla f(\mathbf{m}_t; \mathbf{u}_{k,j}, v_{k,j}) = (W_{t,k,1}, ..., W_{t,k,q})^T \in \mathbb{R}^q$ represent the overall local gradient vector for user k ($k \in \{1,2,...,K\}$) during the t-th ($t \in \{1,2,...,T\}$) communication round, where $\nabla f(\mathbf{m}_t; \mathbf{u}_{k,j}, v_{k,j}) = (\nabla f_1(\mathbf{m}_t; \mathbf{u}_{k,j}, v_{k,j}), ..., \nabla f_q(\mathbf{m}_t; \mathbf{u}_{k,j}, v_{k,j}))^T$ and $W_{t,k,i} = \sum_{j=1}^{S_{\ell,k}} \nabla f_i(\mathbf{m}_t; \mathbf{u}_{k,j}, v_{k,j})$ ($i \in \{1,2,...,q\}$). Following [42], assume that $\nabla f(\mathbf{m}_t; \mathbf{u}_{k,j}, v_{k,j})$ is independent and identically distributed (i.i.d.) and $\nabla f(\mathbf{m}_t; \mathbf{u}_{k,j}, v_{k,j}) \sim \mathcal{N}(0, \sigma_{w,t}^2 \mathbf{I})$, which indicates that $\mathbf{W}_{t,k} \sim \mathcal{N}(0, S_{\ell,k}\sigma_{w,t}^2 \mathbf{I})$. The i.i.d. generated local Gaussian noise

$\eta_{t,k} = (\eta_{t,k,1}, ..., \eta_{t,k,q})^{\mathcal{T}}$ follows distribution $\mathcal{N}(0, \sigma^2 I)$ and is independent of $W_{t,k}$. The edge server aggregates the corrupted local gradient, and it is defined as follows:

$$W'_{t,k} = W_{t,k} + \eta_{t,k}, \tag{7}$$

where $W'_{t,k} \sim \mathcal{N}(0, (S_{\ell,k}\sigma^2_{w,t} + \sigma^2)I)$. The overall local gradients and noise for the t-th round are $W_t = (W_{t,1}, ..., W_{t,q})^{\mathcal{T}}$ and $\eta_t = (\eta_{t,1}, ..., \eta_{t,q})^{\mathcal{T}}$, respectively, where $W_{t,i} = \sum_{k=1}^{K} W_{t,k,i}$, $\eta_{t,i} = \sum_{k=1}^{K} \eta_{t,k,i}$ and $i \in \{1, 2, ..., q\}$. Consequently, from (7), the overall corrupted local gradients for the t-th round are $W'_t = (W'_{t,1}, ..., W'_{t,q})^{\mathcal{T}}$, where $W'_{t,i} = \sum_{k=1}^{K} W'_{t,k,i}$ and $i \in \{1, 2, ..., q\}$. Due to the fact that $W_{t,k}$ and $\eta_{t,k}$ are i.i.d. and independent, the overall corrupted gradients W'_t are i.i.d. and distributed as $\mathcal{N}(0, (S_\ell \sigma^2_{w,t} + K\sigma^2)I)$, where $S_\ell = \sum_{k=1}^{K} S_{\ell,k}$.

Definition 1 (Mutual information privacy [43]). *For every $t \in \{1, .., T\}$, if the mutual information $\frac{1}{q} I(W_t; W'_t)$ during the t-th round is upper bounded by ϵ, namely, $\max_{t \in \{1,..,T\}} \frac{1}{q} I(W_t; W'_t) \leq \epsilon$, the LDP mechanism is said to satisfy ϵ-mutual information privacy for $\epsilon > 0$.*

Definition 2 (Utility [44]). *The utility of W'_t is defined by the distortion between W_t and W'_t, and in this paper, we consider the quadratic distortion $d(W_t, W'_t) = ||W'_t - W_t||^2$. If $\frac{1}{qT} \sum_{t=1}^{T} E(d(W_t, W'_t)) \leq v$, the utility of W'_t is determined by v, where the utility and the distortion have an inverse relationship with each other, .i.e., smaller v corresponds to larger utility.*

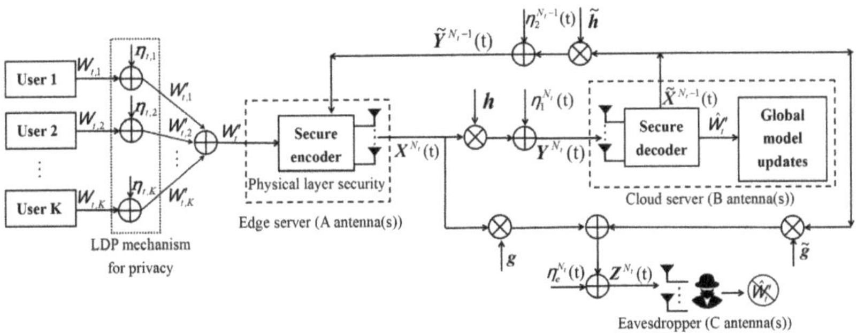

Figure 2. An information-theoretic model of the WHFL system, where the edge server, cloud server and eavesdroppers are equipped with A, B, and C antennas, respectively ($A \geq 1, B \geq 1, C \geq 1$).

2.2.2. Gradient Compression

We employ lossy Gaussian source coding characterized by a quadratic distortion metric, defined as $d(W'_t, \hat{W}'_t) = ||W'_t - \hat{W}'_t||^2$ (the source encoder and decoder are respectively located at the edge server and cloud server), where $\hat{W}'_t$ is the output of the source decoder at the cloud server. Following [45] (Chapter 3.8, pp. 64–65), the edge server's source encoder maps W'_t to $\{1, 2, ..., 2^{qR_t(D)}\}$ and compresses W'_t into an index W''_t that is uniformly distributed over $\mathcal{W}''_t = \{1, 2, ..., 2^{qR_t(D)}\}$. The rate-distortion function $R_t(D)$ is defined as follows:

$$R_t(D) = \begin{cases} \frac{1}{2} \log \frac{K\sigma^2 + S_\ell \sigma^2_{w,t}}{D} & 0 \leq D < K\sigma^2 + S_\ell \sigma^2_{w,t} \\ 0 & D \geq K\sigma^2 + S_\ell \sigma^2_{w,t}, \end{cases} \tag{8}$$

where $\frac{1}{qT} \sum_{t=1}^{T} E(d(W'_t, \hat{W}'_t)) \leq D$. For the cloud server's source decoder, the decoding mapping transforms the indices $\{1, 2, ..., 2^{qR_t(D)}\}$ into $\hat{W}'_t$. Here note that when $R_t(D) = 0$, no message is transmitted, and $\hat{W}'_t$ is set to 0.

2.2.3. Communication Model

At the t-th round, the channel input-output relationships are expressed as follows:

$$Y_i(t) = hX_i(t) + \eta_{1,i}(t), \ 1 \leq i \leq N_t, \quad (9)$$

$$\widetilde{Y}_i(t) = \widetilde{h}\widetilde{X}_i(t) + \eta_{2,i}(t), \ 1 \leq i \leq N_t - 1, \quad (10)$$

$$Z_i(t) = gX_i(t) + \widetilde{g}\widetilde{X}_i(t) + \eta_{e,i}(t), \ 1 \leq i \leq N_t, \quad (11)$$

where the input and output of the feedforward channel are denoted by $X_i(t)$ and $Y_i(t)$, respectively, the feedback channel's input and output are $\widetilde{X}_i(t)$ and $\widetilde{Y}_i(t)$, respectively, and the eavesdropping channel's output is $Z_i(t)$. Note that $X_i(t), \widetilde{Y}_i(t) \in \mathbb{C}^{A\times 1}$, $\widetilde{X}_i(t)$, $Y_i(t) \in \mathbb{C}^{B\times 1}$ and $Z_i(t) \in \mathbb{C}^{C\times 1}$. The average power constraint for the input of the edge server $X_i(t)$ is $\frac{1}{N_t}\sum_{i=1}^{N_t} E[X_i^{\mathcal{H}}(t)X_i(t)] \leq P$, the input of the cloud server $\widetilde{X}_i(t)$ is constrained by $\frac{1}{N_t-1}\sum_{i=1}^{N_t-1} E[\widetilde{X}_i^{\mathcal{H}}(t)\widetilde{X}_i(t)] \leq \widetilde{P}$. The matrices $h \in \mathbb{C}^{B\times A}$, $\widetilde{h} \in \mathbb{C}^{A\times B}$, $g \in \mathbb{C}^{C\times A}$, and $\widetilde{g} \in \mathbb{C}^{C\times B}$ represent the CSI of the feedforward, feedback, and eavesdropping channels, respectively. The channel noises' elements of $\eta_{e,i}(t) \in \mathbb{C}^{C\times 1}$, $\eta_{2,i}(t) \in \mathbb{C}^{A\times 1}$ and $\eta_{1,i}(t) \in \mathbb{C}^{B\times 1}$ are i.i.d. and distributed as $\mathcal{CN}(0,\sigma_e^2)$, $\mathcal{CN}(0,\sigma_2^2)$ and $\mathcal{CN}(0,\sigma_1^2)$, respectively. The input message W_t'' of the edge server is uniformly drawn in the set $\mathcal{W}_t''$, and it is encoded as a codeword of length N_t. Furthermore, the input of the edge server is defined as $X_i(t) = f_{t,i}(W_t'', h, \widetilde{h}, \widetilde{Y}_1^{i-1}(t))$, where $f_{t,i}(\cdot)$ is an encoding function and $\widetilde{Y}_1^{i-1}(t) = (\widetilde{Y}_1(t), ..., \widetilde{Y}_{i-1}(t))$. The cloud server estimates the message $\hat{w}_t'' = \varphi(h, \widetilde{h}, Y^{N_t})$ using the decoding function φ. The input of the cloud server is defined as $\widetilde{X}_i(t) = \widetilde{f}_{t,i}(h, Y_1^i(t), \widetilde{h})$, where $\widetilde{f}_{t,i}(\cdot)$ is an encoding function and $Y_1^i(t) = (Y_1(t), ..., Y_i(t))$. The average decoding error probability $P_{e,t}$ is given by the following:

$$P_{e,t} = \frac{1}{|\mathcal{W}_t''|}\sum_{w_t'' \in \mathcal{W}_t''} Pr\{\varphi(h, Y^{N_t}, \widetilde{h}) \neq w_t'' | w_t'' \text{ sent}\}. \quad (12)$$

Definition 3. *According to [46,47], the CSIs g and $\widetilde{g}$ of eavesdropping channels are defined as follows:*

$$g = \hat{g} + \Delta g, \ ||\Delta g||_F \leq \omega, \ \ \widetilde{g} = \hat{\widetilde{g}} + \Delta\widetilde{g}, \ ||\Delta\widetilde{g}||_F \leq \widetilde{\omega}, \quad (13)$$

where $\hat{g} \in \mathbb{C}^{C\times A}$, $\hat{\widetilde{g}} \in \mathbb{C}^{C\times B}$ are the estimated CSI of g and $\widetilde{g}$, respectively. Δg and $\Delta\widetilde{g}$ represent the legal parties' estimation errors about the perfect CSI of the eavesdropper's channel, and these errors are respectively bounded by parameters $\omega > 0$ and $\widetilde{\omega} > 0$. Here note that $\Delta g = \Delta\widetilde{g} = \mathbf{0}$ corresponds to the situation that the legal parties obtain perfect CSI of the eavesdropper's channel.

Definition 4. *The secrecy level of PLS [48] (the normalized uncertainty of the eavesdropper) is given by*

$$\Delta = \frac{H(W_1'', ..., W_T'' | Z^{N_1}, ..., Z^{N_T}, h, \widetilde{h}, g, \widetilde{g})}{H(W_1'', ..., W_T'')}, \ \ 0 \leq \Delta \leq 1. \quad (14)$$

A transmission rate R is said to be $(\tau, N, \delta, D, v, \epsilon)$ achievable, if for given decoding error probability τ, block length N ($N = \sum_{t=1}^T N_t$), secrecy level δ, $\frac{1}{qT}\sum_{t=1}^T E(d(W_t', \hat{W}_t')) \leq D$, $\max_{t\in\{1,..,T\}} \frac{1}{q} I(W_t; W_t') \leq \epsilon$ and $\frac{1}{qT}\sum_{t=1}^T E(d(W_t, W_t')) \leq v$, there exists a channel code described above such that we have the following:

$$\frac{H(W_1'', ..., W_T'')}{N} = R, \quad \frac{1}{T}\sum_{t=1}^{T} P_{e,t} \leq \tau, \quad \Delta \geq \delta, \quad (15)$$

where $\delta \in [0,1]$, and $\delta = 1$ represents the perfect secrecy. For the WHFL in SISO/SIMO/MISO/MIMO cases, the achievable secrecy transmission rates are respectively denoted by $R_{siso/simo/miso/mimo}(\tau, N, \delta, D, v, \epsilon)$, the channel gains are respectively defined by $h_{siso}, \widetilde{h}_{siso}, g_{siso}, \widetilde{g}_{siso}, \boldsymbol{h}_{simo}, \widetilde{\boldsymbol{h}}_{simo}, \boldsymbol{g}_{simo},$

$\widetilde{g}_{simo}$, h_{miso}, $\widetilde{h}_{miso}$, g_{miso}, $\widetilde{g}_{miso}$, h_{mimo}, $\widetilde{h}_{mimo}$, g_{mimo}, $\widetilde{g}_{mimo}$, *and the CSI estimation errors are defined by* Δg_{siso}, $\Delta \widetilde{g}_{siso}$, Δg_{simo}, $\Delta \widetilde{g}_{simo}$, Δg_{miso}, $\Delta \widetilde{g}_{miso}$, Δg_{mimo} *and* $\Delta \widetilde{g}_{mimo}$.

2.3. Main Results

Theorem 1. *For the MIMO WHFL with K users and T iterations, given that N, τ, v, D, ϵ, δ, and applying the FBL approach in Section 3, the relationship between PLS, privacy, utility, and the noise variance of LDP is characterized by the following:*

$$\max\left\{\underbrace{\max_{\substack{t\in\{1,..,T\},\\ \Delta g_{mimo}}}\left[\frac{D\cdot 2^{\frac{2}{q(1-\delta)}\log\det\left(I+\frac{\hat{g}_{mimo}K_{x_1}\hat{g}^{\mathcal{H}}_{mimo}}{\sigma_e^2}\right)}-S_\ell\sigma^2_{w,t}}{K}\right]}_{\textbf{Secrecy level of PLS}},\underbrace{\max_{t\in\{1,..,T\}}\left[\frac{S_\ell\sigma^2_{w,t}}{K(2^{2\epsilon}-1)}\right]}_{\textbf{Privacy term}}\right\}\leq\underbrace{\sigma^2}_{\textbf{LDP noise variance}}\leq\underbrace{\frac{v}{K}}_{\textbf{Utility term}},\tag{16}$$

where $\hat{g}_{mimo}=g_{mimo}-\Delta g_{mimo}$, $\mathbf{K}_{x_1}=E(\mathbf{X}_1(t)\mathbf{X}_1^{\mathcal{H}}(t))$*. In addition, an achievable transmission rate* $R_{mimo}(\tau,N,\delta,D,v,\epsilon)$ *of our proposed FBL approach is given by the following:*

$$R_{mimo}(\tau,N,\delta,D,v,\epsilon)=\frac{\sum_{t=1}^{T}N_tR_t}{N},\tag{17}$$

where

$$N=\sum_{t=1}^{T}N_t,\quad R_t=\max_{\substack{\sum_{j=1}^{J}P_j=P\\ \sum_{j=1}^{J}\widetilde{P}_j=\widetilde{P}}}\sum_{j=1}^{J}\frac{1}{N_t}\log\left(\frac{3SNR_jd_j^2}{\left[Q^{-1}(\frac{\tau}{8J})\right]^2}\left(1+\frac{SNR_jd_j^2}{\Psi_1\Psi_2}\right)^{N_t-1}\right),\tag{18}$$

$$\Psi_1=1+\xi\frac{d_j^2SNR_j}{\widetilde{d}_j^2\tilde{SNR}_j},\quad \Psi_2=\left(1-\frac{\xi}{\widetilde{d}_j^2\tilde{SNR}_j}\right)^{-1},\quad \xi=\frac{1}{3}\left[Q^{-1}\left(\frac{\tau}{8J(N_t-1)}\right)\right]^2,\tag{19}$$

and $SNR_j=\frac{P_j}{\sigma_1^2}$, $\tilde{SNR}_j=\frac{\widetilde{P}_j}{\sigma_2^2}$, d_j, $\widetilde{d}_j$, P_j, *and* $\widetilde{P}_j$ *are defined in Section 3.*

Proof of Theorem 1. Our FBL approach for the MIMO WHFL is an extension of the classical SK scheme for the AWGN channel with noiseless feedback. The key to this extension is composed of three parts:

- The two-dimensional message mapping method, which maps the message to a complex codeword transmitted over the fading channels.
- An SVD-based pre-coding strategy that divides the MIMO channel into several parallel SISO channels.
- The two-dimensional modulo-lattice operation (MLO) that eliminates the impact of feedback channel noise on the performance of the SK scheme.

Details about the above tools and how to combine these tools to show our FBL approach for the MIMO WHFL are given in the next section, and the formal proof of Theorem 1 is in Appendix A. □

Remark 1. *Here note that in the FBL approach for the MIMO WHFL, we apply an SVD-based precoding strategy to divide the MIMO channel into several parallel SISO channels, which indicates that the FBL approach for the SISO WHFL can be directly obtained since it is a special case of the approach for the MIMO WHFL. The following Corollary 1 proposes an FBL approach for the SISO WHFL and characterizes the relationship between PLS, privacy, utility, and the noise variance of LDP. Since Corollary 1 can be directly obtained from Theorem 1, we omit the detailed proof here.*

Corollary 1. *For the SISO WHFL with K users and T iterations, given N, τ, v, D, ϵ, δ, and using a similar FBL approach to that of Theorem 1, the relationship between PLS, privacy, utility, and the noise variance of LDP is characterized by the following:*

$$\max\left\{\underbrace{\max_{\substack{t\in\{1,..,T\},\\ \Delta g_{siso}}}\left[\frac{D\cdot 2^{\frac{2}{q(1-\delta)}\log\left(1+\frac{|\hat{g}_{siso}|^2 P}{\sigma_e^2}\right)}-S_\ell\sigma_{w,t}^2}{K}\right]}_{\text{Secrecy level of PLS}},\ \underbrace{\max_{t\in\{1,..,T\}}\left[\frac{S_\ell\sigma_{w,t}^2}{K(2^{2\epsilon}-1)}\right]}_{\text{Privacy term}}\right\}\le\underbrace{\sigma^2}_{\text{LDP noise variance}}\le\underbrace{\frac{v}{K}}_{\text{Utility term}},\tag{20}$$

where $\hat{g}_{siso}=g_{siso}-\Delta g_{siso}$. Furthermore, an achievable transmission rate $R_{siso}(\tau,N,\delta,D,v,\epsilon)$ of our proposed FBL approach is given by the following:

$$R_{siso}(\tau,N,\delta,D,v,\epsilon)=\frac{\sum_{t=1}^{T}N_tR_t}{N},\ \ R_t=\frac{1}{N_t}\log\left(\frac{3SNR|h_{siso}|^2}{[Q^{-1}(\frac{\tau}{8})]^2}\left(1+\frac{SNR|h_{siso}|^2}{\Psi_3\Psi_4}\right)^{N_t-1}\right),\tag{21}$$

where $N=\sum_{t=1}^{T}N_t$, $\Psi_3=1+\xi^\star\frac{|h_{siso}|^2SNR}{|\tilde{h}_{siso}|^2\tilde{SNR}}$, $\Psi_4=\left(1-\frac{\xi^\star}{|\tilde{h}_{siso}|^2\tilde{SNR}}\right)^{-1}$, $\xi^\star=\frac{1}{3}\left[Q^{-1}(\frac{\tau}{8(N_t-1)})\right]^2$, $SNR=\frac{P}{\sigma_1^2}$, $\tilde{SNR}=\frac{\tilde{P}}{\sigma_2^2}$, and $|h_{siso}|$, $|\tilde{h}_{siso}|$, $|\hat{g}_{siso}|$ represent the modulus of h_{siso}, $\tilde{h}_{siso}$ and $\hat{g}_{siso}$, respectively.

Theorem 2. *For the SIMO WHFL with K users and T iterations, given N, τ, v, D, ϵ, δ, and using the FBL approach in Section 4, the relationship between PLS, privacy, utility, and the noise variance of LDP is characterized by the following:*

$$\max\left\{\underbrace{\max_{\substack{t\in\{1,..,T\},\\ \Delta g_{simo}}}\left[\frac{D\cdot 2^{\frac{2}{q(1-\delta)}\log\left(1+\frac{||\hat{\mathbf{g}}_{simo}||^2 P}{\sigma_e^2}\right)}-S_\ell\sigma_{w,t}^2}{K}\right]}_{\text{Secrecy level of PLS}},\ \underbrace{\max_{t\in\{1,..,T\}}\left[\frac{S_\ell\sigma_{w,t}^2}{K(2^{2\epsilon}-1)}\right]}_{\text{Privacy term}}\right\}\le\underbrace{\sigma^2}_{\text{LDP noise variance}}\le\underbrace{\frac{v}{K}}_{\text{Utility term}},\tag{22}$$

where $\hat{\mathbf{g}}_{simo}=\mathbf{g}_{simo}-\Delta\mathbf{g}_{simo}$. In addition, an achievable transmission rate $R_{simo}(\tau,N,\delta,D,v,\epsilon)$ of our proposed FBL approach is given by the following:

$$R_{simo}(\tau,N,\delta,D,v,\epsilon)=\frac{\sum_{t=1}^{T}N_tR_t}{N},\ \ R_t=\frac{1}{N_t}\log\left(\frac{3SNR||\mathbf{h}_{simo}||^2}{[Q^{-1}(\frac{\tau}{8})]^2}\left(1+\frac{SNR||\mathbf{h}_{simo}||^2}{\Psi_5\Psi_6}\right)^{N_t-1}\right),\tag{23}$$

where $\Psi_5=1+\xi^\star\frac{||\mathbf{h}_{simo}||^2SNR}{||\tilde{\mathbf{h}}_{simo}||^2\tilde{SNR}}$, $\Psi_6=\left(1-\frac{\xi^\star}{||\tilde{\mathbf{h}}_{simo}||^2\tilde{SNR}}\right)^{-1}$, SNR, $\tilde{SNR}$, N, and $\xi^\star$ are given in Corollary 1.

Proof of Theorem 2. The difference between the approaches in Theorems 1 and 2 is that for the SIMO case, we use a beamforming strategy together with a new pre-coding strategy instead of the SVD-based pre-coding strategy used for the MIMO case. Here the beamforming and new pre-coding strategies respectively transform the feedforward and feedback channels into SISO channels. Then, along the lines of the encoding-decoding procedure in Section 3.1.3, the FBL approach for the SIMO WHFL is obtained, and the detail about this approach is in Section 4. Finally, since the proof of Theorem 2 is included in that of Theorem 1, we omit the formal proof here. □

Remark 2. *Here, note that in the SIMO WHFL, a beamforming strategy transforms the SIMO feedforward channel into the SISO feedforward channel, while a new pre-coding strategy transforms the MISO feedback channel into the SISO feedback channel. Analogously, for the MISO WHFL, first, we apply the pre-coding strategy of the SIMO WHFL to transform the MISO feedforward channel into the SISO feedforward channel, and the beamforming strategy of the SIMO WHFL to transform the SIMO feedback channel into the SISO feedback channel, then along the lines of the encoding-decoding procedure in Theorem 2, the following Corollary 2 for the MISO WHFL is obtained. As the proof follows a similar way to that of Theorem 2, the detailed proof is omitted here.*

Corollary 2. *For the MISO WHFL with K users and T iterations, given N, τ, v, D, ϵ and δ, and using a similar FBL approach to that of Theorem 2, the relationship between PLS, privacy, utility, and the noise variance of LDP is characterized by the following:*

$$\max\left\{\underbrace{\max_{\substack{t\in\{1,..,T\},\\ \Delta g_{miso}}}\left[\frac{D\cdot 2^{\frac{2}{q(1-\delta)}\log\left(1+\frac{||\hat{g}_{miso}||^2P}{\sigma_e^2}\right)}-S_\ell\sigma_{w,t}^2}{K}\right]}_{\textbf{Secrecy level of PLS}},\underbrace{\max_{t\in\{1,..,T\}}\left[\frac{S_\ell\sigma_{w,t}^2}{K(2^{2\epsilon}-1)}\right]}_{\textbf{Privacy term}}\right\}\leq\underbrace{\sigma^2}_{\textbf{LDP noise variance}}\leq\underbrace{\frac{v}{K}}_{\textbf{Utility term}},\tag{24}$$

where $\hat{\boldsymbol{g}}_{miso}=\boldsymbol{g}_{miso}-\Delta\boldsymbol{g}_{miso}$. In addition, an achievable transmission rate $R_{miso}(\tau,N,\delta,D,v,\epsilon)$ of our proposed FBL approach is given by the following:

$$R_{miso}(\tau,N,\delta,D,v,\epsilon)=\frac{\sum_{t=1}^{T}N_tR_t}{N},\quad R_t=\frac{1}{N_t}\log\left(\frac{3SNR||\boldsymbol{h}_{miso}||^2}{[Q^{-1}(\frac{\tau}{8})]^2}\left(1+\frac{SNR||\boldsymbol{h}_{miso}||^2}{\Psi_7\Psi_8}\right)^{N_t-1}\right),\tag{25}$$

where $\Psi_7=1+\xi^\star\frac{||\boldsymbol{h}_{miso}||^2SNR}{||\tilde{\boldsymbol{h}}_{miso}||^2\tilde{SNR}}$, $\Psi_8=\left(1-\frac{\xi^\star}{||\tilde{\boldsymbol{h}}_{miso}||^2\tilde{SNR}}\right)^{-1}$, SNR, $\tilde{SNR}$, N and $\xi^\star$ are given in Corollary 1.

3. An FBL Approach for the MIMO WHFL

For the WHFL in the MIMO case, (9)–(11) can be re-written as follows:

$$\boldsymbol{Y}_i(t)=\boldsymbol{h}_{mimo}\boldsymbol{X}_i(t)+\boldsymbol{\eta}_{1,i}(t),\ 1\leq i\leq N_t,\tag{26}$$

$$\widetilde{\boldsymbol{Y}}_i(t)=\widetilde{\boldsymbol{h}}_{mimo}\widetilde{\boldsymbol{X}}_i(t)+\boldsymbol{\eta}_{2,i}(t),\ 1\leq i\leq N_t-1,\tag{27}$$

$$\boldsymbol{Z}_i(t)=\boldsymbol{g}_{mimo}\boldsymbol{X}_i(t)+\widetilde{\boldsymbol{g}}_{mimo}\widetilde{\boldsymbol{X}}_i(t)+\boldsymbol{\eta}_{e,i}(t),\ 1\leq i\leq N_t,\tag{28}$$

where $\boldsymbol{h}_{mimo}\in\mathbb{C}^{B\times A}$, $\widetilde{\boldsymbol{h}}_{mimo}\in\mathbb{C}^{A\times B}$, $\boldsymbol{g}_{mimo}\in\mathbb{C}^{C\times A}$, $\widetilde{\boldsymbol{g}}_{mimo}\in\mathbb{C}^{C\times B}$, $\boldsymbol{X}_i(t)\in\mathbb{C}^{A\times 1}$, $\widetilde{\boldsymbol{X}}_i(t)\in\mathbb{C}^{B\times 1}$, the elements of $\boldsymbol{\eta}_{1,i}(t)\in\mathbb{C}^{B\times 1}$, $\boldsymbol{\eta}_{2,i}(t)\in\mathbb{C}^{A\times 1}$ and $\boldsymbol{\eta}_{e,i}(t)\in\mathbb{C}^{C\times 1}$ are i.i.d. as $\mathcal{CN}(0,\sigma_1^2)$, $\mathcal{CN}(0,\sigma_2^2)$ and $\mathcal{CN}(0,\sigma_e^2)$, respectively. Here note that the feedforward channel (26) and the feedback channel (27) are both MIMO channels.

In this section, for the MIMO WHFL system, an FBL approach is proposed, which combines the two-dimensional message mapping method, the two-dimensional MLO, and the SVD technique, see the following Figure 3. To facilitate a better understanding of Figure 3, we introduce the two-dimensional message mapping method and the two-dimensional MLO below.

The two-dimensional message mapping method: We first review the message mapping in the classical SK scheme [35] (see Figure 4a). Specifically, for given codeword length n, let the message $W\in\mathcal{W}=\{1,2,...,2^{nR}\}$ and $|\mathcal{W}|=2^{nR}$, where R is the transmission rate. Partition the interval $[-\sqrt{3},\sqrt{3}]$ into 2^{nR} equal sub-intervals, with each sub-interval's midpoint corresponding to a message in $\mathcal{W}$. Let θ denote the midpoint associated with message W, where the variance of θ is approximately 1. This one-dimensional mapping method is shown to be optimal for AWGN channels with real signals. To address the complexity of fading channels, we introduce a two-dimensional message mapping method, detailed as follows:

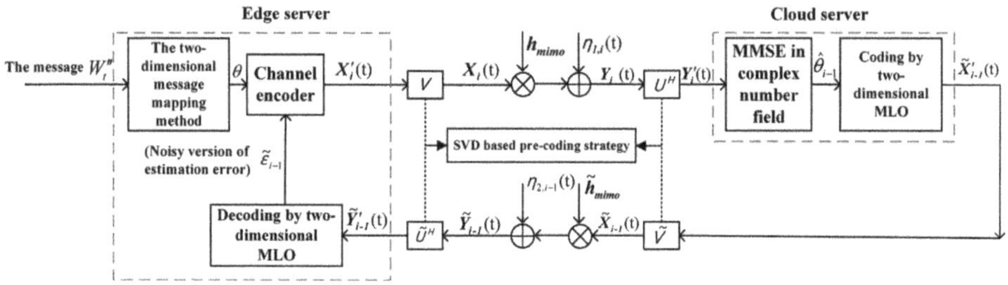

Figure 3. A schematic diagram of the FBL approach for the WHFL over the MIMO channel.

For given codeword length n, let message $W = (W_R, W_I)$, where W, W_R and W_I are uniformly distributed in $\mathcal{W} = \{1, ..., 2^{nR}\}$, $\mathcal{W}_R = \{1, ..., 2^{nR_R}\}$ and $\mathcal{W}_I = \{1, ..., 2^{nR_I}\}$, respectively, and $R_R + R_I = R$. Since the message W is composed of two parts, we place the points (W_R, W_I) in a complex square grid with corners located at $(\pm\sqrt{3}, \pm j\sqrt{3})$ (see Figure 4b). Divide the entire square grid into $2^{n(R_R+R_I)}$ equally spaced sub-grids, and the center point of each sub-grid is mapped to a pair of values in $\mathcal{W} = (\mathcal{W}_R, \mathcal{W}_I)$. Let $\theta = \theta_R + j\theta_I$ be the center point of the sub-grid with respect to (w.r.t) the message $W = (W_R, W_I)$, where θ_R and θ_I represent the real and imaginary components of θ, respectively, and the variance of θ approximately equals 2.

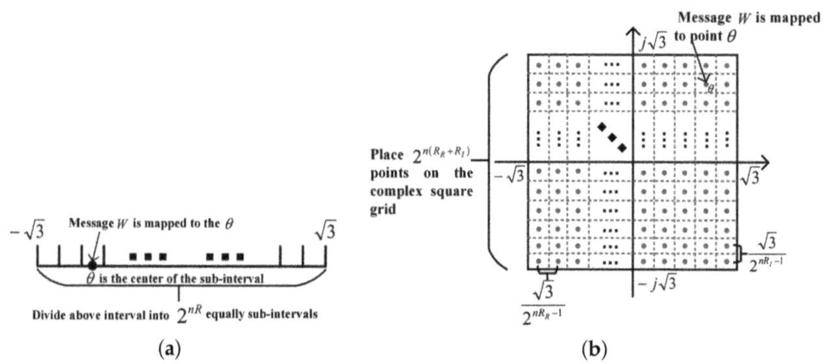

Figure 4. Comparison of the message mapping methods between the classical SK scheme and the scheme in this paper. (**a**) Message mapping of classical SK scheme. (**b**) Message mapping in this paper.

The two-dimensional MLO: The two-dimensional MLO is given by the following:

$$\mathbb{M}_\Lambda[x] \stackrel{\text{def}}{=} x - \mathbb{Q}[x], \tag{29}$$

where the two-dimensional lattice $\Lambda = \Lambda_R + j\Lambda_I$ is a complex plane with $\Lambda_R \in [-\frac{d}{2}, -\frac{d}{2}]$, $\Lambda_I \in [-\frac{d}{2}, -\frac{d}{2}]$, $d > 0$, $j = \sqrt{-1}$, $\mathbb{Q}[x]$ is the nearest neighbor quantization of x w.r.t. Λ, and x is a complex-valued number. Some basic properties of the two-dimensional MLO [39] are listed below.

Proposition 1. *(1). The distributive law* $\mathbb{M}_\Lambda[\mathbb{M}_\Lambda[x] + y] = \mathbb{M}_\Lambda[x + y]$.

(2). If $x + y \in \Lambda$, $\mathbb{M}_\Lambda[x + y] = x + y$, *otherwise, a* modulo-aliasing error *occurred.*

(3). Let the dither signal v *be uniformly distributed on* Λ*, then* $\mathbb{M}_\Lambda[x + v]$ *is uniformly distributed on* Λ*, where* $Var(\mathbb{M}_\Lambda[x + v]) = \frac{d^2}{12} + \frac{d^2}{12} = \frac{d^2}{6}$.

The classical SK scheme does not work in the noisy feedback case, and this is because in such a case, the transmitter cannot accurately obtain the estimation error of the receiver. We show that by applying the two-dimensional MLO to both the feedforward and feedback encoders, the adverse effects of feedback channel noise on the SK scheme's performance can be mitigated, which allows the SK-type scheme to remain effective even in the presence of noisy feedback. The following Figure 5a,b illustrate the differences between the classical SK scheme and the modified SK-type scheme utilizing two-dimensional MLO.

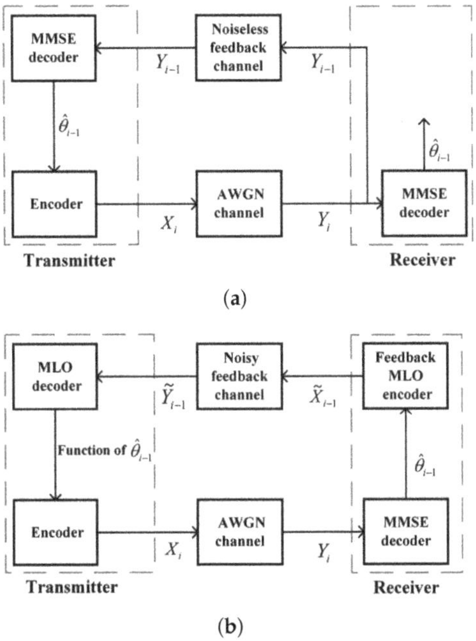

Figure 5. Comparing the mechanisms between the classical SK scheme and the two-dimensional MLO-based SK-type scheme, where $\hat{\theta}_{i-1}$ represents the estimation of the transmitted message θ at time $i-1$. (**a**) The classical SK scheme in a certain round i. (**b**) The two-dimensional MLO-based SK-type scheme in a certain round i.

3.1. An FBL Approach for the MIMO WHFL

3.1.1. Channel Decomposition by SVD

Based on the SVD technique, matrices h_{mimo} and $\tilde{h}_{mimo}$ can be expressed as follows:

$$h_{mimo} = \boldsymbol{U}\boldsymbol{\Lambda}\boldsymbol{V}^{\mathcal{H}}, \ \tilde{h}_{mimo} = \tilde{\boldsymbol{U}}\tilde{\boldsymbol{\Lambda}}\tilde{\boldsymbol{V}}^{\mathcal{H}}, \tag{30}$$

where $\boldsymbol{U}, \tilde{\boldsymbol{V}}^{\mathcal{H}} \in \mathbb{C}^{B\times B}$ and $\tilde{\boldsymbol{U}}, \boldsymbol{V}^{\mathcal{H}} \in \mathbb{C}^{A\times A}$ are unitary matrices. The diagonal matrices $\boldsymbol{\Lambda} \in \mathbb{C}^{B\times A}$ and $\tilde{\boldsymbol{\Lambda}} \in \mathbb{C}^{A\times B}$ have non-negative real number diagonal elements $(d_1,\ldots,d_J)$ and $(\tilde{d}_1,\ldots,\tilde{d}_J)$ [49], respectively, and

$$J = \min(A, B). \tag{31}$$

According to (26) and (30), we have the following:

$$\boldsymbol{U}^{\mathcal{H}}\boldsymbol{Y}_i(t) = \boldsymbol{\Lambda}\boldsymbol{V}^{\mathcal{H}}\boldsymbol{X}_i(t) + \boldsymbol{U}^{\mathcal{H}}\boldsymbol{\eta}_{1,i}(t) \Longrightarrow \boldsymbol{Y}_i'(t) = \boldsymbol{\Lambda}\boldsymbol{X}_i'(t) + \boldsymbol{\eta}_{1,i}'(t), \tag{32}$$

where $\boldsymbol{Y}_i'(t) = \boldsymbol{U}^{\mathcal{H}}\boldsymbol{Y}_i(t), \boldsymbol{\eta}_{1,i}'(t) = \boldsymbol{U}^{\mathcal{H}}\boldsymbol{\eta}_{1,i}(t) \in \mathbb{C}^{B\times 1}$ and $\boldsymbol{X}_i'(t) = \boldsymbol{V}^{\mathcal{H}}\boldsymbol{X}_i(t) \in \mathbb{C}^{A\times 1}$. It is noted that $E(\boldsymbol{X}_i'^{\mathcal{H}}(t)\boldsymbol{X}_i'(t)) = E(\boldsymbol{X}_i^{\mathcal{H}}(t)\boldsymbol{X}_i(t))$ and $E(\boldsymbol{\eta}_{1,i}'^{\mathcal{H}}(t)\boldsymbol{\eta}_{1,i}'(t)) = E(\boldsymbol{\eta}_{1,i}^{\mathcal{H}}(t)\boldsymbol{\eta}_{1,i}(t))$, ensuring that the power constraint of $\boldsymbol{X}_i(t)$ is equal to that of $\boldsymbol{X}_i'(t)$, and the distributions of $\boldsymbol{\eta}'_{1,i}(t)$ and $\boldsymbol{\eta}_{1,i}(t)$ remain the same. As $\boldsymbol{\Lambda}$ is a diagonal matrix, (32) can be decomposed as follows:

$$Y'_{j,i}(t) = d_j X'_{j,i}(t) + \eta'_{j,1,i}(t),\ 1 \le j \le J,\ 1 \le i \le N_t, \tag{33}$$

where $Y'_{j,i}(t)$, $X'_{j,i}(t)$ and $\eta'_{j,1,i}(t)$ denote the j-th components of $\mathbf{Y}'_i(t)$, $\mathbf{X}'_i(t)$ and $\boldsymbol{\eta}'_{1,i}(t)$, respectively.

Similarly, from (27) and (30), (27) can be decomposed as follows:

$$\widetilde{Y}'_{j,i}(t) = \widetilde{d}_j \widetilde{X}'_{j,i}(t) + \eta'_{j,2,i}(t),\ 1 \le j \le J,\ 1 \le i \le N_t - 1, \tag{34}$$

where $\widetilde{Y}'_{j,i}(t)$, $\widetilde{X}'_{j,i}(t)$ and $\eta'_{j,2,i}(t)$ denote the j-th components of $\widetilde{\mathbf{Y}}'_i(t)$, $\widetilde{\mathbf{X}}'_i(t)$ and $\boldsymbol{\eta}'_{2,i}(t)$, respectively, and $\widetilde{\mathbf{Y}}'_i(t) = \widetilde{\boldsymbol{U}}^{\mathcal{H}}\widetilde{\mathbf{Y}}_i(t), \boldsymbol{\eta}'_{2,i}(t) = \widetilde{\boldsymbol{U}}^{\mathcal{H}}\boldsymbol{\eta}_{2,i}(t) \in \mathbb{C}^{A\times 1}$ and $\widetilde{\mathbf{X}}'_i(t) = \widetilde{\boldsymbol{V}}^{\mathcal{H}}\widetilde{\mathbf{X}}_i(t) \in \mathbb{C}^{B\times 1}$. As shown in (32)–(34), applying the SVD technique, the feedforward and feedback MIMO channels can be effectively transformed into J parallel SISO sub-channels.

Power allocating: The edge server assigns power $P_1, \ldots, P_J$ to the J parallel sub-channels for the feedforward channel, where $\sum_{j=1}^{J} P_j = P$. Similarly, the cloud server distributes power $\widetilde{P}_1, \ldots, \widetilde{P}_J$ across the J parallel sub-channels for the feedback channel, where $\sum_{j=1}^{J} \widetilde{P}_j = \widetilde{P}$.

3.1.2. Message Splitting

For given τ, N_t, v, D and ϵ, we define the following:

$$|\mathcal{W}''_t| = 2^{N_t R_t} = 2^{qR_t(D)},\ R_t = \frac{H(W''_t)}{N_t}. \tag{35}$$

Next, the message W''_t is divided into J independent components $(W''_{t,1}, \ldots, W''_{t,J})$, where $W''_{t,j}$ is uniformly distributed over the set $\mathcal{W}''_{t,j} = \{1, 2, \ldots, 2^{N_t R_{t,j}}\}$ and $j = 1, \ldots, J$. Then, we divide each sub-message $W''_{t,j}$ into $W''_{t,j} = (W''_{t,j,R}, W''_{t,j,I})$, where $W''_{t,j,R}$ and $W''_{t,j,I}$ are uniformly distributed over the sets $\mathcal{W}''_{t,j,R} = \{1, 2, \ldots, 2^{N_t R_{t,j,R}}\}$ and $\mathcal{W}''_{t,j,I} = \{1, 2, \ldots, 2^{N_t R_{t,j,I}}\}$, respectively. The rate for each parallel sub-channel is defined as $R_{t,j} = R_{t,j,R} + R_{t,j,I}$. Consequently, the total rate R_t for all J parallel sub-channels during the t-th communication round is as follows:

$$R_t = \sum_{j=1}^{J} (R_{t,j,R} + R_{t,j,I}). \tag{36}$$

3.1.3. An FBL Scheme of Each Parallel Sub-Channel

By using the two-dimensional message mapping method introduced in the last subsection, the message $W''_{t,j}$ is mapped to the center point θ_j of its corresponding sub-grid.

Initialization: At time instant 1, the edge server maps the messages $W''_{t,j}$ to $\theta_j = \theta_{R,j} + j\theta_{I,j}$, and sends the following:

$$X'_{j,1}(t) = \sqrt{\frac{P_j}{2}}\theta_j, \tag{37}$$

Then, the cloud server computes the first estimation $\hat{\theta}_{j,1}$ of θ_j by the following:

$$\hat{\theta}_{j,1} = \frac{Y'_{j,1}(t)}{d_j\sqrt{\frac{P_j}{2}}} = \theta_j + \frac{\eta'_{j,1,1}(t)}{d_j\sqrt{\frac{P_j}{2}}} = \theta_j + \varepsilon_1, \tag{38}$$

where $\varepsilon_1 = \varepsilon_{R,1} + j\varepsilon_{I,1} = \hat{\theta}_{j,1} - \theta_j$ is the estimation error of the cloud server at time instant 1. Define $\alpha_1 = \mathrm{Var}(\varepsilon_1) = \frac{2\sigma_1^2}{d_j^2 P_j}$, $\alpha_{R,1} = \mathrm{Var}(\varepsilon_{R,1}) = \frac{\sigma_1^2}{d_j^2 P_j}$ and $\alpha_{I,1} = \mathrm{Var}(\varepsilon_{I,1}) = \frac{\sigma_1^2}{d_j^2 P_j}$.

Iteration: First, we introduce a shared dither random i.i.d. sequence $\nu^{N_t-1} = (\nu_1, \ldots, \nu_{N_t-1})$, which is perfectly known by both the edge server and the cloud server, and it is uniformly distributed on Λ ($\Lambda = \Lambda_R + j\Lambda_I$ is a complex plane with $\Lambda_R \in [-\frac{d}{2}, -\frac{d}{2}]$, $\Lambda_I \in [-\frac{d}{2}, -\frac{d}{2}]$), and $d = \sqrt{6\widetilde{P}_j}$. Here v^{N_t-1} is independent of all signals transmitted over channels. At time instant i ($2 \leq i \leq N_t$), using the two-dimensional MLO shown in Section 3, the cloud server sends the following:

$$\widetilde{X}'_{j,i-1}(t) = \mathbb{M}_\Lambda[\gamma_{i-1}\hat{\theta}_{j,i-1} + \nu_{i-1}], \tag{39}$$

where γ_{i-1} is a modulation coefficient. From Property (3) of Proposition 1, we have $E(\widetilde{X}'^{\mathcal{H}}_{j,i-1}(t)\widetilde{X}'_{j,i-1}(t)) = \widetilde{P}_j$ (the dither signals guarantee that the codeword transmitted by the cloud server meets the power constraint). Then the edge server computes a noisy version of estimation error $\varepsilon_{i-1} = \hat{\theta}_{j,i-1} - \theta_j$ by the following:

$$\widetilde{\varepsilon}_{i-1} = \frac{1}{\gamma_{i-1}}\mathbb{M}_\Lambda[\frac{\widetilde{Y}'_{j,i-1}(t)}{\widetilde{d}_j} - \gamma_{i-1}\theta_j - \nu_{i-1}] \stackrel{(a)}{=} \frac{1}{\gamma_{i-1}}\mathbb{M}_\Lambda[\gamma_{i-1}\varepsilon_{i-1} + \frac{\eta'_{j,2,i-1}(t)}{\widetilde{d}_j}], \tag{40}$$

where (a) is due to the modulo distributive law in property (1) of Proposition 1. The *modulo-aliasing errors* do not occur in the edge server, if $\gamma_{i-1}\varepsilon_{i-1} + \frac{\eta'_{j,2,i-1}(t)}{\widetilde{d}_j} \in \Lambda$. Hence, the edge server obtains $\widetilde{\varepsilon}_{i-1} = \varepsilon_{i-1} + \frac{\eta'_{j,2,i-1}(t)}{\gamma_{i-1}\widetilde{d}_j}$. Then, the edge server sends the following:

$$X'_{j,i}(t) = \lambda_{i-1}\gamma_{i-1}\widetilde{\varepsilon}_{i-1}, \tag{41}$$

where λ_{i-1} is chosen to satisfy the transmitter's power constraint P_j. Then, the cloud server updates $\hat{\theta}_{j,i}$ by computing the following:

$$\hat{\theta}_{j,i} = \hat{\theta}_{j,i-1} - \hat{\varepsilon}_{i-1} = \hat{\theta}_{j,i-1} - \beta_i\frac{Y'_{j,i}(t)}{d_j}, \tag{42}$$

where $\hat{\varepsilon}_{i-1} = \beta_i\frac{Y'_{j,i}(t)}{d_j}$, and the MMSE estimation coefficient β_i is given by the following:

$$\beta_i = \frac{E(\frac{\varepsilon_{i-1}Y'_{j,i}(t)^{\mathcal{H}}}{d_j})}{E(\frac{Y'_{j,i}(t)Y'_{j,i}(t)^{\mathcal{H}}}{d_j^2})}, \tag{43}$$

which ensures that ε_{i-1} is correctly estimated from $Y'_{j,i}(t)$. Define $\varepsilon_i = \varepsilon_{R,i} + j\varepsilon_{I,i} = \hat{\theta}_{j,i} - \theta_j$, (42) yields the following:

$$\varepsilon_i = \varepsilon_{i-1} - \beta_i\frac{Y'_{j,i}(t)}{d_j}. \tag{44}$$

Further define $\alpha_i = \text{Var}(\varepsilon_i)$, $\alpha_{R,i} = \text{Var}(\varepsilon_{R,i})$, $\alpha_{I,i} = \text{Var}(\varepsilon_{I,i})$. Since ε_i is a CSCG distribution estimation error, we conclude that $\alpha_{R,i} = \alpha_{I,i} = \frac{\alpha_i}{2}$.

Decoding: At time instant N_t, the final estimation obtained by the cloud server is $\hat{\theta}_{j,N_t} = \theta_j + \varepsilon_{N_t}$, where $\varepsilon_{N_t} = \varepsilon_{R,N_t} + j\varepsilon_{I,N_t}$. The cloud server successfully decodes the message $W''_{t,j}$ if $\hat{\theta}_{j,N_t}$ is closest to the message point θ_j, i.e., $\varepsilon_{R,N_t} \in [-\frac{\sqrt{3}}{2^{N_tR_{t,j,R}}}, \frac{\sqrt{3}}{2^{N_tR_{t,j,R}}})$ and $\varepsilon_{I,N_t} \in [-\frac{\sqrt{3}}{2^{N_tR_{t,j,I}}}, \frac{\sqrt{3}}{2^{N_tR_{t,j,I}}})$.

The formal proof of Theorem 1 is provided in Appendix A.

4. An FBL Approach for the SIMO WHFL

For the SIMO WHFL, (9)–(11) can be re-written as follows:

$$\mathbf{Y}_i(t) = \boldsymbol{h}_{simo}X_i(t) + \boldsymbol{\eta}_{1,i}(t), \ 1 \leq i \leq N_t, \tag{45}$$

$$\widetilde{Y}_i(t) = \widetilde{\boldsymbol{h}}_{simo}\widetilde{\boldsymbol{X}}_i(t) + \eta_{2,i}(t), \ 1 \le i \le N_t - 1, \tag{46}$$

$$\boldsymbol{Z}_i(t) = \boldsymbol{g}_{simo}X_i(t) + \widetilde{\boldsymbol{g}}_{simo}\widetilde{\boldsymbol{X}}_i(t) + \boldsymbol{\eta}_{e,i}(t), \ 1 \le i \le N_t, \tag{47}$$

where $\boldsymbol{h}_{simo} \in \mathbb{C}^{B\times 1}$, $\widetilde{\boldsymbol{h}}_{simo} \in \mathbb{C}^{1\times B}$, $\boldsymbol{g}_{simo} \in \mathbb{C}^{C\times 1}$, $\widetilde{\boldsymbol{g}}_{simo} \in \mathbb{C}^{C\times B}$, $X_i(t) \in \mathbb{C}^{1\times 1}$, $\widetilde{\boldsymbol{X}}_i(t) \in \mathbb{C}^{B\times 1}$, the elements of $\boldsymbol{\eta}_{1,i}(t) \in \mathbb{C}^{B\times 1}$ and $\boldsymbol{\eta}_{e,i}(t) \in \mathbb{C}^{C\times 1}$ are i.i.d. as $\mathcal{CN}(0,\sigma_1^2)$ and $\mathcal{CN}(0,\sigma_e^2)$, respectively, and $\eta_{2,i}(t) \in \mathbb{C}^{1\times 1} \sim \mathcal{CN}(0,\sigma_2^2)$. Here note that the feedforward channel (45) is a SIMO channel, while the feedback channel (46) is a MISO channel. Unlike the SVD technique used for the MIMO WHFL that decomposes the MIMO channel into several parallel SISO channels, we use a beamforming strategy to transform the feedforward SIMO channel into the SISO channel, and a new pre-coding strategy to transform the feedback MISO channel into the SISO channel, see the following Figure 6. Further applying the approach for each SISO channel (see Section 3.1.3), the FBL approach for the SIMO WHFL is obtained, and the details are given below.

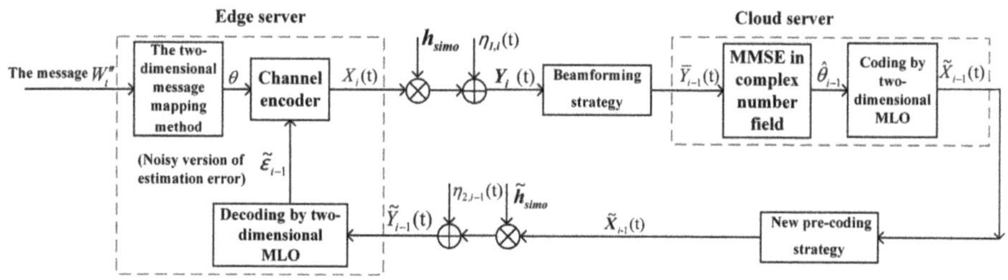

Figure 6. A schematic diagram of the FBL approach for the SIMO WHFL.

Beamforming strategy: The signal received by the cloud server in (45) can proceed as follows:

$$\boldsymbol{h}_{simo}^{\mathcal{H}}\boldsymbol{Y}_i(t) = \boldsymbol{h}_{simo}^{\mathcal{H}}\boldsymbol{h}_{simo}X_i(t) + \boldsymbol{h}_{simo}^{\mathcal{H}}\boldsymbol{\eta}_{1,i}(t) = ||\boldsymbol{h}_{simo}||^2 X_i(t) + \boldsymbol{h}_{simo}^{\mathcal{H}}\boldsymbol{\eta}_{1,i}(t),$$
$$\Longrightarrow \bar{Y}_i(t) = ||\boldsymbol{h}_{simo}||^2 X_i(t) + \bar{\eta}_{1,i}(t), \tag{48}$$

where $\bar{Y}_i(t) = \boldsymbol{h}_{simo}^{\mathcal{H}}\boldsymbol{Y}_i(t) \in \mathbb{C}^{1\times 1}$ and $\bar{\eta}_{1,i}(t) = \boldsymbol{h}_{simo}^{\mathcal{H}}\boldsymbol{\eta}_{1,i}(t) \in \mathbb{C}^{1\times 1}$. Applying (4), the feedforward SIMO channel is transformed into the SISO channel.

A new pre-coding strategy: For the feedback channel (46), allowing the following:

$$\widetilde{\boldsymbol{X}}_i(t) = \frac{\widetilde{\boldsymbol{h}}_{simo}^{\mathcal{H}}}{||\widetilde{\boldsymbol{h}}_{simo}||}\widetilde{X}_i(t), \tag{49}$$

where $\widetilde{X}_i(t) \in \mathbb{C}^{1\times 1}$ and $E(\widetilde{\boldsymbol{X}}_i^{\mathcal{H}}(t)\widetilde{\boldsymbol{X}}_i(t)) = E(\widetilde{X}_i^{\mathcal{H}}(t)\frac{\widetilde{\boldsymbol{h}}_{simo}}{||\widetilde{\boldsymbol{h}}_{simo}||}\frac{\widetilde{\boldsymbol{h}}_{simo}^{\mathcal{H}}}{||\widetilde{\boldsymbol{h}}_{simo}||}\widetilde{X}_i(t)) = E(\widetilde{X}_i^{\mathcal{H}}(t)\widetilde{X}_i(t)) = \widetilde{P}$, which indicates that the power constraint of $\widetilde{\boldsymbol{X}}_i(t)$ is equal to that of $\widetilde{X}_i(t)$. Hence, substituting (49) into (46), we have the following:

$$\widetilde{Y}_i(t) = \widetilde{\boldsymbol{h}}_{simo}\frac{\widetilde{\boldsymbol{h}}_{simo}^{\mathcal{H}}}{||\widetilde{\boldsymbol{h}}_{simo}||}\widetilde{X}_i(t) + \eta_{2,i}(t) = ||\widetilde{\boldsymbol{h}}_{simo}||\widetilde{X}_i(t) + \eta_{2,i}(t), \tag{50}$$

which indicates that the feedback MISO channel is transformed into the SISO channel. Hence along the lines of the encoding-decoding procedure in Section 3.1.3, the FBL approach for the SIMO WHFL is obtained.

Since the proof of Theorem 2 is included in the proof of Theorem 1, we omit the detailed proof here.

5. Simulation Results

5.1. Experimental Settings

The simulation results are derived by averaging 2000 independent channel realizations (i.e., Monte-Carlo simulations). We consider a WHFL system consisting of 10 users, an edge server, and a cloud server, with each user having the same amount of training data. We assume that the channel matrix elements follow an i.i.d. distribution as $\mathcal{CN}(0,1)$ [5,6,17]. Following [47], the maximum normalized estimation errors of the eavesdropper's channel are defined as $\Omega = \frac{\omega}{||g||_F}$ and $\tilde{\Omega} = \frac{\tilde{\omega}}{||\tilde{g}||_F}$, where ω and $\tilde{\omega}$ are defined in (13). The edge server employs Lempel–Ziv–Welch (LZW) source coding [50] to compress the quantized gradients, and the total transmitted data are M bits. The transmission latency for the edge server to upload data is $T_{comm} = \frac{M}{R_{eg}}$ [5], where R_{eg} represents the edge server's transmission rate.

To evaluate the effectiveness of the proposed FBL scheme under real-world conditions, we train a neural network using the MNIST dataset (http://yann.lecun.com/exdb/mnist/, accessed on 20 March 2024), which contains 60,000 training samples and 10,000 test samples of 10 different handwritten digits. The network architecture includes 784 input nodes, a hidden layer containing 20 nodes, and an output layer with 10 nodes. The loss function is cross-entropy, with the hidden and output layers utilizing the ReLU and softmax activation functions, respectively. The neural network contains a total of $q = 15,910$ parameters, and the learning rate is set at $\mu = 0.1$. In the experiments, the following three schemes are compared.

- Benchmark (Perfect HFL): The perfectly aggregated HFL system can be achieved through error-free transmission, which serves as the benchmark accuracy in ideal settings.
- Baseline 1 (Random binning coding scheme (RBCS)-based WHFL [26,28]): The gradient data from the edge servers is uploaded using the RBCS, which is based on traditional low-density parity-check (LDPC) codes with a target bit error rate of 10^{-6}.
- Baseline 2 (Frequency division multiple access (FDMA)-based WHFL with artificial noise (AN) [29]): In the FDMA-based WHFL system with AN, FDMA is employed to transmit gradient data from edge servers to the cloud server, targeting a bit error ratio of 10^{-6}. Additionally, AN is added to the transmitted signals to prevent eavesdroppers from obtaining the true gradient data.

5.2. Experimental Results

We show the results of test accuracy and the cross entropy versus the communication round for SISO/SIMO/MISO/MIMO cases in Figures 7 and 8, respectively. From Figures 7 and 8, we see that if perfect CSI of the eavesdropper's channel is obtained by legal parties, both our proposed FBL scheme, Baseline 1 scheme, and Baseline 2 scheme almost do not affect the learning performance of HFL. This is because Baseline 1, Baseline 2, and our proposed schemes are all capable of transmitting gradient data with a sufficiently low decoding error probability. On the other hand, in our proposed FBL schemes, if imperfect CSI of the eavesdropper's channel is obtained by legal parties, the test accuracy of HFL decreases, and the training loss of HFL increases as the maximum normalized estimation error of the eavesdropper's channel increases. However, note that in such an imperfect CSI case, our proposed FBL schemes still provide the same level of secrecy as that of the perfect CSI case, which shows the robustness of our schemes against imperfect CSI of the eavesdropper's channel. Furthermore, Figures 7 and 8 demonstrate that the eavesdropper cannot obtain the real gradient data when applying our FBL scheme, which indicates that our FBL schemes effectively ensure the PLS of the data.

As depicted in Figure 9, the transmission latency of our FBL scheme is approximately 2 to 5 times lower than that of Baseline 1 and Baseline 2, due to the gain from introducing feedback. Additionally, the transmission latency of our scheme decreases as the number of antennas increases. Furthermore, the transmission latency of Baseline 2 is lower than that of Baseline 1, owing to the gain from introducing AN to counter eavesdropping attacks in Baseline 2. Moreover, Figure 9 shows that the transmission latency of our FBL scheme increases as the maximum normalized estimation error of the eavesdropper's channel increases, and this is because to support the same level of performance, the worse estimation of the CSI of the eavesdropper's channel, the more bits need to be transmitted, which leads to an increase in transmission latency.

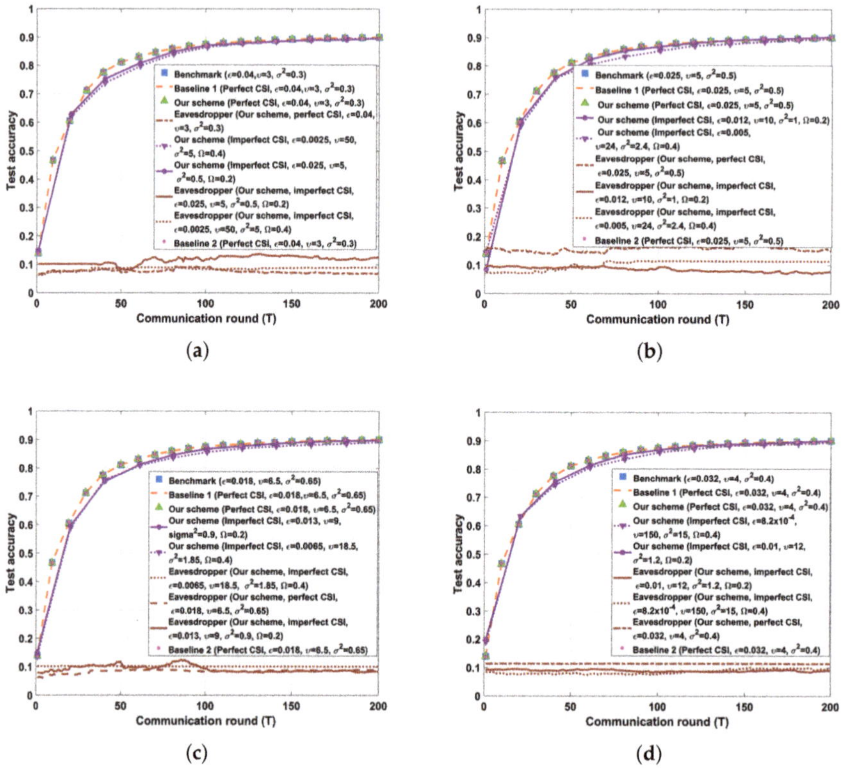

Figure 7. Performance comparison between the different schemes on the MNIST dataset ($K = 10$, $S_\ell = 60,000$, $q = 15,910$, $D = 10^{-4}$, $\tilde{\text{SNR}} = 15$ dB, $P = 10$, $\tau = 10^{-6}$, $\sigma_1^2 = \sigma_2^2 = 1$, $\sigma_e^2 = 2$). (**a**) $A = B = C = 4, \delta = 0.99994$. (**b**) $A = 1, B = C = 4, \delta = 0.99997$. (**c**) $A = 4, B = C = 1, \delta = 0.99997$. (**d**) $A = B = C = 1, \delta = 0.99998$.

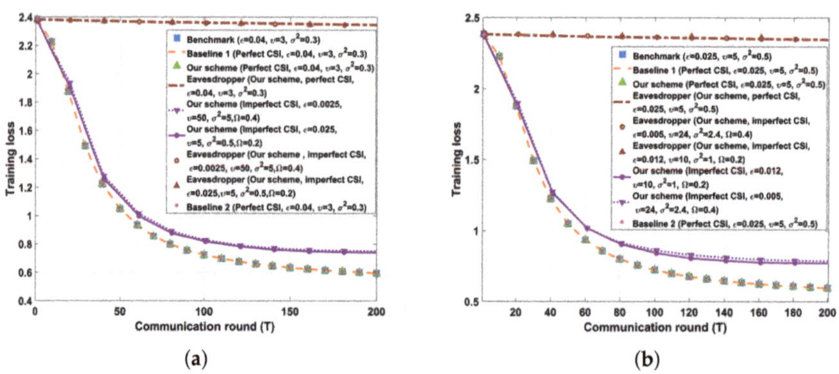

Figure 8. *Cont.*

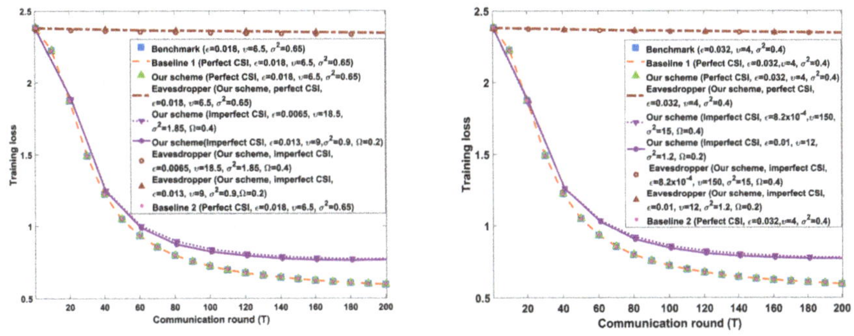

(c) (d)

Figure 8. Performance comparison between the different schemes on the MNIST dataset ($K = 10$, $S_\ell = 60{,}000$, $q = 15{,}910$, $D = 10^{-4}$, $\tilde{\text{SNR}} = 15$ dB, $P = 10$, $\tau = 10^{-6}$, $\sigma_1^2 = \sigma_2^2 = 1, \sigma_e^2 = 2$). (**a**) $A = B = C = 4, \delta = 0.99994$. (**b**) $A = 1, B = C = 4, \delta = 0.99997$. (**c**) $A = 4, B = C = 1, \delta = 0.99997$. (**d**) $A = B = C = 1, \delta = 0.99998$.

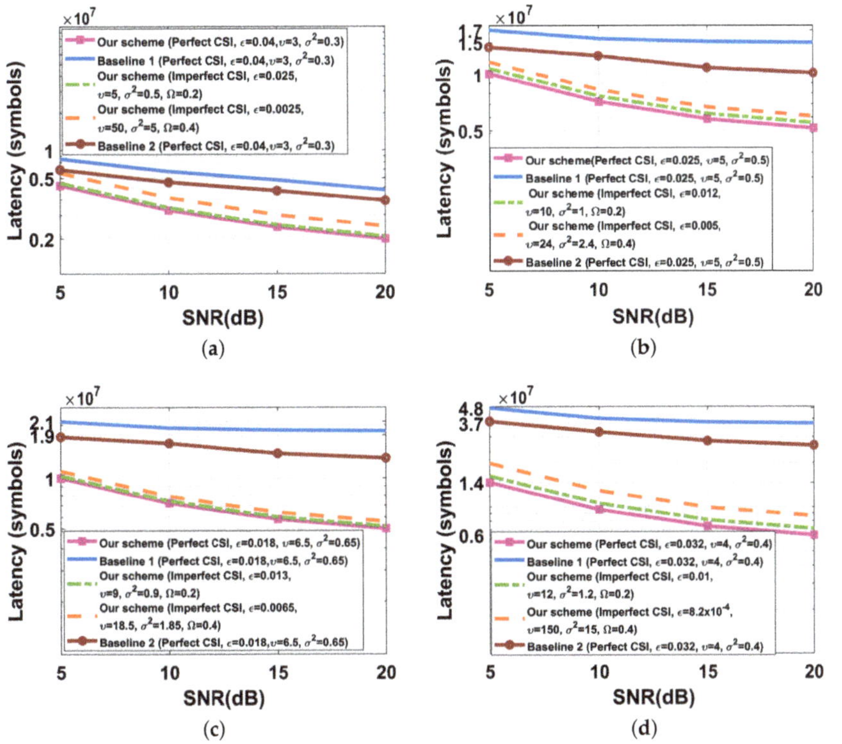

Figure 9. Transmission latency (200 rounds) of the different schemes on the MNIST dataset ($K = 10$, $S_\ell = 60{,}000$, $q = 15{,}910$, $D = 10^{-4}$, $\tau = 10^{-6}$, $\tilde{\text{SNR}} = 15$ dB, $\sigma_1^2 = 1, \sigma_2^2 = 1$, $\sigma_e^2 = 2$, $T = 200$). (**a**) A=B=C=4, $\delta = 0.99994$. (**b**) $A = 1$, $B = C = 4, \delta = 0.99997$. (**c**) $A = 4$, $B = C = 1, \delta = 0.99997$. (**d**) $A = B = C = 1$, $\delta = 0.99998$.

From Table 2, we show that the achievable secrecy transmission rates of our FBL schemes increase with the number of antennas, and the achievable secrecy transmission rates of our schemes are significantly higher than those of Baseline 1 and Baseline 2, due to the gain introduced by feedback in our scheme. Additionally, due to the gain from introducing AN in Baseline 2, its achievable secrecy transmission rate is higher than that of Baseline 1. Furthermore, Table 2 shows that the achievable secrecy transmission rates of our proposed FBL scheme decrease as the maximum normalized estimation errors of the eavesdropper's channel increase, which can be viewed as the price for the worst estimation. From Table 3, we conclude that the achievable secrecy transmission rates of FBL schemes increase as the SNR of the feedback channel increases. Moreover, Figure 10 shows that the transmission latency of FBL schemes increases as the SNR of the feedback channel decreases. Therefore, in our schemes, poorer feedback channel conditions lead to lower achievable secrecy transmission rates and increased transmission latency. However, poorer feedback channel conditions do not directly affect learning performance, as it is primarily determined by the distortion D of lossy source coding, the average decoding error probability τ of channel coding, and the variance of noise introduced by LDP mechanisms.

Table 2. Achievable secrecy rates of the different schemes on the MNIST dataset ($K = 10$, $S_\ell = 60{,}000$, $q = 15{,}910$, $\tilde{\text{SNR}} = 15$ dB, $P = 10$, $\tau = 10^{-6}$, $T = 200$, $D = 10^{-4}$, $\sigma_1^2 = \sigma_2^2 = 1, \sigma_e^2 = 2$).

Number of Antennas	$A = B = C = 4$ (MIMO)	$A = 1, B = C = 4$ (SIMO)	$A = 4, B = C = 1$ (MISO)	$A = B = C = 1$ (SISO)
Our scheme (Perfect CSI)	10.3951(bits/symbol) ($\epsilon = 0.04$, $v = 3$, $\sigma^2 = 0.3$, $\delta = 0.99994$)	4.9928(bits/symbol) ($\epsilon = 0.025$, $v = 5$, $\sigma^2 = 0.5$, $\delta = 0.99997$)	5.1346(bits/symbol) ($\epsilon = 0.018$, $v = 6.5$, $\sigma^2 = 0.65$, $\delta = 0.99997$)	2.6718 (bits/symbol) ($\epsilon = 0.032$, $v = 4$, $\sigma^2 = 0.4$, $\delta = 0.99998$)
Our scheme (Imperfect CSI, $\Omega = 0.2$)	10.3941(bits/symbol) ($\epsilon = 0.025$, $v = 5$, $\sigma^2 = 0.5$, $\delta = 0.99994$)	4.9922(bits/symbol) ($\epsilon = 0.012$, $v = 10$, $\sigma^2 = 1$, $\delta = 0.99997$)	5.1343(bits/symbol) ($\epsilon = 0.013$, $v = 9$, $\sigma^2 = 0.9$, $\delta = 0.99997$)	2.6708 (bits/symbol) ($\epsilon = 0.01$, $v = 12$, $\sigma^2 = 1.2$, $\delta = 0.99998$)
Our scheme (Imperfect CSI, $\Omega = 0.4$)	10.3898(bits/symbol) ($\epsilon = 0.0025$, $v = 50$, $\sigma^2 = 5$, $\delta = 0.99994$)	4.9914(bits/symbol) ($\epsilon = 0.005$, $v = 24$, $\sigma^2 = 2.4$, $\delta = 0.99997$)	5.1338(bits/symbol) ($\epsilon = 0.0065$, $v = 18.5$, $\sigma^2 = 1.85$, $\delta = 0.99997$)	2.6689 (bits/symbol) ($\epsilon = 8.2 \times 10^{-4}$, $v = 150$, $\sigma^2 = 15$, $\delta = 0.99998$)
Baseline 1 [26,28] (Perfect CSI)	4.2827(bits/symbol) ($\epsilon = 0.04$, $v = 3$, $\sigma^2 = 0.3$)	2.1537(bits/symbol) ($\epsilon = 0.025$, $v = 5$, $\sigma^2 = 0.5$)	2.2124(bits/symbol) ($\epsilon = 0.018$, $v = 6.5$, $\sigma^2 = 0.65$)	0.8046(bits/symbol) ($\epsilon = 0.032$, $v = 4$, $\sigma^2 = 0.4$)
Baseline 2 [29] (Perfect CSI)	5.2228(bits/symbol) ($\epsilon = 0.04$, $v = 3$, $\sigma^2 = 0.3$)	2.6265(bits/symbol) ($\epsilon = 0.025$, $v = 5$, $\sigma^2 = 0.5$)	2.6981(bits/symbol) ($\epsilon = 0.018$, $v = 6.5$, $\sigma^2 = 0.65$)	0.9812(bits/symbol) ($\epsilon = 0.032$, $v = 4$, $\sigma^2 = 0.4$)

Table 3. Achievable secrecy rates of our schemes under different feedback channel SNR on the MNIST dataset ($K = 10$, $S_\ell = 60{,}000$, $q = 15{,}910$, $P = 10$, $\tau = 10^{-6}$, $T = 200$, $D = 10^{-4}$, $\sigma_1^2 = \sigma_2^2 = 1, \sigma_e^2 = 2$).

Number of Antennas	$A = B = C = 4$ (MIMO) ($\epsilon = 0.04$, $v = 3$, $\sigma^2 = 0.3$, $\delta = 0.99994$)	$A = 1, B = C = 4$ (SIMO) ($\epsilon = 0.025$, $v = 5$, $\sigma^2 = 0.5$, $\delta = 0.99997$)	$A = 4, B = C = 1$ (MISO) ($\epsilon = 0.018$, $v = 6.5$, $\sigma^2 = 0.65$, $\delta = 0.99997$)	$A = B = C = 1$ (SISO) ($\epsilon = 0.032$, $v = 4$, $\sigma^2 = 0.4$, $\delta = 0.99998$)
$\tilde{\text{SNR}} = 10$ dB (Perfect CSI)	8.0911(bits/symbol)	3.6612(bits/symbol)	3.7589(bits/symbol)	2.2885(bits/symbol)
$\tilde{\text{SNR}} = 15$ dB (Perfect CSI)	10.3951(bits/symbol)	4.9928(bits/symbol)	5.1346(bits/symbol)	2.6718(bits/symbol)
$\tilde{\text{SNR}} = 20$ dB (Perfect CSI)	12.5581(bits/symbol)	6.1146(bits/symbol)	6.3177(bits/symbol)	3.0622(bits/symbol)

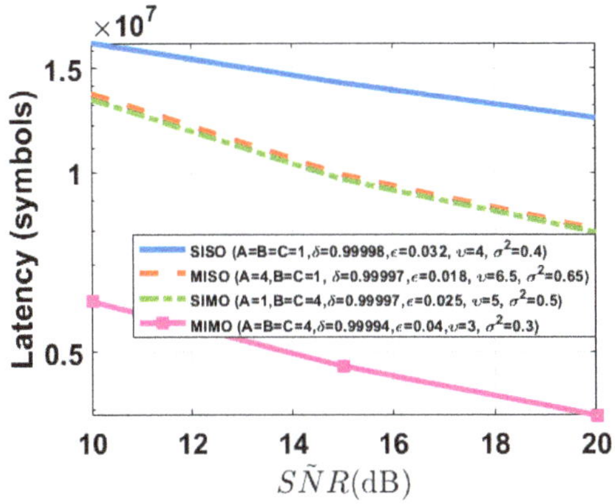

Figure 10. Transmission latency (200 rounds) of our schemes under different feedback channel SNR and perfect CSI on the MNIST dataset ($K = 10$, $S_\ell = 60,000$, $q = 15,910$, $P = 10$, $D = 10^{-4}$, $\tau = 10^{-6}$, $\sigma_1^2 = 1, \sigma_2^2 = 1, \sigma_e^2 = 2$, $T = 200$).

Figure 11 shows the relationship between PLS (measured by the secrecy level), privacy, utility, and the LDP noise variance of proposed FBL schemes. From Figure 11, we conclude that the secrecy level increases as the LDP noise variance increases, and a higher secrecy level leads to a more stringent relationship between privacy and utility (with a smaller ϵ and a larger υ). Apart from this, for a given secrecy level, increasing the maximum normalized estimation error in the eavesdropper's channel results in an increase in the variance of LDP noise, which can be also viewed as the price for the worse estimation.

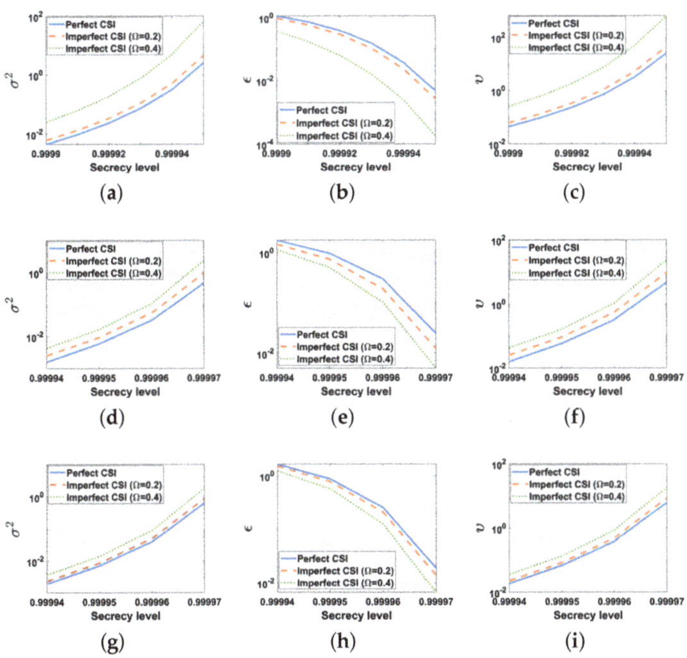

Figure 11. *Cont.*

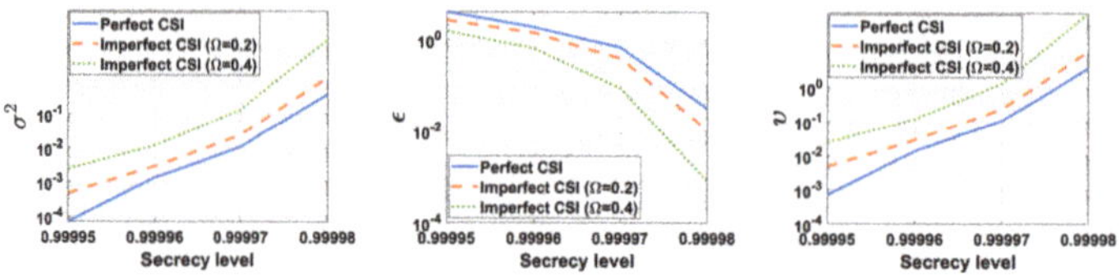

(j) (k) (l)

Figure 11. The relationship between the PLS (secrecy level), the privacy-utility, and LDP noise variance of proposed FBL schemes on the MNIST dataset ($K = 10, S_\ell = 60,000, q = 15,910, D = 10^{-4}$, $\widetilde{\text{SNR}} = 15$ dB, $P = 10$, $\tau = 10^{-6}, \sigma_1^2 = \sigma_2^2 = 1, \sigma_e^2 = 2$). (**a**) $A = B = C = 4$. (**b**) $A = B = C = 4$. (**c**) $A = B = C = 4$. (**d**) $A = 1, B = C = 4$. (**e**) $A = 1, B = C = 4$. (**f**) $A = 1, B = C = 4$. (**g**) $A = 4, B = C = 1$. (**h**) $A = 4, B = C = 1$. (**i**) $A = 4, B = C = 1$. (**j**) $A = B = C = 1$. (**k**) $A = B = C = 1$. (**l**) $A = B = C = 1$.

6. Conclusions and Future Work

In this paper, a practical FBL approach, which is an extension of the classical SK scheme, is proposed for the multi-antenna URLLC-WHFL systems in the presence of PLS. We characterize the relationship between PLS, privacy, and the utility of these WHFL systems, and derive achievable transmission rates of the proposed FBL approach. Simulation results demonstrate that when the edge server has perfect knowledge of the eavesdropper's CSI, our proposed FBL approach not only almost achieves perfect secrecy but also does not affect learning performance. Additionally, simulation results demonstrate that the proposed schemes have robustness even when the edge server has an imperfect eavesdropper's CSI. Apart from this, it has been demonstrated that the transmission latency of our proposed FBL approach is significantly lower compared to traditional RBCS.

Furthermore, this paper focuses on proposing and analyzing a theoretical scheme. The application of this approach in real-world systems still faces practical challenges, such as hardware constraints, power consumption, or synchronization issues. Future work should aim to optimize energy efficiency and address synchronization in more complex multi-antenna systems using the proposed FBL scheme. On the other hand, as the computational complexity of techniques like precoding, beamforming, and SVD increases with the number of devices and communication channels, particularly in multi-antenna systems, further research, and optimization are needed to extend our proposed approach to more complex and large-scale networks. For instance, distributed or hierarchical architectures can allocate the computational load across multiple servers or devices, reducing the burden on individual components. Additionally, low-complexity approximation methods for precoding and beamforming could help lower overall system complexity. Future work will extend our approach to more complex multi-edge scenarios, exploring the impact of interference among edge servers on the WHFL in the presence of PLS.

Author Contributions: H.Z. did the theoretical work, performed the experiments, analyzed the data and drafted the work; B.D. designed the work, performed the theoretical work, interpreted the data for the work and revised the work; and P.X. interpreted the data for the work and revised the work. All authors have read and agreed to the published version of the manuscript.

Funding: This paper was supported in part by the National Key R&D Program of China under grant no. 2022YFA1005000, in part by the National Natural Science Foundation of China under grant no. 62071392, and in part by Chongqing Key Laboratory of Mobile Communications Technology under grant no. cqupt-mct-202302.

Institutional Review Board Statement: Not applicable.

Data Availability Statement: The raw data supporting the conclusions of this article will be made available by the authors on request.

Conflicts of Interest: The authors declare no conflicts of interest.

Appendix A. The Formal Proof of Theorem 1

Appendix A.1. Utility and Privacy Analysis

First, note that since $\mathbf{W}_t$, $\boldsymbol{\eta}_t$, and $\mathbf{W}_t'$ are i.i.d. generated, from Definition 1, we conclude the following:

$$\max_{t\in\{1,..,T\}} \frac{1}{q} I(\mathbf{W}_t; \mathbf{W}_t') = \max_{t\in\{1,..,T\}} \frac{1}{2} \log\left(1 + \frac{S_\ell \sigma_{w,t}^2}{K\sigma^2}\right) \leq \epsilon, \tag{A1}$$

On the other hand, from Definition 2, we conclude the following:

$$\frac{1}{qT} \sum_{t=1}^{T} E(d(\mathbf{W}_t, \mathbf{W}_t')) = \frac{1}{qT} \sum_{t=1}^{T} E(||\boldsymbol{\eta}_t||^2) = K\sigma^2 \leq v. \tag{A2}$$

From (A1) and (A2), the relationship between privacy, utility, and the noise variance of LDP is characterized by the following:

$$\underbrace{\max_{t\in\{1,..,T\}} \frac{S_\ell \sigma_{w,t}^2}{K(2^{2\epsilon} - 1)}}_{\textbf{Privacy term}} \leq \underbrace{\sigma^2}_{\textbf{Noise variance of LDP}} \leq \underbrace{\frac{v}{K}}_{\textbf{Utility term}}, \tag{A3}$$

which indicates that by selecting an appropriate LDP noise to satisfy (A3), both privacy and utility can be ensured.

Appendix A.2. Decoding Error Probability and Convergence Analysis

First, we bound the decoding error probability $P_{e,t}$ of W_t'' transmitted in all parallel sub-SISO channels as follows:

$$P_{e,t} \leq P_{e,t}(1) + P_{e,t}(2) + \ldots + P_{e,t}(J), \tag{A4}$$

where $P_{e,t}(j)$ $(j = 1, \ldots, J)$ represents the decoding error probability of message $W_{t,j}''$, and J is the number of parallel sub-SISO channels, which is defined in (31). Next, we analyze the error events of message $W_{t,j}''$, which consist of the following:

(1) A modulo-aliasing error occurs in the edge server at time instant $i + 1 (1 \leq i \leq N_t - 1)$, and it is defined as follows:

$$E_i = \{\gamma_i \varepsilon_i + \frac{\eta_{j,2,i}'(t)}{\tilde{d}_j} \notin \Lambda\} = \{[\gamma_i \varepsilon_{R,i} + \frac{\eta_{R,j,2,i}'(t)}{\tilde{d}_j} \notin [-\frac{d}{2}, \frac{d}{2})] \cup [\gamma_i \varepsilon_{I,i} + \frac{\eta_{I,j,2,i}'(t)}{\tilde{d}_j} \notin [-\frac{d}{2}, \frac{d}{2})]\}, \tag{A5}$$

where $\varepsilon_{R,i}$ and $\varepsilon_{I,i}$ are the real and imaginary parts of ε_i, respectively, $\eta_{R,j,2,i}'(t)$ and $\eta_{I,j,2,i}'(t)$ are the real and imaginary parts of $\eta_{j,2,i}'(t)$, respectively.

(2) A decoding error occurs in the cloud server at time instant N_t, and it is defined as follows:

$$E_{N_t} = \{\varepsilon_{R,N_t} \notin [-\frac{\sqrt{3}}{2^{N_t R_{t,j,R}}}, \frac{\sqrt{3}}{2^{N_t R_{t,j,R}}}) \cup \varepsilon_{I,N_t} \notin [-\frac{\sqrt{3}}{2^{N_t R_{t,j,I}}}, \frac{\sqrt{3}}{2^{N_t R_{t,j,I}}})\}, \tag{A6}$$

where ε_{R,N_t} and ε_{I,N_t} are the real and imaginary parts of ε_{N_t}, respectively.

Thus, the error probability $P_{e,t}(j)$ is bounded by the following:

$$P_{e,t}(j) \leq Pr\left(\bigcup_{i=1}^{N_t} E_i\right) = Pr\left(\bigcup_{i=1}^{N_t - 1} E_i\right) + Pr\left(\bigcap_{i=1}^{N_t - 1} E_i^c \cap E_{N_t}\right)$$

$$= \sum_{i=1}^{N_t-1} Pr\left(\bigcap_{j=1}^{i-1} E_j^c \cap E_i\right) + Pr\left(\bigcap_{i=1}^{N_t-1} E_i^c \cap E_{N_t}\right)$$
$$= \sum_{i=1}^{N_t-1} Pr\left(\widetilde{E}_i\right) + Pr\left(\widetilde{E}_{N_t}\right), \tag{A7}$$

where E^c is the complement of the set E, and $\widetilde{E}_i = \bigcap_{j=1}^{i-1} E_j^c \cap E_i$. Here, note that $Pr(\widetilde{E}_i)$ $(i \in \{1, ..., N_t - 1\})$ is the error probability that a demodulation error occurs at time instant $i+1$, and no error occurs in all previous times. $Pr(\widetilde{E}_{N_t})$ is the error probability of the final decoding, and no demodulation error occurs in all times. We assume that $P_{e,t} \leq \tau$, which indicates that $\frac{1}{T}\sum_{t=1}^{T} P_{e,t} \leq \tau$ is guaranteed. Then, we choose $P_{e,t}(j) \leq \frac{\tau}{J}$, and we have the following:

$$Pr\left(\widetilde{E}_{N_t}\right) = \sum_{i=1}^{N_t-1} Pr\left(\widetilde{E}_i\right) = \frac{\tau}{2J}, \tag{A8}$$

for simplification, we define the following:

$$Pr\left(\widetilde{E}_1\right) = ... = Pr\left(\widetilde{E}_{N_t-1}\right) = p_m. \tag{A9}$$

Substituting (A9) into (A8), we have the following:

$$p_m = \frac{\tau}{2J(N_t - 1)}. \tag{A10}$$

For the error event $\widetilde{E}_i$, since no demodulation error occurs before time instant $i+1$, according to (43), (44), and the fact $\eta'_{j,2,i}(t)$ is CSCG-distributed, we can conclude the following: $\gamma_i\varepsilon_i + \frac{\eta'_{j,2,i}(t)}{\widetilde{d}_j} \sim \mathcal{CN}(0, \gamma_i^2\alpha_i + \frac{\sigma_2^2}{\widetilde{d}_j^2})$. Hence, we have $\gamma_i\varepsilon_{k,i} + \frac{\eta'_{k,j,2,i}(t)}{\widetilde{d}_j} \sim \mathcal{N}(0, \gamma_i^2\frac{\alpha_i}{2} + \frac{\sigma_2^2}{2\widetilde{d}_j^2})$, where $k \in \{R, I\}$. From (A5) and $d = \sqrt{6\widetilde{P}_j}$, we have the following:

$$\begin{aligned}
Pr(E_i) &= Pr(\{[\gamma_i\varepsilon_{R,i} + \frac{\eta'_{R,j,2,i}(t)}{\widetilde{d}_j} \notin [-\frac{d}{2}, \frac{d}{2})] \cup [\gamma_i\varepsilon_{I,i} + \frac{\eta'_{I,j,2,i}(t)}{\widetilde{d}_j} \notin [-\frac{d}{2}, \frac{d}{2})]\}) \\
&\leq Pr(\gamma_i\varepsilon_{R,i} + \frac{\eta'_{R,j,2,i}(t)}{\widetilde{d}_j} \notin [-\frac{d}{2}, \frac{d}{2})) + Pr(\gamma_i\varepsilon_{I,i} + \frac{\eta'_{I,j,2,i}(t)}{\widetilde{d}_j} \notin [-\frac{d}{2}, \frac{d}{2})) \\
&= 2Q\left(\sqrt{\frac{\frac{3\widetilde{P}_j}{2}}{E(\gamma_i\varepsilon_{R,i} + \frac{\eta'_{R,j,2,i}(t)}{\widetilde{d}_j})^2}}\right) + 2Q\left(\sqrt{\frac{\frac{3\widetilde{P}_j}{2}}{E(\gamma_i\varepsilon_{I,i} + \frac{\eta'_{I,j,2,i}(t)}{\widetilde{d}_j})^2}}\right) \\
&= 4Q\left(\sqrt{\frac{\frac{3\widetilde{P}_j}{2}}{\gamma_i^2\frac{\alpha_i}{2} + \frac{\sigma_2^2}{2\widetilde{d}_j^2}}}\right) = p_m.
\end{aligned} \tag{A11}$$

For simplification, let

$$\xi = \frac{1}{3}\left[Q^{-1}(\frac{p_m}{4})\right]^2 = \frac{1}{3}\left[Q^{-1}(\frac{\tau}{8J(N_t - 1)})\right]^2. \tag{A12}$$

Substituting (A11) into (A12), we have the following:

$$\gamma_i^2\frac{\alpha_i}{2}+\frac{\sigma_2^2}{2\tilde{d}_j^2}=\frac{\tilde{P}_j}{2\xi}. \tag{A13}$$

From (A13), we can conclude the following:

$$\gamma_i=\sqrt{\frac{1}{\alpha_i}\left(\frac{\tilde{P}_j}{\xi}-\frac{\sigma_2^2}{\tilde{d}_j^2}\right)}, \tag{A14}$$

and note that $X'_{j,i+1}(t)$ is subject to the power constraint P_j; hence, we have the following:

$$E[X'_{j,i+1}(t)^{\mathcal{H}}X'_{j,i+1}(t)]=\lambda_i^2E(\gamma_i\varepsilon_i+\frac{\eta'_{j,2,i}(t)}{\tilde{d}_j})^2=P_j. \tag{A15}$$

According to (A13), (A15), we conclude the following:

$$\lambda_i=\sqrt{\xi\cdot\frac{P_j}{\tilde{P}_j}}. \tag{A16}$$

From (43), (A14), and (A16), we have the following:

$$Y'_{j,i+1}(t)=d_j\lambda_i(\gamma_i\varepsilon_i+\frac{\eta'_{j,2,i}(t)}{\tilde{d}_j})+\eta'_{j,1,i+1}(t), \tag{A17}$$

we have the following:

$$\beta_{i+1}=\frac{\lambda_i\gamma_i\alpha_i}{P_j+\frac{\sigma_1^2}{d_j^2}}=\frac{\sqrt{\alpha_i}}{\sigma_1}\frac{\sqrt{\mathrm{SNR}_j(1-\xi\cdot\tilde{\mathrm{SNR}}_j^{-1}\tilde{d}_j^{-2})}}{\mathrm{SNR}_j+d_j^{-2}}. \tag{A18}$$

According to (44), (A14), (A16), (A17), and (A18), we have the following:

$$\alpha_{i+1}=\alpha_i\left(1+\mathrm{SNR}_jd_j^2\frac{1-\xi\cdot\tilde{\mathrm{SNR}}_j^{-1}\tilde{d}_j^{-2}}{1+\xi\cdot\mathrm{SNR}_j\cdot\tilde{\mathrm{SNR}}_j^{-1}d_j^2\tilde{d}_j^{-2}}\right)^{-1}=2d_j^{-2}\mathrm{SNR}_j^{-1}\left(1+\frac{\mathrm{SNR}_jd_j^2}{\Psi_1\Psi_2}\right)^{-i}, \tag{A19}$$

where Ψ_1, Ψ_2, and ξ are defined in (19). From (A6)–(A8), we have the following:

$$Pr\left(\tilde{E}_{N_t}\right)\le Pr\{\varepsilon_{R,N_t}\notin[-\frac{\sqrt{3}}{2^{N_tR_{t,j,R}}},\frac{\sqrt{3}}{2^{N_tR_{t,j,R}}})\}+Pr\{\varepsilon_{I,N_t}\notin[-\frac{\sqrt{3}}{2^{N_tR_{t,j,I}}},\frac{\sqrt{3}}{2^{N_tR_{t,j,I}}})\}\le\frac{\tau}{2J}, \tag{A20}$$

and let

$$Pr\{\varepsilon_{R,N_t}\notin[-\frac{\sqrt{3}}{2^{N_tR_{t,j,R}}},\frac{\sqrt{3}}{2^{N_tR_{t,j,R}}})\}=Pr\{\varepsilon_{I,N_t}\notin[-\frac{\sqrt{3}}{2^{N_tR_{t,j,I}}},\frac{\sqrt{3}}{2^{N_tR_{t,j,I}}})\}=\frac{\tau}{4J}. \tag{A21}$$

Next, we first analyze the term $Pr\{\varepsilon_{R,N_t}\notin[-\frac{\sqrt{3}}{2^{N_tR_{t,j,R}}},\frac{\sqrt{3}}{2^{N_tR_{t,j,R}}})\}=\frac{\tau}{4J}$, .i.e.,

$$Pr\{\varepsilon_{R,N_t}\notin[-\frac{\sqrt{3}}{2^{N_tR_{t,j,R}}},\frac{\sqrt{3}}{2^{N_tR_{t,j,R}}})\}=2Q\left(\frac{\sqrt{3}}{2^{N_tR_{t,j,R}}}\cdot\frac{1}{\sqrt{\alpha_{R,N_t}}}\right)=2Q\left(\frac{\sqrt{3}}{2^{N_tR_{t,j,R}}}\cdot\frac{1}{\sqrt{\frac{\alpha_{N_t}}{2}}}\right)=\frac{\tau}{4J}. \tag{A22}$$

Substituting (A19) into (A22), we have the following:

$$R_{t,j,R} = \frac{1}{2N_t}\log\left(\frac{3\mathrm{SNR}_j d_j^2}{\left[Q^{-1}(\frac{\tau}{8J})\right]^2}\left(1+\frac{\mathrm{SNR}_j d_j^2}{\Psi_1\Psi_2}\right)^{N_t-1}\right). \quad \text{(A23)}$$

Analogously, we can show that $R_{t,j,I} = R_{t,j,R}$. The transmission rate of the j-th ($j = 1, ..., J$) parallel sub-channel in the t-th round is given by the following:

$$R_{t,j} = R_{t,j,R} + R_{t,j,I} = \frac{1}{N_t}\log\left(\frac{3\mathrm{SNR}_j d_j^2}{\left[Q^{-1}(\frac{\tau}{8J})\right]^2}\left(1+\frac{\mathrm{SNR}_j d_j^2}{\Psi_1\Psi_2}\right)^{N_t-1}\right). \quad \text{(A24)}$$

Combining (A24) and (36), and power allocating in Section 3.1.1, (18) in Theorem 1 is obtained. Then, according to $R = \frac{H(W_1'',...,W_T'')}{N}$ in (15), the transmission rate $R_{mimo}(\tau, N, \delta, D, v, \epsilon)$ is given by the following:

$$R_{mimo}(\tau, N, \delta, D, v, \epsilon) = \frac{H(W_1'', ..., W_T'')}{N} \overset{(b)}{=} \frac{\sum_{t=1}^{T} H(W_t'')}{N} = \frac{\sum_{t=1}^{T} N_t R_t}{N}, \quad \text{(A25)}$$

where $N = \sum_{t=1}^{T} N_t$, and (b) is due to the fact that $\mathbf{W}_t'$ is mapped into the uniformly distributed index W_t'' in each communication round, which indicates that $(W_1'', ..., W_T'')$ are independent of each other; hence, (17) in Theorem 1 is obtained. A brief convergence analysis of our FBL approach is given below.

Convergence analysis: Following the convergence proof in [18] and [22], we can prove the existence of convergence in the WHFL system with our FBL scheme when the decoding error probability of our FBL scheme is significantly small. Furthermore, from (A19) and (A22), we conclude that the variance of the estimation error in our FBL approach *converges to zero with double-exponential speed*, which indicates that the final estimation $\hat{\theta}_{j,N_t}$ can always converge to θ_j, and the required coding block length for achieving the desired decoding error probability is significantly short.

Appendix A.3. Security Analysis

First, note that the eavesdropper's equivocation rate, Δ, can be re-written as follows:

$$\begin{aligned}
\Delta &= \frac{H(W_1'', ..., W_T''|\mathbf{Z}^{N_1}, ..., \mathbf{Z}^{N_T}, \mathbf{h}_{mimo}, \widetilde{\mathbf{h}}_{mimo}, \mathbf{g}_{mimo}, \widetilde{\mathbf{g}}_{mimo})}{H(W_1'', ..., W_T'')} \\
&= \frac{\sum_{t=1}^{T} H(W_t''|\mathbf{Z}^{N_1}, ..., \mathbf{Z}^{N_T}, \mathbf{h}_{mimo}, \widetilde{\mathbf{h}}_{mimo}, \mathbf{g}_{mimo}, \widetilde{\mathbf{g}}_{mimo}, W_1'', ..., W_{t-1}'')}{H(W_1'', ..., W_T'')} \\
&\overset{(c)}{=} \frac{\sum_{t=1}^{T} H(W_t''|\mathbf{Z}^{N_t}, \mathbf{h}_{mimo}, \widetilde{\mathbf{h}}_{mimo}, \mathbf{g}_{mimo}, \widetilde{\mathbf{g}}_{mimo})}{H(W_1'', ..., W_T'')} = \frac{\sum_{t=1}^{T} H(W_t''|\mathbf{Z}^{N_t}, \mathbf{h}_{mimo}, \widetilde{\mathbf{h}}_{mimo}, \mathbf{g}_{mimo}, \widetilde{\mathbf{g}}_{mimo})}{\sum_{t=1}^{T} H(W_t'')},
\end{aligned} \quad \text{(A26)}$$

where (c) follows from the Markov chain $W_t'' \to (\mathbf{Z}^{N_t}, \mathbf{h}_{mimo}, \widetilde{\mathbf{h}}_{mimo}, \mathbf{g}_{mimo}, \widetilde{\mathbf{g}}_{mimo}) \to (\mathbf{Z}^{N_1}, ..., \mathbf{Z}^{N_{t-1}}, \mathbf{Z}^{N_{t+1}}, ..., \mathbf{Z}^{N_T}, W_1'', ..., W_{t-1}'')$. The term $H(W_t''|\mathbf{Z}^{N_t}, \mathbf{h}_{mimo}, \widetilde{\mathbf{h}}_{mimo}, \mathbf{g}_{mimo}, \widetilde{\mathbf{g}}_{mimo})$ in (A26) is given by the following:

$$\begin{aligned}
&H(W_t''|\mathbf{Z}^{N_t}, \mathbf{h}_{mimo}, \widetilde{\mathbf{h}}_{mimo}, \mathbf{g}_{mimo}, \widetilde{\mathbf{g}}_{mimo}) \\
&\overset{(e)}{\geq} H(W_t''|\underbrace{\mathbf{g}_{mimo}\mathbf{X}_1(t) + \widetilde{\mathbf{g}}_{mimo}\widetilde{\mathbf{X}}_1(t) + \boldsymbol{\eta}_{e,1}(t)}_{\mathbf{Z}_1(t)}, ..., \underbrace{\mathbf{g}_{mimo}\mathbf{X}_{N_t-1}(t) + \widetilde{\mathbf{g}}_{mimo}\widetilde{\mathbf{X}}_{N_t-1}(t) + \boldsymbol{\eta}_{e,N_t-1}(t)}_{\mathbf{Z}_{N_t-1}(t)}, \\
&\underbrace{\mathbf{g}_{mimo}\mathbf{X}_{N_t}(t) + \boldsymbol{\eta}_{e,N_t}(t)}_{\mathbf{Z}_{N_t}(t)}, \boldsymbol{\eta}_{1,1}(t), ..., \boldsymbol{\eta}_{1,N_t}(t), \boldsymbol{\eta}_{2,1}(t), ..., \boldsymbol{\eta}_{2,N_t}(t), \boldsymbol{\eta}_{e,2}(t), ..., \boldsymbol{\eta}_{e,N_t}(t), \widetilde{\mathbf{X}}_1'(t), ..., \widetilde{\mathbf{X}}_{N_t-1}'(t), \mathbf{h}_{mimo},
\end{aligned}$$

$$\widetilde{h}_{mimo}, g_{mimo}, \widetilde{g}_{mimo}, U, \Lambda, V, \widetilde{U}, \widetilde{\Lambda}, \widetilde{V})$$

$$\stackrel{(f)}{=} H(W_t''|g_{mimo}VX_1'(t) + \eta_{e,1}(t), \eta_{1,1}(t), ..., \eta_{1,N_t}(t), \eta_{2,1}(t), ..., \eta_{2,N_t}(t), \eta_{e,2}(t), ..., \eta_{e,N_t}(t), \widetilde{X}_1'(t), ..., \widetilde{X}_{N_t-1}'(t),$$

$$h_{mimo}, \widetilde{h}_{mimo}, g_{mimo}, \widetilde{g}_{mimo}, U, \Lambda, V, \widetilde{U}, \widetilde{\Lambda}, \widetilde{V})$$

$$\stackrel{(g)}{=} H(W_t''|g_{mimo}VX_1'(t) + \eta_{e,1}(t)) \stackrel{(h)}{=} H(W_t'') + h(\eta_{e,1}(t)) - h(g_{mimo}VX_1'(t) + \eta_{e,1}(t))$$

$$\stackrel{(i)}{=} H(W_t'') - \underbrace{\log\det\left(I + \frac{g_{mimo}K_{x_1}g_{mimo}^{\mathcal{H}}}{\sigma_e^2}\right)}_{\textbf{Information leakage at time 1}}, \tag{A27}$$

where

(e) follows from the fact that conditioning reduces entropy, as shown in (28),

(f) follows from $X_i'(t) = V^{\mathcal{H}}X_i(t), \widetilde{X}_i'(t) = \widetilde{V}^{\mathcal{H}}\widetilde{X}_i(t)$ and $X_i'(t)$ $(i = 2, ..., N_t)$ is a function of $h_{mimo}, \widetilde{h}_{mimo}, \eta_{1,1}(t), ..., \eta_{1,i-1}(t), \eta_{2,1}(t), ..., \eta_{2,i-1}(t)$,

(g) follows from the fact that $\widetilde{X}_i'(t)$ $(i = 1, ..., N_t - 1)$ is only related to ν_i [39] [Chapter 4.1, pp. 61–63], and h_{mimo}, $\widetilde{h}_{mimo}$, g_{mimo}, $\widetilde{g}_{mimo}$, U, Λ, V, $\widetilde{U}$, $\widetilde{\Lambda}$, $\widetilde{V}$, $\eta_{1,1}(t), ..., \eta_{1,N_t}(t)$, $\eta_{2,1}(t), ..., \eta_{2,N_t}(t), \eta_{e,2}(t), ..., \eta_{e,N_t}(t), \nu_1, ..., \nu_{N_t-1}$ are independent of $W_t'', X_1'(t), \eta_{e,1}(t)$,

(h) is due to the fact that (33) and $X_{j,1}'(t) = \sqrt{\frac{P_j}{2}}\theta_j$, and $W_t'' = (W_{t,1}'', ..., W_{t,J}'')$ are mapped into $(\theta_1, ..., \theta_J)$, respectively,

(i) follows from the following:

$$\begin{aligned} h(g_{mimo}VX_1'(t) + \eta_{e,1}(t)) - h(\eta_{e,1}(t)) &\leq \log\det\left(I + \frac{g_{mimo}VE(X_1'(t)X_1'^{\mathcal{H}}(t))V^{\mathcal{H}}g_{mimo}^{\mathcal{H}}}{\sigma_e^2}\right) \\ &= \log\det\left(I + \frac{g_{mimo}E(X_1(t)X_1^{\mathcal{H}}(t))g_{mimo}^{\mathcal{H}}}{\sigma_e^2}\right), \end{aligned} \tag{A28}$$

and $K_{x_1} = E(X_1(t)X_1^{\mathcal{H}}(t))$. Substituting (A27) into (A26), we have the following:

$$\Delta \geq \frac{\sum_{t=1}^T H(W_t'')\left(1 - \frac{\log\det\left(I + \frac{g_{mimo}K_{x_1}g_{mimo}^{\mathcal{H}}}{\sigma_e^2}\right)}{H(W_t'')}\right)}{\sum_{t=1}^T H(W_t'')} \geq \min_{t\in\{1,...,T\}}\left(1 - \frac{\log\det\left(I + \frac{g_{mimo}K_{x_1}g_{mimo}^{\mathcal{H}}}{\sigma_e^2}\right)}{H(W_t'')}\right). \tag{A29}$$

From (8), (35) and (A29), $\Delta \geq \delta$ in (15) is guaranteed if we have the following:

$$\sigma^2 \geq \underbrace{\max_{t\in\{1,..,T\}}\left[\frac{D \cdot 2^{\frac{2}{q(1-\delta)}\log\det\left(I + \frac{g_{mimo}K_{x_1}g_{mimo}^{\mathcal{H}}}{\sigma_e^2}\right)} - S_\ell\sigma_{w,t}^2}{K}\right]}_{\textbf{Secrecy level of PLS}}. \tag{A30}$$

With the assumption that the edge server has an imperfect CSI of the eavesdropper's channel, combining (A30) and Definition 3, (A30) can be re-written by the following:

$$\sigma^2 \geq \underbrace{\max_{t\in\{1,..,T\},\Delta g_{mimo}}\left[\frac{D \cdot 2^{\frac{2}{q(1-\delta)}\log\det\left(I + \frac{g_{mimo}K_{x_1}g_{mimo}^{\mathcal{H}}}{\sigma_e^2}\right)} - S_\ell\sigma_{w,t}^2}{K}\right]}_{\textbf{Secrecy level of PLS}}, \tag{A31}$$

where $\hat{g}_{mimo} = g_{mimo} - \Delta g_{mimo}$. Then, combining (A3) and (A31), (16) in Theorem 1 is obtained.

The proof of Theorem 1 is completed.

References

1. Zhu, G.; Liu, D.; Du, Y.; You, C.; Zhang, J.; Huang, K. Toward an Intelligent Edge: Wireless Communication Meets Machine Learning. *IEEE Commun. Mag.* **2020**, *58*, 19–25. [CrossRef]
2. Yang, Z.; Chen, M.; Saad, W.; Hong, C.S.; Shikh-Bahaei, M. Energy Efficient Federated Learning Over Wireless Communication Networks. *IEEE Trans. Wireless Commun.* **2021**, *20*, 1935–1949. [CrossRef]
3. Amiri, M.M.; Gündüz, D. Federated Learning Over Wireless Fading Channels. *IEEE Trans. Wireless Commun.* **2020**, *19*, 3546–3557. [CrossRef]
4. Jin, R.; He, X.; Dai, H. Communication Efficient Federated Learning with Energy Awareness Over Wireless Networks. *IEEE Trans. Wireless Commun.* **2022**, *21*, 5204–5219. [CrossRef]
5. Zhu, G.; Wang, Y.; Huang, K. Broadband Analog Aggregation for Low-Latency Federated Edge Learning. *IEEE Trans. Wireless Commun.* **2020**. *19*, 491–506. [CrossRef]
6. Zhu, G.; Du, Y.; Gündüz, D.; Huang, K. One-Bit Over-the-Air Aggregation for Communication-Efficient Federated Edge Learning: Design and Convergence Analysis. *IEEE Trans. Wireless Commun.* **2021**, *20*, 2120–2135. [CrossRef]
7. Elgabli, A.; Park, J.; Issaid, C.B.; Bennis, M. Harnessing Wireless Channels for Scalable and Privacy-Preserving Federated Learning. *IEEE Trans. Commun.* **2021**, *69*, 5194–5208. [CrossRef]
8. Wen, H.; Wu, Y.; Yang, C.; Duan, H.; Yu, S. A Unified Federated Learning Framework for Wireless Communications: Towards Privacy, Efficiency, and Security. In Proceedings of the 2020 IEEE INFOCOM Computer Communications Workshops (INFOCOM WKSHPS), Toronto, ON, Canada, 6–9 July 2020; pp. 653–658.
9. Seif, M.; Tandon, R.; Li, M. Wireless Federated Learning with Local Differential Privacy. In Proceedings of the 2020 IEEE International Symposium on Information Theory (ISIT), Los Angeles, CA, USA, 21–26 June 2020; pp. 2604–2609.
10. Yuan, X.; Ni, W.; Ding, M.; Wei, K.; Li, J.; Poor, H.V. Amplitude-Varying Perturbation for Balancing Privacy and Utility in Federated Learning. *IEEE Trans. Inf. Forensics Secur.* **2023**, *18*, 1884–1897. [CrossRef]
11. Kim, M.; Günlü, O.; Schaefer, R.F. Federated Learning with Local Differential Privacy: Trade-Offs Between Privacy, Utility, and Communication. In Proceedings of the 2021 IEEE International Conference on Acoustics, Speech and Signal Process (ICASSP), Toronto, ON, Canada, 6–11 June 2021; pp. 2650–2654.
12. Zhou, J.; Su, Z.; Ni, J.; Wang, Y.; Pan, Y.; Xing, R. Personalized Privacy-Preserving Federated Learning: Optimized Trade-off Between Utility and Privacy. In Proceedings of the 2022 IEEE Global Communications Conference (GLOBECOM), Rio de Janeiro, Brazil, 4–8 December 2022; pp. 4872–4877.
13. Guo, S.; Su, Z.; Tian, Z.; Yu, S. Utility-Aware Privacy-Preserving Federated Learning through Information Bottleneck. In Proceedings of the 2022 IEEE International Conference on Trust, Security and Privacy in Computing and Communications (TrustCom), Wuhan, China, 9–11 December 2022; pp. 680–686.
14. Wang, B.; Chen, Y.; Jiang, H.; Zhao, Z. PPeFL: Privacy-Preserving Edge Federated Learning with Local Differential Privacy. *IEEE Internet Things J.* **2023**, *10*, 15488–15500. [CrossRef]
15. Zhang, N.; Tao, M. Gradient Statistics Aware Power Control for Over-the-Air Federated Learning. *IEEE Trans. Wireless Commun.* **2021**, *20*, 5115–5128. [CrossRef]
16. Liu, D.; Simeone, O. Privacy for Free: Wireless Federated Learning via Uncoded Transmission with Adaptive Power Control. *IEEE J. Sel. Areas Commun.* **2021**, *39*, 170–185. [CrossRef]
17. Yang, K.; Jiang, T.; Shi, Y.; Ding, Z. Federated Learning via Over-the-Air Computation. *IEEE Trans. Wireless Commun.* **2020**, *19*, 2022–2035. [CrossRef]
18. Liu, L.; Zhang, J.; Song, S.H.; Letaief, K.B. Client-Edge-Cloud Hierarchical Federated Learning. In Proceedings of the 2020 IEEE International Conference on Communications (ICC), Dublin, Ireland, 7–11 June 2020; pp. 1–6.
19. Luo, S.; Chen, X.; Wu, Q.; Zhou, Z.; Yu, S. HFEL: Joint Edge Association and Resource Allocation for Cost-Efficient Hierarchical Federated Edge Learning. *IEEE Trans. Wireless Commun.* **2020**, *19*, 6535–6548. [CrossRef]
20. Liu, S.; Yu, G.; Chen, X.; Bennis, M. Joint User Association and Resource Allocation for Wireless Hierarchical Federated Learning with IID and Non-IID Data. *IEEE Trans. Wireless Commun.* **2022**, *21*, 7852–7866. [CrossRef]
21. Wen, W.; Chen, Z.; Yang, H.H.; Xia, W.; Quek, T.Q.S. Joint Scheduling and Resource Allocation for Hierarchical Federated Edge Learning. *IEEE Trans. Wireless Commun.* **2022**, *21*, 5857–5872. [CrossRef]
22. Shi, L.; Shu, J.; Zhang, W.; Liu, Y. HFL-DP: Hierarchical Federated Learning with Differential Privacy. In Proceedings of the 2021 IEEE Global Communications Conference (GLOBECOM), Madrid, Spain, 7–11 December 2021; pp. 1–7.
23. Wainakh, A.; Guinea, A.S.; Grube, T.; Mühlhäuser, M. Enhancing Privacy via Hierarchical Federated Learning. In Proceedings of the 2020 IEEE European Symposium on Security and Privacy Workshops (EuroS&PW), Genoa, Italy, 7–11 September 2020; pp. 344–347.
24. Feng, C.; Yang, H.H.; Hu, D.; Zhao, Z.; Quek, T.Q.S.; Min, G. Mobility-Aware Cluster Federated Learning in Hierarchical Wireless Networks. *IEEE Trans. Wireless Commun.* **2022**, *21*, 8441–8458. [CrossRef]
25. Wyner, A.D. The Wire-Tap Channel. *Bell Syst. Tech. J.* **1975**, *54*, 1355–1387. [CrossRef]

26. Yao, J.; Ansari, N. Secure Federated Learning by Power Control for Internet of Drones. *IEEE Trans. Cognitive Commun. Netw.* **2021**, *7*, 1021–1031. [CrossRef]
27. Wang, T.; Li, Y.; Wu, Y.; Quek, T.Q.S. Secrecy driven Federated Learning via Cooperative Jamming: An Approach of Latency Minimization. *IEEE Trans. Emerg. Topics Comput.* **2021**, *10*, 1687–1703. [CrossRef]
28. Qian, L.; Wu, W.; Lu, W.; Wu, Y.; Lin, B.; Quek, T.Q.S. Secrecy-Based Energy-Efficient Mobile Edge Computing via Cooperative Non-Orthogonal Multiple Access Transmission. *IEEE Trans. Commun.* **2021**, *69*, 4659–4677. [CrossRef]
29. Yan, Z.; Li, D.; Zhang, Z.; He, J. Accuracy-Security Tradeoff with Balanced Aggregation and Artificial Noise for Wireless Federated Learning. *IEEE Internet Things J.* **2023**, *10*, 18154–18167. [CrossRef]
30. Zhang, H.; Yang, C.; Dai, B. When Wireless Federated Learning Meets Physical Layer Security: The Fundamental Limits. In Proceedings of the IEEE INFOCOM Computer Communications Workshops (INFOCOM WKSHPS), New York, NY, USA, 2–5 May 2022; pp. 1–6.
31. Durisi, G.; Koch, T.; Popovski, P. Toward Massive, Ultrareliable, and Low-Latency Wireless Communication with Short Packets. *Proc. IEEE* **2016**, *104*, 1711–1726. [CrossRef]
32. Polyanskiy, Y.; Poor, H.V.; Verdu, S. Channel Coding Rate in the Finite Blocklength Regime. *IEEE Trans. Inf. Theory* **2010**, *56*, 2307–2359. [CrossRef]
33. She, C.; Dong, R.; Gu, Z.; Hou, Z.; Li, Y.; Hardjawana, W.; Vucetic, B.; Song, L.; Yang, C. Deep Learning for Ultra-Reliable and Low-Latency Communications in 6G Networks. *IEEE Netw.* **2020**, *34*, 219–225. [CrossRef]
34. Samarakoon, S.; Bennis, M.; Saad, W.; Debbah, M. Distributed Federated Learning for Ultra-Reliable Low-Latency Vehicular Communications. *IEEE Trans. Commun.* **2020**, *68*, 1146–1159. [CrossRef]
35. Schalkwijk, J.; Kailath, T. A coding scheme for additive noise channels with feedback–I: No bandwidth constraint. *IEEE Trans. Inf. Theory* **1966**, *12*, 172–182. [CrossRef]
36. Gunduz, D.; Brown, D.R.; Poor, H.V. Secret communication with feedback. In Proceedings of the 2008 International Symposium on Information Theory and Its Applications (ISITA), Auckland, New Zealand, 7–10 December 2008; pp. 1–6.
37. Truong, L.V.; Fong, S.L.; Tan, V.Y.F. On Gaussian Channels with Feedback Under Expected Power Constraints and with Non-Vanishing Error Probabilities. *IEEE Trans. Inf. Theory* **2017**, *63*, 1746–1765. [CrossRef]
38. Abadi, M.; Chu, A.; Goodfellow, I.; McMahan, H.B.; Mironov, I.; Talwar, K.; Zhang, L. Deep Learning with Differential Privacy. In Proceedings of the ACM SIGSAC Conference on Computer and Communications Security, Vienna, Austria, 24–28 October 2016; pp. 303–318.
39. Zamir, R. *Lattice Coding for Signals and Network*; Cambridge University Press: Cambridge, UK, 2014.
40. Zhi, K.; Pan, C.; Ren, H.; Wang, K.; Elkashlan, M.; Di Renzo, M.; Hanzo, L.; Schober, R.; Wang, J. Two-Timescale Design for Reconfigurable Intelligent Surface-Aided Massive MIMO Systems with Imperfect CSI. *IEEE Trans. Inf. Theory* **2022**, *69*, 3001–3033. [CrossRef]
41. Schiessl, S.; Al-Zubaidy, H.; Skoglund, M.; Gross, J. Delay Performance of Wireless Communications with Imperfect CSI and Finite-Length Coding. *IEEE Trans. Commun.* **2018**, *66*, 6527–6541. [CrossRef]
42. Chen, Z.J.; Hernandez, E.E.; Huang, Y.C.; Rini, S. DNN gradient lossless compression: Can GenNorm be the answer? In Proceedings of the 2022 IEEE International Conference on Communications (ICC), Seoul, Republic of Korea, 16–20 May 2022; pp. 407–412.
43. Wang, W.; Ying, L.; Zhang, J. On the Relation Between Identifiability, Differential Privacy, and Mutual-Information Privacy. *IEEE Trans. Inf. Theory* **2016**, 62, 5018–5029. [CrossRef]
44. Sankar, L.; Rajagopalan, S.R.; Poor, H.V. Utility-Privacy Tradeoffs in Databases: An Information-Theoretic Approach. *IEEE Trans. Inf. Forensics Secur.* **2013**, *8*, 838–852. [CrossRef]
45. Gamal, A.A.E.; Kim, Y.-H. *Network Information Theory*; Cambridge University Press: Cambridge, UK, 2011.
46. Han, S.; Xu, X.; Fang, S.; Sun, Y.; Cao, Y.; Tao, X.; Zhang, P. Energy Efficient Secure Computation Offloading in NOMA-Based mMTC Networks for IoT. *IEEE Internet Things J.* **2019**, *6*, 5674–5690. [CrossRef]
47. Ng, D.W.K.; Lo, E.S.; Schober, R. Robust Beamforming for Secure Communication in Systems with Wireless Information and Power Transfer. *IEEE Trans. Wireless Commun.* **2014**, *13*, 4599–4615. [CrossRef]
48. Tekin, E.; Yener, A.The Gaussian Multiple Access Wire-Tap Channel. *IEEE Trans. Inf. Theory* **2008**, *54*, 5747–5755. [CrossRef]
49. Tse, D.; Viswanath, P. *Fundamentals of Wireless Communication*; Cambridge University Press: Cambridge, UK, 2005.
50. Welch, T.A. A technique of high-performance data compression. *IEEE Comput.* **1984**, *17*, 8–19. [CrossRef]

MDPI AG
Grosspeteranlage 5
4052 Basel
Switzerland
Tel.: +41 61 683 77 34

Entropy Editorial Office
E-mail: entropy@mdpi.com
www.mdpi.com/journal/entropy

www.ingramcontent.com/pod-product-compliance
Lightning Source LLC
Chambersburg PA
CBHW061605220926
51850CB00005B/429
* 9 7 8 3 7 2 5 8 5 1 1 7 1 *